Will Vaccines Be the End of Us?

BY

John Massaro

FIRST EDITION

JEFFERSONVILLE, NEW YORK

ISBN 978-1-7379660-0-5

Will Vaccines Be the End of Us?
by John Massaro

First edition; first printing Nov. 2021,
second printing October 2025

A book by
EUGENE WALTER PUBLISHERS
BOX 85 • MINERAL POINT • PENNSYLVANIA 15942
A NATIONAL-UNIVERSAL COMPANY

Photo credits: Bettman, dead bodies in Jonestown, GETTY IMAGES; Martha Holmes, young polio victim David Hensley, GETTY IMAGES; Paula Bronstein, Vietnamese boy with deformed head in crib, victim of Agent Orange, GETTY IMAGES; Cambodian girl next to bones of victims murdered by Khmer Rouge, ASSOCIATED PRESS; George Lincoln Rockwell picketing Dallas office of Anti-Defamation League, ASSOCIATED PRESS; Jonas Salk with test tubes, ALAMY

published by

JOHN MASSARO
BOX 45 • JEFFERSONVILLE • NEW YORK • 12748
EMAIL: JOHN@ENDTHESHOTS.COM

Dedication

TO MY BELOVED CHILDREN, SCOTT AND MARY

LIFE HAS dealt you a tough hand — especially you, Mary. Injustice has always been part of the human condition, but why did you have to be the victims of such a cruel injustice — the accident of coming into this world three months before Nature intended?

There are no answers, and there is no turning back the clock. There is only my wish that other parents will look into your innocent faces, and into mine, and trust me when I say that I wrote this book so that they can experience the fullest flowering of the sacred love between parent and child.

You and I have been denied that, and so have millions of mothers and fathers and children whose names and faces will never be known to the world. Children who went to their graves before they could talk, or are serving a life sentence of profound disability; parents for whom life has become a heartache that never ends, who every day suffer the thought, "If only I had known the truth about vaccines."

This book is for them too.

AND TO THE MORE THAN 58,000 YOUNG AMERICAN
MEN WHO FELL IN VIETNAM

TO LIVE in a nation run by gangsters who supervise a global empire means to live with wars — "perpetual wars for perpetual peace," in the words of one astute historian. The Vietnam War was the war of my youth, and had I been born three years earlier, I likely would have been drafted and sent to fight in that pointless and obscene conflict — if I would've had the courage to go. Today, in a nation with no awareness of its own history, the Vietnam War is just two words naming some obscure event. Most Americans living today never knew it and never learned anything about it, or else have forgotten it.

I have not forgotten. I have not forgotten that many young men, my countrymen, went there to fight, and many more did not. I have not forgotten the treachery of the government that sent them to war in Vietnam, and all the phony patriotism that went with it, which I fell for at the time. I now know better. I now know that those men did not die to defend our freedom, nor to stop the spread of communism, nor for any other flag-waving hallucination. They died for absolutely nothing. No, worse than that. They died so that more and more people all over the world would rightly curse and hate America.

Some of us have not forgotten nor will we ever forget your sacrifice. Whether you volunteered for combat with head held high, or dreaded receiving your draft notice, whether you died heroically or trembling with fear, the important thing is this: you went. You didn't get a student deferment, or cook up a medical exemption, or join the National Guard, or flee to Canada: you went. And you went knowing that you were leaving safety behind and walking into extreme danger, because you believed it was the right thing to do.

In going to fight in Vietnam, each and every one of you set an example of courage for the rest of us. Fifty years later, I feel it is my duty to follow your example. Your courage inspired me to write and publish this book, and spread its message, come what may.

Government is not reason; it is not eloquence; it is force! Like fire, it is a dangerous servant and a fearful master. — George Washington

The trade of governing has always been a monopoly of the most ignorant and the most vicious of mankind. — Thomas Paine

The most dangerous man to any government is the man who is able to think things out for himself... without regard to the prevailing superstitions and taboos. Almost invariably he comes to the conclusion that the government he lives under is dishonest, insane and intolerable. — H.L. Mencken

There are people whom it is necessary to detest without compromise. There are people who must be dealt with as enemies of the human race. There are people who have no human heart, and who must be crushed like savage beasts and cleared out of the way.
— Charles Dickens

We are so constituted that we believe the most incredible things, and once they are engraved upon the memory, woe to him who endeavors to erase them.
— Johann Wolfgang von Goethe

When a well-packaged web of lies has been sold gradually to the masses over generations, the truth will seem utterly preposterous and its speaker a raving lunatic.
— Dresden James

You will observe with concern how long a useful truth may be known, and exist, before it is generally received and practiced on. — Benjamin Franklin

Life is short and truth works far and lives long; let us speak truth.
— Arthur Schopenhauer

In times of universal deceit, telling the truth is a revolutionary act.
— George Orwell

Let us dare to read, think, speak and write. — John Adams

Be ashamed to die until you have won some victory for mankind.
— Horace Mann

There is one thing stronger than all the armies in the world, and that is an idea whose time has come. — Victor Hugo

A note to the reader

AFTER MORE than ten years of researching, writing, updating and revising, I said "enough is enough" and in late 2019 submitted this manuscript for final layout work, hoping to publish it shortly thereafter as a regular book for sale, and letting people read it for free on a website I planned to create. Circumstances beyond my control delayed both the printing and the website construction for more than a year and a half. But except for a shifting of text and the addition of one paragraph in March 2020 to the last chapter (headed "Ramblings, Reflections and Conclusions"), and a few grammatical revisions, this book is unchanged since October 31, 2019. This was two months before the first reports of a novel corona virus leaked out of Wuhan, China. Since then, the subject about which I wrote has been shoved aside by the biggest disinformation campaign in world history — the corona virus pandemic, or plandemic, as others have rightly called it. How I failed to grasp the magnitude of this event, in its early stage at least, can be inferred from reading the paragraph referred to above (which begins "Speaking of uncontrollable forces..."). But soon enough, I realized we were seeing the convulsions of a world gone mad. To my knowledge, only one world leader, Alexander Lukashenko, the president of Belarus, had the courage and intelligence to call the spectacle what it was — an "international psychosis."

Notwithstanding the media-generated hysteria over this unremarkable virus, if it even exists (which I doubt), the subject of this book looms larger than ever. Even when the first fake news reports about Covid-19 appeared, people should have been a hundred times more concerned about the danger of vaccines, especially those forced on our children. And now, several months into an intensive campaign to inoculate the great majority of Americans, along with the rest of the world, they should be a thousand times more concerned.

There's not much to be happy about these days, but I'm glad that the publication of this book was put on hold. I can now say that every important point I made, and the title itself, has been validated to the utmost. I believe we're going through one of History's great tribulations. Obviously, I saw the necessity to write an afterword as one last update.

John Massaro
July 17, 2021

Are American parents drinking the Kool-Aid?

ON November 18, 1978, a weird and tragic event, unprecedented in history, occurred at a place called Jonestown in the remote jungle of Guyana, a small country in South America. Jonestown was the site of the People's Temple, a cult that combined elements of conventional religion with a heavy dose of Soviet communism. It was founded and ruled by a creature named Jim Jones, who was called "Dad" or "Father" by his followers, a multiracial fellowship of losers, misfits and dreamers who apparently believed they could create the world's first socialist paradise.

Originally based in Indiana, in 1965 Jones moved his operation to California where he could more easily attract the kind of people he was looking for. In the early 1970s a flurry of newspaper articles critical of the People's Temple, based in part on shocking stories told by defectors, was published in the San Francisco area. Fearing further investigation and a police crackdown, Jones decided to leave for the safety and seclusion of distant Guyana. Nearly one thousand of his followers, many with children, went with him. However, allegations of emotional torment and physical abuse continued to leak out and reached the U.S., prompting California congressman Leo Ryan to pay a fact-finding visit to Jonestown, along with eighteen reporters, photographers and concerned relatives.

The group was not welcome, but after initial tensions subsided, some were allowed to tour Jonestown, a settlement of agricultural plots and simple dwellings. There was no clear evidence of serious problems, but there were hints that things were not right and that some people wanted to leave but were forced to stay against their will. While Ryan spoke to Jones he was attacked by a knife-wielding Temple member, who was quickly subdued. Danger was in the air as the group returned to the airstrip six miles from Jonestown where they had arrived in their small chartered airplane, and were now preparing to leave. They were unaware that they had been followed by several of Jones's armed bodyguards. Reaching the airstrip, where the delegation and a few defectors were waiting to board, the thugs jumped out of their vehicle and opened fire on them. Congressman

Ryan was killed in a hail of bullets; twelve others lay dead or wounded on the tarmac.

When word of the shooting reached the Temple compound, Jones, who had ordered it, knew he was finished. Over towering loudspeakers he told all his followers to assemble in order to commit "revolutionary suicide." Buckets filled with a powdered soft drink similar to the popular supermarket brand Kool-Aid, and spiked with sedatives and deadly potassium cyanide, were prepared. A great majority of People's Temple members obediently lined up, plastic cups in hand, awaiting their turn to drink the poison punch and give it to their children who, of course, had no idea what was going on.

Prior to that fateful day, to test the loyalty of his followers, Jones had held several identical drills during which they were ordered to drink a beverage he told them would kill them, even though he knew it was harmless. Few objected. So this time, after Jones announced that to continue to live would be unbearable, and they were about to pass into a better existence, they again followed his instructions. Forty or fifty, about five percent, had already become disillusioned with life at Jonestown, or sensed that something horrific was about to happen, and tried to slip away into the surrounding jungle. Most were shot or held down and forcibly given a lethal injection. Some did escape and lived to tell their stories.

Shortly after Jones issued his final order, 912 people, including nearly three hundred children, lay dead on the ground. This grisly episode gave rise to the black humor expression "drinking the Kool-Aid," which means never to question established authority, to submit to peer pressure and go along with the crowd even if what they're doing is totally insane.

Some consider what happened at Jonestown a mass suicide, while others call it mass murder. Either way, eerie parallels with mass vaccination exist, if on a much less apocalyptic scale. Vaccines, while rarely killing within minutes, and in most cases not killing at all, nevertheless contain toxic substances and are potentially dangerous for an indefinite period. Yet, most parents, out of a combination of fear, ignorance and social pressure, yield to authority and allow their children to receive scores of them. (Millions, in fact, get the Kool-Aid injected into their own veins when they get their annual flu shot.) They figure it's the right thing to do, simply because everyone else is doing it. And like the pathetic disciples of Jim Jones, their minds are so weak, they are so incapable of thinking for themselves, that they must allow others — the medical "experts" in both government and private practice — to make life and death decisions for their children. There are a few, like the disaffected five percent at Jonestown, who resist, but the other ninety-five percent feel much safer obeying authority and following the crowd.

Is America one big, slow-motion Jonestown?

Or are they more like the people of Cambodia in the late 1970s?

COMMUNISM has many definitions, but the common denominator is that it is the most murderous political system ever conceived. Throughout the twentieth century it erupted in all parts of the world in various forms and degrees, all of them destructive to human life and aspirations. But nowhere was communism more genocidal, on a per capita basis, than in Cambodia, a small country in southeast Asia, where a political party backed by a peasant army known as the Khmer Rouge, and led by a strange man known to the world as Pol Pot, seized power. Pol Pot was driven by a dream to resurrect the ancient glories of the Angkor kingdom — *Khmer* and *Angkor* being names evocative of Cambodia's history — by obliterating all modern influences, including cities. Returning the country to its rural origins was his highest priority. He called his organization *Angkar*.

Riding to power on the convulsions of war, particularly the intensive American air bombing of neutral Cambodia in the region near the Vietnamese border which helped to swell their ranks, the victorious Khmer Rouge army marched into the capital Phnom Penh on April 17, 1975 and immediately began to evacuate the city. Shots were fired as a warning to the people to leave their homes and gather in the streets, where they were told that the Americans were about to bomb the capital and they would be allowed to return home in three days — both lies. Instead, they were forced to march for several days into the countryside, and within a week Phnom Penh, a city of two million, became a ghost town. Smaller cities were emptied in the same manner. Naturally, the people of Cambodia were terrified by this sudden, radical turn of events but they offered no resistance.

Pol Pot renamed the country Kampuchea and set the calendar back to zero. As the mass exodus got underway, people began to suffer, the old and infirm falling by the wayside, soon to be joined by those dying of hunger and thirst. Many more were executed for the most trivial reasons. Families became separated, husbands and wives, parents and children never to see each other again. Reaching their destinations, the people were put to work on grandiose agricultural schemes. Absurd quotas of rice production, impossible to meet, were established. No complaints or protests were tolerated, and all displays of emotion — affection, arguments, crying — were banned. The most timid, fearful questions put to the Khmer Rouge cadres — many of them young teens whose minds had been twisted by heavy indoctrination — would elicit a robotic reply like "The Angkar will take care of you." The slightest deviation from prescribed rules — like giving half a banana to someone seriously ill — would provoke a furious outburst from a guard denouncing the guilty one for questioning the wisdom of the Angkar; to avoid execution, the transgressor would cravenly admit his guilt and pledge to reform his behavior.

People were reduced to the lowest level of existence and died like flies of disease, malnourishment and exhaustion. Anyone considered a burden or a troublemaker was summarily executed, usually by a blow to the head with a farming tool. Sealed off from the outside world, Cambodia sank deeper and deeper into the realm of the unreal. In the words of one survivor named Pin Yathay, "There were no prisons, no courts, no universities, no schools, no money, no jobs, no books, no sports, no pastimes.... There was no spare moment in the 24-hour day. Daily life was divided up as follows: twelve hours for physical labor, two hours for eating, three hours for rest and education, and seven hours for sleep. We all lived in an enormous concentration camp."

The nightmare ended with the overthrow of the regime by the invading Vietnamese army in January 1979. In less than four years, the Khmer Rouge had killed off an estimated two million people, one-fourth of the Cambodian population. Since then, the country has come back to life, but in many ways is still struggling to recover. Shrines filled with the skulls and bones of victims are scattered throughout the country, in stark testimony to what took place here not long ago.

❖

Nature differentiates on its own terms, not man's. Nature may not differentiate at all between the short-lived Khmer Rouge terror and

the quiet, gradual deterioration of health for many spelled out in America's childhood vaccination schedule. Both are extreme examples of delusion and fanaticism run amok. What if the multitude of shots routinely given to most children and adolescents in this country — by age eighteen most will have received between fifty and sixty doses — results in an epidemic of infertility, in addition to so many physical and psychological problems as to prevent twenty-five percent of them from having children of their own upon reaching maturity? Twenty years from now, if the number of healthy children who should have been born to the "shot to death" generation is thus reduced by twenty-five percent, what will the difference be compared to what happened in Cambodia in the 1970s?

Nature's answer: Nothing.

Introduction

I have no doubt whatever that vaccination is an unscientific abomination and should be made a criminal practice. — George Bernard Shaw, 1931

VACCINES are the biggest fraud in the history of medicine. How do I know this? For the same reason I know that former professional football star O.J. Simpson, a black man, savagely murdered his white ex-wife and her male friend on the night of June 12, 1994. There were no witnesses, there was no smoking gun, but every line of evidence and every scrap of common sense pointed to him as the killer. The fact that a jury with nine blacks found Simpson not guilty, in the most publicized courtroom trial in American history, showed that the so-called justice system in this country is as dysfunctional as the so-called health system. Even the dullest minds instinctively understood that this absurd verdict was not the outcome of any rational deliberation but an expression of racial solidarity. (One juror actually gave the clenched fist "black power" salute after it was read.) In the same sense, the glowing "verdict" on vaccines endlessly proclaimed by the overwhelming majority of pediatricians, medical authorities and public health officials, not to mention news media commentators, is a farce. Not only are these people pulled emotionally in favor of vaccines, as were the Simpson jurors in imagining their man was innocent, they have financial and in some cases political interests as well.

The procedure we call vaccination began in 1796, and since then billions have been administered. There is no evidence that one of them has done any good. There are *oceans* of evidence that they do not work, often cause the disease they're supposed to prevent, and carry a long list of terrible side effects, including death. Furthermore, from the beginning, most major figures involved in the development and production of vaccines — those famous and not so famous — have been shameless liars, debased criminals, or both. From the beginning, the premise underlying vaccination theory — that with minimal risk, diseases can be prevented by injecting germs into the body, in order to stimulate natural defenses — has consistently been shown to be false. Vaccines never have done and never will do what their proponents claim, and the practice itself has always been tainted by every form of corruption. For the sake of our children — not to mention people of all ages, and domesticated animals besides — it is imperative to abolish all vaccines and consign them to medical history's junkyard of fallacies and stupidities, where procedures like bloodletting, using heroin to suppress coughs, and treating syphilis with mercury were abandoned long ago.

And yet, despite all the evidence that vaccines have destroyed and continue to destroy so many American children, and also — to be discussed in these pages — that they may well be the origin of polio outbreaks and Down's syndrome, as well as the Spanish Flu and AIDS epidemics, which in the twentieth century probably killed more people than all wars combined — despite all this, we have been going full speed ahead in the opposite direction, especially in the last thirty years. Sometimes it seems as if the hidden schemers who determine vaccination policies and buy out politicians to enact more draconian vaccine laws — always with the connivance of the mass media — are carrying out a secret death wish against humanity. I wouldn't be surprised if some of them are. What we have here is an incredible situation. It justifies mass hangings of liars across the board. It justifies the abolishment of public health bureaucracies at all levels and the demolition of CDC headquarters in Atlanta — or perhaps more fitting, converting it into something like the Tuol Sleng Genocide Museum in Phnom Penh, Cambodia, a testimonial to what the Khmer Rouge did to that country, or the National Holodomor Memorial in Kiev, which commemorates the estimated five million victims of a deliberately engineered famine in the Ukraine in the 1930s.

I thought long and hard about coming up with a more provocative title for this book, one reflecting my belief that parents must be prepared to kill in defense of their children, but decided against it. It's a delicate balancing act these days, trying to get attention and attract readers without crossing an invisible line in a country as lost and leaderless as this one, where faceless bureaucrats and mediacrats are always on the lookout for what they call "hate" and "extremism." Like so many institutions in America today, the mainstream book publishing, reviewing, publicity and distribution network is a travesty. Nearly all books that are politically incorrect in the true sense of that term are self-published, as this book is, and invariably get the silent treatment; there is no mention of them in the media, they are refused distribution through normal channels, bookstores and libraries want nothing to do with them, and the general public is unaware that they exist. So you do what you can, within reason and within the law, to get noticed. And I should make it clear that I'm not trying to glorify violence. I would much prefer to see the objective truth about vaccines be established by calm, honest application of the scientific method. If only American parents were truly free to decide, as they are in deciding to get a flu shot for themselves, whether or not to vaccinate their children — a basic freedom that exists throughout the Western world and elsewhere (though that freedom has recently become imperiled in several countries) — I would never have written this book. Nearly every vaccine zealot, however, firmly opposes this freedom, and even though American children already are the most heavily vaccinated on earth, they favor even more vaccines and tougher laws. In my opinion, these people deserve to die. As we shall see, some of them believe in killing their opponents themselves, and likely already have, hiring professional assassins. This is nothing new in history, though it will come as a shock to complacent Americans who think they live in a nice country. I am not by nature a violent man, but I do believe in the death penalty for heinous crimes, and I recognize that there have been times throughout history when only violence gets the job done. In case anyone has forgotten, our precious freedoms, now mostly a distant memory, were secured only after American colonists took up arms and killed a lot of British soldiers.

Actually, if we lived in a vigorous and sane society, many public health officials, drug company executives and pediatricians — the ones who vaccinate all their young patients, but not their own children — would have been put to death long ago. This is not crackpot or "extremist" rhetoric. In his book *Man, The Unknown*, published in 1935, Nobel laureate Alexis Carrel, a French philosopher, biologist and surgical pioneer — and co-inventor of the world's first artificial heart with the great aviator Charles Lindbergh, also a man of multiple talents and interests — maintained that not only vicious criminals, but leaders in advanced societies who betray the public trust should be executed. Carrel wrote: "Those who have murdered, robbed while armed with an automatic pistol or machine gun, kidnapped children, despoiled the poor of their savings, *misled the public in important matters*, should be humanely and economically disposed of in small euthanasic institutions supplied with proper gasses." (Emphasis added.) Some of these people who are addicted to misleading the public about vaccines also stand convicted of murder in my eyes, though in legal terms "depraved indifference" might be more accurate.

Let me reiterate that the most profound issues are at stake here, which have been addressed by several voices in the wilderness. One such voice was that of the late Dr. Harold Buttram:

Are we a nation of people incapable of recognizing imminent danger signs in the health, welfare, and genetics of our children, and in recognizing these dangers to take corrective actions in their behalf? I think we are capable of taking such actions, but time may be running out. At some unknown future time this process will reach a point of no return in terms of vaccine-induced genetic hybridization that will become incompatible with human reproduction. Mass extinctions are already taking place in plant and animal species, largely due to human encroachments and interventions. Are we soon to follow suit?

❖

This book is not intended to be an introduction to the subject of childhood vaccinations. Those wishing to familiarize themselves with basic, vital facts — facts such as the ingredients in vaccines, how they are made, the endless lies and cover-ups of those who endorse them, the testimony of grieving parents, the collusion between pharmaceutical companies and federal public health agencies, the deadly experiments carried out on children regarded as expendable, and a great deal more — should turn to the bibliography, where I list and summarize those books that were most important to my own education. Valuable websites are also listed at the end. I do touch upon these topics throughout these pages, and occasionally repeat things that already have been written, but essentially this book is an afterword to this repository of accumulated knowledge. Perhaps it is a vain wish on my part that it be the *final* word, especially since, being the perfectionist type, I've always been dissatisfied with what I've written, and already know, as I write these words with much revision yet to be done, that I'll be unhappy with the final product.

This book was a one-man project and part-time "work in progress" for more than ten years. I have revised and updated it many times, have checked statements of fact and often cross-checked them, but towards the end I realized I could not possibly redraft and bring everything up to date. Some chapters have been updated just before publication, others not in five years or more. I also debated whether or not I should write about things that some people would be sure to find offensive, going back and forth numerous times, and finally decided to compromise on this, though how the reader feels about it depends entirely on his or her perspective. Long ago I gave up on trying to shape the chapters that follow into any kind of logical progression; instead, for the most part, this is kind of a scrapbook of ideas and personal stories, quite different from anything else in the anti-vaccine literature.

In creating this book I had to process an enormous amount of information, even though in the end I took some of it out. I also had to resign myself to some informational gaps, including the latest developments in America which have picked up speed considerably in the age of Trump. Nevertheless, I'm certain, as much as I'm certain of anything, that no new turn of events regarding vaccines will nullify the ideas presented in this book, but will only substantiate them. I follow developments as best I can, and the facts about vaccines get worse all the time. Admittedly, due to time constraints, in some instances I relied on my memory and dubious sources like Wikipedia. I have found Wikipedia to be a generally reliable and convenient source of information on subjects that are not contentious, but not to be trusted otherwise. Although I place the highest priority on getting factual details right, there might be a few minor inaccuracies in these pages. But I wish to emphasize that any possible errors of fact are trivial and have no bearing whatsoever on the truth, as I see it, about vaccines.

In the pages that immediately follow, I will say more about the nature of this book, and I will share with you my innermost reasons for writing it. Throughout the book I've taken some detours and include some autobiographical elements and social observations. The connection to vaccines might seem flimsy in places, but to me it all hangs together. Some readers might think otherwise, and I admit there's a personal element involved — perhaps a subconscious desire to write about some things that I know would forever go untold if I don't write about them here. For the most part, these digressions are not at all controversial, and most will find them easy reading. Be warned, however, that on the whole this book is controversial, and unless you're familiar with the underground publishing movement that has quietly persevered for many years in this not so free country of ours, it's unlike any book you have ever read.

Maybe China can teach us a few things

CHINA has always been something of a mystery. The simple fact that Chinese civilization and the race that created it have been around for a long, long time commands respect. And certainly the people and their culture, generally speaking, have admirable traits. But there's also a dark side to the Chinese, and their way of life is definitely not an enviable one. In fact communism, as put into practice beginning in 1949 by Mao Tse-Tung, claimed more lives than in any other country, and was the model emulated by Pol Pot in Cambodia. In part, however, communism in China was a logical extension of the age-old habit of the leadership killing off its unruly subjects by all kinds of cruel methods.

Nevertheless, the leaders of China occasionally do things right — at least in recent decades. After Mao died in 1976, several major reforms were introduced which eased the crushing burdens of daily life to which ordinary people had become accustomed. And while restrictions still exist, China is now a much better place to live than it was under Mao.

As the world leader in the execution of bona fide criminals, including white collar offenders, China has also demonstrated its efficiency in getting rid of its human garbage. In the photograph below, taken on May 29, 2007, Zheng Xiaoyu, head of the Food and Drug Administration, the equivalent of our own federal agency of the same name, is handcuffed by two state security officers moments after a judge of the People's Court sentenced him to death for corruption. Zheng had been convicted of taking bribes totaling $850,000 from eight pharmaceutical companies in exchange for approving substandard medicine that killed or severely sickened hundreds in China as well as other countries to which it had been exported. After losing his first and only appeal, the sentence was carried out, six weeks after it was handed down — the speediness of the judicial process being another thing we can learn from the Chinese.

The gravity of the crime for which Zheng was dispatched has long been considered business as usual by executives at vaccine-manufacturing firms like Merck, and agencies like the Centers for Disease Control (CDC), the Food and Drug Administration (FDA) and others that are supposed to monitor the safety of pharmaceutical products to protect the health of Americans. In large part, however, these corporate and government entities work hand in hand, forming a network of organized crime.

America is terminally corrupt, and until justice and sanity return to this land, parents need to protect their children from the policies laid down by these psychopaths.

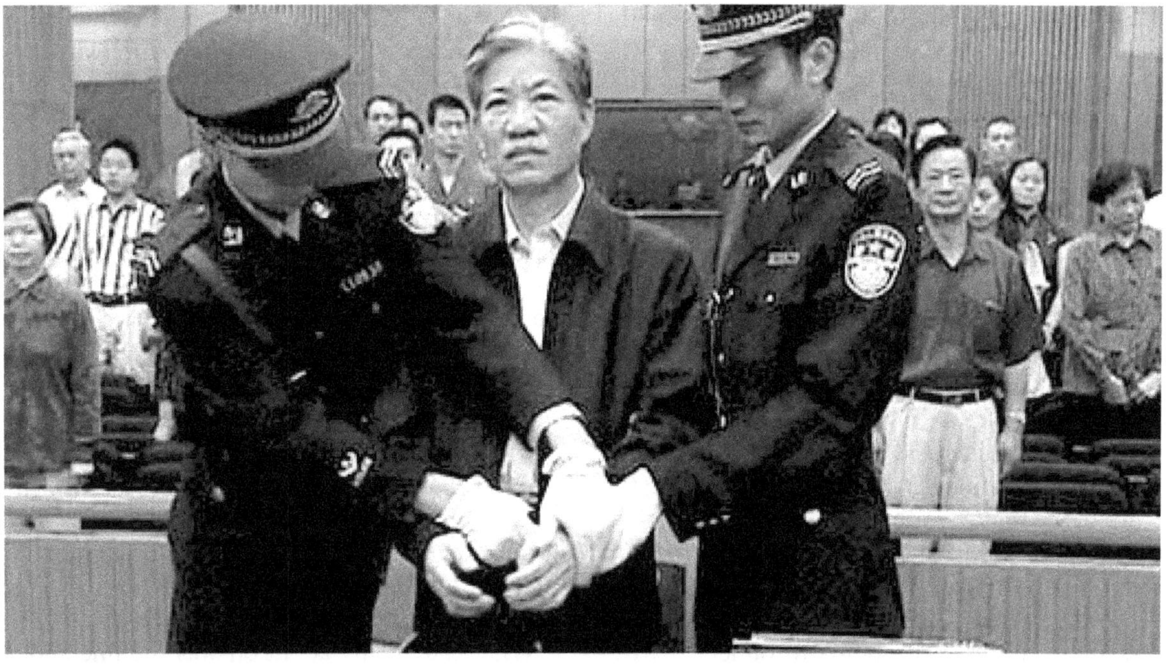

For me, the vaccine debate ended when I was 13

CHILDHOOD vaccination is an extremely emotional issue. That's because doing what's best for one's offspring is the most natural of instincts, the most basic law of life. But what is best? Giving one's child all the mandated vaccines, or refusing to give any of them — or perhaps something in between? There are those who think like I do, and those on the opposite end of the spectrum — and we all say we know what's best for our children. One side, my side, says we're destroying our kids by vaccinating them, while the other side says we would be destroying them by *not* vaccinating them. Is it any wonder that this issue is so explosive? And who is right? Where is the truth in all this?

Since the beginning of time, human beings have believed things that were false, and often ludicrous. The mind is so malleable, so easily seduced by error, so easily swayed by lying scoundrels as well as those who are sincere but mistaken in what they believe. Over the course of thirty years I studied the vaccine issue in my spare time, hearing out both sides, and even now I'm amazed at how wildly contradictory their assertions are and how nothing ever gets resolved. For example, this is what Neil Miller writes in his book *Vaccines, Autism and Childhood Disorders*:

According to WHO [the World Health Organization], *the odds are about 15 times greater that measles will strike those vaccinated against the disease than those who are left alone.*

And this is what Paul Offit writes about the same disease in *Deadly Choices: How the Anti-Vaccine Movement Threatens Us All*:

In 1999, Daniel Salmon and co-workers from the Johns Hopkins School of Public Health found that the risk of contracting measles in five- to nine-year-olds whose parents had chosen not to vaccinate them was one hundred and seventy times greater than for vaccinated children.

Now, simple logic tells us that one of these statements must be wrong; perhaps they're both wrong, but at least one of them *has* to be wrong. Endless examples of this kind of conflicting information can be cited. So again, who's right? It's no wonder that so many parents, anxious to learn the truth of the matter so that they can make the right decisions for their children, don't know where to turn.

So much information about vaccines has been disseminated through so many outlets, particularly in the last thirty years, that even the most tireless researcher with unlimited time at his disposal — which certainly does not describe me — wouldn't be able to examine more than a minute quantity of it. For a while, I collected as much as I could in the way of newspaper and magazine articles, pamphlets, newsletters and material from a wide variety of sources taken from the Internet. This filled a small cardboard box, and in addition I read 37 books on the subject, twelve of them by authors who took a strong pro-vaccine stance. After deciding to write my own book, however, I realized that I'd have to leave a great deal unexamined, and would never be able to keep up with all of the latest developments. But I also felt that I'd made a comprehensive study of this volatile issue, and that any intelligent man or woman who makes an equally inclusive study cannot fail to reach the same conclusion: that in addition to being the biggest hoax in the history of medicine, vaccination — especially compulsory childhood vaccination as practiced in the U.S. — is an enormous criminal enterprise.

From time to time, to refresh my memory about some forgotten detail, I randomly pull something out of my box that I read a long time ago and read it again. Recently I plucked an article from the June 25, 2005 issue of the *New York Times* titled "On Autism's Cause, It's Parents vs. Research." The article is contemptuous towards the anti-vaccine position on a subtle level, as anyone familiar with this newspaper would expect. In it, the two writers briefly discuss the work and publications of a father and son team, Mark and David Geier, who maintained that there is a link between vaccines containing the mercury-based preservative thimerosal and autism. The *Times* writers quote a Dr. Steven Black, director of the Kaiser Permanente Vaccine Study Center in Oakland, California: "The problem with the Geiers' research is that they start with the answers and work backwards," adding, "They are doing voodoo science."

The reporters then quote some federal health officials who dismiss the Geiers' findings, citing two foreign studies that showed the more thimerosal the *less* autism, and a national study that showed no difference. So much for studies proving anything. One can dig up a thousand news articles like this; they all lead nowhere.

What caught my attention was the line about

starting with the answers and working backwards. It struck me that that's what I had done in writing the opening pages of this book. But if the health of a nation's children hinges on the truth or falsity of those answers, what does it matter? What does it matter if the truth, unwelcome though it might to be to some, is written on a napkin, plastered on a billboard, or appears in the beginning, middle, or end of a research paper or book?

With that in mind, I will tell my own story by starting with the answers and working backwards because, in retrospect, I learned the truth about vaccines more than fifty years ago.

❖

I was born in 1953, the oldest of four children. I sometimes laugh about writing this book, because my first memory in life, at the age of four months, was kicking and screaming while being held after receiving my first DPT (diphtheria-polio-tetanus) shot. I don't actually remember the needle going in nor the agonizing pain that must have followed, but I do remember screaming. Many years later, wondering if all this had simply been a bad dream, I asked my mother about it. Yes, she said, I'd had a bad reaction to my DPT shot, and yes, I thrashed and screamed violently in her arms. As best as she remembered, she never brought me in for a second DPT vaccination.

I was also inoculated against polio and smallpox when I was very young, but I have no memory of those shots. When I was in the third grade, our school nurse lined all of us up in the hallway and we were given the new sugar cube polio vaccine to suck on, which we accepted with pleasure. So far as I know, my brother and sister, also born in the fifties, went through the same regimen; as small children, they could not have been vaccinated against any other diseases. In addition to the vaccines just mentioned my younger sister, the youngest in our family, got the measles shot, which came out in 1963, the year after she was born. That was the end of childhood vaccines for all of us; we all received six or seven doses. (I consider each component in a combination vaccine to be one dose, and refer to them as such throughout this book, so that a DPT shot would be three doses.) And that was about the same number that the great majority of us in the "baby boom" generation — loosely defined as those born between 1946 and 1964 — received.

When I was four my parents bought a house in Williston Park, a middle class town twenty miles east of Manhattan. With about 8000 residents and an area less than a square mile, it's one of the smallest towns in Nassau County, which takes up a sizable chunk of western Long Island. It was and still is a mostly "white ethnic" town, with at least half of Italian or Irish ancestry, though there has been an Asian influx in recent years. The entire region, of course, has long been completely developed, the only escape to Nature being a few unspoiled tracts of woods, and some quiet stretches of beach, miles away. Still, it was a nice place to grow up and I had a happy, normal childhood.

There was a strong sense of community in Williston Park, but in the densely packed suburbs, where one town blends into the next with no real boundaries, it sometimes overlapped other towns — to the north, Albertson, also very small, Mineola to the south, more than twice as big, and on the west New Hyde Park, also much larger. Upscale East Williston, on the east side of the railroad tracks, was where the "rich kids" lived, and since they had their own schools and social life, I had nothing to do with them. Nor did I know anyone in New Hyde Park, and very few in Albertson. But the Mineola line was just three hundred yards from our house, and my elementary school was in the Mineola school district, so I grew up with one foot in that town, so to speak.

From kindergarten through the sixth grade, I attended Cross Street School, a five minute walk from home. There were two other elementary schools in town, on Center Street and Park Avenue, which were in the Herricks school district. Both Williston Park and Albertson were evenly parceled between the Mineola and Herricks school districts, the latter of which also covered some of New Hyde Park. In addition there was St. Aidan's elementary school, right next door to Cross Street School, connected with the Catholic church of the same name directly across the street. Williston Park is mostly Catholic, and many parents enroll their children at St. Aidan's, as they did when I was young. It has always been the biggest school in town.

From my first day at Cross Street, I was with the same group of about fifty children, each year divided into two classrooms, so I got to know all of them very well. I was also well acquainted with many other students close to me in age, since we were in the same building together for years. Measles, mumps, chickenpox and German measles (now called rubella) were all common when I was young, especially measles, and most of my classmates came down with one or more of these maladies while in elementary school. We were all familiar with the sight of an empty desk as the day began, and there was never any cause for concern when our teacher announced that Barbara or Jimmy was out with the measles. In fact it always brought out a few snickers because it was funny — like winning the booby prize. Everyone knew, without it ever being explained, that these diseases were just a temporary nuisance, a rite of passage, and that our sick classmate would be back at his or her desk in a week or so. That's how it always happened — without exception. I've even read that kids used to throw "measles parties," deliberately infecting their friends, but that was a little before my time. While in kindergarten my brother came down with chickenpox and my sister, in those early years, both chickenpox and measles; my younger sister avoided all the childhood diseases, as I did.

As a boy I was very sociable, and had a wide circle of friends and acquaintances. I knew almost everyone two years ahead of me and behind me at Cross Street, most of those at St. Aidan's, and got to know many other kids in town, as well as their siblings, through social activities. I was in the cub scouts and boy scouts. Our family belonged to the municipal swimming pool, and many summer days were spent there. As a youngster I played the accordion, and took part in several recitals and competitions. I played Little League baseball beginning at age eight and after that played in the Babe Ruth league. When I got too old for that, I remained active in town for another four years as an umpire and coach, and got to know many more boys who were too young to be my friends, but old enough not to have received any more vaccinations than I did, or at most, one more. As with my own friends, I got to know many of their brothers and sisters too. In this way I learned of only one boy in Williston Park who was partially deaf. He was the younger brother, ten or eleven years my junior, of one of the boys I coached in Little League, and he came to the games wearing a cumbersome hearing aid, a heartbreaking sight.

In addition to organized activities, there was plenty of opportunity for the kind of mild mischief that can be expected of any normal boy growing up, and I was no exception. I came up against some of the rougher kids in town this way, but steered clear of the few really deviant types who, it was said, tortured cats or sniffed glue, the only kind of "drug abuse" I was aware of in the early to mid sixties. Through all the things I did as a child, wholesome or not, by the age of twelve I was acquainted with probably ninety percent of the boys and girls around my age who lived in Williston Park, as well as a fair number who lived in Mineola.

Some kids didn't do well in school, but I knew only one who didn't attend. He was the only child classified as mentally retarded whom I knew. In the unfeeling way that children can be we called him "Joey the retard," though never to his face. I can still see him with his short blond hair and black frame glasses serenely riding his bicycle — he always seemed to be on his bike — staring straight ahead, oblivious to the world around him. Sometimes my friends and I stopped him and bantered with him a bit. I had no idea what he did all day.

I never went to Sunday school, but we did have what was called "religious instruction." Every Thursday afternoon, from the first through the sixth grade, all the Catholic kids at Cross Street, about eighty percent of the student body, were let out an hour early and walked over to St. Aidan's to sit in another classroom under the tutelage of nuns or priests. Although my parents were not really devout, they did go to Sunday mass every week and they made me go with them. There was a Lutheran and an Episcopal church in town, but St. Aidan's had the largest congregation by far, and was pretty much the social nucleus of Williston Park. No matter which priest gave the service, if anything noteworthy in town had happened the previous week — certainly if there was a sudden, tragic death or a child suffering from an unusual illness — he would mention it. But these were rare events, nothing much ever happened in our town, and so the sermon usually revolved around the same old religious patter.

The other source of unhappy local news, when there was any, was the *Williston Times*, which was delivered to our house each week, and which I always thumbed through. Often the *Mineola American* was lying around and I'd glance at that too. There was hardly anything worth reading in either of them — just the usual dull goings-on and tepid articles found in any small town newspaper. But even when something bad happened, I almost always heard about it on the street first: the kid who blew off a few fingers while making a blockbuster firecracker in his basement; the three victims of a fatal house fire; in my later teens, the two my age — one was always in trouble and I just knew he would die young — who overdosed on heroin.

I finished up at Cross Street in 1965, and in September of that year began attending Mineola Junior High School. This was where all the public school kids who lived in the district went to seventh and eighth grade. In addition to all of us at Cross Street, the student bodies of Meadow Drive School in Albertson, and the Willis Avenue, Jackson Avenue and Hampton Street schools in Mineola all funneled into Mineola Junior High. I would estimate the entire student body there back then at 600. I got to know most of my classmates those two years, at least on a superficial level, as well as many students one year ahead and then one year behind me. There was one remedial class, what is now called special education, for nine or ten students who were too slow to keep up with the rest.

Junior high school, now called middle school, is the time in life when boys and girls start becoming interested in each other. This is when I first learned about mononucleosis, the "kissing disease" as it was called. It sounded exotic and scary when I first heard of it, but I soon learned that "mono" was nothing to worry about, although it could take weeks to recover. Mono had a slight notoriety, as if it were a mild venereal disease. Perhaps a dozen kids came down with mono those two years, and we all knew who they were, because the word spread like wildfire.

Sadly, a few young people from Williston Park whom I knew well died in tragic accidents, but it wasn't until I was sixteen that I came face to face with natural death early in life. Two boys, though several years apart in age, died of leukemia around the same time. One was the young son of my former scoutmaster; the other had played on my Little League team. The following year, a very popular boy in town, a senior at Mineola High School, died of complications from pneumonia, although there was speculation that it was connected to dehydration; as a member of the wrestling team, he had been abstaining from food and water in order to lose weight. His death shocked everyone, and a scholarship was established in his name.

Although I don't recall the exact time frame, over the course of those years I do remember that meningitis claimed the lives of two Herricks High School students on separate occasions. This was reported not only in the local papers, but also in the large circulation Long Island newspaper *Newsday*, and it sent ripples of fear through our neighborhood. The victims, who were unknown to me, were not from Williston Park but they couldn't have lived more than two miles away. Insulated as we were from deadly diseases, no one had ever spoken about meningitis — and suddenly here it was, right on our doorstep. Was it contagious? Would others die? Was this the beginning of an epidemic? But both times the days and weeks passed with no one else becoming ill, and both episodes were soon forgotten.

As an aside, after finishing at Mineola Junior High I attended Chaminade High School, an all-boys Catholic school right across the street, from where I graduated in 1971. Although it was in Mineola, very few boys whom I knew went there; nearly all of them came from towns within a twenty mile radius, so naturally I knew nothing of their earlier years nor what happened in their communities. What I will say for the record is that, of the approximately two thousand boys who attended Chaminade for at least one year while I was there — meaning all those from the classes of 1968 through 1974 — I never heard of one who suffered from a so-called vaccine-preventable disease, and in a close-knit school like that, where even the serious illness or death of a student's family member was announced over the PA during homeroom, I certainly would have heard about it.

Despite the fact that I went to a Catholic high school, or perhaps because of it, by the end of my sophomore year I'd become quite skeptical about the teachings of the Church and fed up with going to mass every Sunday. Eventually my parents got tired of arguing with me about it and I stopped going altogether. But they still went and brought home the four-page weekly bulletin which I always picked up and read. I always read the "Pray for the Sick" and "Pray for the Dead" columns first; I was simply curious about who was critically ill and who had recently died in our town. But I never saw any names I knew other than those of the few I have mentioned.

At this point I'll ask the kind reader to forgive me for being a bit nostalgic, and for going a bit overboard in talking about myself — a habit I've never quite been able to break. And perhaps I've slipped up on one or two minor details with the passage of time. But my intent in going into all this detail has been to impress upon you that I was well-informed about the lives of more than one thousand children aged five through thirteen, boys and girls who passed through my life and whom I knew by name, and many others besides, not one of whom suffered any harm from not receiving any of the additional forty or so vaccine doses now pushed by public health authorities from birth through age fourteen — and on top of that twenty more by age eighteen. (Even just a few vaccinations might have caused the leukemia and meningitis deaths I've mentioned, though of course I have no proof.) And more than that, many conditions that children suffer from today were unknown when I was young. I do remember several kids who had allergies to certain foods like chocolate, peanut butter, tuna, whatever, but had no problems at all as long as they avoided them. I never heard of a single case of eczema, asthma, diabetes, digestive tract disease, nor aside from the one case of partial deafness remarked upon, any serious hearing or visual afflictions that are now common. Moreover, the percentage of children in special education programs was only a fraction of what it is today. Certainly there were kids who were undisciplined or "off-key," but I don't recall one with a marked psychological or behavioral disorder. And here I must bring up autism, a word I never heard until sometime in the 1980s. The condition was so rare in the sixties that no accurate records were kept; a widely accepted estimate is 1 in 10,000. In 2016 the rate acknowledged by the CDC was 1 in 68, in 2018 about 1 in 45.

Furthermore, I asked my brother and sister, who were behind me by three and four school grades, respectively, and who, therefore, knew many more children that I did not, and who, incidentally, do not share my strong feelings about vaccines — I asked them if they knew of anyone, while we were growing up, who paid a heavy price after contracting a disease for which vaccines exist now, but were unavailable then. Neither could come up with a single name.

❖

Let's continue working backwards. Let's first consider the Salk polio vaccine, which came along when I was a toddler. As a child and young man, I came into contact with innumerable people ten years or more my senior, who could not have gotten this shot before adolescence. Yet I knew only one who had been stricken by polio, the father of a close friend, whose left leg had been weakened — if my friend hadn't told me about it I wouldn't have known — and who always appeared normal to me. No one ever talked about the bad old days when polio supposedly was a national scourge; no one ever mentioned someone they knew who had been paralyzed or killed by the disease; no one ever told me I was lucky to be born at an ideal time to reap the benefit of the vaccine.

Speaking of the generation that preceded mine, those born in the 1920s, as my parents were, not only missed out on the polio shot when they were young, but the great majority got nothing more than one smallpox inoculation. In reading about the history of vaccines, I came across vague, conflicting information regarding the diphtheria, pertussis and tetanus vaccines, which were in fact developed and licensed in the early twentieth century, though without the fanfare that later attended the polio vaccine. But I could find no estimates as to how many children received them, and nothing I read suggests that they were compulsory, nor that there was any enthusiasm for them. The three-in-one DPT shot came along later, in the 1940s, and that's when these vaccines began to be administered on a mass basis, not before. Diphtheria, even in the years of highest incidence, early in the twentieth century, was always a rare disease, tetanus extremely so. Pertussis (whooping cough) has always been far more common. My mother told me she had it as a child, and I came down with it myself while traveling in Asia in 2013. It lasted for more than three months, and it wasn't much fun, but not then, now, or at any time in the future would I ever consider getting a shot. The whole-cell pertussis vaccine, incidentally, which I received as an infant, and which was later replaced by an acellular version, is widely regarded, even by some vaccine advocates, as the deadliest baby shot ever given — not that the acellular vaccine that children get today should be considered safe.

So it appears very likely that most of our parents, and even many of those born in the 1930s, were vaccinated only against smallpox, a shot that was discontinued in 1972. But to make a final thrust backwards in time, our grandparents, born within range of the turn of the twentieth century, could never have received anything *but* the smallpox vaccine. And if they did, it was in the midst of a firestorm of controversy that had already been raging for a century. No other vaccine has provoked more polemical literature, and more violent resistance — and for a much longer period of time — than the first one.

To return to the present day, we baby-boom parents, more than anyone else, should have raised all kinds of hell against not only vaccinations for the benign diseases we were familiar with as children, but against the onslaught of new vaccines given to our children for diseases we never heard of. Where, for example, did this rotavirus come from? What is it? What about Haemophilus influenza B? Human papillamovirus? Whoever heard of anyone but homosexuals, with their biologically insane perversions, and needle-sharing heroin addicts — and the very rare victim of a contaminated blood transfusion — contracting Hepatitis B?

Half the answer lies with the National Childhood Vaccine Injury Act, to be discussed in this book, and the other half can be inferred from a book I read several years ago titled *Selling Sickness: How The World's Biggest Pharmaceutical Companies Are Turning Us All Into Victims* by Ray Moynihan and Alan Cassels. Not a word about vaccines appears in this outstanding work; instead it documents the juggernaut marketing campaign, which began in the late 1970s, and has never let up, to scare people into believing they are at risk of numerous debilities that exist only in the creative minds found at top advertising firms — and then buying potentially dangerous drugs to treat them. The list is long: high cholesterol levels, osteoporosis, irritable bowel syndrome, erectile dysfunction, social anxiety disorder, premenstrual syndrome and many others. Like all the recently invented Madison Avenue childhood diseases, these make-believe maladies came out of nowhere; prior to the late seventies, they were unheard of. All of them sprouted like mushrooms from the relentless marketing drive of an industry wallowing in a half *trillion* dollars. The difference is that American adults, by their own free choice, have become by a wide margin the world's leading prescription pill poppers, while their children and grandchildren are *not* free, but must comply with state vaccine laws to attend school — laws rammed through by prostitute politicians, their pockets lined by the drug lobby.

In tandem with all this dirty business is the huge number of American children and teenagers, some four million, prescribed mind-altering drugs like Ritalin and Adderall to treat a host of acronymic mental ailments of which attention deficit hyperactivity disorder (ADHD) is perhaps the most commonly diagnosed. If ADHD genuinely exists in certain cases, vaccines are definitely the cause, because there was no such thing when I was growing up. If it doesn't exist, other than in the minds of school psychologists, then in addition to all these new vaccines, rambunctious children today obviously need to be drugged for their own good. What on earth has become of this country? Just what is one to think of parents who go along with this?

Of all the insane policies that have taken hold in America over the course of my life, none is as abhorrent to me as the now routine hospital practice of injecting babies, *within hours of birth*, with the Hepatitis B vaccine. Even if it were safe, which it most certainly is not, the danger that this disease poses to all age groups is nil, except in the high-risk groups mentioned above. Yet not only is this totally unnecessary shot standard procedure in hospitals, but newborn babies are often vaccinated before their parents are notified, if they're notified at all. And what is even more shocking is that very few parents are up in arms about it. Accustomed their whole lives to drinking the Kool-Aid, they accept it as normal.

Reader, if you are still wondering about me, it would go a long way if you share my sense of outrage at this obscene ritual that defiles the miracle of life, the ritual of welcoming a baby into the world with a syringe loaded with Hepatitis B vaccine — and if you share my feeling that something needs to be done about it.

❖

We live in a truth-hating age in which lies have become so pervasive and widely believed that many Americans are almost completely disconnected from reality — a subject I'll take up in these pages. This tissue of lies includes the nonexistent benefit of vaccines. For my part, in dealing with a controversial issue, I've learned to believe without reservation only two types of phenomena: first and foremost, that which I've seen, or haven't seen, with my own eyes; second, facts that might be ignored or covered up by the opposition, but are not disputed. That will be a recurring theme in this book. Of all the children I knew when I was young, I never saw nor heard of one who suffered any ill effects from chickenpox, mumps, measles or rubella, which were common then, nor did I ever see or hear of a child who contracted any of the other diseases for which vaccines began to be piled on in later years. And I have never heard of anyone disagreeing with the claim that today far more children suffer from physical, mental and emotional problems than did children of my generation, because *it's indisputable*. That impresses me far more than statistics or studies of any kind, including those that support my position and that I believe are honest — a very few of which I've cited, in passing, in these pages — and I have no interest whatsoever in what a million vaccine-touting pediatricians, public health officials, scientists and other self-styled experts have to say. I know what I saw and didn't see when I was growing up, and I have also seen, over my lifetime, how harmless diseases have been mendaciously transformed into life-threatening diseases, and how many others sprang up out of nowhere.

Could I be wrong about everything I've written here? Yes. I've been wrong before about a few things I felt strongly about. Maybe there were numerous casualties of childhood diseases who were hidden or sent away by their parents in my town, children who were unlucky to be born a few decades before all these great new vaccines came along. But you're going to have to show me solid proof, and I really don't think that's going to happen.

Vaccines and me

OF COURSE it's only in retrospect that the vaccine debate ended for me when I was thirteen. Before the age of thirty-three, I was completely unaware of any controversy surrounding childhood and all other shots. Like most people, I saw them as an important milestone in mankind's endless quest to wipe out deadly diseases. But the subject didn't interest me and I never gave it any thought. All that I learned until later in life was what my fifth-grade teacher taught us one day. We learned about the great Jonas Salk and his miraculous vaccine which conquered polio and put an end to the constant fear of this dreaded disease. She also told us in a deplorable tone that at the time, this monumental medical breakthrough went largely unnoticed, because the front page headlines were all buzzing about the Hollywood actress Rita Hayworth's latest marital scandal. I later discovered that this statement was absurd, that you had to be living on Mars in April 1955 to escape the media frenzy surrounding Salk and his vaccine. But fifth grade was pretty much the end of my education about vaccines for a long time.

The first time vaccines became an issue with me was during the swine flu scare of 1976, when I was in college. Like millions of other Americans, my parents joined the stampede to the doctor's office to get their shots, and pleaded with me to get one too, but I said no. It wasn't that I knew anything about vaccines; it was more my rebel nature, my refusal to be manipulated by scare tactics and vague rumors of a coming epidemic. As it turned out, the swine flu vaccination campaign was a disaster, with lots of finger-pointing and egg on the faces of health bureaucracy "experts." No swine flu ever broke out, hundreds were killed or seriously injured by the shots, and the program was quietly dropped. Actually, it was a scenario that has been repeated countless times with vaccines, the only difference in this case being the intensive news coverage leading up to it with the president himself, Gerald Ford, leading the country by rolling up his sleeve and getting the shot. But at the time I figured it was just a fluke, a well-intentioned plan that somehow went wrong, and I quickly forgot about it.

A few years later I began driving an oil truck, which has been my livelihood ever since; I deliver heating oil to homes and businesses on Long Island. Before marriage and children came along, I worked only five or six months during the cold weather, and satisfying a thirst for adventure and a desire to see as much of the world as I could, I spent two or three months traveling abroad as cheaply as possible each summer. Each year from 1981 to 1988 I traveled to a different part of the Third World. In 1981 I went to western Africa. I had learned that some countries in the region required proof of vaccination for yellow fever and cholera, and since I was concerned about my health, I read a book titled *The Traveler's Health Guide* by Dr. Anthony Turner, a British Airways medical officer with long years of experience in the world's undeveloped areas. He strongly recommended not only the required vaccinations, but others as well. So two months prior to departure, I went to the medical building at Kennedy Airport in New York and got shots for yellow fever and cholera, which were recorded and stamped in a small booklet in English and French that they gave me. Before giving me the yellow fever shot, which was said to be effective for ten years, the nurse on duty asked me if I was allergic to eggs. I told her I wasn't. That was the first and last time anyone ever asked or told me anything prior to sticking a needle in my arm. Later that month I returned for a typhoid vaccination as well as a cholera booster.

Over the next five years, before leaving home,

I got more cholera and typhoid vaccinations, and as an added precaution against hepatitis A, which is common in Africa, Asia and South America, I requested some gamma globulin injections as well. In all, I got about 15 "travel shots" in the 1980s. I figured they couldn't hurt, they could only help, so why not get them? I'd never heard anything bad about vaccines, no one giving me the shots had ever mentioned any risks or side effects, and other than my DPT shot as an infant, I never had an adverse reaction to any vaccination.

The first time I heard "the other side of the story" was in 1987, while traveling in Nepal. I'd met two Australian women who, like me, were interested in trekking for four or five days in the Himalayas, so we hired a guide and some porters through a travel agency in Kathmandu, the capital, and off we went. One evening, while chatting and sipping tea around a campfire, the subject of health came up, as it often does among travelers to Third World countries, and I casually mentioned that I'd gotten a gamma globulin shot before leaving home. They replied that they chose not to get one because gamma globulin was made from other people's blood, and there was a controversy in Australia as to whether one could be infected with the AIDS virus from this shot. *Other people's blood!* When I heard these three words I nearly fainted. Up until then I had never given any thought to the contents in that needle, nor that there was danger of any kind associated with vaccines, the swine flu episode notwithstanding. Now, since the AIDS virus was said to have a long incubation period, and since this was my third or fourth gamma globulin shot, I wondered if I was a dead man walking. But 32 years have elapsed since that conversation — after which I never let anyone stick a needle into me again — and I'm happy to report that I'm still here.

Five years later I got married. Since both my wife and I were well into our thirties we tried to have children right away, but without success. After three years with no results, we started going to fertility specialists and having tests done to identify the problem. They couldn't find anything wrong with her, but I was told that a microscope examination had revealed antibodies on my sperm cells, and left with the speculation that this was why she had failed to become pregnant. I mentioned this to a friend who worked in the health field, and he told me it was the strangest thing he'd ever heard. When I asked the doctor how this could have come about, he told me he had no idea. It would be years before I learned that vaccines can impair fertility, and provoke other conditions in which the body's immune system goes haywire, attacking its own cells as if they were foreign pathogens. The possibility of fertility impairment and many other side effects is mentioned in fine print on the product inserts of several vaccines that few people, including doctors, bother reading. Incidentally, I later learned firsthand of another man with the same condition. He had received numerous vaccinations during his years in the Navy, and was the only infertile one in a family of seven.

In 1993 my wife and I began trying to have a baby through in vitro. By this time my wife, who had heard some bad things about childhood "immunizations," as they are deceptively called, had gotten me interested in the subject since, of course, I was planning to be a father. She listened to Gary Null, who often denounced vaccines on his weekly radio program. I started listening to him myself. Null, an advocate of holistic medicine, has been around a long time, and while his occasional rudeness to his guests irritated me, I felt that he was really on to something when he ripped vaccines. It was his show that prompted my wife to order Neil Miller's *Vaccines: Are They Really Safe And Effective?* I read this little book in one sitting and it struck me like lightning. We agreed, as we agreed on almost nothing else, that our children would never be vaccinated.

I became obsessed with this subject, and started reading everything I could. And the more I read, the more incensed I became. Throughout this period, I began to observe something else, both before and after I learned that "routine" vaccinations can damage the auditory nerve. When I first started delivering oil, I noticed, at most, three or four "Deaf Child Area" signs on the many residential streets I drove on. A few years later it was eight or ten, then a few years after that twenty-five or thirty, and eventually seventy or eighty. They were popping up everywhere, and looking back, their steady increase corresponded to the ever increasing number of childhood vaccines. Over the years I've driven on, at most, ten percent of the streets on Long Island, and I can only guess that these signs represent a small fraction of the children in Nassau and Suffolk counties who have suffered serious hearing loss or total deafness. As the years passed, signs indicating blind and autistic children living nearby also began to appear.

I won't go into detail regarding our experience with in vitro, other than to warn you to do your homework thoroughly prior to undergoing any medical procedure, and to never blindly trust what doctors tell you. After two failed attempts, my wife did become pregnant, but we were given poor advice, which led to our making an unsound decision, not informed about certain risks, and as a result she gave birth to twins three months prematurely. My daughter went into shock right after she was born, my son needed an emergency operation to keep him alive, and they both remained in the neonatal intensive care unit for about three months. Now in their twenties, both have mild cerebral palsy, and my daughter is also severely autistic, having been diagnosed at the age of three. She had suffered heavy brain bleeding after birth, and a scan taken shortly before she left the hospital showed an area of dead cerebral tissue.

For a little icing on the cake, I went through more than three years of bitter divorce litigation, which began when my children were five — a battle not of my choosing. Combined with the despondency over my daughter's condition, and losing my job — a risk I had taken after months of arguing to please an unbalanced woman whom I never should have married — it was about all I could endure. In writing these words, I'm not venting my sorrows and misfortunes; that's bad form. I only wish to make it clear that the idea of writing this book was born in the crucible of those years. I look back with few regrets and much satisfaction on a life well spent — particularly my adventures around 97 countries in every part of the world. And I have long since resumed a normal existence. My daughter is in a home, well taken care of, my son, though dependent on the care of my former wife, is doing well; although they live more than a hundred miles from me, I see both of them regularly, and have shared many wonderful times fishing, camping and traveling with my son. You adjust mentally, and you put your trials and tribulations in their proper perspective. Many others have had it far worse than I, not to mention the hundreds of millions who have perished under murderous governments, and died or been disfigured in war. Tragedy is part of the human condition; some just get hit harder than others, and in the end we all get an equal dose.

But there were low points during this period, and even later, when I struggled to find some meaning, some purpose to my life. Many a night I sat in my recliner, alone in my apartment, reflecting on the events in my life, good and bad, that led to the present, and wondering what I hope to accomplish in my remaining years. Most parents never think about this; their purpose in life is to do the best they can for their children, to see them grow to their full potential and go on to have children of their own. But for parents of handicapped children the situation is entirely different: while some may feel otherwise, for me, it was a living death taking care of my daughter after she was diagnosed as autistic. And while my son is better off, there is still the void of knowing he will never be able to live independently, and the uncertainty of what will become of both my children when my former wife and I are gone.

The more I reflected on my situation, the more I was driven by an inner need to leave something of myself behind — if not in the grandchildren that I'll never have, then in a small flame to illuminate the hard, upward path taken by the best men and women. I would transform my tragedy into something good, something that would derail the approaching tragedies of other families. I would use the gift of a rebellious intelligence bestowed on me to expose one of the most catastrophic ideas ever hatched by misbegotten human brains. Thus, I wrote this book because I *had* to. And although the manner in which vaccines upended my life is unusual, maybe even unique, my feelings are exactly the same as those of so many other parents whose children were maimed or killed by them. In the words of a woman who lost her infant son, an only child, ten hours after receiving the DPT, Hib and oral polio vaccines, "Now I will channel all of my anger and my ultimate, unconditional love for my precious baby into this cause [exposing the danger of vaccines]. No one should have to feel and know this living nightmare."

Monuments to medical madness

OTHER than route drivers who cover a wide area, very few know of the abundance of signs on Long Island — and probably in many areas of the country — warning of a disabled child in the vicinity. There has been a clear correlation between their increase and the steep increase in childhood vaccinations. Plenty of parents would start asking questions about vaccines if they knew of these signs.

In 2008, around the time I first thought about writing this book, I began making a list of the streets where these signs randomly appeared on my rounds making oil deliveries. (There were also a few near my home.) Most of the first signs, I noticed years earlier, advised of a deaf child, later blind children as well, though for a long time I was ignorant of the connection between vaccines and ocular problems, and neglected to record them. Thus, there were fifteen or twenty I didn't photograph. The signs alerting drivers to an autistic child began to appear later. Over the years, I occasionally delivered oil in areas new to me, and consequently came across signs I hadn't seen before. I wrote all of them down, and took all the photographs seen here over a two year period beginning in 2015. When I finally got around to doing this, I noticed about thirty signs had been removed, apparently because the child had grown into adolescence or moved, and I also noticed that on quite a few signs, the word "child" was replaced by "person." These I did not photograph.

So what do these signs prove? In a court of law, probably nothing. But to my mind, these signs are a stockpile of circumstantial evidence that vaccines destroy lives. Of course I could've gone around knocking on doors locating the homes of these children, and getting their parents' opinions. I didn't do this because I felt it was an invasion of privacy, and it would have been very time-consuming. Also, there frequently is a delayed onset between vaccination and visible injury, and uninformed parents might not recognize a link, or even deny it. Nevertheless, it would be interesting to find out how all these parents feel. It also would be interesting to find out if, among all these children, a single one was unvaccinated.

These signs undoubtedly represent a small percentage of the disabled children among us. For lack of knowledge or personal reasons, many parents never request them. I know of two autistic children in different towns with no signs on their streets, and I've seen countless cars with the familiar "autism awareness" sticker parked in driveways on streets where no sign stood.

There are approximately 38,000 residential streets in Nassau and Suffolk counties, which cover all Long Island east of the New York City boroughs of Brooklyn and Queens. The great majority of them I've never seen. Some towns I've worked heavily, many others not much at all, and in the case of eastern Suffolk, never. There are no readily available statistics for how many handicapped children live on Long Island, much less statistics of confirmed vaccine injuries. I accept that a small amount of blindness and deafness may be due to genetic factors, but in all or nearly all cases of autism, I indict vaccines. I should also point out the absence of signs indicating that a child with asthma or diabetes lives on the block, these conditions also caused by vaccines, and that the great majority of vaccine injuries are not severe but rather mild to moderate, for example dyslexia or impaired visual acuity. Taking all this into account, it would be unwise to offer an estimate of the number of children and teens in this area whose lives have been impeded or ruined by vaccines. My guess would be in the many tens of thousands, if not more than one hundred thousand. My guess is that the misfortune of more than ninety percent of the children represented by these 197 signs can be laid at the door of routine vaccinations. They are listed here in alphabetical order:

ALBERTSON: Funston Ave, Taft Pl; AMITYVILLE: Brewster La, Cedar St, Maple Dr, Penndale Dr; BALDWIN: Bertha Dr, Carman Pl, Grove St, Jefferson St, Merrick Rd, Prospect St, Seaman Ave, Waddell St; BAY SHORE: Longshore St, Louise Dr, Montauk Dr, Pine Dr; BELLMORE: Bellmore Rd, Howard Rd, Newbridge Rd, Oswego St, Riviera La, Russell St, Wallen Ln, Warren Ave; BETHPAGE: Avoca Ave, Brenner Ave, Davis Pl, Stewart Ave; COPIAGUE: Dante Ave; EAST MEADOW: Alder Ave, Ava Dr, Avis Dr, Bellmore Rd, Bruce Dr, Earl Pl E, Eighth St, Lloyd Ct, Newbridge Rd, Oakdale Rd, Starke Ave, Stuyvesant Ave, Sussex Rd, Wilson Rd; EAST ROCKAWAY: Cambridge Rd, Nicholas Ave, Oxford Rd; ELMONT: Clay St, John Ave, Kirkman Ave, Lehrer Ave, Lydia Ave, Meacham

Ave, Oakley Ave; FARMINGDALE: Iriquois Pl, James St, Midwood St, Regina Rd, Taylor Dr; FRANKLIN SQUARE: Adonia St, Daffodil Ave, Fendale St, Hancock St, Maxwell St, Naple Ave, Rule St, Willow Rd; FREEPORT: Frankel Ave, Grand Ave, Morton Ave, Overton St, Rutland Rd; GARDEN CITY: Sackville Rd; GLEN COVE: Andover Pl; HEMPSTEAD: Belmont Pkwy, Jefferson Pl, Weir St, Whitson St, Yale St (2 different signs); HEWLETT: Broadway, Harvard Ave, Hewlett Pkwy; HICKSVILLE: Bobwhite Ln, First St, Halsey Ave, Shari Ct, Willoughby Ave; LEVITTOWN: Blacksmith Rd, Chase Ln, Gardenia Ln, Haymaker Ln, Meridian Rd, N Bellmore Rd, Orchid Rd, Squirrel Ln, Stonecutter Ln, Universe Dr; LINDENHURST: Alhambra Ave E, N Sixth St, S Fifth St; LYNBROOK: Devine St, Ocean Ave, Ruth Pl; MANHASSET: Dennis St; MASSAPEQUA: Bernard St, Camp Rd, Chester Ave, Cheryl Rd, Elizabeth St, Ford Dr S, Francine Ave, McKinley Pl, Nancy Pl, N Elm St, N Idaho Ave, Pembroke Dr, Pocahontas St E; MASSAPEQUA PARK: Cartwright Blvd, Maryland Ave, Oakdale Pl, Pennsylvania Ave, Westwood Rd S; MERRICK: Bayview Ave, Bedford Ave, Decker Ave, Florence St, Meadowbrook Rd, Park Ave, Taft Ave, Willis Ave; MINEOLA: Cleveland Ave, Coolidge Ave, Marcellus Rd, Seward Ave; NEW HYDE PARK: Falmouth Ave, Herbert Dr, S Seventeenth St; OCEANSIDE: Cornwell Pl, Dambly Ave, Davison Ave, Evans Ave, Oceanside Rd, Pearl St, Waukena Ave; OLD BETHPAGE: Haypath Rd, Maggio Ln; OYSTER BAY: Glen Cove Rd, W Oak Hill Rd; ROCKVILLE CENTRE: Concord St, Grand Ave, Judson Pl, N Kensington Ave, Raymond St, Rockville Ave; ROOSEVELT: W Centennial Ave (2 different signs), W Fulton Ave; SEAFORD: Darby Ln, Ionia St, Washington Ave, Willoughby Ave; SYOSSET: Chelsea Dr, Terrehans Ln; UNIONDALE: Belmont Pl, Birch St, Hempstead Blvd, Newport St; VALLEY STREAM: Cornwell Ave, E Dover St, Jedwell Pl, Pilgrim Pl, Putnam Ave, Sloan Dr S, Virginia St, Wyngate Dr W; WANTAGH: Cornelius Ave, Francis Dr, James Rd, McDonald Ave, McLean Ave, Morgan Dr; WEST BABYLON: Arnold Ave, Belmont Ave, Burgess Ave; WEST HEMPSTEAD: Dogwood Ave, Eagle Ave, Eighth St, Junard Blvd, Knollwood Dr; WESTBURY: Bowling Green Dr, Cypress Ln W, Grand Blvd, Middle Ln; WILLISTON PARK: Prospect St, Sheridan Ave

On the following pages are the pictures I took of these signs.

DEAF CHILD AREA
AUTISTIC
20
BLIND CHILD AREA
AUTISTIC CHILD AREA
BLIND CHILD AREA
DEAF CHILD AREA
AUTISTIC CHILD AREA
AUTISTIC CHILD AREA
BLIND CHILD AREA
AUTISTIC CHILD AREA
AUTISTIC CHILD AREA
AUTISTIC CHILD AREA
DEAF/BLIND CHILD AREA
AUTISTIC CHILD AREA
DEAF CHILD AREA
DEAF CHILD AREA
VISION IMPAIRED CHILD AREA
NO STOPPING
AUTISTIC CHILD AREA
AUTISTIC CHILD AREA
AUTISTIC CHILD AREA
AUTISTIC CHILD AREA
AUTISTIC CHILD AREA
AUTISTIC CHILD AREA
AUTISTIC CHILD AREA
AUTISTIC CHILD AREA
DEAF CHILD AREA
DEAF CHILD AREA
AUTISTIC CHILD AREA

DEAF CHILD AREA
DEAF CHILD AREA
NO
AUTISTIC CHILD AREA
DEAF CHILD AREA
BLIND CHILD AREA
AUTISTIC CHILD AREA
AUTISTIC CHILD AREA
WATCH COMMUNITY
AUTISTIC CHILD AREA
DEAF CHILD AREA
DEAF CHILD AREA
DEAF CHILD AREA
DEAF CHILD AREA
DEAF CHILD AREA
DEAD END
DEAF CHILD AREA
DEAF CHILD AREA
Gladmore St
AUTISTIC CHILD AREA
DEAF CHILD AREA
AUTISTIC CHILD AREA
CHILDREN AT PLAY
BLIND CHILD AREA
BLIND CHILD AREA
AUTISTIC CHILD AREA
DEAF CHILD AREA
BLIND PERSON
DEAF CHILD AREA
DEAF CHILD AREA
BLIND CHILD AREA
AUTISTIC CHILD AREA
AUTISTIC CHILD AREA
AUTISTIC CHILD AREA

AUTISTIC CHILD AREA
AUTISTIC CHILD AREA
AUTISTIC CHILD AREA
CHILDREN AT PLAY
AUTISTIC CHILD AREA
AUTISTIC CHILD AREA
DEAF CHILD AREA
DEAF CHILD AREA
SLOW
CHILDREN
AUTISTIC CHILD AREA
DEAF CHILD AREA
AUTISTIC CHILD AREA
DEAF CHILD AREA
AUTISTIC CHILD AREA
DEAF CHILD AREA
AUTISTIC CHILD AREA
AUTISTIC CHILD AREA
DEAF CHILD AREA
DEAF CHILD AREA
DEAF CHILD AREA
DEAF CHILD AREA
DEAF CHILD AREA
NO
DEAF CHILD AREA
TOWN
DEAF CHILD AREA
DEAF CHILD AREA
DOG WASTE
DEAF CHILD AREA
DEAF CHILD AREA
AUTISTIC CHILD AREA
DEAF CHILD AREA
DEAD END
DEAF CHILD AREA

AUTISTIC CHILD AREA
STOP
DEAF CHILD AREA
DEAF CHILD AREA
DEAF CHILD AREA
AUTISTIC CHILD AREA
AUTISTIC CHILD AREA
AUTISTIC CHILD AREA
DEAF CHILD AREA
DEAF CHILD AREA
AUTISTIC CHILD AREA
AUTISTIC CHILD AREA
AUTISTIC CHILD AREA
DEAF CHILD AREA
AUTISTIC CHILD AREA
AUTISTIC CHILD AREA
AUTISTIC CHILD AREA
AUTISTIC CHILD AREA
DEAF CHILD AREA
AUTISTIC CHILD AREA
AUTISTIC CHILD AREA
DEAF CHILD AREA
AUTISTIC CHILD AREA
AUTISTIC CHILD AREA
AUTISTIC CHILD AREA
AUTISTIC CHILD AREA
DEAF CHILD AREA
DEAF CHILD AREA
AUTISTIC CHILD AREA

AUTISTIC CHILD AREA
AUTISTIC CHILD AREA
DEAF CHILD AREA
AUTISTIC CHILD AREA
DEAF CHILD AREA
DEAF CHILD AREA
SLOW
DEAF CHILD AREA
TRAFFIC
DEAF CHILD AREA
DEAF CHILD AREA
DEAF CHILD AREA
DEAF CHILD AREA
AUTISTIC CHILD IN AREA
AUTISTIC CHILD IN AREA
DEAF CHILD AREA
NO
DEAF CHILD AREA
DEAF CHILD AREA
DEAF CHILD AREA
DEAF CHILD AREA
AUTISTIC CHILD AREA
DEAF CHILD AREA
River Av
AUTISTIC CHILD AREA
AUTISTIC CHILD AREA
AUTISTIC CHILD AREA
AUTISTIC CHILD AREA
AUTISTIC CHILD AREA
DEAF CHILD AREA
AUTISTIC CHILD AREA
CHILDREN AT PLAY
DEAF CHILD AREA

DEAF CHILD IN AREA
NO STOPPING HERE TO CORNER
NO TRUCKS OVER 4 TONS GROSS
DEAF CHILD AREA
DEAF CHILD AREA
DEAF CHILD IN AREA
DEAF CHILD AREA
DEAF CHILD AREA
DEAF CHILD AREA
DEAF CHILD AREA
DEAF CHILD AREA
DEAF CHILD AREA
CHILD AREA
DEAF CHILD AREA
DEAF CHILD AREA
DEAF CHILD AREA
AUTISTIC CHILD AREA
DEAF CHILD AREA
DEAF CHILD AREA
DEAF CHILD AREA
DEAF CHILD AREA
DEAF CHILD AREA
AUTISTIC CHILD AREA
AUTISTIC CHILD AREA
DEAD END
WORK AREA AHEAD
DEAF CHILD AREA
AUTISTIC CHILD AREA
AUTISTIC CHILD AREA
DEAF CHILD AREA
30
DEAF CHILD AREA

AUTISTIC CHILD AREA
AUTISTIC CHILD AREA
AUTISTIC CHILD AREA
AUTISTIC CHILD AREA
AUTISTIC CHILD AREA
DEAF CHILD AREA
DEAF CHILD AREA
AUTISTIC CHILD AREA
AUTISTIC CHILD AREA
DEAF CHILD AREA
DEAF CHILD AREA
20
DEAF CHILD AREA
CHILD AREA
DEAF CHILD AREA
DEAF CHILD AREA
DEAF CHILD AREA
DEAF CHILD AREA
DEAF CHILD AREA
CHILDREN AT PLAY
DEAF CHILD AREA
DEAF CHILD AREA
DEAF CHILD AREA
AUTISTIC CHILD AREA
AUTISTIC CHILD AREA
SLOW
AUTISTIC CHILD AREA
20 M.P.H.
DEAF CHILD AREA
25 M.P.H.
DEAF CHILD AREA
DEAF CHILD AREA
AUTISTIC CHILD AREA
SLOW

Foundation of quicksand: the smallpox vaccine

It would seem to be impossible for a rational mind to conceive that a filthy virus derived from a smallpox corpse, the ulcerated udder of a cow, or the running sores of a sick horse's heels, and cultivated in scabbed festers on a calf's abdomen, could fail to have disastrous effects when inoculated into the human body. — Dr. Beddow Baily, 1936

HOW do crazy medical ideas get started in the first place, then gain momentum? For the same reasons the current cholesterol scam, and a dozen others, came out of nowhere and are now raking in billions of dollars: there are plenty of swindlers in the world; most journalists are morons or prostitutes or both; people are gullible; they look to authority figures for guidance, even though most of them don't know what they're talking about. I've already mentioned Gerald Ford, who went on national television to issue dire warnings about a possible swine flu epidemic, and was photographed, broadly smiling, getting a shot from his personal physician; a deeply moved Dwight Eisenhower honored Jonas Salk at a special ceremony at the White House rose garden, days before the trickster's phony polio vaccine started killing American kids; Ronald Reagan awarded the National Medal of Science to Merck's top vaccine developer, the seamy, foul-mouthed Maurice Hilleman; First Lady Rosalynn Carter, a dear friend of Jim Jones, was a tireless campaigner for childhood vaccines, and endorsed books written by Paul Offit who, like Salk, will receive his due in a later chapter. One might think that Mrs. Carter would've learned something after Jonestown.

But let's go back to 1796 when Edward Jenner, an English doctor of sorts - he seems to have bought his medical degree for 15 pounds — performed what is recognized as the world's first vaccination. At that time in rural England, some held the dubious belief that a person could not catch smallpox if he first came down with cowpox, an innocuous disease caught by milking the infected udders of cows. (Full-blown smallpox was a dreadful disease which was often fatal and sometimes left surviving victims permanently scarred; there are so many conflicting statistics as to its prevalence in bygone centuries that I will not cite any here, other than to say that, while not rare, it does not appear to have been as common as generally supposed.) Jenner picked up on this belief and decided to experiment. He cut the arm of a boy named James Phipps and inserted cowpox matter. The boy caught cowpox. Several weeks later he injected him with smallpox residue. When he didn't contract smallpox, Jenner declared success. But other doctors disputed this, claiming they knew of dairymaids who had caught cowpox, and in later years smallpox. Jenner backed off and admitted that he too knew this was true.

Actually, Jenner's experiment was merely a modification of a procedure known as variolation, which had been practiced for hundreds of years, originating in China then making its way west. Variolation involved inserting the powder of smallpox scabs into the superficial scratches of an uninfected person to induce a mild case of the disease, which theoretically would result in lifelong immunity. The wife of the British ambassador to the Ottoman Empire came across variolation in 1717 and introduced it to England four years later. But there were problems and uncertainties, not to mention fatalities, with variolation, and by the end of the eighteenth century many were disillusioned with the method. Variolation transitioned into vaccination, the difference being that now diseased scrapings from cows, rather than humans, were being used.

Jenner was a shady character who stole the cowpox vaccination idea from a Welsh farmer named Benjamin Jesty. Two years after being caught lying about his achievement, he came up with the idea that men who milked cows soon after they tended to horses suffering from "grease" — a disgusting disease characterized by oozing pus near their hooves — were *truly* immune from smallpox. In a paper he published titled "An Inquiry into the Causes and Effects of the Variolae Vaccinae" he suggested injecting children with a new and improved vaccine — cowpox infected with horse pus. While most people, including his medical colleagues, were horrified by this idea, some parents actually let Jenner experiment on their children with this vile concoction. The vaccine proved to be a failure.

So Jenner went back to plain old cowpox, even though he himself had admitted, years earlier, that it couldn't guarantee immunity from smallpox. Despite all the evidence that his vaccinations didn't work, he remained fixated on the idea, and in 1807 petitioned the government for financial support, claiming his product had "the singularly beneficial effect of rendering through life the person so inoculated perfectly secure from the infection of smallpox," even though he knew this to be false.

A sidebar to this story which highlights the flaws of even the keenest minds, and makes it easier to understand the credulity of people in general, involves Thomas Jefferson. As a young man, Jefferson had submitted to variolation which, as in England, had become increasingly unpopular in the American colonies. Jefferson saw vaccination as a big improvement in conferring immunity from smallpox, so much so that he had many of his slaves, relatives and neighbors vaccinated. In a cordial letter to Edward Jenner dated May 14, 1806, he congratulated him on his discovery, which he considered a huge medical advance. Of course, in those days the flow of information across the ocean took a long time and was severely limited by modern standards, so it's possible that Jefferson was unaware of all the denunciations heaped on Jenner in his native England.

In any event, convinced of Jenner's sincerity, Parliament granted his request and mass inoculation campaigns followed. Pretty soon vaccinated people started coming down with smallpox. How could this be? It couldn't be, so Jenner and his supporters denied it. When it became impossible to deny, he changed his tune and said that if his vaccination did not prevent smallpox, it at least resulted in a milder case — a refrain that's still heard today. When these milder cases started dying, Jenner came up with a new line: those who died had been injected by "spurious" cowpox; those who recovered had received the real thing!

Jenner died in 1823. By that time three kinds of his vaccine were in use, including the cowpox-horse pus variety. With the vaccination fad spreading among a segment of the population, and Jenner gone, new excuses were made to explain away their ineffectiveness and danger. One puncture isn't good enough, some doctors said; two, or even three or four were recommended. The "needle" of those days was a sharpened point of ivory dipped in pox brew, and driven through the skin. "Mothers nearly always protest," noted a chief surgeon, who referred to the ivory devices as "real instruments of torture." But the multiple injections failed too, so the vaccinators came up with yet another story. Now people were told they had to come back for more and more injections "until vesicles cease to respond to the insertion of the virus." This didn't work either.

All that was left to the vaccination cult was outright fraud. People who were admitted to hospitals with smallpox were registered as unvaccinated, even though their vaccination scars were plain to see. Or else the smallpox they were suffering from was given a different name, which led George Bernard Shaw, the great Irish playwright and man of letters, to remark: "During the last epidemic at the turn of the [20th] century, I was a member of the Health Committee of London Borough Council. I learned how the credit of vaccination is kept up statistically by diagnosing all the re-vaccinated cases as pustular eczema, varioloid or whatnot — except smallpox." This "change the name game" is another trick that has come down through the ages and still employed to mask the observable fact that vaccines do not work.

❖

It's incredible to think that such a flimflam with such a terrible track record could take hold in the world's most advanced countries. Yet take hold it did. By the late nineteenth century, several European countries had mandatory vaccination laws on the books. The results were devastating. One million vaccinated Germans contracted smallpox in 1870-71; thousands died. "The hopes placed in the efficacy of the cowpox virus as a preventative of smallpox have proved entirely deceptive," said the German chancellor. In Italy, a leading medical professor wrote, "The epidemics of smallpox we have had [from 1887-89] have been so frightful that nothing before the invention of vaccination could equal them." A terrible epidemic took place in the early 1870s in England, which had a 98% vaccination rate; surviving hospital records show that about 90% of the victims had been vaccinated and re-vaccinated.

This epidemic gave a strong push to the anti-vaccine movement in England, which gained traction in 1853, when smallpox vaccinations became compulsory. Those who disobeyed the law were prosecuted, their homes and property confiscated; parents who tried to protect their children from the shots were fined and jailed. Shortly after the epidemic subsided, Mary Catherine Hume published *150 Reasons for Disobeying the Vaccination Law by Persons Prosecuted Under It*, which advocated open defiance. She wrote:

Vaccination is poisoning of the blood... capable of sending inwards to the vital organs the blood impurities which might otherwise have been safely got rid of by means of the natural disruptive process.... The Vaccination Acts — for rendering compulsory on the citizens of a so-called free country — are tyrannical, unconstitutional, and in the true sense illegal; Godless and insane; idolatrous, being acts for the continual offering up of human sacrifices; odious to all lovers of freedom; fraudulent, as evading and violating another existing Act of Parliament; demoralizing to the whole community; and, on the direct and indirect testimony of their own supporters and advocates, an utter and absolute failure.

But it wasn't until 1907 that the law was changed and vaccinations became entirely voluntary in England which, as in most other Western countries, is the case today.

By the turn of the century, smallpox vaccination was widespread in the United States, and mandatory in several states. As in England, there was plenty of opposition accompanied by sporadic violence and lawsuits contesting the legality of forced vaccination. The results on both sides of the Atlantic were often tragic, as indicated by the following testimonials, which were sometimes published with photographs of the victims in life and in death:

Margaret Ann, the only daughter of Mr. and Mrs. Donald W. Gooding, was pronounced a perfect baby by the doctor when she was born. This beautiful and healthy infant was vaccinated at the age of four months. The first two injections didn't take, so a third vaccination was given, after which inflammation of the brain developed within five days. She was taken to the hospital where she remained for many weeks. At the age of 13 months she was blind and could not learn to walk. She also developed digestive disturbance and convulsions.

Mona Stevenson was vaccinated at the age of five weeks with the official glycerinated calf-lymph. After five weeks of suffering in which the child's face and arm was partly eaten away by the vaccine disease, the child died.

My wife was in good health until she was vaccinated November 1, 1922. On the 9th of November, she broke out with a pronounced skin rash all over her body. On November 29, she became seriously ill. Sores appeared in place of the rash. The doctor said it was smallpox. She died on December 5.

On November 13, 1922, my wife submitted to vaccination in order to hold her position with the telephone company. Up to the time of the vaccination she was in good health. On December 8, she became ill with enlarged lymphatic glands under her arms. She died on December 10.

Horace Capewell was a fine, healthy baby with beautiful eyes. He was vaccinated when five weeks old. Nine days later his body was a mass of sores. His eyes became affected and at the age of five months he was totally blind.

Ernest Cheeseman was vaccinated at the age of nine weeks with the standard glycerinated calf-lymph. Up to that time, he was healthy and normal. Five days after his vaccination a serious form of syphilitic skin disease developed. The body was covered with eczema-like eruptions and the feet were drawn up, all out of shape. The sores that broke out around the mouth looked like burnt meat and stood out at least an inch. Death relieved him of his suffering.

While most doctors of the era had an almost religious belief in vaccinations, as they do today, some did not, and spoke out forcefully against them:

I have been a regular practitioner of medicine in Boston for 33 years. I have studied the question of vaccination conscientiously for 45 years. As for vaccination as a preventative disease, there is not a scrap of evidence in its favor. The injection of virus into the pure bloodstream of the people does not prevent smallpox; rather, it tends to increase its epidemics, and it makes the disease more deadly. Of this we have indisputable proof. In our country cancer mortality has increased from 9 per 100,000 to 80 per 100,000 or fully 900 percent increase within the past 50 years, and no conceivable thing could have caused this increase but the universal blood poisoning now existing. — Dr. Charles E. Page, Boston practitioner

I believed that vaccination prevented smallpox, or if it did not absolutely prevent it in every case, it modified the disease in some cases, and I believed that re-vaccination, if only frequently enough, gave absolute immunity. Experience has driven all that out of my head; I have seen vaccinated persons get smallpox, and persons who had been re-vaccinated get smallpox, and I have seen those who have had smallpox get it a second time and die of it. — Dr. J.C. Ward, Royal College of Surgeons, England

After collecting the particulars of 400,000 cases of smallpox, I am compelled to admit that my belief in vaccination is absolutely destroyed. — Professor Adolf Vogt, chair of Vital Statistics and Hygiene at Berne University

To affirm that there never has been any scientific warranty for a belief in the alleged protective virtues of vaccination and that its practice is backed by ignorance and indifference, is a sorry charge to make against the medical profession... but the charge, I regret to say, is only too true. I know whereof I affirm, for I, too, must plead guilty of the charge. Before discovering my mistake I had vaccinated more than 3000 victims, ignorantly supposing the disease I was propagating to be a preventative of smallpox. Having taken for granted what my teachers had asserted, I was a staunch believer in the alleged efficacy of vaccination as a prophylactic against smallpox. I remained in this blind and blissful state of ignorance for several years, and not until I acquired a little experience in the school of observation and reflection did I discover that my faith was pinned to a shameful fraud. — Dr. J.W. Hodge, practitioner

How is it that smallpox is five times as likely to be fatal in the vaccinated as unvaccinated? How is it that, as the number of people vaccinated has steadily fallen, the number of people attacked with variola has declined and the case mortality has progressively lessened? The years of least vaccination have been the years of least smallpox and least mortality. These are just a few points in connection with the subject which are puzzling me, and to which I want answers. — Dr. L.A. Parry, practitioner

Never in the history of medicine has there been produced so false a theory, and such fraudulent assumptions, such disastrous and damning results as have followed the practice of vaccination; it is the ultma thule [extreme limit] of learned quackery, and lacks, and has ever lacked, the faintest shadow of a scientific basis. The fears of the people have been played upon as to the dangers of smallpox, and the promise of sure prevention by vaccination, until nearly the whole civilized world has become physically corrupted by its practice. — Dr. E. Ripley, Connecticut practitioner

Many doctors also saw a connection between vaccination and cancer, a matter best summed up by the observation of Indiana doctor W.B. Clark: "Cancer was practically unknown until cowpox vaccination began to be introduced. I have had to do with 200 cases of cancer and I never saw a case of cancer in an unvaccinated person." This is an intriguing statement which, however, our esteemed cancer experts have never followed up on, because like vaccination, cancer is a big business that requires lots of customers.

❖

As the twentieth century wore on, the debate within the medical community became red-hot. Simply stated, the smallpox vaccine was so dangerous and ineffective, and this fact had become so obvious, that an impasse was reached over the continuation of its use. Gradually, many countries eased their requirements. The last naturally occurring case of smallpox in America was recorded in 1949; in 1972 smallpox vaccinations ended in this country.

But vaccinations continued in impoverished parts of the world. WHO, the global health bureaucracy, estimated there were 15 million smallpox cases worldwide, and announced they would launch a massive vaccination campaign to wipe smallpox off the face of the earth — which many questioned since it was now common knowledge that the vaccine *caused* the disease. Most of their efforts were concentrated in Africa. In October 1977, WHO identified Ali Maow Mallin, a Somali cook, as the last person to contract a natural case of smallpox, and less than three years later held a ceremony at their headquarters in Geneva declaring that smallpox had disappeared from the planet. Considering that, as today, there were billions of human beings living in poverty in 1977 who could not possibly have been examined by WHO's surveillance teams, such a declaration was preposterous.

Through the 1980s and 1990s, little was heard of smallpox. Supposedly the only remaining known samples were stored in secure government laboratories in Atlanta and the Siberian city of Novosibirsk. Nevertheless, small batches of the vaccine continued to be manufactured and used for no apparent reason. A backpage news brief beneath the heading "Unneeded Smallpox Shot Causing A Serious Illness" appeared in the September 19, 1981 issue of the *New York Times*. It stated that one million American military personnel had been unnecessarily vaccinated each year, and many were suffering from a "sometimes fatal" condition. This was just one of countless outrages associated with vaccination in that year alone, and either underreported or ignored by the American news media.

A glaring example of this censorship was a bombshell front page story in the May 11, 1987 issue of the *London Times* indicating that the AIDS epidemic in Africa was triggered by the mass smallpox vaccination campaign of the previous decade — not one word of which ever appeared in the *New York Times* or any other major media outlet here. The *London Times* exposé was written in response to a tip from an adviser to WHO who had been asked to investigate the suspicious link between WHO's ambitious vaccination program in central Africa and the sudden appearance of AIDS in precisely the same countries. The advisor concluded that there was indeed a connection and filed his report. When WHO buried it, he contacted the *Times*.

The late Dr. William Douglass, a colorful and outspoken publisher and author, part Georgia hillbilly and part world traveling medical pioneer, went further. In a pamphlet he wrote titled *WHO Murdered Africa*, he claimed that the smallpox vaccine didn't *trigger* AIDS, it *deliberately introduced* this horrible disease. AIDS, he wrote, was "a cold-blooded, successful attempt to create a killer virus [by splicing deadly sheep and cow viruses, which in their natural state are harmless to man] which was then used in a successful experiment in Africa." Quoting from WHO's own publications, he charged that this organization, in collaboration with the National Cancer Institute, made the virus, used the smallpox vaccination campaign as their vehicle to spread it, then sat back to see what would happen. Never one to mince words, Douglass suggested "closing down all laboratories in this country that are dealing with these deadly retroviruses" — he singled out Fort Detrick, Maryland — "then we must sort out the insane, irresponsible and traitorous scientists involved in these experiments and try them for murder."

I realize that statements like this sound wacky to many people who are accustomed to reading only material found on newsstands or at the public library. They will begin to make sense as you make your way through this book.

❖

Although smallpox burst into the news after the events of September 11, 2001, and stayed there for some time, strange developments were taking place before that date. In 1997, four years before the World Trade Center and the Pentagon were hit, Dynport, a vaccine company based in Frederick, Maryland, the location of Fort Detrick, was awarded a contract by the Defense Department to produce a new smallpox vaccine. The CDC contracted with another firm, Acambis, to manufacture the same vaccine in 2000. Since smallpox supposedly had been eradicated, and it appeared likely that the last two known samples would be destroyed, scientists and public health authorities wondered out loud about these actions. "A lot of people thought this was a crazy idea, to make a new vaccine when the disease didn't exist," said Dr. Margaret Hamburg of the Department of Health and Human Services (HHS).

Three months prior to September 11, a study code-named "Dark Winter," which simulated an outbreak of smallpox in the U.S., was conducted by a team referred to as bioterrorism specialists. The study predicted that within two months, three million people would be infected, the country's infrastructure would break down, and there would be riots nationwide. But according to another study made by the CDC, the assumption that one person would infect ten others, and each of these another ten and so on, was a wild exaggeration. Statistics compiled from previous outbreaks showed that smallpox was not nearly that contagious. Nevertheless, Dark Winter coordinator Donald Henderson, who also had led WHO's supposed smallpox eradication campaign in Africa, concluded that at least 100 million doses of the vaccine were necessary to safeguard America. And shortly after September 11, Vice President Dick Cheney was shown a video of Dark Winter and urged to speed up production of the vaccine.

The following month, the CDC unveiled legislation called "The Model State Emergency Health Powers Act." Section 504a gave public health officials and state governors the authority, among other things, to arrest and forcibly vaccinate anyone deemed to be unprotected, or a threat to spread, an infectious disease. Furthermore, it specified conditions under which property could be seized, and the Constitution and Bill of Rights suspended. In addition, it repealed laws allowing medical or religious exemptions to vaccinations, and cleared police, National Guard and government authorities of any liability arising from harm or death due to a forced inoculation. According to the website publichealthlaw.net, last updated on January 27, 2010 (but since taken down), "38 states and DC have passed a total of 66 bills or resolutions that include provisions from or closely related to [MSEHPA]."

Remember that there hasn't been a single case of smallpox in this country since 1949, vaccination against this disease was discontinued here in 1972, and in 1980 WHO proclaimed that smallpox no longer existed anywhere on earth, outside of two laboratories. Then the rumors started. It was said that impoverished Russian scientists in Novosibirsk had sold off smallpox cultures to foreign terrorists to be used against America, and that the usual boogeymen, Iraqi leader Saddam Hussein, and North Korea's Kim Jong-Il, had secret stocks in their possession and might be preparing to saturate the U.S. with smallpox (see, for example, a *Washington Post* article of November 5, 2002 and an ABC news report of January 7, 2006). Not a shred of evidence supported these absurdities, and they quietly faded away.

Dire predictions that there might not be enough vaccine to protect Americans alternated with warnings that the vaccine was too dangerous to use. One week a shrill warning would be issued that the bioterrorists could strike at any moment; the next week the likelihood of a smallpox attack was said to be almost nil. As local and state health authorities stood by waiting for directions from the federal government — directions that never came — pharmaceutical firms jostled each other for lucrative contracts to manufacture the vaccine. Then there was disagreement over who should be held liable for the inevitable casualties of mass vaccination — the drug firms or the government. Then politicians disagreed among themselves as to whether or not victims should be compensated, and if so, how much. Editorial writers around the country urged both caution and action — sometimes in the same editorial. President George Bush, on a visit to CDC headquarters in Atlanta, expressed concern about people dying from the vaccine; two weeks later he postponed indefinitely the destruction of America's last smallpox culture.

What no one denied was that severe injury and death would accompany widespread vaccination. Even the American Medical Association (AMA), usually in the front lines of any vaccination drive, backed off, saying the vague threat of a biological attack just wasn't worth the risk — a risk that no one could clearly quantify. Undoubtedly, many arguments took place behind closed doors. On many fronts, no one could predict what would happen if an executive order came down requiring that all Americans be vaccinated against smallpox. Bureaucrats knew there would be sheep-like compliance among millions considering it their patriotic duty to get the shot, but there would also be a violent backlash in a country where a lot of people own guns. This was probably on George Bush's mind when he finally came up with a compromise in the midst of his contrived war plans and scare stories about Saddam and smallpox; there would be no national vaccination program *but* a half million health care workers would have to submit to the smallpox shot.

Bush's plan ran into immediate stiff resistance. More than ninety percent of targeted health care workers flatly refused to be vaccinated. In addition, seven people with heart conditions suffered severe reactions, and two died, prompting the CDC to suspend the shots for those with a history of heart problems. Pretty soon almost everyone was refusing them, and by the summer of 2003 the plan was scrapped, while its architects congratulated each other that the vaccine wasn't quite as deadly as they had feared.

❖

This, then, is a capsule history of the disease and vaccine that started it all. It's a microcosm of the vaccine story that's been repeated over and over — a story that always brings out the worst in man, and the worst kind of men. The book has been closed on smallpox for 16 years now, but not entirely. In just the last nine years, as I write, there have been outbreaks in Uganda, the Congo, India and the country of Georgia in the Caucasus. There is an obvious reluctance to diagnose these cases as smallpox, because this would erase the legend so beloved of all the alphabet soup health agencies that this disease was conquered by vaccines. Instead we're told about orthopox and monkeypox, and that WHO has determined that in Uganda the outbreak was "likely" chickenpox, the lesions of which are clearly distinguishable from those of smallpox, whereas monkeypox appears to be smallpox by a different name. I am neither a virologist nor a taxonomist, but this seems to be a typical example of the name-switching game started by Jenner.

The March 2011 outbreak in rural Jharkand state in India, whatever it was, was written off by local health authorities as a superstitious belief of villagers. I can only suspect that bribes were offered or arms twisted to induce them to make this report. It's reminiscent of Heather Whitestone, the first woman with a disability to be crowned Miss America, in 1994. It was initially reported that she had become deaf at eighteen months following a DPT shot. A week later the story changed dramatically, following input from the American Academy of Pediatrics (AAP) and her pediatrician; now it was a Hemophilus influenza infection that had caused the nerve damage to her ears (which, if true, was probably provoked by the shot, a common occurrence). I can just imagine what went on behind the scenes.

The smallpox vaccine is the most important one to study in terms of seeing the big picture. It has been around for more than 220 years, much longer than any other vaccine, and it is the theoretical foundation upon which all vaccines have been created. In all that time it has never ceased to be highly controversial; every claim of success by its adherents has been relentlessly attacked and convincingly refuted by its opponents. Nevertheless, since it's no longer given to children, and despite recent deliberations about administering it at gunpoint, if necessary (current laws in several states will allow this if a national health emergency is declared in Washington), it gets little discussion these days. The permanent spotlight falls instead on the polio vaccine. That's the big success story, the vaccine that saved so many lives, the one everyone "knows about," the one that gives the word *vaccine* such a positive spin. It amazes me that so few people have taken a good, hard look at this incredible fraud, which we will do in the next chapter.

The 1872 and 1911 articles from the *New York Times* reproduced below provide us with some perspective on the smallpox vaccine as it was seen over a century ago.

SMALL-POX.

Vaccination the Only Defense Against the Disease—Views of the New-York Academy of Medicine on the Question.

The following is the report of a Committee on Vaccination made to the New-York Academy of Medicine Feb. 1, 1872, and now published in accordance with a resolution of that body, which accepted and approved the report and authorized its publication:

REPORT.

The undersigned, Committee of the New-York Academy of Medicine, having been appointed to report upon the questions relating to small-pox and vaccination which concern the medical profession and the public, respectfully submit the following statement of facts to encourage and aid the work of thorough vaccination: The fatal prevalence of small-pox in numerous cities and towns in this country and throughout most of the civilized world at the present time, has caused some alarm and skepticism in regard to the protection afforded by vaccination. Thousands and even millions of people still neglect the only sure means of protection against the scourge they dread, yet while from doubt and delay, the duty is thus neglected by vast numbers of people. The daily observations and experience of physicians have, for seventy years, uniformly proved that in their personal exposure to the sick of small-pox, and by the severest testing of the attendants of the sick in hospital, sick-room and ambulance thorough vaccination continues to give full protection against the contagion, for they are freely exposed to its presence, and even to its touch without harm. Thorough vaccination is thorough protection against small-pox, and as effective now as in the time of JENNER. It is generally admitted, and clearly proved, that thorough vaccination not only protects against small-pox without danger to life or diffusing any infection. Scarred and disfigured faces are rarely met with at the present day, except in persons who, having neglected to be vaccinated, have taken small-pox in the natural way. In proof of this protective influence of vaccination, and the liability to infection without it, we may mention that in a severe epidemic in Norwich, England, of 306 persons in several families, there were among 91 vaccinated persons, 2 cases of mild varioloid, and no deaths; and of 215 unvaccinated persons, 200 took the disease, and 46 of them died.

REVACCINATION.

By the medical profession, revaccination is recommended, not only as a measure of renewed security against small-pox, but as a test of the completeness of the first vaccination. In most persons a first re-vaccination succeeds in a certain degree; some may be found wholly insusceptable to it—thoroughly protected against contagion—and in some persons it will take barely enough to show that the virus was not wholly inefficient; but, sometitimes, with a degree of energy which very closely resembles that of a primary vaccination. Revaccinnation is necessary, and a duty, to test the completeness of protection and to exhaust the receptivity of small-pox, and of vaccination itself. It is generally supposed that the suceptibility to vaccinia and similarly to small-pox infection, is proportionate to the success attained in the revaccination, and that they in whom the success of it is almost perfect, were but imperfectly protected by the first. Such was the opinion of JENNER himself, who thought that the security which the first vaccination imparted was in direct proportion to the degree of perfection of the vaccine process, and advised that vaccination should be repeated as long as any effect was produced. He who is insusceptible to one or more revaccinations, with fresh virus, may generally deem himself as secure as vaccination, or even an attack of small-pox itself, can make him.

SMALL-POX AFTER VACCINATION.

Small-pox, it is true, occasionally occurs in persons who have been once vaccinated; but its occurrence is very rare in those recently vaccinated, or revaccinated once or oftener, or in those in whom the first vaccination was thoroughly good, as judged by the Jennerian rules for vaccination.

In these instances the disease manifests itself oftenest, in mitigated form, known as varioloid and the protection afforded by vaccination is found not only to be equal to that which is conferred by a previous attack of small-pox It may in truth be said that secondary small pox is oftener more severe, malignant and fatal than that which occurs after vaccination—that very rarely destroys life or disfigures the person. It is believed that a revaccination will destroy any susceptibility to small-pox infection which may remain from incompleteness or imperfection of the primary vaccination. In proof of the correctness of this opinion it may be stated that in the Army of Bavaria revaccination has been compulsory since 1843, and from that date until 1857—a period of fourteen years—not even a single case of (unmodified) small-pox occurred nor a single death from the disease. The vaccine virus now in use in New-York City is justly regarded by the most experienced physicians as trustworthy and effective. There are various methods of renewing vaccine virus other than obtaining it from the human arm. Without discussing the relative merits of vaccine virus derived from the several sources, it may safely be said that the vaccine virus in use in this City properly taken and properly applied, is trustworthy and effective, and affords all the necessary protection.

VACCINE VIRUS NOT A CARRIER OF OTHER DISEASES.

It is a prevalent idea in the public mind, that vaccine matter may communicate directly various cutaneous and constitutional diseases with which children may suffer after vaccination. The virus is then alleged by ignorant persons to be "bad," or "impure," and in view of these contingencies, this is a point on which most persons are extremely solicitous, and which tends to deter many from vaccination. There is one disease, syphilis, which some persons have supposed may be transmitted from a child laboring under it, to a healthy one, by vaccination, but admitting that this might happen, under circumstances of gross carelessness, it would bear no appreciable proportion to the vast multitudes of vaccinations which have given perfect protection against small-pox without any suspicion of harmful effects from vaccine. It is believed never to have been seen in the practice of the most experienced vaccinators or surgeons especially conversant with that disease, nor has a single case been met with in the systematic inspections of vaccination in England, which have already extended to millions of persons, and the risk, indeed, of any such occurrence is so infinitesimally small that, for all practical purposes, we may regard it as non-existant. It is, however, with the appearance of other cutaneous eruptions and consequences which follow vaccination at longer or shorter intervals, that the public is chiefly concerned. These are blotches and rashes, erysipelas and eczema and enlargement and suppuration of the glands of the neck, and other indications of disease called scrofulous.

Such complications of vaccination are undoubtedly witnessed occasionally, but they are mostly the results of a pre-existing tendency. The vaccine disease is attended with a febrile movement and changed condition of the blood, which, when the predisposition to scrofulous affections exists in the system of the person vaccinated, may tend to excite them into activity, as many other causes do.

The diseases which occasionally follow vaccination are not themselves communicable or contagious from person to person, nor are they produced in each of the persons vaccinated with the same matter. They cannot be communicated at will, or at the pleasure of the vaccinator. No new disease has been known to follow vaccination; all had pre-existed, and they occur not only after it, but after small-pox itself, natural or inoculated, measles or scarlatina, &c., or without these, and when no vaccination had been performed. Scrofula, when developed has always been latent in the system of the person vaccinated, dependent on a peculiar state of the blood and the constitution. To a neglect of reasonable precautions, and to carelessness in the performance of the first vaccination are due the unmerited discredit into which this greatest boon of a merciful Creator has occasionally fallen, and the fatal objection of certain classes to its employment. With due precautions, no other disease than that of vaccinia will be communicated. The distinguished English vaccinators, Dr. Marson, in 40,000 vaccinations; Dr. Leese, in as many more; Sir Wm. Jenner, in 13,000 sick children and adults in London, and Dr. West, of the Children's Hospital, in 26,000 children, all concur in saying that they have never seen any other diseases communicated with the vaccine, and that they disbeleive the popular reports that diseases are so communicated; nor is the child from whom the lymph is taken found to be suffering from the disease it is said to have imparted. Fresh lymph is always to be preferred when it can be obtained, but crusts may be used if such lymph is not to be had and there is immediate necessity for vaccination. When practicable it is desirable to vaccinate from arm to arm, with virus taken on quill-slips directly from a healthy individual to the arm of another.

VACCINATION TENDS TO DIMINISH DISEASE AND MORTALITY.

Small-pox greatly tends to promote the development of scrofula and consumption. Vaccination, by preventing small-pox, also indirectly diminishes scrofulous diseases: and the death-rate of most constitutional diseases, to say nothing of small-pox itself, has materially decreased as the practice of vaccination has increased; and not only has mortality been diminished, but life has been lengthened and population augmented. The exact reverse is true of the former practice of inoculation of small-pox.

AGE FOR VACCINATING; AND NUMBER OF VESICLES.

There is no age at which a person should or should not be vaccinated, when a necessity exists for immediate protection against small-pox; but it is best done in early childhood, for then the irritation is less, and the child unable to relieve itself by scratching the part, which, with the irritation of the clothes, often interrupts the progress of the vesicle, causes it to discharge prematurely and become inflamed; whereby a vaccination which might have run its course naturally becomes altered, and an inflamed an angry sore is produced, leaving a deformed, and ill-looking scar and a diminished protection. Matter, too, taken from such a sore is worse than valueless. The age of three months is, on the whole, to be preferred. The most distinguished English authorities in vaccination, SIMON, MARSON, SEATON, and others, recommend, as preferable, and as ensuring greater ultimate protection, four or five separate good-sized vesicles; and of these, in taking virus for the purpose of vaccinating others, one, at least, should remain untouched.

THE FIRST VACCINATION.

The utmost care should be taken in the performance and management of a first vaccination, to make it as nearly perfect, regular and exhaustive as possible, for upon this depends very much the after insusceptibility to small-pox; and it is doubtful whether any amount of subsequent revaccination will fully compensate for the deficiency of the first.

THE PERIOD FOR REVACCINATION.

It may be left very much to be determined by circumstances; it can never be done too early or too promptly in persons exposed to small-pox, and it should be performed in every person soon after the age of puberty, however frequently it may be repeated at earlier or later periods of life. It may succeed at one time when it would not at another; indeed, it appears to be true that revaccination, as well as primary vaccination, actually takes more readily in some years than in others. "One thoroughly good vaccination to start with," says SEATON, "and one careful revaccination after puberty, are all that is necessary for protection as complete as any known proceeding can give against small-pox." After the twenty-fifth year of age, the liability to small-pox in the well vaccinated is very small. The effects of a revaccination are occasionally as perfect in appearance and the symptoms produced as are those of a primary vaccination of infancy, and just as a second attack of small-pox, may be as complete as the first; but oftenest the revaccination produces only slight and brief symptoms and appearances, and sometimes no effects whatever—which could not be the case if the protective influence of vaccination steadily wore out in the human system with advancing age. Of the nearly 50,000 revaccinations in the Prussian Army in the year 1833, only about one-third were perfectly successful; resembling, that is, the results of a primary vaccination so closely as scarcely to be distinguished from it; the remaining two-thirds being more or less modified or failing entirely. In other trials on a large scale, the success has not been as great. It should be remembered that secondary small-pox occurs on an average in about one per cent. of cases; and that while we urge revaccination, both as a test of remaining receptivity to it, and of further exhaustion of what may remain, it is well known that a single well-performed vaccination, perfect in all respects, does, with few exceptions, suffice to secure for life an individual subjected to it. The small-pox has lost none of its malignity and virulence. And while unvaccinated persons are allowed to accumulate, it is vain to hope for any exemption from epidemics of greater or less extent. When every child shall be carefully vaccinated, and revaccinated at later periods, when the errors, ignorance and prejudices with which the subject is now invested shall be dispelled, and the human family shall become willing to place itself unhesitatingly under the "ægis of JENNER," vaccination will become even more deserving of the confidence and encomium of its friends and the public, less open to objection and refusal, and small-pox, if it does not disappear entirely, will become less frequent and fatal. In times of epidemic prevalence of small-pox, all should encourage vaccination. Vaccination and revaccination are positive duties. It is to be regretted that any persons should be found so ignorant and prejudiced as, by precept or example, to oppose or discourage this safe and simple means of protection. In order to give complete and assured protection against small-pox, every person, not recently and thoroughly vaccinated, should be at once revaccinated, and in subsequent life should repeat this duty as often as every five or six years, until in adult life the repeated revaccination cease to have effect. The perpetual exhaustion of each individual's susceptibility to vaccinia and small-pox should be regarded as a duty more imperative than that of maintaining a policy of insurance, or the necessary apparatus and well-drilled methods of protection against accidental fire or death. For small-pox is a calamity that must, if possible, be prevented, even at the trouble and expense of an excess of precaution. In view of the dangers of delay in the duty of vaccination, and of the obligation of physicians and all others to secure prompt attention to this duty, the members of this Committee would urgently recommend that vaccination should be promptly secured to every child; also, that revaccination should not be neglected in any person at the period when the change from childhood to adult life renders the individual specially susceptible to contagion, and that in the presence of small-pox every community should adopt measures to provide for general vaccination and revaccination as a public duty.

[Signed.]

EDMUND R. PEASLEE, M. D., President New-York Academy of Medicine.
ELISHA HARRIS, M. D., Chairman of Committee.
WM. C. ROBERTS, M. D., Vice-President Academy of Medicine.
ALFRED C. POST. M. D., Ex-President Academy of Medicine.
STEPHEN SMITH, M. D., Commissioner of Health.
JONAS P. LOINES, M. D., Vaccine Physician Eastern Dispensary.
FRANK P. FOSTER, M. D., Vaccine Physician City Dispensary.
EDWARD H. JANES, M. D.
GOUVERNEUR M. SMITH, M. D.
Committee on Vaccination and Small-pox, New-York Academy of Medicine,
NEW-YORK, Wednesday, Jan. 31, 1872.

The New York Times
Published: February 10, 1872

CALL VACCINATION A MEDICAL SHAM

Speakers Tell Brooklyn Philosophical Association It Doesn't Bring Immunity from Smallpox

DANGER IN IT, THEY THINK

Physician Defends the Practice and Offers to Join in a Novel Test of Its Efficiency.

The Brooklyn Philosophical Association met yesterday afternoon, with the usual platform favorites on hand to make it warm for the lecturers. The topic was "Vaccination." Joseph F. Rinn, President of the association, had the subject "Does Vaccination Grant Immunity from or Lessen the Severity of Smallpox?" Harry Weinberger, a lawyer, who recently had a case on Staten Island where a father refused to allow his children to be vaccinated by the school authorities, spoke on "Should Compulsory Vaccination Be Abolished?" They contended that vaccination was in great part a medical sham.

Mr. Rinn said it was daring on the part of medical men to insist that the laity should not be permitted to pass judgment upon the evidence offered by the facts as to the value or lack of value of medical remedies. He quoted an article in the Bulletin of the Academy of Medicine of Pittsburg, which gave the results of an investigation by a physician into what is taught at medical colleges in this country. This article, he said, showed that there was amazing lack of standardization of studies and that men "with hardly any knowledge of medical science were let loose on the community to gain experience on humanity."

"In spite of this laxity on the part of the medical profession," he said, "these half-trained students, who gain a smattering of knowledge out of books, taught to them in many instances by others that had gained their knowledge out of books, receive certificates as doctors from these colleges, and immediately assume an attitude of dogmatism to the public on subjects such as vaccination, of which they have but little knowledge."

Mr. Rinn then mentioned his offer of a prize of $1,000 to any hospital that would provide five doctors who would submit to vaccination with vaccine virus supplied by the New York Board of Health, and who, after having passed on the vaccination as perfect, would let doctors he chose inoculate smallpox matter into their bodies. If three out of the five did not contract smallpox, Mr. Rinn said, he would forfeit to the hospital or any charity the sum named.

"The fact is," he said, "the medical profession knows little or nothing about the vaccine it uses, and hardly a man among them could conscientiously assert that he knew what kind of virus he was using on his patients. They call Jenner the founder of vaccination and in all their medical works praise him, yet I challenge the medical profession to show that any vaccine factory in this country or any Board of Health is manufacturing virus according to principles laid down by Jenner in his book published in 1801, entitled 'An Inquiry Into the Cause and Effects of Variola Vaccinae.'

"What are they using for vaccine virus? Horse grease, cowpox, as first advocated by Jenner, or pure horse grease, as he later in life adopted? Jenner distinctly proved that 'spontaneous cowpox' was useless, and warned the profession against its use, and yet the advertisement of prominent vaccine firms state that their seed virus is from cows having 'spontaneous cowpox.' Other firms are and have been using virus from cows that were inoculated with pure smallpox from human beings. Now, if Jenner is the father of vaccination and the medical profession has never disavowed belief in it, by what right or by what discovery since made do they depart from the Jennerian principle of vaccine virus, and then enforce it upon us by law?

"The medical profession is constantly quoting 'pure calf lymph from healthy calves.' How can calves be healthy when they have to be diseased in order to get the virus?

"It is a direct violation of the principle of asepsis in surgery, which is advocated by scientific medicine and stands for this, that no foreign organism shall be permitted to go into any cut part on the human body, and aseptic medicines are used in all operations to prevent what is done in vaccination. Any surgeon who would put matter from a sore into a freshly cut body in an operation would now be considered a fit subject for a lunatic asylum, and yet they do this very thing in vaccination, and never seem to realize its dangers. When accidents and deaths occur from vaccination they always contend that the vaccination was improperly done or that the person was not clean, but always deny that the vaccination itself could cause the fatal result, but I challenge them to produce vaccine virus free from bacterial taint."

Mr. Weinberger complained of the Board of Education for its stand on compulsory vaccination, and reviewed the case of Herbert A. Thorpe of Staten Island, who had threatened to shoot anybody who attempted to vaccinate his children. He said he had asked the Board of Education to provide a school in every borough where children whose parents objected to vaccination could send them. He also had a word for "doctors who take their profession into politics." He said that if they did not endeavor to have laws passed through the Legislature like that for compulsory vaccination, appropriations like those for Boards of Health would have to be cut down and many young doctors would lose their places. He said he was against compulsory vaccination because rich children were favored—"and the rich," he said, "can always get their rights, can always hire lawyers who may get them even a little more."

He advocated sanitation instead of vaccination. "Tear down the tenements and pay the landlords for them."

When the Chairman, Dr. William H. Weston, opened the free discussion a white-haired man mounted the rostrum and said:

"When I was 8 years old I suffered from smallpox so badly that it was thought I was doomed to blindness for life. Every one in the little village where I lived had had it. One child would get it, and it would spread through the whole place. That was seventy-seven years ago. Forty-five years ago I went back to my birthplace, and, mindful of the old plague, noticed the faces about me carefully. There was no visible sign that any one had had it, although these were not lacking in the old days. I asked one of three persons who remembered me as a child about the disease. I was told there was no more smallpox; that vaccination had wiped it out."

A doctor in the audience said he had been in the Department of Health and had vaccinated about 5,000 children in the public schools. He said he had seen only five cases of smallpox in the district he was assigned to and that there were no difficulties arising from vaccination itself.

"Let Mr. Rinn," he said, "choose ten persons who do not believe in vaccination, and I will take ten who do. We will then all be exposed to a malignant case of smallpox. To three out of his ten I will promise the best of medical attention for the malady they contract, and to the other seven the most gorgeous of funerals."

The New York Times

Published: February 6, 1911

In 1915 a man from upstate New York named James Loyster, whose 11-year-old son died after being vaccinated against smallpox, printed and circulated a 40-page pamphlet, with photographs, summarizing 51 vaccine disasters, 30 of which were fatal. Thirteen were reprinted in a book published in 1920 titled *Horrors of Vaccination: Exposed and Illustrated* by Charles Higgins. This work was dedicated to President Woodrow Wilson, and naively implored him to intervene and put an end to the practice. It contained many photographs of victims young and old. Reprinted on the following pages is the title page and introduction to Loyster's pamphlet, along with four case histories.

Figure 5 in the introduction is a photograph, taken after death, of an English infant named Mona Stevenson, who succumbed on August 4, 1908 to the effects of vaccination, her face and body covered by smallpox pustules.

"VACCINATION RESULTS IN NEW YORK STATE IN 1914"

SHOWING PHOTOS AND PARTICULARS OF MANY CHILDREN KILLED BY VACCINATION IN 1914

By JAMES A. LOYSTER
CAZENOVIA, N. Y.

In Memory of his Son

Lewis Freeborn Loyster

who Died September 21, 1914
as a sequence of
Vaccination

INTRODUCTION

THE object of this pamphlet is to publish the result of a painstaking inquiry into the effects of vaccination in the State of New York in 1914.

My immediate personal interest in the subject is due to the death of my only son as a result of vaccination.

I have been a believer in and advocate of vaccination. I was myself vaccinated in childhood by the arm-to-arm method without ill effects. It was in accordance with this belief and in an honest effort to comply with the law that I had my son vaccinated. Even his death did not entirely shake my faith in the practice, but it led me to make an investigation of the results of vaccination in New York State in 1914. Owing to the difficulty of making a canvass in the great cities, no effort was made to collect statistics in New York and Buffalo, and but little in Rochester, Syracuse or Albany. My investigations were, therefore, practically confined to the rural or semi-rural portions of the State.

The result has been the gathering of such an appalling story of death and illness as to completely shatter my belief in the wisdom of enforced vaccination.

It should be understood that I am not a physician. This leaves me free to write with greater frankness in certain instances than would be permitted to a doctor by the ethics of his profession.

I desire to be understood as seeking nothing but the common good of humanity. In the pages that follow I have attempted to write without bias as far as is humanly possible. Only established facts are presented; extravagant statements have been avoided; in no single instance has a quotation been made from an anti-vaccination source. I have felt that my case is so strong that I could afford to be generous in my arguments.

The reader is invited to go over the facts as presented and draw his own conclusions as to the accuracy of the deductions made.

JAMES A. LOYSTER.

Cazenovia, N. Y., Jan. 5, 1915.

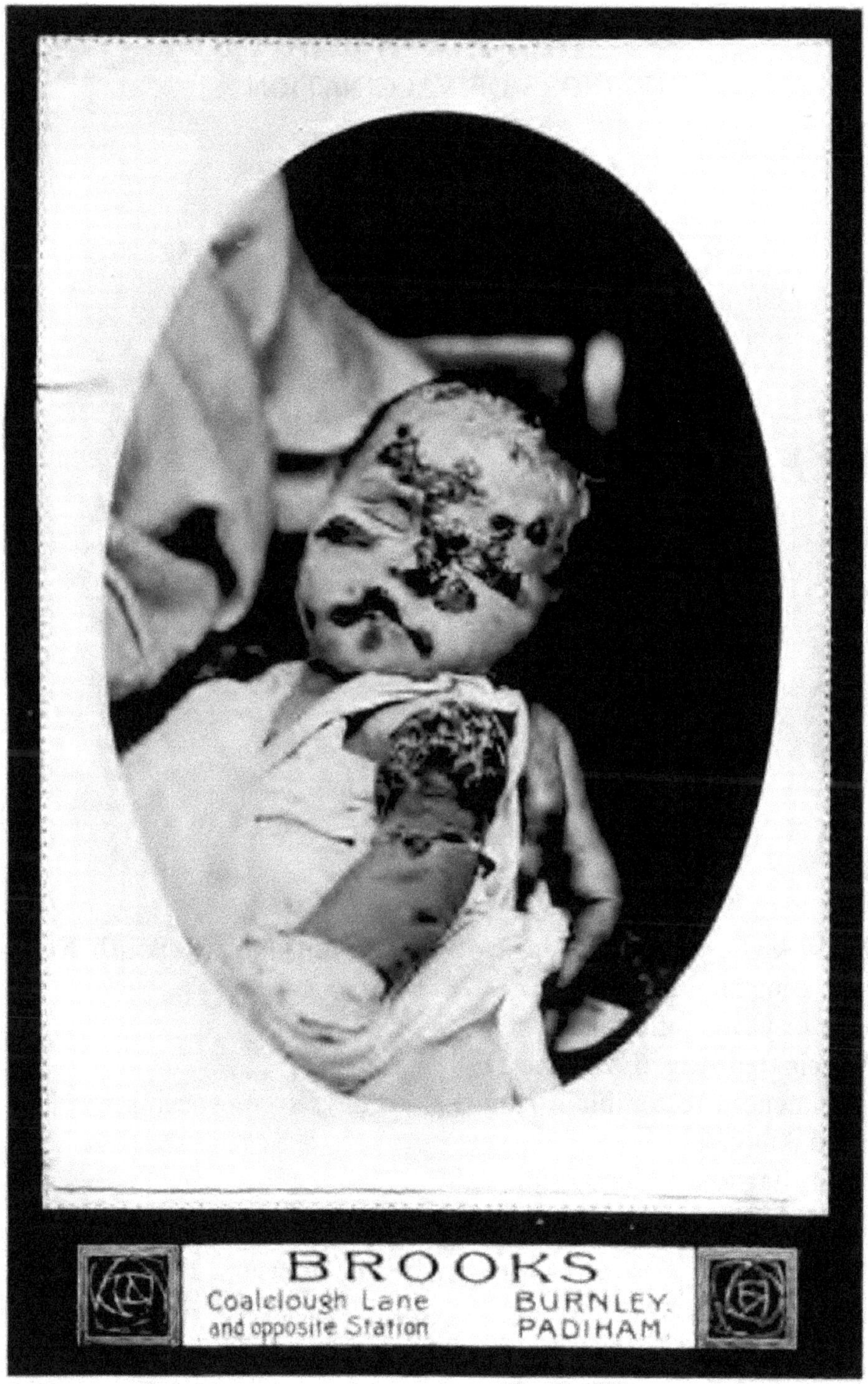

Fig. 5. Photograph of a little English baby two months old, taken after death, killed by vaccination in thirty-six days. This is a very severe and frequent form of fatal vaccination known as "Generalized Vaccinia," where the vaccine sore spreads all over the body in a series of big confluent pustules very like confluent smallpox, being thus clearly an aggravated case of pus infection and septicemia.

HORRORS OF VACCINATION

CASE No. 18

Edward Johnson

EDWARD JOHNSON, son of J. Johnson, No. 65 South St.. Binghamton, N. Y.

Age 11 years 7 months.

Vaccinated Aug. 6, vaccine "E."

Commenced to complain Aug. 14,—8 days from vaccination.

Died Aug. 26.

First diagnosed "Blood Poison."

Subsequent diagnosis "Typhoid Meningitis."

Patient had been in perfect health prior to vaccination. He was ordered to hospital for an operation to which parents would not consent. There were no convulsions. He was unconscious about a week. Paralysis of the throat a prominent symptom. Breathing much labored.

Both this case and No. 17 were subjects of sharp controversy among doctors and others in the city of Binghamton. There is no lack of medical opinion as to vaccination having been the primary cause of illness, but this is stoutly denied by other physicians. Considered collectively with the similar cases herewith reported, it seems that there can be little doubt of their being similar to the rest of the vaccination fatalities.

PETITION TO THE PRESIDENT

CASE No. 19

BELLE HINMAN, daughter of Prof. M. C. Hinman, Tully, N. Y.

Age 7.

Vaccinated July 25, vaccine "H."

Commenced to complain Aug. 7,—12 days from vaccination.

Died Aug. 10.

First diagnosis, "Acute Indigestion."

Final diagnosis, "Cerebro-spinal Meningitis."

This little girl was also in perfect health and unusually well developed for age, both physically and mentally. She complained of feeling ill on Aug. 7 with symptoms indicating a digestive disturbance. The local physician called it acute indigestion. She soon lapsed into unconsciousness and never rallied. Expert counsel was called and case pronounced cerebro-spinal meningitis. A lumbar puncture disclosed in spinal column a clear fluid under normal pressure. Cultures were sterile; intra-spinous injection of the Flexner serum without beneficial effects. Paralysis was extensive, involving the throat; breathing labored and death due to respiratory insufficiency. Temperature elevated during entire illness. All symptoms were so exactly parallel to those of Cases 24 and 25 as to warrant the suspicion that this, too, was infantile paralysis.

BELLE HINMAN

PETITION TO THE PRESIDENT

CASE No. 24

WILBUR DOYLE

WILBUR DOYLE, son of Ivan Doyle, New Woodstock, Madison County, N. Y.

Age 8.

Vaccinated Aug. 28, vaccine "E."

Commenced to complain Sept. 9,—11 days from vaccination.

Died Sept. 14.

Diagnosis, "Infantile Paralysis."

This boy became unconscious Sept. 9. Temperature nearly 106. Kidneys involved; necessary to use catheter. Temperature dropped to about normal for few hours. General paralysis appeared. Throat paralyzed Sept. 13. Could not swallow. Temperature again became extremely high, reaching 107.4 rectal at 3 A.M. on the 13th. Remained unconscious until death, which occurred at 10:35 P.M. Sept. 14.

Note. The make of vaccine virus used in the different fatal cases shown herein is indicated by the Code letters "E" and "H." E indicates the virus made by the Mulford Company of Philadelphia, and H the virus made by Parke Davis & Company of Detroit. These were the two manufacturers whose virus is officially alleged to have caused the epidemic of Foot and Mouth Disease in 1908. See page 99.—C. M. H.

HORRORS OF VACCINATION

CASE No. 25

Lewis Freeborn Loyster

LEWIS FREEBORN LOYSTER, son of James A. Loyster, Cazenovia, N. Y.

Age 11.

Vaccinated Aug. 29, vaccine "E."

Commenced to complain Sept. 10,—11 days from vaccination.

Died Sept. 21.

Diagnosed "Infantile Paralysis."

This boy was the picture of health. To quote the attending physician, he was "a splendid specimen" physically. On the night of Sept. 10 he had a bad headache. The next morning at six was no better. At ten he was found by his mother unconscious. He was very constipated and slightly nauseated. Vomited once. Examination of urine showed indican in considerable quantities. The initial diagnosis was acute indigestion. Blood examined and found normal except for an excess of white corpuscles, explained as a natural sequence of vaccination. Unconsciousness continued. On Sept. 13 lumbar puncture made. Fluid from spinal cord not under pressure; perfectly transparent, subject to microscopic and culture tests; proved absolutely sterile. Case diagnosed as infantile paralysis of the cerebral type. Temperature very high, ranging from 104 to 106 rectal. About Sept. 14 throat became paralyzed. For five days could not swallow a drop. Food administered through rectum; medicine hypodermically. Paralysis of right leg and arm appeared about Sept. 17. Lungs filled with thick mucus. Respiration labored. Slight cyanosis. Small amount of oxygen administered continually after Sept. 16. Death occurred at 10:10 P.M. on Sept. 21 from paralysis of respiratory muscles. Temperature about 107 at death. Microscopic examination of spinal cord after death confirmed diagnosis as infantile paralysis.

The true, suppressed, and disgraceful story of the polio vaccine

THE polio vaccine is the one that everybody "knows about." Almost every American adult "knows" that polio was a dreaded disease until the great Dr. Jonas Salk came along with his miracle vaccine and wiped it out overnight — one of medicine's greatest success stories. It's *the* vaccine that people always point to when you begin raising doubts as to why a child should be vaccinated. "Look at what happened with polio," they invariably reply. "Nobody has polio anymore. How can you say that vaccines don't work?"

Few subjects have been drenched in as much disinformation as polio — the disease, the vaccine, its creator. As we will shortly see, the Salk polio vaccine myth is nothing more than the residue of an out of control news media binge that lasted two weeks in April 1955. By the end of that month the vaccine had proved to be such a disaster that some media factions did an about-face and began to report many of the troubling facts about it, though in a much more subdued manner. With a little bit of time and effort, you can find and read the deflating stories that appeared, mostly in late April and early May 1955, in many American news publications, but soon thereafter were buried and have remained buried to the present day. Additionally, with the passage of time and the facilitation of the Internet, an abundance of information has come to light attesting to the vile character of this man Jonas Salk.

What is truly amazing in the face of all these damning facts is that so few are aware of the heated controversy surrounding the inception of the Salk vaccine, and the Big Lie endures, as entrenched as ever. Since it's the main pillar of the vaccine dogma, and the source of more brainwashing than all the other childhood vaccines combined, it merits a detailed and lengthy discussion.

Before delving into the medical aspects of polio and the story behind the vaccine, let's ground ourselves with a few little-known facts about which I've never encountered any disagreement:

- Polio was always a relatively rare disease in America. There never was any nationwide "epidemic" nor anything close to one.
- More than 90% of human beings are naturally immune to the polio germ and will not develop symptoms under any conditions.
- When the Salk vaccine came out, many prominent doctors and scientists expressed grave doubts about it and refused to vaccinate their own children with it.
- On May 7, 1955, just weeks after the announcement of Salk's wondrous achievement and the start of a nationwide inoculation campaign, Surgeon General Leonard Scheele temporarily suspended the program after more than a hundred children contracted a severe case of polio from the shots.
- Because of the obvious danger of the vaccine, many European countries rejected it. Yet polio in these countries continued to fade away to the point of virtual nonexistence.
- In 1960, it was discovered that millions of doses of polio vaccine, which was (and still is) grown on diced monkey kidneys, had been contaminated with a cancer-causing monkey virus known as SV-40.

Many other claims *have* been contested, such as that the incidence of polio had been declining for many years in the U.S. and Europe prior to 1955, and that reported cases of the disease in the New England states were five to ten times higher during the year *after* the inauguration of mass vaccination than the year *before* it. I have no reason to doubt these claims, but as I have no wish to become entangled in statistics, I leave further investigation to the reader.

❖

Poliomyelitis, or polio as we all know it, is caused by an intestinal virus that attacks the central nervous system. It is contagious and can be spread through food, water and contact with contaminated feces — changing an infected baby's diaper, for example. Polio can also be brought on by other shots, a phenomenon that became evident earlier in the twentieth century with the advent of the pertussis and diphtheria vaccines. Many biologists working independently noticed this link and published papers on it. *Why* it happens is unclear; *that* it happens has been admitted even by vaccine proponents.

Excessive sugar ingestion appears to be another culprit. Polio always struck with the greatest frequency in the summer months, when consumption of soft drinks and ice cream was at its highest. This connection was demonstrated by an American doctor and nutritionist, Benjamin Sandler, who in the spring of 1949 warned the residents of North Carolina, through local newspapers, to cut way back on these products. As a result, sales fell 90%, as did the rate of polio throughout the state. However, this movement lost steam the following year and sales of sugary refreshments rebounded, as did cases of polio. Furthermore, Sandler compiled data showing that in China, despite its huge population that lagged far behind the West in health standards, polio was almost unheard of; he attributed this to the very low consumption of sweetened food products in that country, where sugar was a scarce commodity.

Another causative factor may well have been the widespread use, beginning in 1945, of the insecticide DDT, which by the 1960s was widely criticized due to environmental concerns, and was eventually banned in most of the world. Ironically, it was used here mainly to kill flies, which were thought to spread polio. The "experts" of the day reassured parents that it was perfectly safe not only to spray in areas like playgrounds and beaches where children gathered, but also, as incredible as it sounds, to sprinkle it in powder form on their food. It's possible that many cases of acute DDT poisoning were misdiagnosed as polio. We now know that toxic substances can cause profound changes in cell structure, a classic example being the connection between asbestos and mesothelioma.

❖

Polio is a baffling disease; its epidemiology is not well understood. Decades ago at least, it seems to have been more common in Western nations than in the world's most impoverished regions. And it appears there were more cases of polio in America in the twentieth than in the nineteenth century. One gets the feeling that had other vaccinations been halted, sugar intake lowered, and DDT never used, polio would have ceased to exist here. But of course such a hypothesis was never tested.

Misconceptions about polio abound, the most common being that anyone who contracts the disease will die or become paralyzed. This is not true. Symptoms vary in intensity, but usually include fever, headache, sore throat, weakness, neck stiffness and joint pain. Most polio cases in this country went undiagnosed because very often the symptoms were indistinguishable from other common ailments that clear up on their own, such as a bad cold. Even when paralysis does occur, a complete recovery often follows.

I don't mean to trivialize the fate of those unfortunate people whose lives were devastated or ended by polio in its most virulent form, which obviously is a terrible disease, capable of paralyzing the limbs or respiratory muscles — the latter condition requiring the use of the "iron lung," a medical device invented in 1928 that forces the diaphragm to contract and expand. Nevertheless, to put things in their proper perspective, polio was never common in this country. What was common was publicity about it. In the 1940s and 1950s, when the race was on to produce a vaccine, ten times as many children died in accidents as from polio; cancer claimed three times the number. Although exact figures are impossible to pin down, I would estimate that less than 1 in 3000 Americans was ever seriously affected by polio. In my life, I have known only one such person, mentioned earlier. Quite frankly, polio never deserved more than a fraction of the attention it got in the medical research community, and had it been ignored, it likely would have disappeared on its own, as it in fact did in other advanced countries.

Why, then, was it thrust into the spotlight? Because of its most famous victim: Franklin Roosevelt. Struck down by polio at thirty-nine, desperately trying every therapy to regain use of his legs and failing, confined to a wheelchair for the rest of his life, Roosevelt became obsessed with the idea of defeating the disease. Before and after he became president in 1932, he plowed his immense wealth and political clout into his obsession. In 1938 he teamed up with a friend and former law partner named Basil O'Connor — a man with ambitions in philanthropy as well as medical projects — to form a nonpartisan group called the National Foundation for Paralysis. Its objectives, Roosevelt proclaimed, were to find a cure for polio and provide the best treatment available for its victims. The National Foundation, as it came to be called, would evolve into the largest voluntary health organization of all time.

The crusade against polio was revolutionary in the annals of medicine. Spanning two decades, it employed media spokesmen, politicians, movie stars, public relations agencies, fund raisers, doctors and local chapter organizers. It gave birth to the "poster child." It changed the way the government licensed and tested new drugs. It mobilized the entire nation with a community spirit. Through the tireless efforts of volunteers, donations eventually soared into the hundreds of millions of dollars, a fair chunk of it from pocket change — the famous March of Dimes campaign, which was conceived in 1938 by the vaudeville entertainer Eddie Cantor and approved by President Roosevelt (which is why his profile began appearing on the dime in 1946, the year after his death). Many famous people, often with the best of intentions, lent themselves to the cause well into the 1950s — people as diverse as Bing Crosby, Joe DiMaggio, Richard Nixon and the Lone Ranger. Everything from church suppers and square dances to presidential birthday balls and Hollywood extravaganzas were employed to generate publicity and research funds to support the quest for a vaccine. Some critics began to mock polio as a "celebrity disease," a justified reaction considering that the money and effort lavished on it were ridiculously out of proportion to its actual danger, while more common and deadly diseases went underfunded.

It was all doomed to fail, and fail it did because vaccination, it bears repeating, rests on the false premise that injecting a human being with dead or weakened amounts of a disease germ will stimulate the body to produce antibodies that will prevent the disease, thus enhancing health. Life doesn't work that way. It had been proven by all the vaccines that came before polio, and it would be proven by all the vaccines that came after. The best minds that enlisted in the war on polio were no match against the laws of Nature.

Moreover, many of those who took up the fight against polio were scoundrels to begin with. Although the fact is never taught in school nor mentioned on television, Franklin Roosevelt was detested by a large number of Americans who saw him as the consummate lying politician, especially in the way he maneuvered his country into World War Two. And while I can understand and sympathize with his ambition to wipe out polio, it was a reckless undertaking. Basil O'Connor, his hand-picked director of the National Foundation, was a man of limited intelligence and unlimited vice, who years later, even when serious problems with the new vaccine had become evident, would introduce legislation to make vaccination compulsory, allegedly to get rid of surplus stock.

And then there were the men and women in the long white coats, feverishly working in their laboratories to create an effective vaccine, elbowing each other for fame, fortune and delusions of scientific immortality. They were vain, rotten people and some of them were criminals in a very real sense. Two names from this rogues' gallery would emerge and endure to the present day: Jonas Salk and Albert Sabin.

Salk and Sabin were bitter rivals and would remain so for forty years, right up to their deaths in the 1990s. They represented different schools of thought, and along with their respective colleagues, followed separate paths to the same elusive goal. Salk believed that a vaccine made of virus killed by formaldehyde (a main ingredient in embalming fluid and a known carcinogen) was the answer, while Sabin felt that only a live polio virus, weakened (in medical terms, *attenuated*) in the laboratory, also by toxic chemicals, would be effective. This debate over which was the better approach actually predated the work of these two men; it went back to the 1930s when rudimentary polio vaccines were tested on a small scale with deadly results. In fact, the debate has never ended. The debate is irrelevant because both vaccines are proven failures and both opened a Pandora's box of new and serious medical problems. However, experience has shown that the live virus vaccine is more dangerous.

Even those who believe in vaccination must admit that it is now and always has been a crapshoot. What concentration of formaldehyde should be used to kill the virus, yet keep it intact enough so as to stimulate the immune system to develop antibodies? Or, if you want to keep the virus alive but safe, how much would be needed to lower its potency yet still trigger an antibody response? Would one shot do, or were two or more needed? What is the best medium on which to grow viruses to prepare them for vaccines? Was mineral oil a safe adjuvant (antibody stimulant)? If so, why did some virologists believe that it caused swelling at the injection site? Would it be better to use water, even though it lacked adjuvant properties? Would use of any of the three known virus strains in the vaccine confer immunity to all three types of polio, or was an exact match needed? Was a three-in-one vaccine too dangerous? To those who had committed their lives and careers to a polio cure, there were no ready answers to these and many other questions. *These issues could be resolved only through experimentation.*

In his book *Immunization: Theory vs. Reality* Neil Miller defines vaccinations simply as "crude experiments conducted on innocent people." That is exactly what they are, and this is true of *all* vaccinations. But without a doubt, polio vaccines have surpassed all others with respect to the number of innocent people — the vast majority of whom were children — experimented on.

❖

In researching this topic, I was taken aback by something I had never heard of. Over the years, about 100,000 monkeys from India and the Philippines, as well as chimpanzees from central Africa, were captured in the wild then flown to the U.S., where they were used to test the polio vaccine before being destroyed. Invariably they arrived here malnourished and miserable, and a number of them attacked their human handlers in the lab, at least once with fatal results. Some of them endured intense pain before being put to death. This brings up the touchy subject of using animals in experiments for the ostensible purpose of benefiting humans. There is a wide range of opinions on vivisection — the practice of harming live animals in conducting scientific research — and the question will always be raised, who's to say what's right and what's wrong? It's all quite subjective. Personally, even though experimentation on lower animals like mice and hamsters bothers me less than experiments performed on primates, *all* vivisection disturbs me. Then again, I'm sure there are those who would consider me cruel for being a hunter, though I find such a sentiment ridiculous because man is a natural predator and meat eater. In my experience most hunters, like me, are sensitive to the suffering of animals and strive for the quickest kill. In any case, to my way of thinking, there is something fundamentally wrong with people who have no qualms about using vast numbers of monkeys in dubious vaccine experiments, then just killing and throwing them in the trash like potato peels. Such people strike me as demented.

From monkeys it's just a short step to children. More than a million American children were used to test both the live virus and killed virus vaccines. Most had a choice, or at least their parents did; the rest were treated like laboratory rats. In fairness to the polio vaccine fraternity, all of whom are now deceased, there was a wide spectrum of opinion on this most sensitive issue. Some were troubled by the idea of exposing so many children to unknown risks, if not vehemently opposed to it. Others, however, were not. Among the earliest of the latter group was John Kolmer, a Philadelphia pathologist, who in 1935, working independently, vaccinated more than 10,000 youngsters with his own secret recipe containing live viruses. This resulted in at least a dozen cases of paralytic polio, nine of them fatal. Kolmer claimed to have injected himself and his two sons with his concoction, and to have been granted consent by the children's parents. At a medical convention later that year, a government health official came close to accusing Kolmer of murder, to which he replied, "Gentlemen, this is one time I wish the floor would open up and swallow me," and sat down without another word. So it appears that Kolmer, at least in the end, possessed a glimmer of a conscience.

The same cannot be said for Hilary Koprowski, a scientist at Lederle Laboratories. In 1950, Koprowski fed his live virus vaccine, diluted in a glass of chocolate milk, to twenty handicapped children in Letchworth, an institution located in New York's Hudson Valley. This was carried out in secret, with only one health official at Letchworth, an acquaintance of Koprowski's who supported the experiment, aware of what was taking place. According to Koprowski, all the children developed polio antibodies and none contracted the disease.

Six years later, after the rise and fall of Salk's killed virus vaccine (which we will come to), Koprowski was invited to Belfast, Northern Ireland for a guarded testing of his product. Stool samples of inoculated children revealed that the weakened virus had actually become stronger while passing through the body, a most alarming observation. Fearing a polio outbreak, the Irish director of the experiment, a virologist named George Dick, whose own daughter had been vaccinated, stopped it. "Had it continued," Dick said, "I have no doubt at all that we would have paralyzed a number of children." He questioned Koprowski's data and expressed a deep disappointment in him, a feeling widely shared by Koprowski's American colleagues who viewed him as brash, dishonest and lacking a conscience.

Koprowski then moved on to the Congo where he tested his vaccine on nearly a million rural Africans, the great majority of them infants and toddlers. The vaccine was administered in liquid form, by squirting it into their mouths. The youngest children received fifteen times the adult dosage. Koprowski apparently ran into problems with WHO (the UN's World Health Organization), whose support he claimed he had, although WHO denied backing him. No satisfactory results were achieved — one can only guess at the human toll — and Koprowski's vaccine was never licensed for use in any country.

Unaware of Koprowski's Letchworth venture at the time, an up and coming microbiologist at the University of Pittsburgh named Jonas Salk was thinking along the same lines. In a letter to the research director of the National Foundation, which was sponsoring his work, Salk wrote: "I think the time has come for these experiments to be carried out in man.... I have investigated the local possibilities for such an experiment and find... there are institutions for hydrocephalics and other similar unfortunates. I think we may be able to obtain permission for a study."

Salk did obtain the permission he wanted from two establishments in western Pennsylvania, one of them a grim state institution where the hopeless and forgotten lived out their lives. Subjects were used with and without consent, the vaccine was deemed safe and potent, and this led to the famous Francis Field Trials of 1954.

"Field trial" is a euphemism for experiment, and this was a major medical experiment the likes of which America had never seen. It was supervised by Thomas Francis, Salk's fatherly mentor and close associate. Although concerns and apprehensions were raised throughout the country and a small percentage of parents refused to give consent or later withdrew, the experiment, fueled by a huge sum of money and intense publicity, proceeded as planned and in the spring of 1954, 1,080,680 schoolchildren were injected with the inactivated Salk vaccine, while 749,236 were given a placebo. For nearly a full year in this pre-computer age, Francis, assisted by an army of volunteers and paid statisticians, collected and tabulated the results.

In 1959, following a visit by Soviet virologists to his laboratory in Cincinnati and then an invitation that took him to Leningrad (now St. Petersburg), Salk's arch-rival Albert Sabin pulled off the biggest medical experiment in history: 77 million — everyone under the age of 20 in the Soviet Union — were given vaccines containing his live polio virus strains. Although Stalin had been dead six years, a totalitarian regime still ruled the country, and needless to say, neither these young people nor their parents had any say in the matter. Sabin's Soviet counterparts proclaimed the experiment a great success, as one would imagine, but what really happened behind the Iron Curtain remains unknown to this day. If one extrapolates from events in the U.S., it's likely that the vaccine killed or paralyzed thousands of Russian children.

But it was the Francis Field Trials alone that led to The Myth. This experiment was undertaken to determine the feasibility of vaccinating all American children; if it worked for a million, it would work for all. From the summer of 1954 to the spring of 1955, the nation held its collective breath while data from 211 counties in 44 states poured into Ann Arbor, Michigan, where Thomas Francis had set up a vaccine evaluation center.

Yet during this time frame, and in the months that preceded it, there was confusion and unease behind the scenes. Some monkeys were still being infected by vaccines that had been presumed safe. Tense, unpublicized meetings over safety issues took place between officials of the National Institute for Health, a federal agency, and the National Foundation. Since Salk had elected to use what was called the Mahoney strain in his vaccines — the most virulent of the three known polio viruses — there was particular worry that not all the viruses had been killed. The technicians at Parke-Davis, a Detroit-based pharmaceutical company selected to manufacture the vaccine for the trials, could not duplicate Salk's product and were confused by his instructions, which in any event were constantly changing while Salk strove to perfect his vaccine. And in evaluating the data, Francis had to deal with the deaths of hundreds of children, most of them from accidents, but a significant percentage from natural causes, and some from polio itself.

The Salk Myth was born on April 12, 1955. On that day, and for the next two weeks, Salk was catapulted to the status of a scientific god. A tidal wave of hero worship swept over the land, a tidal wave that has drowned the truth to this day. In the weeks leading up to it, the media had been dropping heavy hints with news headings like "Tracking a Killer" and "Closing in on Polio." Actually, the whole thing had been orchestrated to fall on April 12, the tenth anniversary of FDR's death. As flashbulbs popped at a packed news conference, Thomas Francis, with Salk standing at his side waiting to speak, made the announcement that so many people had waited so long to hear: the vaccine worked, and it was safe. Polio was now a disease of the past.

What followed can only be called a media-choreographed fit of national insanity. As the news spread like wildfire around the country, car horns honked and church bells pealed. People wept with joy. Salk was cheered and toasted as a true hero and benefactor to mankind — a humble, soft-spoken scientist who had persevered for years in an ultimately successful quest to wipe out a horrible disease. Envelopes addressed to Salk containing donations, thank-you notes and pleas to find a cure for cancer poured into the University of Pittsburgh by the thousands. Two brand new luxury automobiles arrived. Elementary schools sent giant posters signed by the entire school body, inscribed "We Love You, Dr. Salk." Salk was approached by agents from three major Hollywood studios for the exclusive rights to his life story. Honors were bestowed on him from many quarters, including the U.S. House and Senate, which gave him the Congressional Gold Medal, the nation's highest civilian award. President Eisenhower arranged a Rose Garden ceremony, where in a voice trembling with emotion he profusely thanked The Great One for his contribution to medicine which would spare so much suffering, and promised to give the vaccine to every country that wanted it. "POLIO IS CONQUERED." screamed the front page of the *Pittsburgh Press*; *Newsweek* extolled "A Quiet Young Man's Magnificent Victory;" Salk's profile adorned the cover of *Time*. Viewers nationwide heard TV's top newsman, Ed Murrow, begin his live broadcast from Ann Arbor with these words: "Today a great profession made a giant step forward and the news that came out of this room lifted a sense of fear from the homes of millions of Americans."

But it was all a rush to judgment — actually a badly aimed *missile* to judgment — and by the end of the month, after the missile had reached its peak, it came crashing down to earth in resounding tragedy. The truth of the matter, behind all the media hoopla, was that the vaccine was swamped with unanswered questions and unresolved problems. Foremost was the issue of safety and adequacy. Some of Francis's deductions, for those who cared to examine them, were conjectural and his conclusion that the vaccine was 60 to 70 percent effective against the most prevalent polio strain was hardly a ringing endorsement. The illnesses and deaths went largely unexplained. It appeared also that the trials were riddled with fraud, since, critics pointed out, all cases of polio contracted within a month after vaccination were not tallied on the pretense that they were "pre-existing."

In her hard-hitting 1993 book *Vaccination* (subtitled *100 Years of Orthodox Research Shows that Vaccines Represent a Medical Assault on the Immune System*), Australian doctor and researcher Viera Scheibner wrote: "The authors of the [Francis] report stated that there were 57/100,000 cases of polio among the vaccinated and 54/100,000 among the unvaccinated. So the rate of polio was slightly higher in the vaccinated than in the unvaccinated.... Based on their own figures, the trial showed a total failure of the Salk polio vaccine to protect against poliomyelitis." Thus, one must ask, on what basis did Thomas Francis — who had a reputation as one of the more cautious vaccinologists — pronounce the trial a success?

To all this was now added the problem of the vaccine's manufacture and distribution throughout the country. Following the dispute with Parke-Davis, the National Foundation invited five other pharmaceutical houses to sign on, which they did. But not enough vaccine had been produced prior to the release of the Francis Report because the government had not yet licensed it for manufacture, and prior to April 12 the companies had no way of knowing what the demand, if any, would be. Thus, the vaccine was in short supply at a time of overwhelming demand, with the warm weather polio season approaching, no less. Rumors and charges of price gouging began to fly. Moreover, there was disagreement as to the role of the federal government; President Eisenhower had assumed the big pharmaceutical companies would work out among themselves any shortage problems that might arise, while editorials railed against the incompetence of his administration, and in particular the newly created Department of Health, Education and Welfare (HEW). If that weren't enough, the triple testing safety procedures in vaccine manufacture which had been instituted eighteen months earlier were now largely abandoned, simply because there was little oversight, government or otherwise, in the mad rush to produce the vaccine and get it to market.

Two weeks after the delirious Ann Arbor press conference, telephones started ringing in public health offices around the country, and then in Washington D.C.: *children were coming down with polio*. The cases piled up. Something had gone very wrong. Still, thousands of children continued to be vaccinated every day. The government's top scientists held an all-night emergency meeting in the nation's capital. Polio specialists from around the country, both critics and supporters of the vaccine, were awakened by phone calls, briefed on the situation, and asked their opinions and advice. The batches of vaccine causing the disease were narrowed down to Cutter Laboratories in Berkeley, California, although suspicions arose that other companies also had distributed improperly produced vaccine containing live virus. As health officials began conducting random tests of vaccine lots, they continued to grapple for a response, and more children contracted polio. Finally, Surgeon General Leonard Scheele suspended the vaccination program pending a review of the six drug companies. The campaign was resumed shortly thereafter, but now many people had become disillusioned with the Salk vaccine and kept their children away from it. Yet Salk was never stripped of his sainthood and never pulled down from his pedestal, where he remains to this day.

The Salk vaccine continued to be used through the late 1950s, but it had fallen from grace, and interest now shifted to Sabin's live virus vaccine, which gradually displaced its predecessor. Theoretically the attenuation process was supposed to knock out the virus while keeping it alive, but in truth it was a matter of hit or miss tinkering under the microscope. But within a year of his Soviet experiment, with the Salk vaccine foundering, Sabin was given the green light by the U.S. government to mass produce his oral live-virus vaccine. This time there was no media razzle-dazzle. Taken in a sugar cube, it was easier to deliver, cheaper to make, and in 1963, two years after it was sanctioned by the American Medical Association over Salk's objection, it became the sole polio vaccine in this country. (As I mentioned earlier, I remember the school nurse pulling us out of our third grade classroom, lining us up, and giving us our little treat. I don't believe there was any parental consent.)

No Cutter incident would ever tarnish the Sabin vaccine. Instead, over the years it failed and in many cases inflicted polio but without much publicity. Like Salk's vaccine, it too was stained with fraud. For example, to make it look more effective, the number of polio cases required to define an "epidemic" was nearly doubled. Symptoms of paralysis had to be exhibited for 60 days before qualifying as polio, whereas formerly it had been only 24 hours. And aseptic meningitis, a condition closely related to polio and barely distinguishable from it, was now counted as a separate disease although it had been reported as polio before the introduction of the live vaccine.

At least outwardly, polio continued to decline in America, but there is no evidence that either vaccine played a role in the process. Other advanced nations steered clear of both vaccines because of all the controversy, but polio disappeared there as well. It didn't take long before the Sabin vaccine had become a prime suspect in the handful of polio cases occurring each year in the U.S. In 1976 Jonas Salk testified before Congress that Sabin's vaccine was "the principal if not sole cause" of all reported polio cases in this country since 1961, the last year Salk's vaccine was exclusively used. Several lawsuits were brought against Lederle Laboratories, the leading manufacturer of the oral polio vaccine, by parents of vaccine-damaged children. By 1992, the CDC got around to admitting that every case of polio in the U.S. in the previous thirteen years had been caused by the Sabin vaccine. (The logical question is, "Then why on earth continue to vaccinate against polio?")

The CDC maintained that these polio cases amounted to about ten a year. However, an independent analysis of the federal government's own vaccine database over a five-year period in the 1990s revealed 13,641 reports of adverse events following use of the vaccine, including 6,364 emergency room visits and 540 deaths. The fallout from these tragedies prompted health bureaucrats to reconsider their recommendations, and in 1996 a "mixed" schedule which used both vaccines was adopted by most pediatricians. When this didn't solve the problem the CDC turned the clock back forty years, endorsing a full return to the Salk vaccine — though it should be noted that the CDC's own publication *Polio: What You Need To Know* warns the few parents who read it that the inactivated (Salk) polio vaccine can cause "serious problems or even death...."

Meanwhile, in the topsy-turvy world of vaccines, Finland was going the opposite way. Following a small outbreak of polio in that country among people who had been injected with the Salk Vaccine (as reported in the February 8, 1985 issue of *American Medical News*) Finnish doctors recommended a campaign to inoculate all adults with the Sabin vaccine!

There are those who believe that polio never disappeared at all, and the evidence backs them up. In what seemed to be an eerie replay of Koprowski's aborted Belfast experiment, only on this occasion not stopped in time, a polio outbreak among recently vaccinated children in Egypt in

1983 revealed that the three live virus strains in the vaccine had recombined into a previously unknown fourth strain. This happened again in the Dominican Republic in 2000 amid numerous cases of paralysis, an incident of particular concern as it was believed that polio had disappeared in the western hemisphere.

Dr. William Douglass, who published a monthly newsletter, *Second Opinion*, took this issue further. In an August 1996 article headed "Chronic Fatigue Syndrome: The Hidden Polio Epidemic," Douglass, a vociferous opponent of vaccines, asserted that polio never went away, but now manifests itself as chronic fatigue syndrome, adult paresis, myalgic encephalomyelitis, multiple sclerosis, amyotrophic later sclerosis, Guillain Barre syndrome, idiopathic epilepsy and a host of other neurological conditions with fancy names that were rare or unknown a hundred years ago. Drawing on a paper published in the *Annals of the New York Academy of Sciences*, Douglass convincingly argued that the three original polio strains, altered as they were while passing through the human body in the vaccination process, have mutated over the years into at least 72 viral strains that can cause polio-like diseases.

❖

The animosity between Jonas Salk and Albert Sabin adds a bit of humor to this sad saga. Their feud began around 1950 and continued right up to their deaths in 1995 and 1993, respectively. In personality they were complete opposites. Salk was quiet, preferring to avoid direct confrontation with Sabin, and for that matter with others who didn't like him. Sabin was clearly the aggressor, making no secret of his contempt for his rival, often referring to him as a "kitchen chemist." "You could go into the kitchen and do what he did," Sabin once sneered. Whenever the two were in the same room, such as at a meeting or symposium, Sabin would throw a barb Salk's way, while Salk bit his tongue. Envious of Salk's acclaim at first, Sabin's vaccine would eventually enjoy a much longer reign. In later years Salk's son Darrell, also a doctor, went after his father's tormentor. In 1984 he testified on behalf of a Kansas man who sued Lederle Labs, maker of the oral vaccine marketed as Orimune, after contracting paralytic polio from his recently vaccinated infant daughter. (A jury awarded the man $10 million on the basis of an interoffice memo written by a Lederle doctor that discussed "the possibility of reduced Orimune sales if the company took steps to inform doctors of the risks associated with administering the drug.") And more than once the younger Salk was adamant in calling for a switch back to his father's killed-virus vaccine, which he regarded as clearly superior. "He doesn't know what he's talking about," Sabin scoffed. [His work is] completely out of focus, distorted....Just a chip off the old block."

This jousting sometimes erupted between the proponents of the two different vaccines and even the manufacturers. I have a brochure called *What You Need to Know About Polio and its Prevention* published by Connaught Laboratories, which makes what is now called the enhanced inactivated polio vaccine (eIPV). The brochure falsely credits both vaccine types with eliminating polio from much of the world, then, in a thinly veiled attack on its competitor, twice implies that the oral polio vaccine (OPV) alone carries the risk of inflicting polio. This is also false (although the risk *is* greater with OPV). This brochure appeared right after the CDC came out with the new mixed sequence vaccination schedule. The previous year, under the heading "Lederle Lobbies to Stop Polio Vaccine Policy Change," the National Vaccine Information Center, a Virginia-based organization that exposes vaccine dangers, ran a related article in their newsletter *The Vaccine Reaction*. From the January/February 1996 issue:

In a surprising game of hardball, Lederle Lab came out swinging after the Advisory Committee on Immunization Practices (ACIP) voted at its June 1995 meeting to make a transition from the live oral polio vaccine (OPV) to the inactivated polio vaccine (IPV) in order to prevent cases of OPV-associated polio in the U.S. As the sole provider of OPV in the U.S., Lederle is fighting to retain its monopoly market and has been vigorously lobbying members of Congress and HHS officials to prevent ACIP from going forward with its recommendation. Lederle engaged a high profile Washington D.C. law firm which sent a letter to ACIP challenging its right to make policy change without publishing it first in the Federal Register and charged ACIP members with violating conflict of interest rules. A law firm retained by Connaught Labs, a manufacturer of IPV vaccine, sent a letter to ACIP blasting Lederle for their "desperate attempt... to maintain its monopoly at the expense of the public health.

Lawyers, money-grubbing drug companies, backbiting scientists. Do you really expect the truth to emerge from all their brawling? Do you really think the health of our children is their top priority?

❖

In the end, the most tragic chapter of this story might be one that surfaced relatively late, and casts a long shadow over the present and future. I refer to the contamination of countless millions of polio vaccine doses in the 1950s and 1960s with viruses from monkey kidney tissue on which the polio viruses were grown. This is old news; some excellent pieces of investigative journalism on this topic appeared decades ago in secondary mainstream publications like *Science*, *Money*, *New York* and *Rolling Stone*, as well as in renowned medical journals, not to mention alternative health publications.

As early as 1954, researchers had discovered scores of viruses in this medium used for polio vaccine production. The fortieth of these simian viruses, known as SV-40, would later come to the forefront because of its virulence. But it wasn't until 1959 that a conscientious government scientist named Bernice Eddy discovered that viruses harmless to one mammal could cause cancer in another. Experimenting, she injected extracts of monkey kidney used for polio vaccine production into hamsters. They developed fatal tumors. She knew, however, that she was overstepping her bounds. Five years earlier, she had tested polio vaccine lots submitted by Cutter and quickly had paralyzed monkeys on her hands. She alerted her superiors, but in the jubilation surrounding Ann Arbor, nothing was done. "They just went ahead and released the vaccine anyway, a lot of it," she recalled. "The monkeys, they just discarded."

Now she was watching hamsters die, and again she blew the whistle. And again her concerns were ignored. Out of frustration, while at a meeting on polyoma viruses in New York, she mentioned that one could also get cancer from Rhesus monkey kidney cell extracts — the kind used to grow polio viruses. This enraged her boss and upset everyone in attendance, including Albert Sabin, whose vaccine at the time was on a meteoric rise. Eddy's reward was a gag order, a padlock on her laboratory door, and a demotion.

Shortly thereafter, Eddy was vindicated by Maurice Hilleman, the chief Merck vaccine researcher. Hilleman was the first to isolate and identify SV-40. Sabin's vaccine was loaded with SV-40, meaning that tens of millions of children in America alone had these viruses pumped into their bloodstreams. The public was never told. "I don't think anybody along the way was irresponsible," Hilleman said many years later. "It was important not to convey to the public [this] information because you could start a panic. They had already had production problems with people getting polio. If you added to that the fact that they found live [monkey] virus in the vaccine, there would have been hysteria." (Hilleman's statement about no one being irresponsible is plainly contradicted by Eddy's experience.)

Studies published in eminent medical journals around the world have established a causal connection between the polio vaccine and an alarming rise in certain types of cancer, especially bone, lung and brain cancer, in addition to leukemia. Numerous microscopic examinations of cancerous tissue have revealed the distinctive DNA pattern of SV-40; the only logical explanation is that the viruses were transmitted to the victims through polio vaccination.

Of course, SV-40 is just one of many viruses, not to mention other biological debris, that made its way into polio vaccines, and probably still does, since today's vaccine continues to be developed and manufactured in much the same way as earlier versions. Yet another concern is the suspicion that these viruses, which are not naturally found in human beings, have become incorporated in human DNA and are being passed in the womb from mother to child. Cancer researchers are at a loss to explain why SV-40 has been found in the brain tumors of children who presumably did not receive tainted vaccines, and for that matter, also found in a significant percentage of blood and semen samples taken from healthy individuals. What this seems to boil down to is unintentional genetic engineering on a massive scale. What this portends for the future, it's too early to say. My gut feeling is that the eventuality will not be good, and might be grave.

❖

More than any other disease — indeed, more than any other branch of medicine — the polio vaccine is inseparable from the supposed greatness of its two most celebrated champions, Jonas Salk and Albert Sabin. Salk is by far the "greatest" in the entire field; he is the savior of children, the avatar of vaccine victory over disease. Salk is to medicine what Galileo is to astronomy, what Leif Erikson is to exploration. That, at any rate, is what just about every American, including me, was brought up to believe. But what did Salk and Sabin do to earn our admiration? What kind of men were they *really*? If what their colleagues thought of them is any indication, then there must be a special place in hell for both men, with Salk occupying the hotter spot. Sabin, at least, had some admirers within the vaccine fraternity, though I don't understand why. Even his friends, who thought highly of his skill behind the microscope, conceded that he was obnoxious and abrasive, at times even cruel. As one put it, "All of us who know Albert admire him and detest him at the same time for the same reasons." Maurice Hilleman, *not* a friend, went further: "Sabin was a mean goddam bastard. Smart. But a brainsucker. He went from one field to another, always sneaking in." Quite a revelation, coming from a self-described mean bastard himself.

Sabin also was a rank hypocrite. When, at a roundtable discussion, Hilary Koprowski admitted that he had tested his live virus vaccine on handicapped children, Sabin lashed out at him. "How dare you! Why did you do it? Why? Why?" When Koprowski replied that someone had to cross the line, Sabin shot back, "You're not sure of this, you're not sure of that. You may have caused an epidemic!" Yet this was the man who went on to conduct the largest experiment on children in history — in a bleak communist dictatorship where no one could refuse his own highly questionable live-virus vaccine.

Furthermore, Sabin was the type who could never admit he was wrong. This is understandable, such as when, at a conference, Bernice Eddy implied that children could get cancer from his vaccine; it would have meant renouncing ten years of work. But considering the consequences, that's the path an honorable man would have taken. Likewise, he could not accept that his vaccine was anything less than perfect, and totally rejected the evidence that children were coming down with polio from it. "He was so strong-willed, he thought he could will it away," a colleague said.

Most of us know a few people like Albert Sabin. There's no mystery about them; they're simply the kind of people we like to avoid. A Jonas Salk, with his quiet, serious demeanor, is much harder to figure out unless you get to know him well. That might explain, in part, why the media glorified him, while reserving milder praise for Sabin. At one time or another, everyone involved in polio research had unkind things to say about Salk. Sabin's rants could be written off as professional jealousy, but Salk operated differently. Jonas Salk was a weasel. Little by little, his sly behavior made him a despised and permanent outcast among his peers.

Ann Arbor on April 12, 1955 was the climax. While America was in the throes of hero worship that would turn to dismay in two weeks, those on the inside already had a jaundiced view. After being introduced to a standing ovation by Thomas Francis, to whom he owed everything, Salk quickly alienated those closest to him. In thanking a long list of people who had helped him in different capacities, he didn't even mention his own laboratory workers who had toiled for years under him. Then, expounding on Francis's comment that the vaccine used in the field trials was shown to be 60 to 70 percent effective, he said that his newest vaccine might theoretically lead to 100 percent protection. For the rest of that day — indeed, for the rest of his life — Salk would bask in his instant fame and divine status. Very few were aware of what the news writers and television cameras hadn't recorded. "After Jonas finished talking," Francis recalled, "I went over to him, sore. 'What the hell did you have to say that for,' I said. 'You're in no position to claim 100 percent effectiveness. What's the matter with you?'" Tom Rivers, a top virologist deeply involved in polio, was infuriated at Salk's comments. "To my mind, it was an implied criticism of the way Francis had run the field trials, and nothing should have detracted from the kudos that Tommy received that day." "The bedlam was disgusting," another scientist remarked of the carnival atmosphere. "It was as if four supermarkets were having their premieres on the same day." Researcher Paul Clark wrote his good friend Thomas Francis: "I am deeply concerned, as are many others, with all the hysterical publicity — Polio is licked, Salk the miracle man stuff. The public is gullible.... I am tempted to get out my sharpest pen and stick it into the balloon as far as I can." A National Foundation insider added, "We could see that success... would make a public god of him, distorting the meaning of his work, crediting him with achievements that belonged to Enders and Bodian and so many others, and lousing him up with other scientists. We could see... but it was not our headache."

The most deeply hurt were his own staff, who sat in the packed audience without hearing a word of gratitude for their years of dedicated service. Even David Oshinsky, the scrupulously neutral author of *Polio: An American Story*, from which I've extracted these quotes, briefly takes sides here, writing of Salk: "One of his great gifts was a knack for putting himself forward in a manner that made him seem genuinely indifferent to his fame, a reluctant celebrity, embarrassed by the accolades, oblivious to the rewards.... [Those close to him knew] a man who cultivated the press with the same care he cultivated viruses, crafting his image with a film director's eye." Part of that image was a fake concern for the well-being of children, as in his famous quote which even anti-vaccine author Neil Miller fell for: "When you inoculate children with a polio vaccine you don't sleep well for two or three weeks."

Salk's carefully crafted image did not escape the eagle eye of Australian researcher Viera Scheibner who, in her 1993 book *Vaccination*, quotes at length from an article Salk published in the *Journal of the American Medical Association* (JAMA) in the aftermath of the Cutter incident, in which he expressed an earnest commitment to safety:

146 cases of paralytic polio developed in vaccinated children and their contacts within a short period of time. This demanded a very intensive re-examination of the theoretical and practical implications of vaccine preparation, testing, and use.... It has been realized always that the preparation of a safe poliomyelitis vaccine would, at the beginning at least, require adherence to detail such as is not demanded for the preparation of any other immunizing agent. It was recognized, too, that there must be incorporated into the vaccine preparation process itself a test for safety, which we refer to as the "margin of safety," of such degree that the most sensitive tests... would be negative for living virus.... The relative infrequency of severe paralysis under natural circumstances requires, above all else, that the vaccine must be free, insofar as it is possible to create such a preparation, of the capacity to induce the disease that it is intended to prevent; nor should a vaccine for poliomyelitis cause such side effects as would make it undesirable.... The objective in the preparation of a poliomyelitis vaccine cannot include the knowing or willful acceptance of a risk that is tangible, or measurable to any degree. Any risk that is involved, so long as it is recognized, must be corrected, whatever may be its cause.

Scheibner then comments:

Every point in this proclamation has been violated and there is no guarantee that it does not continue to be violated. Children receiving any of the polio vaccines continue contracting paralysis from the vaccine. The widespread incidence of chronic ill health, an endless stream of respiratory infections, usually resisting all treatment, occurring in small children and the continuing high incidence of child leukemia and cancer are themselves the evidence that not all is healthy in the kingdom of vaccines. The continued contamination of polio vaccines with animal viruses is of special concern.

The most damning revelations about Jonas Salk came from his chief lab assistant, Julius Youngner. Of the daily work atmosphere, Youngner recalled:

There was no personal warmth — I mean none. The first rule we learned was to call him "Dr. Salk," never "Jonas." He would speak to us through a wall of notes and memos. He refused to teach. We were the only lab that didn't hold a seminar, not even a bag lunch. Here was a guy who could always find an hour to brief some reporter at the local Chinese restaurant, but could never find the time to sit down with his own people.

One incident that occurred in 1954 forever diminished Youngner's estimation of Salk, and led to their falling out. Youngner had co-authored a paper about a color test he had developed to measure the concentration of polio virus, and gave a draft to his boss to read. The following week, Salk told him he had lost it, but not before jotting down some notes while reviewing it:

"I was incredulous. If there were those who could be scatterbrained or disorganized enough to 'lose' a manuscript, Jonas was not among them. Quite the contrary; he was meticulous and disciplined and I knew of no instance in which he behaved in such an irresponsible manner. Holding my tongue, I waited to see what he would come up with."

A few days later, Salk handed Youngner a reconstructed draft which contained all the original data. When Youngner inquired about this, Salk explained that he had found the tables but not the text. Even more disturbing was the new title page on which Salk's name appeared for the first time — *above* those of Youngner and his partner, even though Salk had contributed absolutely nothing to the color test. "When I questioned the change, [Salk] said that since he had to reconstruct the entire paper it was only fair that his name go first. I was dumbstruck and realized that this was a substantive issue that would break our relationship if I carried the argument further...."

Salk's behavior on the eve of the Cutter incident went far beyond hypocrisy in light of his statement quoted by Scheibner; it was blatant criminal neglect. Shortly after the Ann Arbor press conference, Youngner was invited to visit Cutter's vaccine plant. He was horrified at the sloppiness of their operation and reported back to Salk that the employees at Cutter didn't know what they were doing, adding that he intended to notify the proper authorities so that polio vaccine manufactured by Cutter would never be injected into children's veins.

Jonas was unexpectedly calm through my recital. He agreed that it was a serious situation with terrible potential consequences. For this reason he suggested that it would be better if the letter came from him.... I was completely taken in — to my knowledge the letter was never written. Jonas never gave me a copy of it, and he never mentioned the matter to me again.... When the Cutter incident began to unfold... I was immobilized. I realized that Jonas probably had done nothing — but neither had I. My guilt at being taken in by him was oppressive, but what to do? Silence was my response.

As a schoolboy, I had been brainwashed by a photograph of Saint Jonas gazing into his miracle vaccine jar. Almost every American has seen one of these photos. Youngner enlightened me on this score: "All the photographs of Jonas 'in the laboratory.' All the shots of Jonas in his white coat, surrounded by lab equipment, microscopes; Jonas intently holding up and looking at culture bottles — all were set up either in his office or an empty room before the photographers came.... Jonas was his own press agent. He leaked like a sieve."

Why should we believe Youngner? Because he had no reason to lie. Because he felt what any decent human being would feel. Because it validated what other acquaintances of Salk said about him. Because it formed a consistent pattern with the way Salk behaved throughout his life, including his early years as a left-wing activist and vocal opponent of Hitler's rise to power in Germany, who nevertheless wangled a draft deferment and spent World War Two in a Michigan research lab.

After Sabin's vaccine became the vaccine of choice in the early sixties, Salk faded from view, but not entirely. He retreated to California and established the Salk Institute, a jumble of ugly slabs overlooking the Pacific Ocean. His avowed purpose was to bring together the country's brightest minds in pursuit of a higher level of human existence that combined the cutting edge insights of medicine, evolution and philosophy. Founded in 1963, it has given the world nothing, and my impression, after scanning the website www.salk.edu, is that of a sanctuary for New Age eggheads and dilettantes. By the late 1980s, other motives had surfaced. The Salk Institute founded a department that worked under contract with the Pentagon, and in 1988 concluded a $32 million deal with the Army to produce vaccines and biological reagents. (The military, it should be noted, is wedded to the vaccine industry.) Nor was this the first indication of Salk's greed. After commencing work on an AIDS vaccine — another scheme that went nowhere — Salk co-founded the Immune Response Corporation to produce and market the vaccine that never was, which through his manipulation netted him ownership of stock shares worth over $3 million. By this time few scientists took Salk seriously, and his critics were too numerous to count. Still, he vainly plodded on with his AIDS vaccine. "There have to be people who are ahead of their time," he said. "And that is my fate."

Salk managed to ride the crest of one delirious press conference to the end of his life. In 1977 he was awarded the Presidential Medal of Freedom — the year after publicly throwing his support behind the disastrous swine flu vaccination campaign. President Reagan honored him in 1985, designating May 6 "Dr. Jonas Salk Day," and urging Americans to pay tribute to a man who had supposedly saved so many lives. *Time* would later proclaim him one of the top 100 scientists and thinkers of the twentieth century, and put him on the cover once again. In his later years, Salk continued to rack up awards and honors. And in small ways, Americans would continue to adore him. Airline pilots would announce he was on board and passengers would burst into applause. Restaurants and hotels would accord him VIP treatment. And opinion polls revealed that, in the eyes of the public, Salk remained a giant in the field of medicine.

But the facts speak loud and clear. Jonas Salk was a media invention, nothing else. He was an ambitious, mediocre, sleazy little man who discovered nothing and accomplished nothing, whose only real legacy is a trail of broken lives. This was a man who blamed pharmaceutical companies for his own flawed production technique; who stole ideas from others and passed them off as his own; who estranged most of his colleagues because of his total lack of integrity; who convinced people he couldn't sleep at night worrying about vaccinated children, and publicized his pious concerns about the safety of his vaccine in a prestigious medical journal — then secretly refused to stop the distribution of 400,000 doses he knew to be *un*safe. Nimbly navigating a river of fraud his entire career, he actually believed the current would carry him to scientific immortality, stupidly unaware that the truth always wins out in the end, even if it takes centuries. It's just a matter of time before Salk comes crashing down from his pedestal and his vaccine, along with all other vaccines, will be seen as just another disaster among several disasters in the checkered history of medicine.

For now, the Big Lie of the miracle Salk polio vaccine continues to flourish, but Salk got his earthly comeuppance in 1993, two years before his death, when he was invited to attend the unveiling of his portrait in the medical complex auditorium at the University of Pittsburgh — a final opportunity to loaf in his hollow glory. Before the ceremony, he went to see his former assistant Julius Youngner, now an elderly, distinguished professor at the school. The two men hadn't seen or spoken to each other in over thirty years. To Salk it was a courtesy call, or perhaps a desire to make amends; Youngner had other things in mind. Releasing the pain and bitterness bottled up for so long, he asked Salk:

"Do you still have the speech you gave in Ann Arbor in 1955? Have you ever reread it? We were in the audience, your closest colleagues and devoted associates, who worked hard and faithfully for the same goal that you desired.... Do you remember whom you mentioned and whom you left out? Do you realize how devastated we were at that moment and ever afterward when you persisted in making your co-workers invisible? Do you know what I'm saying?"

Salk answered that he did. But Youngner wasn't finished. "I [told him] I also was disturbed by his behavior [in] the Cutter incident and that I still had not forgiven him for this. Jonas was clearly shaken by these memories and offered little response. There is no doubt in my mind that he knew what I referred to."

When Salk died of heart failure at the age of eighty-one, the reverential newspaper obituaries were as predictable as the setting sun, but they were not the final tribute. The Greatest Press Conference in History continues to be celebrated on milestone anniversaries. What I have alluded to above bears repeating: Jonas Salk and his polio vaccine have been thoroughly discredited in books, including even mainstream books, in magazine articles, and all over the Internet — yet the Myth lives on. It is a subject more relevant to mass psychology than biology.

Of course, not everyone was taken in by this swindle. Then, as now, there were the marginalized ones, the crackpots, the conspiracy theorists — that is, the voices of sanity who ripped Salk and the very idea of a polio vaccine to shreds, before and after Ann Arbor. They wrote underground articles in a cheeky style that has all but vanished, with brash headings like "Childslaughter, Inc.," "New England and Salk's Offal," and "The Salk Monkey Kidney Juice." Several of these articles — along with brief stories of youngsters victimized by Salk's deadly mischief, as they appeared in small-town newspapers throughout the country — were reprinted in Eleanor McBean's *The Poisoned Needle*, published in 1957, which captured the dissident spirit of the times, and proved once again that the truth is timeless. A few of these stories are reproduced on the following pages.

Caution was cast to the winds in this country and our U.S. Government and medical profession had no hesitancy in using the *unproved* vaccine and repeating their blundering tests on millions of human beings even after the large number of deaths and paralysis had *proved* it to be not only a failure but a dangerous killer. Most of the experiments were made on children too small to fight back.

A PARTIAL LIST OF DEATHS FROM SALK VACCINE

Susan Pierce (age 7), Pocatello, Idaho, died April 27, 1955
Ronald Fitzgerald (age 4), Oakland, Calif., died April 27, 1955
Allen Davis Jr. (age 2), New Orleans, La., died May 4, 1955
Janet Kincaid (age 7), Moscow, Idaho, died May 1, 1955
Danny Eggers (age 6), Idaho Falls, Idaho, died May 10, 1955

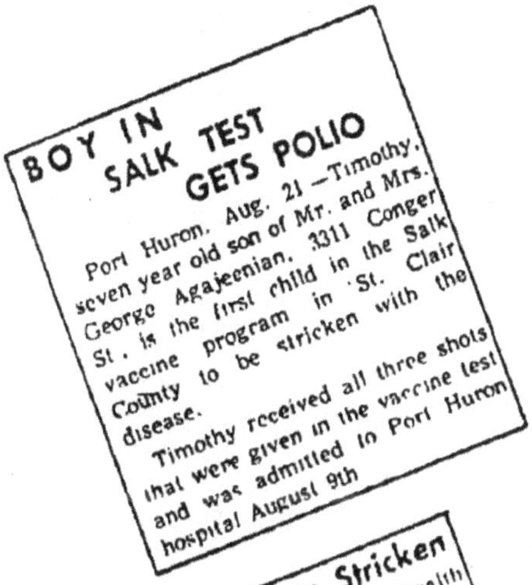

BOY IN SALK TEST GETS POLIO

Port Huron, Aug. 21—Timothy, seven year old son of Mr. and Mrs. George Agajeenian, 3311 Conger St., is the first child in the Salk vaccine program in St. Clair County to be stricken with the disease.

Timothy received all three shots that were given in the vaccine test and was admitted to Port Huron hospital August 9th.

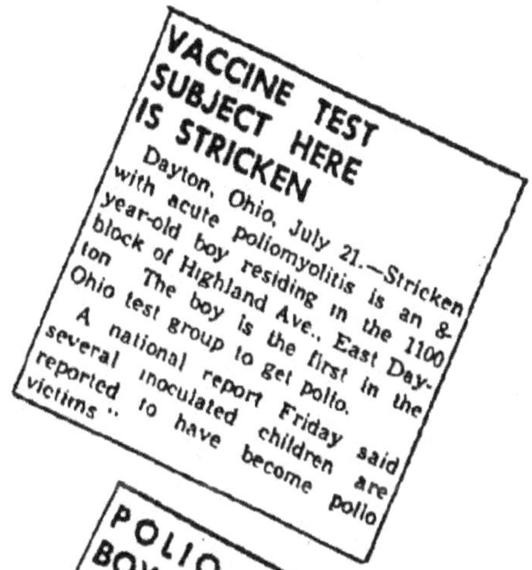

VACCINE TEST SUBJECT HERE IS STRICKEN

Dayton, Ohio, July 21.—Stricken with acute poliomyolitis is an 8-year-old boy residing in the 1100 block of Highland Ave., East Dayton. The boy is the first in the Ohio test group to get polio.

A national report Friday said several inoculated children are reported to have become polio victims.

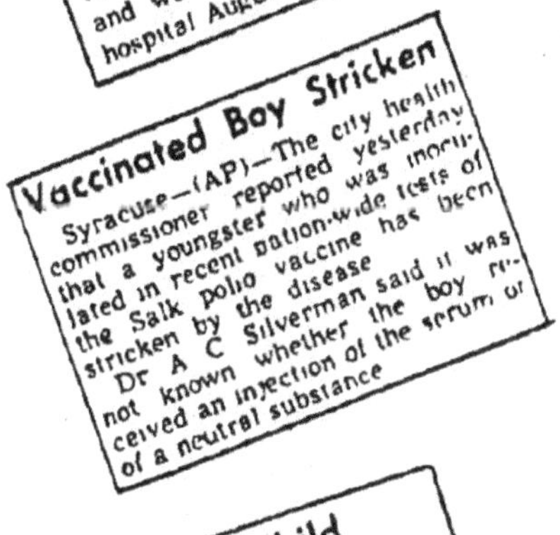

Vaccinated Boy Stricken

Syracuse—(AP)—The city health commissioner reported yesterday that a youngster who was inoculated in recent nation-wide tests of the Salk polio vaccine has been stricken by the disease.

Dr. A. C. Silverman said it was not known whether the boy received an injection of the serum or of a neutral substance.

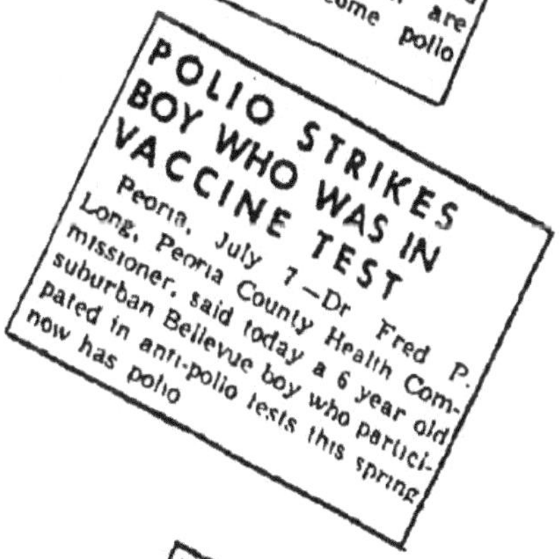

POLIO STRIKES BOY WHO WAS IN VACCINE TEST

Peoria, July 7—Dr. Fred P. Long, Peoria County Health Commissioner, said today a 6 year old suburban Bellevue boy who participated in anti-polio tests this spring now has polio.

Second Tulsa Child Succumbs To Polio

Tulsa, May 29—Patricia Reddrick, 8 year old Tulsa second grade pupil, died late last night of polio to become the second victim of the disease here this year.

Her parents, Mr. and Mrs. Earl Reddrick, said Patricia was among the Tulsa County second graders in the polio vaccine trials.

Dr. T. P. Haney, health director said 'some deaths will occur whether the vaccine has been used or not.'

INOCULATED CHILD HIT BY POLIO

Pontiac, Mich., June 13.—A second Michigan child who participated in the recent Salk polio vaccine tests has developed polio, and her mother also has been hospitalized for observation.

The newly stricken polio patient is Linda Brooks, 7, of Farmington township. She was admitted to hospital June 4.

GIRL TAKES POLIO AFTER VACCINATION

Houston, Texas.—A seven-year-old girl who took all three shots of the Salk polio vaccine was stricken Friday with polio and admitted to Hedgecroft Clinic.

Jeannette is the daughter of Mr. and Mrs. Elton Kirchenwitz. A spinal fluid test taken showed Friday she was suffering from polio.

POLIO HITS TEST CASE

Seattle, June 16. — A Kitsap county youngster taking part in polio vaccine tests in Washington state, has contracted the disease, Dr. W. R. Giedt, state epidemiologist, said today.

BOY WITH 'SHOTS' CONTRACTS POLIO

Corpus Christi.—A 7-year-old boy who took the series of Salk polio vaccinations has contracted polio, the city-county health unit reported Monday.

He was one of the Corpus Christi second graders who took part in a nation-wide test of the new polio vaccine developed by Dr. Jonas Salk of Pittsburgh.

The State Health Dept. said in Austin there had been "one or two" similar cases in Texas.

SALK TESTER, 8, GETS BULBAR POLIO

Bradford, Pa., Sept. 10. — Dr. H. J. McGhee, McKeen County Health Director, reported today that a second grader who was inoculated with the Salk vaccine is a victim of bulbar-type polio.

Eight-year-old Terry F. Tessena, of Smethport, is in Bradford Hospital in serious condition and has been in an oxygen tent for two days. Bulbar polio is the most serious type.

SALK VACCINE PATIENT GETS POLIO ATTACK

Fort Wayne, Ind., July 28.—The first case of polio among the Indiana children who received the Salk vaccine injections last spring has developed at Fort Wayne.

Dr. Walter Kruse, Fort Wayne Board of Health, said the illness of John Erb III, 8, has been diagnosed as polio and he was one of the second grade children given the vaccine injection.

Shot Recipient Admitted Here As Polio Case

Oklahoma.—Larry Eulert, 8, of 1039 N. Delaware Ave., who, according to hospital attendants, received all three of the shots of Salk vaccine last spring, was admitted to the polio division at Hillcrest Medical Center Thursday.

The boy, son of Mr. and Mrs. Luther Eulert, was placed in an iron lung after a tracheotomy was performed to facilitate his breathing.

Oklahoma State officials report 3 other children contracted polio following their participation in Salk vaccine tests.

VICTIM OF POLIO HAD RECEIVED SALK VACCINE

Topeka, Aug. 21.—A 7-year-old Topeka girl who received the Salk polio vaccine as part of a nation-wide vaccination test is hospitalized here with polio this week.

She is Barbara White, daughter of Mr. and Mrs. Woodrow White.

Everything published by the CDC pertaining to vaccines, such as the two-sided handout reproduced below, is rubbish. All vaccine information served for public consumption by the government-medical-pharmaceutical cartel is written at the Sesame Street level. Compare the capsule history of Jonas Salk and the polio vaccine, shown on page 67 (following the two-sided handout), taken from the CDC website, with the appalling and undeniable facts that anyone can easily find on the Internet.

| DISEASES and the VACCINES THAT PREVENT THEM |

INFORMATION FOR PARENTS

Polio

Last updated July 2011

What is polio?
Polio is a disease caused by a virus. It can cause lifelong paralysis (can't move parts of the body), and it can be deadly. But, the polio vaccine can protect against polio virus.

What are the symptoms of polio?
Most people who get polio do not have any symptoms.

A small number of people (4 to 8 people out of 100) have minor symptoms that may include the following:
- Fever
- Tiredness
- Headache
- Nausea and vomiting
- Sore throat
- Stiff neck and back
- Pain in arms and legs

These symptoms usually last 2 to 10 days and then go away on their own.

How serious is polio?
Even though most people have no symptoms or minor symptoms, the risk of lifelong paralysis is very serious. Even children who often seem to recover fully can develop new muscle pain, weakness, or paralysis as adults, 30 or 40 years later.

About 2 to 5 children out of 100 who have paralysis from polio die because the virus affects the muscles that help them breathe.

About 1 out of 100 people who get polio have weakness or paralysis in their arms, legs, or both. This paralysis or weakness can last a lifetime.

How does polio spread?
Polio spreads easily, usually from the stool (feces) of an infected person to the mouth of someone else through hands or objects, like toys, that have small amounts of the stool on them. It can also spread from the mouth of an infected person, as through kissing.

Benefits of the polio vaccine (IPV)
- Saves lives.
- Protects young children from serious disease and lifelong disability.

Side effects of the polio vaccine (IPV)
- The most common side effects are usually mild and include redness and pain from the shot.

What is the polio vaccine or IPV?
IPV is a type of polio vaccine. IPV stands for inactivated (killed) polio vaccine. It is given by a shot.

The polio vaccine protects children by preparing their bodies to fight the polio virus. Almost all children (99 children out of 100) who get all doses of IPV will be protected from the polio virus.

Why should my child get the polio vaccine?
The polio vaccine prevents polio. Even though no polio cases have originated in the U.S. in 20 years, the disease still occurs in some parts of the world. It would only take one traveler with polio from another country to bring polio back to the U.S.

When should my child get the polio vaccine?
Children should get four doses of IPV at the following ages for best protection:
- One dose each at 2 months and 4 months;
- A third dose at 6 through 18 months; and
- A fourth (booster) dose at 4 through 6 years of age.

It is safe to get IPV at the same time as other vaccines.

DEPARTMENT OF HEALTH & HUMAN SERVICES USA

CDC CENTERS FOR DISEASE CONTROL AND PREVENTION

AMERICAN ACADEMY OF FAMILY PHYSICIANS
STRONG MEDICINE FOR AMERICA

American Academy of Pediatrics
DEDICATED TO THE HEALTH OF ALL CHILDREN™

Is the polio vaccine safe?

IPV is very safe, and it is effective at preventing polio disease. Vaccines, like any medicine, can have side effects. But severe side effects from IPV are very rare.

If my child does not get the polio vaccine, will he get polio?

Without the vaccine, polio spreads very easily. Before the polio vaccine, there were more than 20,000 cases of polio in the U.S. each year. Today, thanks to the vaccine, there are no cases of polio in the U.S. But if people stopped vaccinating, we could see cases of polio again.

How can I learn more about the polio vaccine?

To learn more about the polio vaccine or other vaccines, talk to your child's doctor.

Call **800-CDC-INFO** (800-232-4636) or go to ***http://www.cdc.gov/vaccines*** and check out the following resources:

- Vaccines and Preventable Diseases—Polio Vaccination: ***http://www.cdc.gov/vaccines/vpd-vac/polio/default.htm***
- Common Questions Parents Ask about Infant Immunizations: ***http://www.cdc.gov/vaccines/spec-grps/infants/parent-questions.htm***
- Vaccines website for parents: ***http://www.cdc.gov/vaccines/parents***

What can I do to protect my child from polio?

✔ Vaccinate your child on time.

✔ Talk with your child's doctor if you have questions.

✔ Keep a record of your child's vaccinations to make sure your child is up-to-date.

The Centers for Disease Control and Prevention, American Academy of Family Physicians, and American Academy of Pediatrics strongly recommend all children receive the polio vaccine according to the recommended schedule.

DISEASES and the VACCINES THAT PREVENT THEM
Updated February 2013

Polio

Polio Pioneers

More than 50 years ago, polio held U.S. families in a grip of terror. Especially during the summertime, when polio seemed most likely to circulate, parents feared they would hear in the news or from neighbors that someone in the community had polio. "People tried to keep their children safe from the potentially paralyzing disease by keeping them out of public places such as pools, parks, and theaters," explained Dr. Anne Schuchat, director of the National Center for Immunization and Respiratory Diseases at the Centers for Disease Control and Prevention (CDC).

The nation came together like never before in an effort to create a vaccine to protect children from polio. Millions of Americans raised funds in their communities for research. Much of the funding came through the National Foundation for Infantile Paralysis (presently the March of Dimes Foundation), founded in 1938 by President Franklin D. Roosevelt, himself paralyzed by polio in the prime of his life. Before the March of Dimes drew national attention to the search for a polio vaccine, two attempts to develop a polio vaccine had failed—neither produced immunity and some deaths were blamed on one of the vaccines.

In 1952, Jonas Salk and his team at the University of Pittsburgh created the first effective polio vaccine. By 1954, it was time to test the Salk vaccine widely. Thomas Francis Jr. at the University of Michigan led the nationwide test, the scale of which had never been seen before. More than 1.8 million school children across the United States participated. Thousands of health care professionals and other volunteers administered the vaccine and collected results.

Everyone had the same goal: victory over polio. In 1955 the results were proclaimed: the Salk vaccine was "safe, effective, and potent!"

The Salk vaccine was made by killing the poliovirus, and it was given as a shot. The second polio vaccine licensed for use in the United States, created by Albert Sabin, was an oral polio vaccine, which was made by using a live weakened version of the poliovirus. By 1963, a formulation of this vaccine that prevented three strains or types of polio—like the Salk vaccine before it—was available.

"Polio was eliminated in the United States in 1979," said CDC's Dr. Greg Wallace. "But, because polio still circulates in other parts of the world, we need to continue vaccination in the United States."

"Scenarios for polio being introduced into the United States are easy to imagine, and the disease could get a foothold if we don't maintain high vaccination rates," explains Dr. Wallace. "For example, an unvaccinated U.S. resident could travel abroad and become infected before returning home. Or, a visitor to the United States could travel here while infected. The point is, one person infected with polio is all it takes to start the spread of polio to others if they are not protected by vaccination."

It's important to remember that in the 1950s, protecting the public from polio was, in the truest sense, a national project. Every effort was made to see that the vaccine would be widely available to all children and polio would be wiped out.

Vaccinating each child in the United States today remains a priority. As some countries have seen in the last decade, without widespread vaccination, the disease rapidly returns and people must once again work to eliminate it.

Symptoms of Poliovirus Infection

Poliovirus infection, while greatly feared and sometimes dangerous, is usually not obvious. Most people infected with poliovirus have no apparent symptoms. But unfortunately, anyone infected with poliovirus can spread it.

About 4% to 8% of people infected with poliovirus have minor symptoms that don't last long. They may have a sore throat, fever, tiredness, nausea, headache, or stomach pain. About 1% to 5% of people infected with poliovirus may feel stiffness in their back and neck with a severe headache or pain in their arms and legs. These symptoms usually go away within 2 to 10 days.

In Rare Cases, Poliovirus Infection Can be Very Serious

Unfortunately, even children and adults who have been healthy all their lives can get seriously ill from poliovirus infection. About 1 out of 100 people infected with poliovirus develops polio disease and becomes paralyzed for life.

CDC

AMERICAN ACADEMY OF FAMILY PHYSICIANS
STRONG MEDICINE FOR AMERICA

American Academy of Pediatrics
DEDICATED TO THE HEALTH OF ALL CHILDREN

Jonas Salk poses for the camera.

America was brainwashed about the polio vaccine by the clever use of images, like Salk gazing into his test tubes and flasks and the polio-stricken boy soliciting donations. In reality, the Salk vaccine *caused* more severe cases of polio than it prevented — if it prevented any at all. Other pictorial propaganda was less subtle, like the "Wellbee" poster of 1963, and the cover of a 6-page "Fact vs. Fiction" pamphlet printed by vaccine maker Pasteur Merieux Connaught, which was as childish as the content. Was there a strategy here in targeting the less brainy, or are most Americans really this stupid?

In the wake of the media consecration of St. Jonas of Pittsburgh, many places were renamed in his honor, including a junior high school (now called a middle school) in Levittown, on Long Island. Levittown has three school districts, and

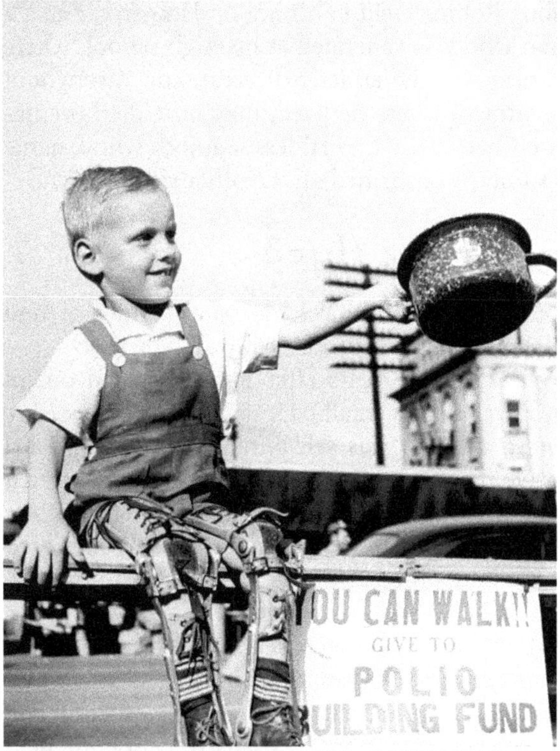
YOU CAN WALK!!
GIVE TO
POLIO
BUILDING FUND

"Wellbee" says
BE WELL!
take
ORAL
POLIO
VACCINE
• tastes good
• works fast
• prevents polio

TAKING THE MYSTERY
OUT OF VACCINES:
FACT VS
FICTION

Emotional appeals and kindergarten-level indoctrination are the hallmarks of vaccination propaganda.

all three have strong baseball programs, a fact know to the present writer who was a Nassau County high school baseball umpire for 24 years. Two miles from the Salk Middle School, also in Levittown, is Division Avenue High School, whose varsity baseball coach for thirty years was Doug Robins, who compiled the second all-time winningest record among New York state high school baseball coaches. I worked many of Doug's games over the years, and always looked forward to them. Not only were you assured of seeing high school baseball at its best, but Doug was that rare kind of coach who could question a call, even get a bit angry at times, yet always remain respectful towards us umpires. We all liked Doug, and it was easy to see that all his players as well as their parents who came to watch their sons play looked up to him. He was a genuinely nice man who led by example.

Doug died of brain cancer in 2007 at the age of 58. He was five years older than me, and as a child was almost certainly injected with Salk's polio vaccine. We now know that millions of doses of that vaccine were contaminated with the SV-40 virus that has consistently been found in cancerous human brain tissue. While it might seem improbable to suggest a vaccine-cancer link with the passage of fifty years, some cancer agents are very slow-acting, and brain cancer is much more common now than it was three generations ago. However, unless a biopsy is performed to detect the presence or absence of SV-40, the cause can only be surmised.

Shortly before Doug passed away, the baseball field at Division Avenue High School was named Doug Robins Field in his honor. How ironic that a man who was venerated at his high school, where he also was the athletic director, and throughout Levittown, where he lived, may have died prematurely because of a worthless scumbag whose name is similarly enshrined on the other side of town.

What kind of country is this?

EZRA POUND, born in Idaho and raised in Pennsylvania, was a renowned poet and an American original. Unconventional in every way, he was adamantly opposed to his country's entry into World War Two. Pound greatly admired Benito Mussolini and his fascist system of government, settling down in Italy long before the beginning of hostilities. Even before American soldiers landed in Europe, he made radio broadcasts from Rome on a regular basis, in down to earth language salted with homegrown slang, blaming Franklin Roosevelt, Jewish schemers and international finance for the conflict. Pound was not alone in his accusations. Prior to the Japanese attack on Pearl Harbor, which was no surprise but rather the intended outcome of Roosevelt's war plotting, many prominent Americans, most notably Charles Lindbergh, head of the America First Committee, had publicly spoken out against Jewish influence and the build-up of war propaganda, though in a much more civil manner. Joseph Kennedy, father of the future president and then U.S. ambassador to Great Britain, made the same charges in private. So did many others, in hushed voices that in large part fell silent after America entered the war. Pound, however, remained outspoken and on the air right up to the end of the fighting. When the war in Europe ended in May 1945 he was arrested by U.S. forces, imprisoned and gawked at in a tiny outdoor cage for three weeks, then transferred to a military camp in Pisa. Shipped back in chains to the Land of the Free in November, he faced a real possibility of being executed for treason, but instead was declared mentally unfit and committed to St. Elizabeth's Hospital in Washington D.C., a decrepit institution where he languished for more than twelve years. Released in 1958, he wasted no time in returning to Italy, where he quietly spent his remaining years, dying in 1972 at the age of 87. Upon arriving by ship in Naples, a reporter asked him how it felt when he was let out of the insane asylum in America after being locked up for so long. Pound replied, "I never was. When I got out of the hospital I was still in America, and all America is an insane asylum."

In 1964, six years after Pound walked out of St. Elizabeth's, a small paperback titled *None Dare Call it Treason* was self-published by a Missouri native named John Stormer. In less than a year it went through twenty printings and sold more than seven million copies, even though it was ignored by the mass media. I never heard of it until 1976 and will always remember it as the first book I read that cut through fifteen years of fog and allowed me to see world events much more clearly, even though the author left much unsaid and I still had a lot to learn. This book exhaustively documented the fact that, far from being "the leader of the free world," our own government, with the support of all our major institutions, was responsible for bringing to power and propping up nearly every communist regime on earth. Chapter One begins with the question "Have We Gone Crazy?" in large print. John Stormer was a religious, conservative Republican, the very personification of mainstream America. Nothing he wrote suggests that he was opposed to the American military role in the Second World War, even though he often mentions the secret deal Roosevelt made in February 1945 with Joseph Stalin at Yalta, in the Ukraine, which sold out all eastern Europe to the slavery of communism. In personality and lifestyle Stormer had nothing in common with Ezra Pound. Yet, by different routes, both men reached the same conclusion: a streak of insanity runs through America. I wholeheartedly agree. And this country has gotten a lot crazier since 1964, and continues to get crazier by the day, as more and more have come to see. But a solid majority are still severely out of touch with reality — the definition of insanity. Many still wave the flag and talk about the freedom we enjoy. What freedom are they talking about? The freedom to get out of bed in the morning? The freedom to cross the street?

What I've just written, and what I've written below, will appall some readers. It would've appalled me as a younger man. I was brought up in a basically conservative family — my father was a World War Two veteran, and his older brother, an uncle I never met, was killed in action in that conflict — and instilled with a sense of patriotism both at home and in school, soaking up the myths about America being the greatest country the world has ever seen, a beacon of freedom to people everywhere, our soldiers always being the good guys who could do no wrong, and so on. Amidst the suffocating liberal atmosphere of my college, I remained defiantly patriotic. So proud was I to be an American that the first time I traveled to Europe, in 1978, getting around for six weeks on a rail pass with some hitchhiking mixed in, I sewed a small American flag on my backpack because I wanted everyone to know my nationality. This, however, brought no compliments but only an occasional sneer from people with whom I'd struck up a conversation, who told me in polite terms, "You should get rid of that thing. Nobody wants to look at that." This hurt and confused me; it was the beginning of a long, sobering education that only a combination of independent travel and reading can provide.

It's natural to love one's country, to be patriotic, but not when hubris, ignorance and terminal stupidity masquerade as patriotism. It's right to detest and take up arms against any enemy that poses a military threat to one's native land, but not when that "enemy" is no enemy at all but an invention of conspirators behind the scenes who have their own agendas, as has been the case with every foreign conflict this country has been involved in, beginning with the Spanish-American War of 1898 — subversive forces that George Washington warned against in his Farewell Speech, in which he counseled friendly neutrality towards all nations, special favors and political entanglements with none. That was the basis of American foreign policy for more than a century, and it served this nation very well.

Now and then, on a sudden impulse to capture the essence of what America once was, I go to my bookshelf and pull down a book about the development of this nation in its infancy. When I do, I never fail to be inspired by the men who created America — and above all, Washington, whose leadership not only won the Revolutionary War, but held the country together when it threatened to fall apart shortly after victory on the battlefield. In particular I enjoy reading about all the constructive energy that went into hammering out the Constitution, one of mankind's loftiest achievements in efficient, strictly limited government. And despite the arguments, the personality conflicts and the private interests that sometimes animated them, despite the fact that, in the final analysis, a nation cannot be saved by words written on a scrap of paper, perhaps at no other time in history was such a great task accomplished by a gathering of men with so much intelligence and integrity.

All the Founding Fathers were learned, practical men and keen students of history, and most of them had a pessimistic view of human nature. They knew that good governments invariably go bad and that tyranny is the end result. They knew the republic they were establishing was something fragile that might not last. They were right about this, of course, but they could not foresee that regional differences and destructive ideas would lead to a fratricidal civil war, much less what would become of their country in two hundred years, any more than the most perceptive among us today can visualize what this land will look like in 2080, or at the dawn of the twenty-third century. What anyone with a basic knowledge of history *should* know is that America, as it once was, what it once stood for, is gone forever. There is no comparison between the America of 1800 and the America of 2000, only a world of contrast. A country is only as good as its people; simply said, a country *is* its people. In that respect, perhaps no other nation got off to a more promising start, and in historical terms, went downhill faster. A country populated by five million sturdy, self-reliant men and women, most of British stock — the descendants of those brave and fit enough to make a daring trans-Atlantic voyage, then survive the rigors of a strange new land — has nothing in common with the hodgepodge of 325 million that America has become. This land is now a magnet for unassimilable aliens from every corner of the planet escaping their mostly failed societies, among them millions of scroungers and misfits. This situation, which has created chronic tension over and above the problems inherent in our large native black population, with its huge criminal and welfare underclass, has no parallel in history and cannot possibly last. The white population alone overflows with deadbeats who keep reproducing. To these must be added the drones and gangsters employed by government, especially at the federal level, and their corporate partners in crime — what has come to be known at the "Deep State." Even our better men and women have been reduced to the level of automatons portrayed in George Orwell's *Nineteen Eighty-Four*.

Years ago I came across a quote on the Internet that caught my attention: "Just look at us. Everything is backwards; everything is upside-down. Doctors destroy health, lawyers destroy justice, universities destroy knowledge, governments destroy freedom, the major media destroy information, and religions destroy spirituality." It was attributed to a Michael Ellner, whom I know nothing about. Whoever he is, whatever else he might believe, he accurately described America and furnished a starting point, although the rot and insanity spreads far beyond the institutions he mentioned. Since this book is about vaccination it focuses on the first charge, doctors destroying health, but this ties in with almost everything else that's backwards and upside-down in America.

This chapter is neither a history lesson nor a treatise on everything that is wrong with America. Rather it's an attempt, within a limited space, to show that the current insane vaccine situation can best be understood by facing some hard facts about this country, the people who run things, and its citizens. I can only pick and choose a few such facts. Perhaps it's impractical for me to try to do this in just a few pages, as it took me a good ten years after graduating college to unlearn nearly everything I had been taught in school from age six, to cleanse my brain of all the poison pumped in by the news media, and then to discover hundreds if not thousands of suppressed facts I had never been aware of. Hopefully some readers will find this discussion illuminating, and will be inspired to learn more on their own.

Let's start with one overarching and irrefutable fact: Since World War Two, primarily by means of the aerial bomb, the U.S. government has murdered more innocent people outside of its borders than any nation in world history. By any definition, including the one in the dictionary, deliberately killing innocent people in order to accomplish certain objectives is terrorism, which means that the government in Washington D.C. is the world's bloodiest and longest running terrorist organization, with the Pentagon its base of operations. Actually, this government has frequently terrorized its own citizens since the Civil War. Few people realize just how savage the Civil War was, one example being Union general William Sherman's infamous "march to the sea" in Georgia in 1864. After burning the capital city of Atlanta to the ground, Sherman's 60,000 troops marched for five weeks in a column thirty miles wide to Savannah, 250 miles away, annihilating everything in their path — putting the torch to all homes and buildings, destroying all crops and foodstores, and killing all livestock. This was far from being the only atrocity against southern civilians. William Brownlow, a pro-Union Methodist preacher from Tennessee, provides a fine example of the spirit of the times. Listen to this excerpt of a speech given by this man of God at a New York convention in 1862:

If I had the power, I would arm every wolf, panther, catamount and bear in the mountains of America, every crocodile in the swamps of Florida, every Negro in the South, every fiend in hell, clothe them in the uniforms of the Federal army and turn them loose on the rebels of the South and exterminate every man, woman and child south of the Mason-Dixon line. I would like to see especially the Negro troops, marching under Ben Butler, crowd the last rebel into the Gulf of Mexico and drown them as the Devil did the hogs in the Sea of Galilee.

This white-hot fanaticism flourished even years before the first shot was fired at Fort Sumter, the Missouri-Kansas border being a chronic flashpoint. Death squads on both sides, bitterly divided over the issue of slavery, and as brutal as any Central American drug gang, carried out numerous massacres in this area, an episode referred to as Bleeding Kansas. Very few schoolchildren learn about the details of this terrible conflict, and fewer still learn about Reconstruction, the twelve-year reign of terror over the defeated South that began when the war ended, and gave birth to the Ku Klux Klan, which in its early days was led by men of the highest caliber who waged a successful guerilla war against barbaric Federal occupation and restored normalcy to the southern states. Is there a people on this planet who know less about their own history than Americans?

Since the Civil War, American history has constantly been stained by violence and injustice. I'd be the last one to stoke guilt feelings among white people for the deeds of their forefathers — God knows there's been enough of that — but the fact remains that the Indians were killed off in vast numbers during the westward expansion that took place in the latter half of the nineteenth century, and the great herds of bison upon which they largely depended for their survival were nearly driven to extinction, millions of them shot simply for sport. Much of this was bound up with the laying of railroad track from coast to coast, and the raw materials of railroad construction and operation were in turn bound up with the mines, mills and factories where white laborers, including women and children, often toiled long hours in horrifically unsafe conditions. Many men whose families depended on their salaries suffered crippling injuries, and when this happened they received no compensation and were left to fend for themselves. The American labor movement overflowed with violence and does not make pretty reading. Then, as now, unbridled political power flowed from the coffers of the ruthless super-rich. Then, as now, man's inhumanity to man, American-style, reigned, and so did the rotten System that today throws children's lives on the vaccine roulette wheel while the casino always wins. Industries change but human nature doesn't. Jack London, an extraordinary personality and magnificent writer, portrayed all this in his novel *The Iron Heel*, published in 1908. For a man who lived only forty years, London's literary output was astounding, as was the range of his adventures and experiences. Jack knew human struggle and the hard side of life firsthand, and as with most of his novels and short stories, worked these themes into *The Iron Heel*. It was fact in fictional form, and it's sad but enlightening to see how only the window dressing has changed since long ago, how the boundless wealth of the industrial magnates back then bought off politicians, judges, professors, police, journalists and ministers, while those whose physical labor built this country were crushed under the ruling oligarchy — under "the iron heel." This echoes down to the present day with recent abominations like NAFTA, which threw hundreds of thousands of Americans in the manufacturing industries out on the street as their jobs were shipped to poor countries where labor costs are much lower. Then, as now, it's all about corporate greed; then, as now, the attitude of the plutocrats is "let them eat cake." In his novel London prophesied an America torn apart by all manner of chaos and violence, including bombings, assassinations, and wholesale bloody reprisals in which the nation's oppressed workingmen rose against the rich and ruthless and their entrenched federal agents. Since then, only the names and circumstances have changed. A similarly violent novel with the theme of total race war, *The Turner Diaries*, was written by William Pierce, founder of the white nationalist organization the National Alliance, under the pseudonym Andrew MacDonald. First published in 1978, it has sold close to a million copies to date. Good luck finding it in any bookstore or public library.

America's decline took a sharp downward turn with the election of Woodrow Wilson. 1913, the first year of his presidency, was a fateful year that saw the inception of the income tax, the privately run Federal Reserve Bank, and the Anti-Defamation League (ADL) of B'nai B'rith, which has grown into a powerful police state spy and smear agency and has corrupted numerous law enforcement agencies, most importantly the FBI. In 1916, Wilson, who exhibited signs of clinical insanity, and whose dementia worsened over time, ran for a second term on the slogan "He kept us out of war." No sooner was he re-elected than he started talking about "a war to make the word safe for democracy" and "a war to end all wars." Ah yes, a war to end all wars. World War One created the conditions that led to World War Two, ushering in the modern era of what the eminent historian Harry Elmer Barnes called "perpetual war for perpetual peace," and what Professor Revilo Oliver, one of America's great twentieth century minds, called "contrived wars for hallucinatory ends." These wars continue to the present day. The First World War also ushered in the draft, which was activated in the next three contrived wars, though happily abolished after the Vietnam War. I can't imagine a worse mockery of freedom than being ordered, against your will, and under threat of imprisonment, to report for military training in order to be shipped off to a foreign war based on false flags — and this most certainly includes World War Two — depriving you of your freedom so you can defend freedom, as it were, while in reality always begetting more war crimes and more human misery.

Every foreign war America has fought has been accompanied by lies and idiotic propaganda, but the First World War was the worst. For 140 years America held Germany in high esteem as a natural friend, as one of Europe's and the world's most cultured and dynamic nations. Whipped to a frenzy overnight by the press, Germans now became barbaric Huns. It's difficult to believe that some of the illustrations from this period, printed on large posters, were taken seriously — German soldiers depicted as gorillas wearing spiked helmets, impaling Belgian babies on bayonets, and such. In an orgy of perverted patriotism, egged on by the fledgling motion picture industry, Americans responded. Nationwide, German-Americans suspected of disloyalty were beaten, arrested, tarred and feathered; rumors of secret German spies and incursions by German submarines and aircraft abounded; words of German origin were dropped from conversation so that sauerkraut became "liberty cabbage" and hamburgers "liberty steaks." There were other manifestations of national madness compared to which Iraqi soldiers throwing Kuwaiti babies out of hospital incubators and Kim Jong-Un targeting California with nuclear warheads seem quite credible.

Since I'm one of just a few thousand Americans to have visited North Korea on an ordinary tourist visa when it was legal — as of September 1, 2017 it's illegal as deemed by our government, not theirs — I wish to say more about this country. One man, Air Force General Curtis LeMay, deserves special mention. With the press of a button to release bombs, it appears that LeMay personally killed more human beings than anyone who ever lived. Truly a genocidal maniac, "bombs away LeMay," as his subordinates called him, lived for one thing: roasting innocent people. He perfected his talent in the skies over Germany, but mostly Japan, dropping incendiary bombs on 64 Japanese cities, including the March 10, 1945 raid on Tokyo, which left behind some 85,000 charred corpses, more than were killed in either of the atomic bombings of Hiroshima and Nagasaki, even though most Americans have never heard of it. LeMay also headed the Strategic Air Command during the Korean War. At the end of World War Two, it was our government, in the person of an obscure military bureaucrat named Dean Rusk, who later served as Secretary of State in the Kennedy and Johnson administrations, that casually split the ancient nation of Korea in half at the 38th parallel. The north was handed over to Joseph Stalin, whose point man in Korea was Kim Il-Sung, grandfather of the current leader Kim Jong-Un. Rusk, who by his own admission knew little about the country, made this decision after studying a map of Korea in *National Geographic* magazine. This is recounted on page 124 of his memoirs *As I Saw It*. Five years later, Kim Il-Sung's army invaded the south in a bid for a total communist takeover. This initiated the Korean War, which lasted three years and ended in a stalemate. The U.S. Air Force flattened literally every town and city in North Korea by conventional bombing and the use of napalm. A tremendous amount of napalm was used, and by most estimates between 1.5 and 2 million civilians died. So let's get this straight. First we allow Stalin, one of history's greatest mass murderers and our great ally in World War Two, to foist a particularly harsh version of communism on the people living in the northern half of Korea, who had no say in the matter, a failed system which endures to this day; then, while pretending to eradicate the very regime we were responsible for creating, we wipe out about one-fourth of the civilian population. Imagine filling all of our thirty major league baseball stadiums to capacity with innocent men, women and children, drenching them with gasoline (napalm is jellied gasoline), then tossing a lit match on them. That gives you some idea of what we did to that country. I never knew any of these things myself until I did some reading in 2013 shortly before I visited North Korea — a totally safe and welcoming country as long as one follows a few simple rules that a child can understand, contrary to all the scaremongering propaganda put out by the State Department and television.

The greatest one-day massacre of human beings in all history, incidentally, carried out by us and our partners in terror, the British, took place in Dresden, Germany on the evening of February 13-14, 1945, near the end of the war when Germany's military defeat was assured and that undefended, beautiful baroque city was overflowing with refugees fleeing westward from the raping, pillaging, murdering Mongolian hordes that our ally Joseph Stalin had transported from the Soviet Union's Asian provinces. No one really knows how many died in this firestorm. A conservative estimate of 135,000 asphyxiated or burned alive was made by British historian David Irving. Other estimates run much higher. Prior to the publication of Irving's book *The Destruction of Dresden* in 1963, few people outside of Germany were aware of this most unspeakable of all war crimes; even some British pilots who had dropped the bombs were shocked when Irving showed them photographs of the carnage. To this day most Americans have never heard of it. The Dresden holocaust is just one of hundreds of suppressed facts about World War Two, the crumbling foundation of so many monstrous lies that most Americans are addicted to. For those who wish to rise above their ignorance, I can think of no better antidote than M.S. King's *The Bad War*, published in 2015, a fast-moving book of mini-chapters packed with facts and photographs that even I, a longtime student of revisionist history, had never come across.

❖

Right through all the the futile foreign wars that America has fought, to the present day, medical corruption has flourished here at home. It's impossible for me to stay abreast of events as they unfold, but for our purpose this is hardly necessary because medical despotism predates the Civil War. Interestingly enough, Benjamin Rush, a signer of the Declaration of Independence and a doctor himself, voiced his concern about a future medical tyranny in America. But his warning was ignored, and this tragic oversight, when a government charter for the new nation was being drawn up, planted the seeds of the poisonous tree that is current CDC vaccination policy, and a great deal more. The long history of the arrogant disdain for the health and well-being of ordinary people on the part of the ruling elite, beginning in 1847 with the founding of the American Medical Association, is wittily told in *Murder by Injection: The Story of the Medical Conspiracy Against America* by Eustace Mullins. In his inimitable style, Mullins applies his dry sense of humor to a very serious subject, which had me chuckling much of the time while reaping the fruits of the forty years of research he put into this work. The second longest chapter is titled "The Profits of Cancer," which speaks for itself. While many Americans are aware that the standard approach to cancer of surgery, radiation and chemotherapy is often barbaric and useless, few know just how relentlessly those who have offered alternative treatments have been persecuted. The sordid and often secret medical crimes carried out against the citizens of this country with the approval of interlocking huge corporations and federal and state public health agencies, are far too numerous to go into here. You can read about many of them on the Wikipedia page headed "Unethical Human Experimentation in the United States." This information appears to be quite accurate based on what I've read elsewhere, but it's noteworthy that not a single vaccine experiment is mentioned. This is understandable given that the editors at Wikipedia, who claim to be neutral, are very fashion-conscious and sneer at any idea critical of vaccines.

In recent decades the unholy alliance of mega-corporations, particularly the pharmaceutical giants, and regulatory agencies, has all but declared war on the health food industry, directly targeting purveyors of vitamins, supplements, raw milk and other products. At the forefront of this campaign is the FDA, the match of any federal agency for corruption and brutality. An Internet search conducted by typing in the key words will uncover FDA horror stories galore, right to the present day. Their methods first came to my attention with the May 6, 1992 raid on the Seattle clinic of Dr. Jonathan Wright, a renowned author, public speaker and former editor of *Prevention* magazine. Acting in concert with the Kings County police department, an FDA SWAT team smashed down the front door, and with guns drawn, terrorized patients and staff, shouting orders and occupying and ransacking the premises for several hours. Dr. Wright's crime? He had been injecting his patients with high-dose B vitamins, a safe and beneficial procedure. No one had complained about him; his patients and colleagues held him in the highest esteem. No charges were ever filed against Dr. Wright.

In an outright campaign of terror and intimidation the following year, the FDA, joined by heavily armed agents of the Drug Enforcement Agency (DEA), IRS, U.S. Customs and U.S. Postal Service, launched nearly forty commando-style raids on health food stores, vitamin companies and natural health clinics throughout Texas. In none of these raids were there any reasons to expect resistance. People in the health food and alternative medicine fields are not violent. This is simply the same kind of out-of-control government thuggery about which America's founders repeatedly warned posterity.

A System that behaves so viciously towards people whose views on health and healing are far apart from conventional allopathic medicine will naturally target medical dissidents perceived as a threat to the vaccine industry. Nor should it come as any surprise that the U.S. is the least free nation in the Western world, and probably one of the least free in the entire world, when it comes to childhood vaccinations. Specifically, parents must show proof of a long list of shots in order for their children to attend school. For a long time, only Mississippi and West Virginia refused to allow religious exemptions to mandated vaccines, but since 2015, three more states — California, New York and Maine — have been added to the list and there will likely be more after this book is printed. The more damage that vaccines do, the more that vaccine fanatics want to force them on us. The relative free choice remaining in most states is being chipped away. And just because you comply with your state laws doesn't mean a busybody from Child Protective Services, flanked by two detectives, won't be knocking on your door to ask questions. Just because you follow the rules in submitting a statement requesting exemption doesn't mean it's going to be granted. In most cases it is; in my case, it was. But in a country as upside-down as this one, where bureaucrats and judges routinely make up their own laws or interpret existing laws using the wildest logic, parents have been dragged into court over the exemption issue, and in quite a few cases jailed for child abuse. But it gets much worse. I've read some shocking stories about parents who objected to having their infants vaccinated, but gave in to scare tactics or veiled threats, only to see their precious children suffer a severe reaction or even die, after which they were accused of "shaken baby syndrome" to protect some stupid doctor or nurse, and charged with grave felonies. Can you imagine losing a child like this, then being arrested and dragged into court with the prospect of facing a long prison sentence? What a truly criminal practice vaccination is. What a truly horrible nation America has become.

It appears we've now reached the point where gangsters in the government-corporate medical cabal put out contract killings on individuals they don't like. In just one three-day period, from June 19 to June 21, 2015, three doctors whose practices were grounded in holistic health died suddenly in mysterious circumstances — two in Florida and one in North Carolina. The first, Jeffrey Bradstreet, a thorn in the side of the vaccinators, was widely known and dearly loved by the parents of autistic children whom he treated. Bradstreet's own son had become autistic after an MMR shot. He was found floating in a river with a gunshot wound to the chest, days after FDA agents had descended on his office. His death was ruled a suicide. Can anyone doubt that Dr. Bradstreet was murdered? Two days later, on Father's Day, two fathers, Bruce Hedendal and Baron Holt, were found dead. Hedendal, who had had some run-ins with the FDA, was slumped over the wheel of his car. On April 3, 2018, the body of a black CDC epidemiologist, Timothy Cunningham, was found floating in the Chattahoochee River in Atlanta, not far from CDC headquarters, nearly two months after he had mysteriously disappeared. Speculation swirled as to whether he had found that black babies are more susceptible to vaccine damage and autism than white babies, and as to how he died. Jumping into a frigid river seems to me an unusual way to end it all, but suicide was the coroner's ruling. As one anonymous commentator wrote about these deaths: "You may call me a conspiracy theorist if I may call you a coincidence theorist."

Shortly after the Cunningham case was closed, Dr. Suzanne Humphries, a well-known author and outspoken critic of vaccines, received a series of death threats, including one that detailed exactly how she would be killed during a mass shooting at a public event. An image of her dead body covered with semen stains with her eyes gouged out was also sent to her. Terrified, as anyone would be, she contacted the FBI. The FBI then tried to frame her for fabricating threats against herself. Probably the only reason this didn't happen is that it was publicized, with all the details of the FBI's sloppy lies and maneuvers, on naturalnews.com.

What is one to make of a country whose top law enforcement agency employs so many criminals who have committed so many crimes — and is never held accountable? I'm sure there are good people in the FBI, and I'm equally certain that there are many liars, perverts, killers, cover-up and frame-up artists in this agency. If you doubt that, you need to read *The Franklin Cover-Up* by John DeCamp. It's the most horrifying book I've ever read. It's about the kidnapping, abuse and murder of children in America that feeds the sexual appetites of many highly placed people in business and government, and their protectors, including those in the ranks of the FBI. DeCamp was a Vietnam veteran, lawyer, Nebraska state senator and father of four. Some young victims who revealed facts and some of his partners who were investigating these crimes were themselves murdered. DeCamp names names, lots of them. Only one person sued him for libel — and lost. I warn you, read this book only if you have a strong stomach and if you really want to know what America has come to. If you don't read it, at least read some of the 298 reviews on Amazon. Actually, you may already know that the FBI, assisted by members of the Army's Delta Force, laid siege to the compound of the Branch Davidian religious sect near Waco, Texas on April 19, 1993, engulfing it in flames and burning alive 76 innocent people, including 21 children. This atrocity, after a 51-day standoff, may have been premeditated mass murder and was the only time in U.S. history that tanks were used in a military attack against its own people. Millions of Americans watched in horror as it unfolded on live television. Those who were too young to understand or born after it happened should watch the video of the assault on the Internet to gain some insight into the federal government. A government capable of doing something like this and getting away with it is capable of doing *anything*, including abducting the children of anti-vaccine parents and killing anyone who resists.

America isn't the land of the free and the home of the brave; it's the home of the hired assassin and the land of the frightened sheep. Taking a short break from writing this chapter, I decided to jot down some names of people I'd read about who died before their time. The certainty that they were murdered, in my mind, approaches one hundred percent, and in a few cases there isn't the slightest doubt. The most famous of these would be George Patton, the audacious World War Two general who through the war, led his troops and fought with fanatical aggression, having swallowed whole the big lie that Hitler was a menace to America and the whole world, but who later came to see that he had been fighting on the wrong side and that Soviet communism was the real menace. For much of 1945 he spoke and wrote with contempt towards Jews and with admiration of the German people and the fighting qualities of their soldiers. Despite mounting opposition against him, he refused to keep his mouth shut or put his pen down. When a staged car accident did not come close to killing him, his brief stay in a hospital, while recuperating from minor injuries, did on December 23, 1945. A strange car accident on November 13, 1974 was successful in killing Karen Silkwood, a technician at the Kerr-McGee Fuel Fabrication plant in Cimarron, Oklahoma. The site processed plutonium, a dangerous radioactive element, and she became concerned with the lack of safety precautions and disregard for public health, and decided to go public with her information. Heavily contaminated with plutonium herself, she had made arrangements to turn over documents to a newspaper reporter when she died on a lonely country road. On August 10, 1991 a cleaning lady who worked at the Sheraton Hotel in Martinsburg, West Virginia discovered the naked body of Danny Casolaro in a bathtub in room 517. Casolaro, a freelance author, had been working on a book titled *The Octopus*, which tied together several strands of an international cabal that included the Inslaw software theft, the Department of Justice, Iran-Contra, and money laundering. He had been lured to Martinsburg to meet someone who, he believed, would provide substantial inside information for his book. Casolaro's wrists had been deeply slashed about ten times and there was blood everywhere. The local police did not want to get involved. His death was ruled a suicide. Suicide also was the cause of death attributed to a black police officer on the Oklahoma City force, Terrance Yeakey, who was among the first to arrive and help the badly injured at the scene of the Alfred Murrah Federal Building after it was blown up on April 19, 1995, killing 168. Seeing several federal agents standing around, he immediately sensed that something fishy was going on, and later came to believe in a massive cover-up, the official account contradicting what he had witnessed. (Look at a photograph of the devastated building and you will see debris blown outward, indicating that one or more bombs had exploded inside, an impossible result of an outside truck bomb detonation.) Yeakey was deeply troubled by the whole affair and never stopped talking about it. He and his family were constantly threatened by anonymous phone callers. His body was found in a field on May 8, 1996. Yeakey was a huge man who must have violently resisted his murderers, which might explain the multiple stab wounds, strangulation marks, and single bullet fired through his skull. Some suicide. Stanley Meyer was an inventor who had created what he claimed was an automobile engine that could run on water which, were it to replace the internal combustion engine, would obviously pose a huge threat to the profits of the petroleum industry. Two men posing as Belgian investors met with him in a restaurant in Grove City, Ohio on March 20, 1998. Taking a sip from a glass of cranberry juice, he grabbed his throat and ran outside where he died on the sidewalk. His last words were "They poisoned me!" Cerebral aneurysm, the police concluded after a three-month investigation. I can only wish there were more FBI men like Ted Gunderson, a special agent who had supervised FBI offices in Memphis, Dallas and Washington D.C. Like John DeCamp, who mentioned him in *The Franklin Cover-Up*, he was committed to protecting children from the vast pedophile network that operates throughout this country and reaches into the highest corridors of power, including the FBI, as he well knew. Despite the clear dangers, and numerous threats against him, he continued his investigations well past his retirement. He died of bladder cancer on July 31, 2011. Three years earlier, he revealed that he had tested positive for arsenic and cyanide poison. His fingers had turned black, a telltale sign of arsenic poisoning. These, as I said, are just a few names that popped into my head, after which I reviewed the details on the Internet. I've read about many, many more over the years.

Readers unfamiliar with the names above may still have heard of the famous "Clinton body count." Anyone who has looked into the background of Bill Clinton knows that, as governor of Arkansas in the 1980s, he turned a blind eye to the transport of drugs on a huge scale in and out of a small airport near Mena, a sleepy town in the southwestern corner of his state. This is what paved his road to the White House. The Clinton body count refers to the more than 100 people connected to the Clintons in some capacity, intimately or from a distance, who died in suspicious circumstances. Many, obviously, were murdered, either on direct instructions from Bill or Hillary Clinton, or more likely, from the invisible wirepullers who engineered their climb to political power and have ready access to a stable of assassins. With the exception of White House counsel Vince Foster's supposed suicide on July 20, 1993, none of these deaths received any substantial news coverage. One of the dead was Jerry Parks, who had worked closely with Foster and who, as a longtime member of Bill Clinton's security team when he was governor, knew many repugnant details of his personal life. After learning of Foster's murder, Parks grimly told his family, "I'm a dead man." In the early evening of September 26, 1993, Parks was stopped for a red light at an intersection in Little Rock, when another car pulled beside his, a man jumped out with a 9 mm handgun, and pumped several rounds into him. The best compilation of the Clinton body count, with all the details, is found in the "murder" volume of the trilogy titled *Hillary (and Bill)* by Victor Thorn. (The other two volumes are "sex" and "drugs.") Thorn was a brave investigative reporter who wrote many excellent books, one of his most important being *9-11 Evil: Israel's Central Role in the September 11, 2001 Terrorist Attacks*. He was found dead of a gunshot wound to his head on August 1, 2016, his fifty-fourth birthday. Just one more convenient suicide.

So again I ask, "What kind of country is this?" Taking what I've written above into account, is it any wonder that we have such a destructive childhood vaccination policy? Are all the puzzle pieces of the American nightmare beginning to fit?

Many Americans, oblivious to reality, go right on believing all the absurd myths they were raised on. I know them. Most of them are nice people and I take no pleasure in insulting them. They're the kind of people who fly the flag on the front porch and have never read a thought-provoking book in their lives. They seem to think they're free because they're not in jail, and because their refrigerators are full and there's plenty of booze and beer on tap and lots of entertainment and leisure activities to be enjoyed. They have a nice salary or a nice pension and life is good. Sure, the country has problems but we can get through it. We're still the freest people on earth, aren't we? We need to get our children vaccinated because that's how we protect them and protect our society from disease, right? It's part of being patriotic, isn't it? Don't you know that vaccines wiped out polio? Haven't numerous studies proven that vaccines don't cause autism? If vaccines were really dangerous, surely our government would tell us about it, wouldn't they?

Actually, there's little personal freedom here beyond that which exists in nearly every impoverished and backward country, like the freedom of religion, the freedom to travel, and the freedom to change one's residence. Americans have the freedom to vote for Tweedle-Dee or Tweedle-Dum every four years, and the freedom to send the puppet of their choice to congress who will rarely if ever do anything in their interests. They have no say nor are there any referendums to change government policies that they object to, that is to say, nearly all of them, the most disastrous being the deliberate flooding of this land with people of all races and cultures which has made America one of the world's most artificial nations. There are many onerous restrictions on individual freedom that have multiplied over the years as government has reached leviathan proportions, and with it the empowerment of legions of stupid and arrogant bureaucrats, backed up by increasingly militarized police forces that have been conditioned to look upon anyone with an independent streak as a threat. Politically, America can best be described as a soft tyranny, where life is still normal for most in a material sense, but also one with a growing army of disaffected citizens, a nation that at its core is bankrupt and whose future portends revolution, secession or civil war. History does not stand still.

In reality, there are only two categories of personal freedom in America that carry any weight: the right to free speech and the right to keep and bear arms, as enumerated in the first and second amendments, respectively, to the Constitution, that estimable but ultimately defective blueprint for a higher statecraft. Paradoxically, at least on paper, Americans are still free to voice opinions that in other countries would be prosecuted on various grounds, often "hate speech." America has become an uncertain place, however, and anyone who openly espouses naughty ideas, or offends the wrong people, does not know if he will be ignored, monitored or harassed, or if a nice little accident or suicide or heart attack or conviction carrying a long prison term will be arranged for him. In some professions, you stand an excellent chance of being fired even though the government will leave you alone, so free speech comes with the freedom of losing your livelihood. If you're a high school social studies teacher, for example, and you deviate from the established version of World War Two, you'll likely be on the unemployment line the following week. But as long as you don't pose a real threat to the System, you probably won't be arrested. Some people have told me, not jokingly, that they would never subscribe to a politically incorrect publication, or log on to a politically incorrect web site, for fear of getting on a government list. This is unfounded and pathetic, though I suppose understandable, given that records of all electronic communications of U.S. citizens now are kept under lock and key by the National Security Agency (NSA) in a huge data storage facility near Bluffdale, Utah, and high-tech, unlimited surveillance, both secret and in the open, has advanced to a level that Orwell never dreamed of.

Likewise, while the second amendment states that "the right to keep and bear arms shall not be infringed," it is blatantly infringed in some states and municipalities, though less so or hardly at all in others. And like the first amendment, which certain individuals would love to curtail using their pet phrases, there are many who would love to ban certain types of guns as a first step toward their ultimate goal of confiscating all firearms, a chilling dress rehearsal of which was seen when New Orleans was placed under martial law in the wake of Hurricane Katrina in 2005. Unfortunately for them, however, guns are here to stay. Gun ownership is a deeply ingrained American tradition and there simply is no way that people are going to voluntarily surrender their weapons en masse as they did, for example, in Australia in the late 1990s. There are about 300 million privately owned firearms in this country, a unique historical phenomenon that has never existed anywhere else. American citizens are by far the most heavily armed people on earth, and this presents a huge problem to thugs at every government level who will have to think twice before going door to door in order to forcibly vaccinate people which, as I noted in the chapter on smallpox, was, and probably still is, a dormant law on the books in 38 states. What remains to be seen is how far state or federal government will go before people start shooting.

Somebody once described America as the Roman Empire on steroids. It's a good description. Rome, the glory and might of the ancient world, its military forces spread far and wide, arose, declined and disintegrated over the course of 1500 years. In America the process has taken place much faster. America is now collapsing under the weight of its lies in much the same way as the old Soviet Union did with its oppression and extended empire. America is no more a free nation than the Soviet Union was a worker's paradise. Flying the flag, singing *God Bless America*, the "salute to our military" ritual that often takes place before huge crowds at televised sports events — none of this cardboard patriotism puts a dent in reality, the reality that America is death to the world and to itself, the reality that every day American children are fed into the maw of this beast when they get vaccinated.

This has been a very negative chapter, so I'll try to end it on a positive note. I'll repeat what I stated above, that a country is only as good as its people. Well, there are still a few million really good people scattered around America — men with hearts of gold and fists of iron and women who stand behind them. The message of this book rings loud and clear with them. I know they're out there because I've met them — and so have many Europeans who have told me how truly wonderful Americans are when you meet them in their own country, as opposed to the poor impression they make as tourists abroad. As a young man, I hitchhiked all around the country and had many great conversations, and was invited into many homes by people who, in picking up a stranger with his thumb out, showed they were a little braver and kinder than most. Later I drove at least 30,000 miles, over numerous trips, in my own car. Everywhere. And I've met plenty of other nonconformists in out of the way places. Interesting, kindhearted people who think for themselves and despise everything about Washington D.C. The comment boards of iconoclastic blogs and web sites sizzle with their opinions. Here's one I came across on the topic of vaccines: "Put a needle in my arm and I'll put a bullet in your brain." They're not people who vegetate in front of the television set. Not for a moment did they believe the media lie that the Waco holocaust was "an apparent mass suicide." They're the kind who know there is something profoundly wrong with America, even if they can't put their finger on it — the kind who are fed up with all the lies, wars and corruption, with the flood of non-white invaders who keep pouring in, with the total lack of leadership, with political dog and pony shows that pit a witch like Hillary Clinton against a clown like Donald Trump. The only thing that makes me feel good about being an American these days is my bond with these people. And even if they've never weighed the vaccination issue before, they'll know what I'm talking about when they read this book. One thing they're *not*, is politically conservative, and that's what I'll take up in the next chapter.

The fraud of conservatism

A FEW words are in order here about the canard that this is a free country with plenty of competing ideas and opinions, as in liberal versus conservative, Left vs. Right, Democrat vs. Republican. This distracting soap opera has been playing for a long time, and I'm hardly the first one to point it out. Way back in 1968, third party presidential candidate George Wallace famously quipped that "there's not a dime's worth of difference between the Democrat and Republican parties." And much further back still, Ernest Everhard, the hero of Jack London's *The Iron Heel*, confronted Congress in one scene with these words: "There are no Republicans nor Democrats in this House. You are lick-spittlers and panderers, the creatures of the Plutocracy." As the French proverb goes, the more things change, the more they remain the same.

Although these distinctions have become irrelevant, I occasionally use the terms "Left" and "liberal" as they're commonly understood, but "conservative" no longer means anything to me. I write these words as one who used to proudly called himself a "staunch conservative," but many years ago came to see that conservatism is a bowl of mush, a synonym for stupidity, hypocrisy and cowardice. There was a time when the word really meant something, when it meant defending and *conserving* everything precious in Western civilization that has come down through the ages. Genuine conservatism was exemplified by Charles Lindbergh. An intensely private man of sterling character who loathed the press, Lindbergh nevertheless went public in alerting his countrymen to the danger of Jewish influence in the newspaper and film industry, as previously mentioned, and he also warned of the terrible squandering of the white race's best genes that would inevitably result from another world war. The issues Lindbergh raised in his speeches and in his wartime diary, which later was published in book form, are as fundamental as ever, but no self-styled conservative since 1945 has dared to bring them up. To do so would be career suicide in a country where media executives determine the boundaries of permissible speech and opinion, beyond which those of all political stripes know not to stray. Lindbergh himself was relegated to near non-existence after the war because of his views. World War Two, a cataclysm for every nation involved, has become the bedrock of lies on which the modern world teeters, yet our hollow military victory in that terrible conflict, which brought communism to half of Europe as well as China and Korea, is considered a glorious triumph of good over evil by all respectable conservatives.

In our times, Ron Paul, the former Texas congressman — and incidentally, a retired obstetrician who favors parental freedom of choice on vaccination — has been the only public figure I can think of who espouses the conservatism embodied in the men who created America. As principled a man as Paul is, however, he has always avoided the core issues of race and Jewish power. His integrity was rewarded by media obscurity, in addition to Republican Party dirty tricks that nullified his presidential ambitions in 2008 and again in 2012.

Pat Buchanan is the grand old man of American conservatism; many would agree that he's the best that conservatism can offer. This professional chameleon deserves mention only because he has been around forever and has occasionally made sense, at times in the past even angering media bosses and milquetoast conservatives with mild criticism of the Zionist lobby and our endless futile wars. But he's an old hand at switching sides, and has always landed back on safe ground in the Establishment camp, having a string of books on the *New York Times* bestseller list to show for it — the hallmark of every well-known conservative, and the reward for playing the game. This is what one would expect of an insider who worked for Richard Nixon and Ronald Reagan in their presidential years — both of whom wore a conservative hat but never did one thing to put their country on the right track. Nixon was nothing more than an ambitious and pragmatic politician, while Reagan was a basically nice man but totally lacking in substance, a Hollywood actor who played his conservative role to the hilt. What Buchanan offers is a kind of schizophrenic conservatism. Quoting from his writings many years ago, independent researcher Michael Hoffman showed that Buchanan has an ingrained habit of contradicting himself and compromising the truth, that his modus operandi is "to mix very eloquent insights and very well-written and memorable and defiant turns of phrase with one or two sentences crammed with weasel words....[and to] adapt himself to certain illusions so as not to appear too weird and far out to Joe Sixpack." Buchanan's half-hearted bid for the presidency, in which he resorted to cheap tactics to distance himself from the "far right," deeply disappointed his supporters, not once but twice. A tired old gasbag forever stuck in political kindergarten, a profile in mediocrity whose accomplishments in life add up to zero, he continues to blab and write dust. On vaccination he has taken no position that I know of. The best that can be said for him is that he has more self-control than the rest.

Conservatism has become undefinable. The voting records of politicians who get themselves elected by singing a conservative tune are often indistinguishable from those of their liberal counterparts. Americans who call themselves conservatives don't even know what they believe, preferring to let the tough talkers do their thinking for them. In addition to Fox News, that media fortress of belligerent couch potatoes, there are plenty of vapid conservative web sites and obnoxious radio talk show hosts who have no interest in honest debate but rather in maximizing their ratings by insulting anyone who disagrees with them — if they're allowed on the air in the first place. They've been playing this game for as long as I can remember with their silly outbursts and their complaining about the way things are, lashing out at "liberals" and "the Left" all the way to the bank, the same way their adversaries lash out at "right-wing extremists" and "white supremacists" (with even some conservatives now, like Sean Hannity, bashing "white supremacists"). It's a carefully supervised farce, even if most of them don't realize it. Virtually all these conservative media puppets, not to mention politicians, have three things in common: they're rich, they're strong supporters of Israel, and they endorse American military aggression around the world. Some are pretty bloodthirsty about it. Most of them also avoided military service during wartime, some going to great lengths to shirk their patriotic duty while others did the fighting and dying. The Vietnam War is a case in point, and Trump a typical chickenhawk. With the draft looming, Donald Trump got four student deferments, then, despite playing three collegiate sports, claimed he had heel spurs, his fifth deferment.

Nothing illustrates the meaninglessness of contemporary political jargon better than the subject of this book. There is no "conservative" or "liberal" position on vaccination. The late, liberal Democratic senator Edward Kennedy was an advocate of compulsory vaccination, as was the equally slimy conservative Republican senator Bill Frist, a doctor with strong ties to the drug industry. Yet Kennedy's nephew, Robert Kennedy Jr., also a steadfast liberal who supported Hillary Clinton for president, has become an articulate spokesman exposing CDC corruption and ripping the vaccine industry (though sadly he insists that he's pro-vaccine). The children of pacifist hippie types are as likely to be unvaccinated as the children of heavily armed survivalist types. We associate the first group with the Left, the second with the Right, but both loathe the government. As the political sands have shifted, the distinctions between Left and Right have largely evaporated, except in street fighting where the lines are clearly drawn. To my mind, the only difference between liberals and conservatives these days is that the political IQ of the latter is, on average, five points higher, with both in the low double-digit range. Conservatism, in the old sense of the word, no longer exists, having been hijacked by "neo-conservatives," a grossly misleading term. This is the subject of the excellent and effectively banned book *The Judas Goats* by the late Michael Collins Piper, which traces the fraud of modern conservatism back to its roots — the Trotskyite branch of communism in the old Soviet Union, heavily Jewish in origin, whose followers emigrated to the U.S. and metamorphosed into the fanatically pro-Israel neo-cons, from whom nearly all well-known conservatives today take their cues. According to Piper, "The modern-day disciples of Trotskyism are key figures among The Enemy Within, twisting old-fashioned conservatism into a divisive and destructive force that is utilizing America's military might, the blood of its children, and its national treasure to enforce a global Zionist imperium — in short, a New World Order." And Piper was right on target in slamming "conservative" Fox News as a controlled pseudo-opposition — a seemingly sensible alternative on domestic issues to the liberal slop dished out by the other major networks, and one that consequently attracts a large audience, but in reality a blaring propaganda platform for endless war and American and Israeli terrorism.

The Judas Goats, published in 2006, is as timely as ever, but the proliferation and growing influence of conservative web sites and some new players merits additional comment since they have added to the confusion. A good example is Tucker Carlson, who came on the scene after Piper's work was published and who, like Pat Buchanan, is more sensible and likable than the rest of the crew. But he too knows which side his bread is buttered on, and expounds accordingly. Perhaps the most fraudulent of the conservative web sites is breitbart.com, which is not only heavily staffed by Jews and stridently pro-Israel, but was actually conceived in that country. These people sometimes take stances, probably insincerely, on certain issues, like gun control and Third World immigration, that strike a chord in white Americans and this confuses them. When the mainstream media refer to such entities or individuals as "ultra-conservative" or "far right" they confuse the issue even more, which is their intention.

❖

Donald Trump was elected president with the overwhelming support of the white working class and a solid chunk of the white middle class — conservatism USA. He appealed to the hopes and dreams of so many frustrated people in this country who for so long have yearned for real leadership, who were absolutely starved for it. Was he not the embodiment of conservatism and the American dream himself, a billionaire businessman with an American flag pinned in his lapel? He was going to make America great again — whatever that was supposed to mean. He had a vision and a concrete plan to straighten things out after eight years of Obama's Left-liberal misrule. He was going to drain the swamp. I, for one, was not swept along by the huge wave of enthusiasm generated by the crowds that turned out for Trump's campaign rallies. Even though, at the time, I knew nothing of his history of schemes and failures as a business tycoon, inherited from his equally sleazy father, I had an uneasy feeling about him. On the other hand, I admit to being taken by his style. I found his choice of words most refreshing, and I was especially delighted by his pledge to reduce our military presence abroad. In my lifetime he was the only presidential candidate to speak in the straight-shooting language of the American workingman. I truly believed there was a chance that he would do at least some good.

I am revising this in February 2020, shortly after Donald Trump was acquitted in the play-acting impeachment hearings, which bored the country to death, and which I knew would not culminate in his removal from office — not that that would have changed anything. It has been more than three years since this man became president, and it has become painfully obvious to me and a lot of other people, including many of the 63 million who voted for him, though probably not most of them, that he is the biggest impostor ever to slither into the White House. He broke all records in reversing the number of promises he made as candidate once he became president, as well as the speed with which he reversed them. Even before he was sworn in, most of his key cabinet appointments were snakes dredged from the swamp he had promised to drain. America is now in the kind of disarray that I pictured a Clinton presidency would bring, and whether or not Trump finishes his first term, or is re-elected — and in spite of the living fossils who will stand by him no matter what, imagining that he must be doing everything right because he has been relentlessly attacked by "liberals" and "Democrats" over the most frivolous matters imaginable — conservatism has reached its inevitable dead end. And who better to stand on its corpse than this impulsive, narcissistic man-child of no moral, emotional or intellectual depth, of no core beliefs or values other than a love of money, cheap sex, Israel and military might, whose life never had any purpose other than to claw his way to the top of the heap. And now that he's there he can watch his country sink faster than under any previous administration, while continuing to live in a world of make-believe, as evidenced by his countless lies and idiotic "tweets" and public statements. The acrimony between Trump, his remaining supporters, and his many critics in the government, military, media and entertainment has become a theatre of the absurd, a spectacle of divisiveness on the level of a school cafeteria food fight between children, with all the players vying for top honors in stupidity and irrelevance. I fail to see any qualitative difference between Trump haters on one side and his diehard defenders on the other side.

Donald Trump's tattered presidency is the logical sequel to his earlier years. Anyone who takes the trouble to examine his record as a New York City real estate developer will discover, as I belatedly did, that he is not a man of the people but a man of the gutter. And it's in the gutter where so many proponents of conservatism dwell. One of Trump's former friends and supporters, who wrote a flattering book about him but later turned against him, though he may have flipped again, is San Francisco-based talk show host Michael Savage, host of *The Savage Nation*, which at one time was syndicated on 400 radio stations with a weekly listening audience of more than ten million. He also runs a web site, michaelsavage.com. Savage is a Bronx Jew whose real name is Michael Weiner. He is widely described as a conservative, and has taken many stands popular with conservatives. Of late the equally nauseating conservative pseud Ben Shapiro has moved in on some of his air time. Weiner's opinions are all over the map. He is noted for saying outrageous things on his radio show, even outdoing the cackling idiocy of Ann Coulter, another of this bunch who is often tagged with the ridiculous label "far right." This was Weiner's take on autism, as heard on July 16, 2008:

I'll tell you what autism is. In 99 percent of the cases, it's a brat who hasn't been told to cut the act out. That's what autism is. What do you mean they scream and they're silent? They don't have a father around to tell them, "Don't act like a moron. You'll get nowhere in life. Stop acting like a putz. Straighten up. Act like a man. Don't sit there crying and screaming, idiot.

As the father of an autistic child, I can't even say that this offends me. Rather, it saddens me that this country has sunk to such a level of mental retardation that a throwback like this still has a large following among listeners who call themselves conservatives. This tirade did result in a backlash which saw some advertisers pull out of his program, but no such backlash occurred following his April 17, 2006 show when he called for the killing of 100 million Moslems. But this tough talker was always shy about grabbing a gun and doing any killing himself. While in his twenties and with a war being fought in southeast Asia, Weiner was palling around with LSD guru Timothy Leary and that poster boy of degeneracy, pedophile "poet" Allen Ginsberg.

Weiner's Talmudic fantasy came on the heels of a statement made the previous month, on March 8, by Bill O'Reilly of *The O'Reilly Factor* fame, that "in a sane world, every country would unite against Iran and blow it off the face of the earth." This call to genocide undoubtedly pleased his Jewish paymasters. Earlier in life, however, this studio warrior showed a marked preference for the college campus over the rice paddies and jungles of Vietnam. For two decades, before his excessive libido led to his dismissal at Fox News, the pompous O'Reilly, who had fobbed himself off as a down-to-earth guy with working-class roots, was America's favorite TV conservative commentator. O'Reilly actually thinks he's smart. He's so shallow that he thinks all the books he has written are really great because they all became *New York Times* bestsellers. He's not so dumb, however, that he doesn't recognize which side of an issue is the safe side to be on. On the February 2, 2015 edition of *The O'Reilly Factor*, he was on camera with Megyn Kelly — who, in an amusing turn of events, and joined by other women, later tattled about the off-camera lewd behavior of this slob who quietly spent more than $40 million to settle several sexual harassment lawsuits. While this bimbo talked a mile a minute, sounding like the usual broken record, about the wonderful measles vaccine, and how her three children are all up to date on their shots, and so on and so forth, O'Reilly nodded and nattered in agreement.

It happens that O'Reilly and I attended the same high school (he graduated just before I started) and, owning an alumni directory, I know his address. He lives on a private road overlooking Manhasset Bay — quite a climb up the socioeconomic ladder from Levittown, where he grew up. It also happens that Manhasset Bay used to be a favorite fishing spot of mine, and through a neighbor who had worked as a security guard in the area, I had permission to park in a private driveway and walk down to the water where I would cast from shore for striped bass. O'Reilly's home was almost directly across the street. Mr. Working Class lives in a mansion that not one workingman in America would ever be able to buy unless he wins the lottery. His latest gig is something called the Great American Wealth Project. Completely out of touch with unemployed blue collar Americans, he wants to see them get rich, which is the meaning of life for all good conservatives. Since becoming a liability because of his carnal misconduct, he started his own online news show, which is nothing but the usual conservative blather.

For sheer hypocrisy and hot air, however, no one has surpassed Rush Limbaugh, America's most listened to radio personality and most famous conservative (still at work after a diagnosis of lung cancer). Anal cysts prevented Limbaugh from fighting in Vietnam, but like the others, he discovered his manhood later in life behind a microphone, beating the war drums in lockstep with the media mantra of the day. It's hilarious that Limbaugh so often denounces the very media that have long employed him and made him one of America's wealthiest celebrities. Rush makes about $40,000 an hour, somewhat more than the average salary of his dumbshit fans. A few times a year I turn the radio dial to listen to the braying of this jackass for a few minutes, and each time I do I'm left speechless as to how anyone can take him seriously. This includes his scorn of the fact that vaccines cause autism, and his pummeling of the British doctor Andrew Wakefield, about whom more later. With the Internet, though, it's gratifying to see that there are intelligent Americans out there who see through all this nonsense, and every day are calling out Limbaugh, O'Reilly, and their fellow racketeers. "Conservatism Incorporated" is the fitting term that intelligent Americans have stamped on the whole sorry business.

But for now business is still good. Among the most recent to join the pack is Charlie Kirk and his Turning Point USA scam. Three cheers for Donald Trump for being the first president ever to appoint an open homosexual, Richard Grenell, to a cabinet post, Kirk recently crowed on social media. This was eight months after our family values president held an LGBT rainbow flag and called for the decriminalization of homosexuality around the world; two months after our pro-free speech president, signed an executive order cutting off federal funds to colleges which allow students to criticize Israel; two years after our pro-gun ownership president, of whom the NRA leadership is so enamored, buckled under media pressure after the Parkland, Florida school shooting and said, "I like taking guns away early.... Take the guns first, go through due process second." Even Clinton and Obama didn't go this far.

❖

Whether in government or media, conservatives just react to the leading news stories each day, sounding off within their prescribed boundaries. Step over the line and you're sent packing, as Joe Sobran, who wrote a column for *National Review* under the withering eye of conservative poseur William F. Buckley, found out. Whether out of ignorance or expedience, they are accomplices to the routine censorship employed by the media. A case in point is the dispossession and violence directed by the new black rulers of South Africa against white people, especially farmers and their families in outlying districts. Gruesome photographs of some of the thousands murdered, many tortured to death, can be found on the Internet. This has been going on for more than twenty years now amidst a near total global news blackout — a blackout matched by the delirium over the supposed "horrors of apartheid" under white rule — a favorite theme, incidentally, of that conservative blowhard Rush Limbaugh. Anyone familiar with the facts on the ground in Africa, as I am, having traveled far and wide and read about fifty books on all aspects of that continent, knows that the situation in South Africa under apartheid as projected by the news media was a lie of monstrous proportions. Back in the 1980s, when hardly a week passed without the television screen misrepresenting everyday life in South Africa, pretending that the whole country was going up in racial flames, only a few conservatives spoke in feeble tones in defense of that nation. There were even some "conservative" Republican congressmen — I think here of former Speaker of the House Newt Gingrich, the consummate sewer rat politician and yet another slacker who ran away from the Vietnam War — who voted to tighten the screws on that beleaguered country with punitive economic sanctions that hurt ordinary blacks more than anyone else, as I discovered during my 1988 visit to South Africa.

Over the years, on any number of issues, the conservative position has shifted to the left in acquiescence to the ground rules laid down by the mass media bosses. To cite one more example, the so-called civil rights movement of the 1960s, with all its destructive rioting in more than a hundred American cities, is now looked upon as social progress by many conservatives. It was the wimpy conservative icon, Ronald Reagan, who in 1983 gave a rousing speech and signed into law a national holiday commemorating the birthday of Martin Luther King, a moral leper whom two generations of American schoolchildren have been taught to revere as a very great man, an apostle of nonviolent resistance to injustice. This is the same Michael King — his real name — who provoked disorder or violence everywhere he went, who plagiarized much of what he wrote, and who was such a thoroughgoing fraud and sexual deviant that the record of his private life, as compiled by federal investigators, was sealed by court order in 1977, not to be opened until 2027, though happily some of it has already leaked out.

❖

Once upon a time, I thought that personal freedom, as long as it didn't encroach upon the lawful rights of others, was one of the main tenets of conservatism. I can't imagine a more basic freedom than the freedom of devoted parents to make a medical choice for their child. Yet I don't know of one prominent conservative — aside, perhaps, from Ron Paul — who has taken a strong and consistent stand in support of this choice, or expressed indignation at the status quo in regard to childhood vaccination. To do so would not be wise. While professing to believe in free speech, conservatives, as I have mentioned, know there are limits to what they can say and write, and they adjust their sales pitch accordingly. Vaccination is a sensitive issue, a political minefield where it's better not to tread for too long.

Every now and then conservatives say something I agree with. From time to time some of them have raised lukewarm objections to certain aspects of vaccines or compulsory vaccination, while tiptoeing around the dogma that puts vaccines on the highest medical pedestal. This even includes Weiner, as well as that most servile of footlickers, Sean Hannity. Even Trump, as a private citizen, strongly voiced his opinion that autism was connected to vaccination, and after winning the 2016 election, met with Robert Kennedy Jr. and pledged to appoint him to head a commission on vaccine safety. So typical of the man, he turned around and appointed two ardently pro-vaccine directors, Scott Gottlieb and Brenda Fitzgerald, to the FDA and CDC respectively, and forgot about Kennedy. Two years later, when the media cooked up another artificial crisis over an upsurge of measles, blaming anti-vax parents, Trump once again showed what a spineless, brainless jellyfish he is by going with the flow: "They have to get the shots," he said. "The vaccinations are so important."

Those I have named in this chapter are but a few of the well-known conservatives of our day who have led so many Americans into the swamp of no return. There are many, many more in government, entertainment, political activism, on the radio, on television, and on the web. As lost in space as conservative Americans are, it's hard for me to believe that an appreciable number of them will ever again buy the kind of snake oil that Trump successfully peddled in 2016. But we'll see. The 2020 presidential election will certainly be interesting, and guaranteed to widen the cracks in a country that's already cracked up. No matter who wins, expect nothing progressive on the childhood vaccination front or anywhere else from anyone identifying with this moribund word "conservative." Conservatism is for losers. Conservatism is finished.

America's gravest problem

The media and educational system have dumbed the people down to a level hitherto unknown in the civilized world. They are modern-day zombie populations, led around by the nose — mentally so manipulated that they cannot think straight, much less act in their own self-interest, either as individuals or as societies and states. Both in spirit and in reality, they have become the tax-paying cash cows and playthings of an alien oligarchy. — Ernst Zundel

IN THE foregoing pages I have made scattered references to the news media, an institution that deserves a closer examination because of its unrestricted and awesome power, including the power to mold popular views about vaccines and sway public health authorities to initiate draconian vaccination policies. Of all the myths that Americans are addicted to, perhaps the most ludicrous is that they are privileged to have a "free press" — that they are free to examine all sides of an issue, the substance of which is readily accessible as "news" conveyed through various outlets of the mainstream media. The truth is very different. The truth is that Americans are as misinformed, lied to and brainwashed as are the subjects of the world's most totalitarian societies, past and present. In reality, most Americans have been reduced to the intellectual level of insects based on what they read in the daily newspaper or weekly magazine, what they hear on the radio, and above all what they see on television — not to mention books, movies and the Internet.

The American news and entertainment media constitute the most sophisticated mass mind control apparatus ever created. Not my words, but an excellent description. The mind control techniques are far more insidious than the crude and blatant propaganda messages, with illustrations, that I once saw in the communist dictatorships of eastern Europe, and more recently in North Korea. Americans are told, without being aware of it, what is socially respectable to believe, with some variety and phony controversies tossed in to give the appearance of a free marketplace of ideas. Man is a social animal, and the pressure to conform to this fake reality that the media bosses have invented is overwhelming. Few dare tread in the territory designated for "hate groups" or "neo-Nazis" or "extremists" or "white supremacists" — buzzwords used like prods to keep the cattle under control. "Conspiracy theory" is my personal favorite. As if there's no such thing as conspiracies. As if men have never plotted behind closed doors. As if history isn't full of political conspiracies. As if the world's jails aren't filled with people who conspired to commit crimes and got caught. It's certainly true that some people have overheated imaginations and visualize conspiracies without supporting evidence, but only an imbecile would take this to mean that conspiracies don't exist. "Conspiracy theory" is just another catchphrase endlessly thrown out to belittle those few who can think for themselves, who intuitively know that what they read in the newspapers and see on television is a muddle of lies — to make the weak-minded think that anyone with ideas that challenge the media line is somewhat eccentric, if not a total nut. It's an utterly meaningless term parroted by the fashionable and clueless, and even by some who should know better. All of this is an endless game of psychological warfare. Everything — literally every single issue that has shaped the times in which we live, here and abroad — every prominent personality, event and topic going back a century or more — has been deliberately twisted or censored by the mainstream media. As a result, by the time the average American reaches the age of 30 or so, his or her brain has been marinated in so much nonsense that it can no longer function normally, any more than a pickle can be turned back into a cucumber.

Disseminating what is called "news" in this country has a long, lamentable history, even before the early twentieth century, by which time it had become a nearly total Jewish enterprise. Thomas Jefferson spelled it out in 1807 when he wrote, "Nothing can now be believed which is seen in a newspaper. Truth itself becomes suspicious by being put into that polluted vehicle." The following year he referred to "the putrid state into which our newspapers have passed, and the malignity, the vulgarity, and mendacious spirit of those who write for them."

The situation is by no means unique to America, of course. Oswald Spengler described it in his classic work *Decline of the West*, published during the chaos and despair that swept across his native Germany near the end of World War One, long before the age of television:

What is truth? For the multitude, that which it continually reads and hears.... What the Press wills, is true. Its commanders evoke, transform, interchange truths. Three weeks of press work and the truth is acknowledged by everybody. No tamer has his animals more under his power. Unleash the people as reader-mass and it will storm through the streets and hurl itself upon the target indicated, terrifying and breaking windows; a hint to the press staff and it will become quiet and go home. The Press today is an army with carefully organized arms and branches, with journalists as officers, and readers as soldiers. The reader neither knows, nor is allowed to know, the purposes for which he is used, nor even the role that he is to play. A more appalling caricature of freedom of thought cannot be imagined.... And the other side of this belated freedom — it is permitted to everyone to say what he pleases, but the Press is free to take notice of what he says or not. It can condemn any "truth" to death simply by not undertaking its communication to the world — a terrible censorship of silence, which is all the more potent in that the masses of newspaper readers are absolutely unaware that it exists.

In *The Eleventh Hour*, published in 1998, the late British patriot John Tyndall updated and expanded on the role of media subversion in his country, which mirrors our own state of affairs:

Daily, the media feed the people with a diet of degeneracy, combined with censored news and comment, all with the purpose of reducing them to a herd of cattle lacking either the wit or the will to resist the "brave new world" that their masters have planned for them, which world is now in an advanced stage of construction. Daily, there is the constant stream of poison directed to undermining all the finer human qualities that go to the building and preservation of successful nations. Along with non-stop political propaganda, in its direct and indirect forms, there is an unending avalanche of products encouraging moral and sexual license, homosexuality and every other imaginable depravity and perversion, while the same methods are used to denigrate morality, manliness, loyalty, patriotism, pride of race and the soldier qualities. A nation subject to this conditioning over the period of one or two generations becomes a nation ripe for conquest and enslavement, whether these are achieved by military or political means. The media are worth 100 divisions to any would-be enemy of war — so long as they are controlled and directed in the manner that operates in contemporary Britain.

Ironically, Donald Trump deserves credit for popularizing the term "fake news," and it's amusing that the real purveyors of fake news, those in the mainstream media, have turned the label around and pinned it on the alternative media — some of which, I must say, lacks credibility as well. I say ironically because Trump is very selective in which media entities he denounces as fake, too thickheaded to see that "conservative" oriented outlets, of which his favorite seems to be Fox News, are just as fake as the rest of them. Ironically because Trump won the 2016 presidential election thanks to all the free publicity he got from the very newspapers and television stations he accused, correctly, of propagating fake news, which struck a resounding chord in the hearts of so many down-to-earth American voters. The presidential campaigns of Ron Paul received so little news coverage that it amounted to outright censorship. Trump would have been ignored in the same manner if he had any honor which, as a pliant tool of the real power brokers behind the scenes — some of whom have now abandoned him — he most certainly does not. Everything in America is smoke and mirrors, and it has reached the point where the paid liars in the media have become so entangled in their web of lies that they now resemble cartoon characters. For those of us interested in ascertaining the truth, the whole thing has taken on a comic opera aspect, with a growing pushback from ordinary people who are fed up with being lied to, but unable to figure out what's really going on. In some instances *no one* can figure out what's really going on, as in America's military meddling in the Mideast, and domestic mass shootings. A case in point is footage that ABC aired on both its *World News Tonight* and *Good Morning America* programs on October 13-14, 2019, purporting to show Turkish military forces slaughtering Kurdish civilians at night with tremendous firepower in the Syrian border town of Tal Abyad. A viewer noticed that the video clip was actually from a 2017 machine gun shoot at a popular bi-annual event at the Knob Creek range in West Point, Kentucky which attracts firearms enthusiasts from around the country. ABC was forced to apologize and admit its "mistake." Also, in an age when certain "news" events are likely contracted out to companies that can supply adoring fans, protestors or "victims," and openly advertise their services (see, for example, California-based crowdsondemand.com, and crisiscast.com with headquarters in London and a "presence" on four continents), the line between reality and unreality easily disappears.

One of the most incisive exposes on the media I've ever read is the *National Vanguard* research report titled "Who Rules America?" which is on the Internet. Not only does this essay explain how the news and entertainment media falsify the reality to which most Americans adjust their opinions, it reveals the interlocking corporate gears of this behemoth, and discusses its immense political power — the power, among other things, to instigate race riots in American cities, make or unmake American presidents, foment war and revolution abroad, topple governments, destroy nations. And it names more than 50 individuals, along with several photographs, of these mostly unknown, shadowy media executives. While the lineup has changed over the years, at all times the overwhelming majority have been Jews, a group that comprises a mere three percent of the U.S. population. This corrosive Jewish influence is literally a matter of life and death, and no sticking one's head in the sand or rolling one's eyes, no cries of "anti-Semite" or "Jew hater" will change that. In a later chapter I address this subject in detail, since vaccination is one of its many tentacles. In this chapter I only touch upon it, taking note that these Jewish media executives and other immensely powerful Jews have a private agenda that few others share, even if there are some personal feuds within this group, which probably accounts for a few sexual predators and financial pirates among them being openly shamed and thrown to the sharks. "News" involving countries like Syria, Iran, North Korea, and in the past, South Africa, and on many other issues, is little more than a refined expression of the malignant hatreds of these media moguls, or in some cases, like Israel, their warm devotion, dished out to the public by the mental midgets who work for them. The few mainstream journalists who possess brains and integrity end up quitting or getting fired.

❖

For most of my life I've subscribed to various publications darkly characterized, when they are mentioned at all, as "hate sheets" by the media mind rapists. I've also read hundreds of books similarly classified and effectively banned, including those mentioned throughout this work — books never displayed during the charade known as "Banned Books Week" sponsored every September by the American Library Association. With one exception, I have never seen any of this literature for sale on any newsstand or on the shelf of any bookstore or public library. Nevertheless, there will always be some restless minds and free spirits who seek this kind of material. The situation in America is much like that of the communist nations of eastern Europe whose regimes fell, one after the other, near the end of the twentieth century. There, the intelligent and rebellious, the ones who could not stomach the lies that most of their countrymen swallowed without protest, secretly wrote, printed and circulated, at great personal risk, literature considered subversive by the authorities. The Russian word for this movement was *samizdat*. We have our own version of *samizdat*, the one difference — and it's a big difference — being that it's legal to publish and distribute writings that violate existing taboos. Yet this freedom is crippled when such literature is denied the usual channels of publicity and distribution. Genuinely banned books can be purchased online or directly from their publishers, but the damage has been done when only a small minority know about them. And even if many more Americans were aware of such books, most people, programmed by decades of Stalinist education and media lies, would be frightened at the prospect of opening their minds to forbidden ideas.

Then there's the Internet, which has been called "the printing press of the second American Revolution." The Internet has ripped a huge gash in the iron curtain that concealed essential truths from Americans for so long. It's the only forum of truly free speech we have, though in the past few years the monopolistic tech giants — and these too are heavily Jewish at the executive level — have taken bold and disturbing steps in the direction of censorship, always under the pretense of fighting "hate." There are, of course, pools of disinformation floating around the net, some of which tax one's abilities to establish the truth on controversial subjects. There are trolls, and there are plenty of nitwits who use it as a soapbox. Still, it's an amazing innovation that those of us born two generations ago could never have foreseen. With a few words and a mouse click, one can instantly connect with intelligent ideas and opinions light years removed from the mainstream media. And because of the Internet, the army of informed and angry Americans has steadily grown. Yet here too, a majority of citizens avoid politically incorrect Web sites, both out of fear of being monitored, or because it's too painful to entertain ideas that family and friends disapprove of, or that might contradict beliefs long held dear.

One of the more amusing complaints of the pro-vaccine people is that the media are guilty of blowing up vaccine risks out of proportion by putting out unsubstantiated scare stories. Having combed through hundreds of newspaper and magazine articles, I know the opposite is true. Journalists are the pro-vaxxers' best friends. As far back as I can determine, they have consistently blown up the supposed risks of *not* getting vaccinated. It's true, however, that the media position on this subject is not monolithic. From time to time, on banal TV chat shows, and in sound bites sometimes featuring celebrities, vaccines have been given a bad rap. And over the years some major exposés, few and far between, have appeared on television. Perhaps the most influential was "DPT: Vaccine Roulette" which aired on April 19, 1982 on WRC in the Washington D.C. area and then on NBC-affiliated stations around the country. This show provoked an avalanche of phone calls to doctors' offices and public health departments around the country. But people forget. How many older folks remember that program? How many people younger than fifty have ever heard of it? It's the *pro*-vaccine viewpoint that is repeatedly hammered — at times softly and subtly, at times forcefully — into impressionable American heads. Mainstream reporters who get too big for their britches don't last long, as the case of Valeri Williams makes clear. In 2003 Williams, an investigative reporter at WFAA-TV in Dallas, exposed the mercury-autism link, the dangers of the pneumococcal pneumonia vaccine Prevnar, and the shenanigans of the federal approval process of vaccines in general, which shook up government health regulators and vaccine manufacturers. The station started feeling the heat, but Williams was not intimidated. Then suddenly — poof! — she was gone from WFAA, and all her broadcasts on vaccines were deleted from the station's Web site. Business as usual in the USA.

I don't know what goes on in the executive suites of major media headquarters. Undoubtedly, personal matters come into play now and then. For all I know, the decision to air "DPT: Vaccine Roulette" was made after a shouting match involving one party whose child or grandchild became autistic after being vaccinated. Also, I can't explain why a pivotal book like *Evidence of Harm* was published by a major firm, St. Martins Press, and shot to the top of the *New York Times* best-seller list after being favorably reviewed in that newspaper, the most powerful of any media organ, whereas the great majority of books critical of vaccines are self-published or printed by obscure companies and totally ignored by the book review segment of the media. In any case, it's obvious that the media barons are not shouting about vaccine risks from the rooftops. They could if they wanted to. They could, if they so chose, spark an instant revolt by printing big block headlines like KILLER VACCINES: CHILDREN DIE WHILE PROFITS SOAR, or otherwise agitating the public. They have that power, and have used it over and over to wreak havoc around the world. There are several reasons why they will never do it in regard to vaccine dangers. Foremost is the symbiotic relationship between media conglomerates and the pharmaceutical giants; the former rake in billions in advertising revenues, while the latter rake in billions in resulting sales. The wealth and power of these corporate titans are intertwined, as evidenced by their membership in conspiratorial organizations like the Bilderberg Group and the Trilateral Commission. Furthermore, the media bosses, bottom feeders that they are, instinctively sense that the anti-vaccination movement is genuinely progressive, and therefore they are driven to destroy it. Lastly, I suppose that some of them actually believe that vaccines are beneficial. Most rank and file journalists, airheads by definition, surely believe this.

If there's any doubt as to the media's entrenched position on the vaccine issue, a quick scan of editorials that came in the wake of a measles outbreak traced to a "patient zero" at Disneyland in California in December 2014 should remove it. Some 125 were said to be infected by this carrier, whose identity and vaccination status were never established, nor were there any statistics on the vaccination status of those infected. In other words, there was no solid information to go on, and it's possible that many or most of the infected had been vaccinated against measles, demonstrating the failure of the vaccine. Nevertheless, this was an opportunity not to be missed for the vaccine pushers to jump all over their opponents, as conveyed in these headings: *Forbes*: "Measles Can Kill, And It's Spreading. Sue Parents Who Didn't

Vaccinate? Absolutely" (January 28, 2015), and "Anti-Vaccine Doctors Should Lose Their License" (January 30, 2015); *Time*: "Unvaccinated Families' Addresses Should Be Made Public" (February 11, 2015) and "Facebook Must Shut Down The Anti-Vaxxers" (February 19, 2015); *The Atlantic*: "Fight The Anti-Vax Movement With Lawsuits" (February 4, 2015); *USA Today*: "Jail Anti-Vax Parents" (January 28, 2015). None of this came out of the blue. Well before this phony measles "crisis" came along, the media elites had occasionally shown where they stand, as in these lines from an October 12, 2006 *New York Times* editorial: "The pendulum has swung too far toward letting parents opt out. States need to work harder at educating parents about the value of vaccination and should get tougher in granting exemptions." (The *Times* got its wish in 2019 in its home state when Governor Andrew Cuomo, a quintessential political whore, signed a bill ending religious exemptions in the wake of what the media sensationally reported as a new crop of measles cases at some private Jewish schools.)

Surpassing everything was the editorial "Preying On Parents' Fears," printed in the May 8, 2017 issue of the *Boston Herald*, a daily tabloid. This was a response to what the media claimed was another measles flare-up in the Somali community of Minneapolis. Setting aside here another critical issue — just who it is that's settling unwanted immigrants from Somalia in Minnesota at taxpayers' expense — this was another fake crisis that had to be milked. The editorial ended with these words: "These are the facts: Vaccines don't cause autism. Measles can kill. And lying to vulnerable people about the health and safety of their children ought to be a hanging offense." This was an open declaration that we anti-vaxxers deserve to be killed. It provoked an immediate widespread and furious response, so furious that it seems the *Herald* scrubbed it from their archives: I was unable to retrieve it.

I read all this with pleasure. It's good that the media have dropped all pretense of objective reporting. It's good that they're making more and more enemies with their malicious stupidity. It's good that they've revealed themselves for the snarling little communists they are. Now everyone with eyes to see and ears to hear knows that we're engaged in a real war.

❖

It would probably be unreasonable on my part to ask readers to do what I did to ring in the new century — take my television set out to the curb with the rest of the trash. Although I miss the escape of a ballgame or a nature show now and then, I've never looked back. Nor do I read any daily newspaper. I turn WCBS radio on for ten minutes each day at 7 AM to hear the "top stories," just to get a rough idea of what's happening in the country and around the world, dreading to hear that another U.S. aerial slaughter has begun. Ten minutes is all I can stand. For uncensored news and comment I go to my favorite hate sites on the Internet, which also keep me informed of all the lies recently reported by the mainstream media.

Much of what I've discussed in this chapter obviously has nothing to do with vaccines, though it has plenty to do with the System we live under that mandates childhood vaccination. I wrote it in the hope of unshackling the minds of those who can still *think*. I wrote it to give the reader a sense of perspective, to show that the mainstream media is the *last* place anyone should expect to find the truth about vaccines. What it really comes down to is the slow drip mental poison, like all the rest of the media mental poison, instilling the message that only morons or weirdos reject vaccines for their children — and nobody wants to be seen as a moron or weirdo. Are you strong enough to resist this poison? Or will you follow the rest of the cattle over the cliff?

The white man's Mau Mau

I USED to think that white people, on average, were much more intelligent and civilized than black people. I also used to think that blacks, particularly the natives of Africa, were much more ignorant than whites. I still believe these things, but I no longer use the word *much*. Sometimes, it seems to me, the most barbaric of human traits have simply become more refined in the white race. For example, it's common in Africa, during tribal wars or political uprisings, to kill one's enemies by hacking off heads or limbs with a machete, the African weapon of choice, or by burning, boiling or engaging in other methods of torture for personal pleasure. This kind of behavior is considered so deviant among armed white men in any circumstances that I cannot recall ever reading a single account of it (although I don't doubt it has happened on rare occasions). Yet white American airmen, in times of war, have incinerated millions of innocents with phosphorus or napalm, a death no less horrible than the cruelest African torture, the difference being that these killings are carried out from miles above in sophisticated flying machines and the killers never hear their victims scream in agony nor do they see a single blackened corpse.

As for ignorance, generally speaking white Americans are on a par with denizens of the Congo rain forest. Most Americans know as much about the movements toward independence in African colonies as a naked Dinka tribesman in the southern Sudan knows about the wars this nation waged in Korea and Vietnam, events that took place in the same time frame. Colonialism refers, of course, to the political control and exploitation by various European powers of much of the world inhabited by the colored races. This often brought stability and benefited them in a material sense, but it also brought social friction, and among many native peoples a smoldering resentment as well. This was certainly true on the continent of Africa, most of which was colonized in the twentieth century by the French and British.

In the two decades following World War Two, several African demagogues, spurred in large part by communist propaganda, appealed to the worst impulses of their people in a bid to drive out the white man. One such demagogue was Jomo Kenyatta, who led an underground movement called Mau Mau (pronounced to rhyme with "now now"), and who went on to become the first president of Kenya, when that east African colony became independent of Great Britain in 1963. In the 1950s, the Mau Mau sect, embraced by Kenyatta's Kikuyu tribe, struck terror into the hearts of everyone in Kenya — whites, Asians, blacks of other tribes, and above all Kikuyus who remained loyal to the British crown and actively opposed them. Thousands were butchered. The insurrection was discussed in detail in a book published in 1955 titled *Inside Africa* by John Gunther, an American journalist who spent nearly a year traveling around the Dark Continent. While in Kenya he moved around at considerable peril, escorted by a British constable, through an area known as a hotbed of Mau Mau activity. I quote at length to offer a broad picture of some of the unpleasant realities of Africa, which exist today the same as they did six hundred years ago, when the first Portuguese explorers penetrated the west African coast and recorded what they saw:

Mau Mau methods of murder are, as everybody knows, peculiarly atrocious. Aside from stolen firearms, the terrorists have three chief weapons: the panga, *a long, heavy knife like a machete; the* simi, *a double-edged sword shaped like a scythe; and the* rumgu, *a nail-studded wooden club. When the Mau Maus kill with* pangas, *the victim may be literally sliced to pieces or chopped to bits. One reason for this is that every member of the attacking gang must join in the kill by actual participation with a weapon, so that all are equally guilty and nobody can inform on anybody else. Often the eyeballs of the victim are removed, so that he cannot "see" after death who his killers were. The Kikuyus are profoundly superstitious people.*

Of all things connected with Mau Mau, the most exceptional — as well as horrible — is the oath. When a Kikuyu enters into the organization, or is forcibly inducted into it, he takes an oath, which is one reason why the movement has such a compulsive hold on people. If a member attempts to give up his allegiance, thus breaking his oath, other Mau Maus will do their best to track him down and murder him. Oath taking has always been part of the tradition of the Kikuyu tribe, and oaths per se do not have any sinister connotation. Methods of administering the hideous Mau Mau oath vary in different districts, but the general procedure is something like this, Candler told us. The candidate for admittance is brought at night to a lonely forest hut, and crawls through an arch of banana leaves, muttering an incantation seven times. The number seven is important, and refers to the seven orifices of the human body. Then the eyes are gouged out of a sheep, and placed over the candidate's eyes, temporarily blinding him, while he licks a mixture of blood, excrement, and juices from the sheep's eye. He repeats the oath, which is administered by a witch doctor or elder of the tribe, phrase by phrase; this commits him among other things to murder anybody who may be assigned to him as a victim, even if it is a close relative or his best friend. Other ceremonies take place too obscene for mention. An incidental point is that there is a sizable initiation fee, ranging up to 62.50 shillings ($8.75), the equivalent of a month's wages or more. This makes the oath doubly serious, and collection of these fees gives considerable economic power to the Mau Mau organization.

Somewhere in my voracious reading of all things African back in the 1980s, when I made three trips to that continent, I came across some of the ceremonies that Gunther considered "too obscene for mention." In one of these, the Mau Mau inductee was required to have anal intercourse with a mutilated, dying sheep. On other occasions, he was ordered to drink the "Mau Mau cocktail," a blend of human semen and menstrual blood. In some cases, he was forced to dig up the remains of his parents and eat their desiccated flesh. The unstated but obvious goal of these subhumans was to establish total domination over their fellow man by forcing him down to the most abysmal depths of degradation.

Like the Mau Mau initiation ceremony, the childhood vaccination is an exercise in submission and degradation, backed by the threat of force — and it too comes with a hefty fee. It, too, is an oath, unspoken perhaps, but an oath invoking supernatural powers just the same: "When I inject this potion into your child, he will be protected from terrible diseases." Or, more likely to be actually spoken: "If you refuse my shot, your child may be struck down by polio" (or measles or tetanus, etc.). There are differences, of course. The vaccination ritual doesn't take place in a filthy, smoky hut but in the spotless surroundings of the pediatrician's examination room. On the other hand, the contents in that little vial reserved for our children are, if anything, more disgusting than the ingredients mentioned above that were used in the Mau Mau oath. These contents, past and present, include mouse brains pulverized in a blender; emulsions of aged, infected rabbit spinal cord tissue; chicken embryo proteins; monkey kidney; calf fetus blood; dried powder of calf belly scabs; lung cells from aborted human fetuses; genetically engineered cow-human viruses; and horse serum antibodies. Added to this abbreviated list are traces of anti-freeze, embalming fluid, detergent and other chemical substances.

The other major difference is that foreign biological matter was administered to Mau Mau initiates by mouth, not by needle. When you swallow something harmful, the toxins pass from the intestinal wall to the liver, where they are neutralized if the amount isn't overwhelming, before returning to the intestine through the bile duct. Eventually they're excreted. But this natural purification system is bypassed when putrescent particles from other animal species, in addition to chemicals, are injected directly into the bloodstream. Of course, both practices are dangerous and unspeakably revolting. But, theoretically at least, the Mau Mau oath is less of a health threat than the vaccination oath, especially in the long run. And while nearly everyone recognizes bestiality for the abomination that it is, vaccination takes this crime against Nature one step further by introducing the genetic material of animals into healthy human cells.

❖

There are those who claim that Jomo Kenyatta had no involvement with Mau Mau, that it was a British kangaroo court that convicted him for terrorist activities and sent him to prison for six years. I seriously doubt this, but it's beside the point. No one disputes that Mau Mau slaughtered thousands of innocents in cold blood, and the reminiscences of Kenyatta's closest associates, together with his public appearances, all preserved on film, make it clear that he was not a nice man.

It really would have been interesting, in the 1950s, to have witnessed a panel discussion with Jomo Kenyatta and some of the polio vaccine chasers. Kenyatta was no savage, at least not on the surface. He attended college in London, wrote a book, married a white woman (one of four wives) and became a smooth-talking politician. Indeed, he was a polished African gentleman, greatly admired by many white liberals — the same kind of idiots who in later years worshiped the Marxist revolutionary Nelson Mandela. Kenyatta would have fit right in sitting at a table discussing human progress with the likes of Salk, Sabin, and Koprowski. Yes, it would have been most interesting listening to their dialogue, observing their mannerisms and discerning their underlying motives.

Of course, this is just an idle musing on my part. But I'm sure of one thing: the deeds of Salk, Sabin and their brethren have been far more destructive to humanity than anything a missing link like Jomo Kenyatta ever dreamed up.

A deadly dogma

Note: this chapter compares aspects of the vaccine dogma to other dogmas that many people hold dear. Some may find my ideas upsetting or infuriating. Please read no further unless you consider yourself totally open-minded.

WE HUMAN beings are part of the natural world. We are animals and much less unique in our behavior than we would like to believe. One characteristic that does distinguish us from the rest of the animal kingdom is our highly developed brain. This has allowed a few of us to accomplish magnificent feats that the highest ape could never come close to duplicating. However, the human brain is a two-edged sword that has given others free rein to do all kinds of horrible things far beyond the capacity of any other life form. The human brain is the source of endless destruction of our natural environment, and the needless deaths of so many living beings, including our own. It has also has made us vulnerable to falsehoods and delusions of every description, some of which fall under the heading of dogma.

Superstition is a mild form of dogma. It's not as rigid, not as universal, and not empowered by bad government, as dogma often is. Astrology is a good example of superstition. It maintains that the positions of the moon, stars and planets at one's birth predestines that person to a particular temperament and to certain twists and turns throughout life. Conceived four or five thousand years ago, it was based on an image of the physical universe that was totally wrong. Nevertheless, it survives and even flourishes to the present day, kept alive by millions of horoscope readers, and even by men and women in search of a compatible mate born under a certain zodiac sign. But at no time did astrology ever ascend to an official position of social or governmental influence; it remains what it always has been — harmless nonsense shunned by most rational people.

It's when superstition escalates into dogma that trouble starts. Dogma runs deeper and stronger, is imposed by established authority, and enforced, at the very least, by the code of social approval, and as often as not throughout history, by the lawman or the executioner. Except for those rare individuals who are capable of thinking flexibly and independently, dogma also provides a frame of reference to understand the inner and outer world, which is a powerful human need. Through the ages and among all human societies, a wide variety of dogmas have existed, in some cases at odds with other dogmas. Despite the fact that all dogmas are by definition falsehoods, most people need *some* fixed idea to give them their mental bearings. Put another way, most people believe what they believe not because they're interested in the truth, but because it makes them feel good. This is often conditioned by fear of punishment for thinking forbidden thoughts. Attempting to deprive people of their most cherished beliefs is a cruel act that can shatter their equilibrium, provoking anger, scornful laughter or frozen silence, depending on the individual personality. Yet it has been those philosophers, scientists, inventors and other rebels who were able to think *un*dogmatically — to think outside the convention of their times — who were responsible for raising man to the fragile plane we call civilization. It is they who have always posed a threat to the upholders and enforcers of dogma — a threat not only to their power and prestige but to their income as well. And since all dogmas are false, they always require some modification when the heat of critical inquiry is turned up high enough by those brave few willing to risk unpleasant consequences.

Western man's greatest dogma, of course, has been Christianity, the core of which is the divinity of Jesus Christ. (For what it's worth, I believe in evolution not creationism, and I do believe that a man of extraordinary character named Jesus walked the earth in ancient times, but was as mortal as the rest of us.) It may well be that organized religion, however fallacious its doctrines, exerts a positive, perhaps even an essential stabilizing influence on every society. It also has inspired some of mankind's most beautiful creations, as exemplified by Europe's soaring Gothic cathedrals, Bach's cantatas, and Michelangelo's paintings and sculptures. But here I set these issues aside. The issue before us is how an idea so illogical, even after mutating into countless schisms over the past two thousand years, and being stripped of all its legal might these past two centuries — after spawning so much neurosis, psychosis and bloodletting, which reached its peak in the Thirty Years' War of 1618-1648 pitting Catholics against Protestants across nearly the entire continent of Europe, which wiped out half the population in some areas — still endures in various dogmatic forms. For example, some forty million Americans, mostly in the Baptist denominations, actually believe that the Jews literally are God's Chosen People, and a fair number of these even believe that destroying the perceived enemies of the state of Israel with nuclear weapons would be fulfilling biblical prophecy. The notion of Israel's sanctity has long infected the minds of several high-ranking people in government and the military; President Ronald Reagan, who really believed that we were living in the "end times" as foretold in the Book of Revelation, was a prime example.

Most Christians are not zealots or fanatics. For the majority, Jesus's divinity serves as a kind of security blanket, something to wrap oneself in when life becomes hard to bear. What is instructive here, however, is mankind's incredible inertia. Ever since that collection of fairy tales, massacres and half-baked history took shape as the Holy Bible, Christians have had the opportunity to read something that contradicts itself more than a hundred times, but how many have actually read it? Furthermore, those who believe that the Bible is the infallible word of God, and Jesus his earthly emissary, need to explain why, when he lost his temper with a fig tree just because it had no figs at the time, he cursed it and caused it to wither to its roots (Mark 11: 12-14, 20-21), and worse, why he said of those who rejected him as their ruler, "Bring them here and slay them before me" (Luke 19: 27). Is that any way for a Son of God to talk? And of course there are many more uplifting passages in the Old Testament, such as "Happy shall be he who takes your little ones and dashes them against the rock" (Psalms 137: 9). In fact, the Good Book is replete with accounts of mass murder on the scale of Hiroshima, carried out against the enemies of the ancient Hebrews.

The great astronomer Galileo, and Darwin two hundred fifty years after him, dealt Christian dogma a one-two blow which sent it reeling. It's still reeling. Yet, even though the dogma has lost its punch, it survives. As recently as 1859, when Darwin's *On the Origin of Species* was published, his ideas were violently attacked by learned men throughout the Western world, not because they were wrong, but because they refuted the creationist myths in the Book of Genesis, which at the time all respectable citizens were expected to believe. And even though it is no longer taboo to believe in evolution, the firestorm of controversy that Darwin ignited smolders to this day. Most people who adhere to any of the world's major religions don't want to hear about it.

It's equally remarkable that, at the dawn of the seventeenth century, a scientist of Galileo's stature had to go underground to escape the clutches of the Roman Catholic Church. So did Copernicus, whose observations and brilliant mathematical calculations forced him to conclude that the earth was *not* the stationary center of the universe around which all celestial bodies revolved, as the Church taught. Copernicus kept a low profile, deferring the publication of his findings until after his death. Galileo built magnificently on the foundation that Copernicus laid, but his work unfortunately coincided with the Inquisition. He constantly lived in fear of Church authorities and often found himself in hot water. In later life he was charged with heresy and put on trial. Knowing he would face certain imprisonment, and possible torture or execution, he recited an apology in which he "abjured, cursed and detested" his earlier writings in which he had contradicted the official Catholic dogma of how the solar system worked. As a reduced punishment, he spent the last eight years of his life under house arrest.

Giordano Bruno, a contemporary of Galileo, paid the supreme price for defying that dogma. This fiery, restless spirit traveled widely around Europe and England, studying, writing, lecturing, meeting royalty, fascinating some and infuriating others with his revolutionary concept of the universe. Bruno went much further than Copernicus, completely rejecting the static, Aristotelian explanation of outer space, which the Church had incorporated five hundred years earlier, and even lambasted Aristotle as "the stupidest of all philosophers." (Aristotle, for his part, fled Athens in 323 B.C. for upsetting the authorities there with his "lack of reverence for the gods," mindful that not long before him the ever-questioning Socrates had been forced to drink poison hemlock, his death sentence for committing a similar crime.) God, asserted Bruno, was not the thundering tyrant of the Old Testament, but a primordial energy pervading a universe of infinite space and heavenly bodies, endowing everything with freely developing forces of change and growth. Jesus, he said, was just a magician. This kind of insolence alarmed the authorities, who considered it very bad for business.

Fifty years earlier, Martin Luther, outraged by the rampant corruption of Church officials who were making personal fortunes selling "ticket to Heaven" indulgences to sinners, initiated the first Protestant sect of Christianity bearing his name, at the same time stemming the flow of illicit profits. But the Church's thriving trade in baptisms, marriages, tithes and real estate continued unabated.

Bruno became a wanted man. Eventually he was betrayed to the Inquisition by a malevolent nobleman named Mocenigo, who envied his phenomenal memory and intellect. Bruno was thrown into a filthy Vatican dungeon where, clad in rags and malnourished, he languished for seven years. From time to time he was dragged out and given a chance to recant, but unlike Galileo, he refused. Finally tiring of this impossible man, nine resident cardinals put him on trial for heresy, found him guilty, and sentenced him to death. On February 17, 1600, Giordano Bruno was burned at the stake on the Campo Dei Fiori in Rome.

There is a world of difference between the power and influence of the Catholic Church during the Renaissance and that of the contemporary Christian churches, but some things, including the profit motive, have not changed. Bruno was a very real threat to the priestly monopoly as well as the authority of dogma in his time; that dogma has run out of steam over the past four hundred years, and much of what remains is more social gospel than religious gospel, with a pronounced leftward tilt. But churches still compete with each other for customers — there are about 800 listings in the Suffolk County yellow pages — and when not ensconced in their mansions, America's top televangelists continue to preach about the Armageddon that never comes while flying around the country in their Lear jets.

It's foolish to imagine that men have outgrown the backwardness and closemindedness exhibited in ages past, other than to grant that modern day heretics are no longer burned alive. Dogma and heresy are still very much with us, and even though Christian doctrine has become totally disconnected from political power, new dogmas have stepped in to fill the vacuum. To really get a feeling of dogmatic authority and the fear and intolerance it engenders, we need look no further than the so-called Holocaust, which has replaced Christianity as the number one dogma in the Western world. What we are seeing here is the birth of a new religion, which should be obvious from the universal capitalization of the letter H in this word, a word which began appearing only in the 1970s to signify the alleged attempt by Adolf Hitler to kill all the Jews of Europe.

As the late professor Revilo Oliver pointed out in his book *The Origins of Christianity*, Christianity was a copycat religion modeled in almost every detail on Zoroastrianism, an ancient Persian religion which was based on a real man named Zoroaster who, like Jesus, was transformed by his followers into a savior. Now we see Holocaustianity imitating Christianity the same way — the nation of Israel as the Jewish people's resurrection, Hitler as Satan, Holocaust shrines and museums sprinkled like chapels across the Western landscape, the sense of collective guilt for allowing the Jews to be "crucified," the social pressure to conform to the tall tale of the gassed six million, the persecution of heretics (euphemistically called "deniers"), and as with all major dogmas, the establishment of a lucrative racket, which has reaped billions in films, books, research grants and reparation schemes of all kinds, the prime beneficiary being the state of Israel itself.

All dogmas crumble under the weight of serious inquiry. And it is the simplest questions that cause the first cracks. Even as a child, upon being taught in weekly Catholic religion class that all the good people in the world who had been born before Jesus came to Earth were not "saved" and had to spend eternity not in Heaven but in a neutral place called Limbo, I remember thinking, "What kind of all-knowing, all-understanding God could be so unfair? Why should all those people be deprived of eternal happiness in Heaven because of something they had no control over?" (I imagine the Church has replaced this and a lot of other hooey I was taught with more persuasive hooey.)

So it is with the Holocaust. We might ask, for example, if the Germans, so renowned for their efficiency, really intended to exterminate all the Jews in Europe, why were so many of them in German-occupied areas still alive at the end of the war? Arthur Butz, author of the landmark study *The Hoax of the Twentieth Century*, asked this question many years ago. It has never been answered. British historian David Irving has asked many more troublesome questions. The multilingual Irving has written more than thirty books on World War Two, though none focus specifically on the Holocaust. He has spent much of his life studying the archives of that era in institutions throughout Europe and America, and has an encyclopedic knowledge of the personal lives of famous figures associated with the Second World War, particularly from Germany. Why, he has asked, are there documents with Hitler's signature authorizing the summary execution of commissars on the eastern front, and as a temporary measure, the euthanasia of Germany's institutionalized mentally incurable at a time when hospital beds were urgently needed for wounded soldiers — yet not one that mentions the mass killings of Jews? Why, furthermore, in all the recorded comments about Jews made to his inner circle, such as those published in the book *Table Talk*, does he speak only of expulsion and never extermination? The late French revisionist scholar Robert Faurisson once challenged the dogmatists in the simplest way. He said, "Draw me a Nazi gas chamber."

No one ever did. One rebuttal, by an establishment historian named Jean-Claude Pressac, was attempted, which in turn was demolished by Faurisson. For those interested, the minutiae can be found online.

No revisionist historian, incidentally, would argue that many Jews, though certainly not all, were uprooted and transported to work camps, and that many died of starvation or disease, as did a far higher number of Germans, after Germany's infrastructure was destroyed by sustained Allied carpet-bombing, and that Hitler was determined to expel most Jews from Europe and eventually transport them either to Russia or Madagascar — much as they had been expelled from European countries or territories nearly fifty times in the past thousand years. What these modern heretics dispute is the dogma that Jews were the victims of an attempted wholesale extermination, that mass execution gas chambers were employed by the Nazis, and that anything close to six million perished.

Most Americans, who take our First Amendment for granted, or don't even know what it means, are unaware that it's a felony in several European countries and also in Canada to defy Holocaust dogma. This in itself is a strong hint as to where the truth lies. For an act of heresy committed in 1989, Irving was arrested in Austria on November 11, 2005 and imprisoned over a year. He could have been locked up for twenty years in that country, where the penalty for heresy is severe. He's not the only one. For openly challenging the dogma, thousands have been prosecuted on charges of "inciting racial hatred" and "defaming the dead" in Austria and Germany, where the Alice-in-Wonderland courts hold that "truth is no defense," and some defense lawyers have actually been threatened with prison for defending their clients too vigorously! Robert Faurisson, though not jailed, was dragged into court and fined several times. Exhausted from the struggle for historical truth, the late Ernst Zundel, a Canadian citizen and an incorrigible heretic, endlessly harassed by legal authorities and brought to trial in 1985 and again in 1988, moved to rural Tennessee, thinking America was the land of the free. On February 5, 2003 he was arrested by federal agents on the pretext of missing an immigration hearing, shipped back in chains to Canada where he was locked up for two years, then flown to Germany, the country of his birth, where he spent five more years in prison before and after being tried and convicted. Such is the treatment in store for the Brunos of our day; and such is Jewish media censorship that most Americans know nothing about these events.

While dogma is enforced from above with stiff legal language, on the street level it's enforced with violence. Faurisson was savagely beaten by three Jewish thugs while walking his dog in broad daylight in a Paris park on September 16, 1989. The home of Ernst Zundel, who endured endless death threats, was firebombed, and he and his supporters were spat on, screamed at and punched while making their way to the Toronto courthouse for his trial. On July 4, 1984, the southern California warehouse of the Institute for Historical Review, a major publisher of Holocaust heresies, went up in flames and most of its inventory was reduced to ashes, after repeated threats and harassment of its staff. No one was ever arrested for this crime, which calls to mind the burning and sacking of the library in Alexandria, Egypt, the ancient world's greatest repository of knowledge, at the hands of Christian fanatics in 391 A.D.

All dogmas are also characterized by a contradiction of physical laws. For example, revisionist historians have pointed out that the cremation of a single human body is a slow process, and it would have taken thirty years or more to burn millions of corpses in the limited crematoria facilities that existed — just as they still exist in all developed countries — in the concentration camps. Other phenomena, as related by supposed eyewitnesses, are reminiscent of the silly stories that abound in the Bible. A former Auschwitz inmate who testified at the Zundel trial claimed that he could tell who was being cremated by the color of the flames rising from the chimneys; green flames meant Hungarian Jews, blue flames Polish Jews, and so forth — aside from the fact that chimneys emit smoke, not flames. The late Nobel Peace Prize winner Elie Wiesel, who more than anyone else popularized the term "Holocaust" and spent much of his life spreading the word about that alleged event, claimed that geysers of blood from murdered Jews spurted out of the ground for months. This buffoon — and a vicious buffoon at that, who openly called for the imprisonment of deniers — was exposed as a shameless liar in a blog called "Elie Wiesel cons the world," and in the devastating book *Holocaust High Priest* by Warren Routledge. Wiesel spouted so much nonsense that in a sane world he would have been laughed into oblivion. That six consecutive U.S. presidents felt compelled to grant him a personal audience, and grovel at his every word, speaks volumes about Jewish supremacy in America.

Although sidestepping the question of whether or not the Nazis murdered millions of Jews, Norman Finkelstein, in his book *The Holocaust Industry*, exposed the Holocaust for what it is, namely an international free-for-all of swindlers and shakedown artists. His book did not endear him to his fellow Jews. Another courageous American Jew, David Cole, wearing a yarmulke and posing as a curious little *schlemiel*, interviewed the curator of the Auschwitz museum, Franz Piper, also Jewish, and in an unguarded moment Piper admitted that the so-called gas chambers were actually built by the Soviets in the 1950s. This was consistent with my own visit to Auschwitz in 1991, during which I saw several crude and ridiculous atrocity exhibits but nothing specified as a gas chamber, the alleged existence of which was purposefully left vague for gullible tourists. Cole's narration of his trip to Auschwitz and his interview of Piper, which can be seen and heard on YouTube (or could when this was written; the Jewish-owned platform is increasingly censoring its users, many of whom have fled to Gab or Bitchute), is a dagger in the heart to the dogma. Everyone should watch it. Furthermore, in response to the onslaught of sound revisionist critiques of the Auschwitz legend, and shortly after my visit, a plaque commemorating the supposed murder of four million at this camp was taken down and replaced by another plaque stating that about one and a half million innocents were executed — which is still light years from the truth. Yet the hallowed figure of six million Jews killed by the Nazis remains, and woe to he who challenges it! This shows that the primary feature of dogma is fear and intimidation, certainly not reason — and since dogma, like death and taxes, is inevitable, fear and intimidation to varying degrees have always cast an ominous shadow over human events.

As far as major historical dogmas go, the fable of the six million gassed Jews is a very young one, but it became engraved in stone throughout the Western world shortly after its inception due to the near-total Jewish control of judicial and educational institutions, and practically all organs of the mainstream news and entertainment media, particularly in America. All that's left standing of the ancient lineage of Christian dogma, which in medieval times contained numerous components but over the centuries has splintered off in so many muddled directions, is a dim belief that Jesus was a flesh and blood embodiment of the force governing the universe that people call God.

Not only do all dogmas fall apart when closely examined, they shrivel under the naked critical eye. I have already mentioned my trip to Auschwitz and the lack of any installation designated as a gas chamber there or at the ancillary camp of Birkenau, which I also visited. Earlier, in 1978, I visited Dachau as well, located just outside of Munich, and was quite surprised to read, on one of the signposts, something to the effect that while the Nazis had planned to exterminate Jews here, the plan was never implemented. This, in fact, was my first inkling that there was something screwy about the Holocaust story because I distinctly remember being taught in elementary school that Dachau and Auschwitz were the worst of the Nazi death camps. In fact, there was an official U.S. Army photograph, widely circulated and reprinted after the war in many school textbooks, including mine, showing an American G.I. standing before the door of a clothing disinfection chamber, used to kill disease-spreading lice, always a concern in the camps — a device transformed by propagandists into an execution gas chamber — on which a skull and crossbones appeared, with the hours of operation, and the words (in German): "Caution! Gas! Life Danger! Do not open!" The caption under the photograph read: "Gas chambers, conveniently located to the crematory, are examined by a soldier of the U.S. Seventh Army. These chambers were used by Nazi guards for killing prisoners of the infamous Dachau concentration camp."

But, I later learned, the dogma was quietly revised in 1961. The new and improved dogma — despite all the lurid atrocity stories that already had been spread around the world, despite all the confessions that had come out at the Nuremberg Trials (many of them extracted by torture, or offered in desperation to avoid the noose) — was that the Jews had been exterminated only in German-occupied Poland, not in Germany proper. These days the dogma is once again on the run. With so many untenable stories about Auschwitz having been exposed, the focus has shifted to imaginary death camps in places like Treblinka and Majdanek, names I never heard when I was young.

Likewise, many layers of Christian dogma have long been cut loose, though at a much slower pace. Back in the thirteenth century, St. Thomas Aquinas, a towering intellect in his time — though also a rather intolerant one, who wrote that heretics should be put to death — codified the dogma in his *Summa Theologica*, a 21-volume treatise on God and all creation, which synthesized divine revelation and Aristotelian logic, and upon which Catholic doctrine is largely based to this day. His ideas came to him while sitting within the stone walls of a Roman seminary. Nearly four hundred years later they were blown to dust by what Galileo saw through his powerful and magnificent telescope, and by what Darwin discovered, not in the gloom of a secluded study, but during his five-year voyage aboard the *H.M.S. Beagle*, especially his excursions in the coastal and interior regions of South America, where the record of life, past and present, existed in visible profusion. *On the Origin of Species* is remarkable in its totally open-minded and *un*dogmatic line of reasoning, Darwin never hesitating to air the nagging doubts and problems which he anticipated would come up against his theory of evolution, some of which he admitted that he himself could not resolve.

❖

In the final analysis, the two dogmas I've been discussing aren't going anywhere anytime soon. The heart of Christian dogma common to all denominations that has remained unchanged for two millennia — Jesus Christ as savior of mankind, as the only physical manifestation of God ever to appear on our planet — will continue to fill the spiritual needs of some two billion earthlings. And barring a revolution that will put an end to Jewish control of the boob tube and all other mechanisms of mass mind control, the six million Holyhoax will continue to reign as the supreme dogma of Western man.

However, when people can physically see or otherwise strongly sense a pack of dogmatic lies that's been foisted on them, such lies rest on a shaky foundation. The dogma of Communism, to be discussed further on, was so plainly untruthful, such a gross insult to even average minds, that it could be enforced in most countries only by terror, by the constant fear of imprisonment or execution — and still it was watered down or totally evaporated nearly everywhere in a relatively short span of time as world dogmas go.

The dogma of the equality of the races is probably the second most powerful one in America and in much of the West today, and it's an amusing one because while nearly everyone pays lip service to it in public, in private a large number of white people don't believe it. What makes it even more amusing is that in Africa, very few natives have heard of this dogma and even fewer would take it seriously. On the Dark Continent, whites supervise virtually all the schools, hospitals, factories, airports, industries, and utilities where they exist — I saw this everywhere in my travels through eighteen countries in sub-Saharan Africa — and not only is this situation not resented by the vast majority of blacks there, it's accepted as perfectly natural. The famous German missionary doctor Albert Schweitzer, who spent most of his long life tending to the sick at a remote jungle hospital in the west African nation of Gabon, and who knew Africa as well as anyone, put it best when he said, "The superiority of the white man is so obvious to the African native that it ceases to be taken into account." That's one of the truths about modern Africa you'll never hear.

Of course, you don't need to travel to Africa to see through this dogma. Nor is it necessary for me to belabor the fact that numerous comparative anatomies have shown that the Negro brain is smaller, lighter and less complex than the Caucasian brain. Unless you have tunnel vision, all you need to do, if you live near a black area, is go there and have a look. If you live here on Long Island, a twenty minute stroll with eyes open from Garden City to Hempstead, or Merrick to Roosevelt — to offer just two examples — provides visible proof that the races are not equal in any sense. As everyone around here knows, it would not be wise to walk through Hempstead, Roosevelt and other black towns on Long Island after dark. So daffy are the times in which we live, incidentally, that the previous sentence would be construed by most journalists and educators as racial hatred. Having been acquainted with many black people in my life, I freely admit that many are deserving of respect, even though I choose not to associate with them. And certain exceptional individuals — I think here of Moise Tshombe, a political leader in the Congo during that country's descent into chaos and savagery in the 1960s — stand higher in character and intelligence than 99% of the white population anywhere. None of this changes the fact that blacks, on average, are less intelligent, have a lower capacity for civilization, and are more prone to violent behavior than whites, and probably all other races for that matter.

The dogmatists claim that racial differences are only skin-deep, and all races are inherently equal. For them, environment means everything and heredity means nothing, so that a nation of Zulus growing up in Elizabethan England would have given the world a Shakespeare, a Drake, and a Newton, only with black skins. Conversely, by their own logic, an English population originating in Africa should be living in mud huts and hunting with spears, which was hardly the case of the English, as well as the Dutch, in South Africa, who, in conditions of bare subsistence built a modern nation fully the equal of the one their ancestors left behind.

Not only are racial distinctions plain to the eye, they stir powerful biological forces that are the root cause of social instability in countries where two or more widely dissimilar races live next to each other, as any American not in a catatonic trance should know. Even within what is loosely called the white race, there are noticeable differences, both physical and temperamental, between what anthropologists refer to as the Nordic and Mediterranean subraces, which correspond to northern and southern Europe, respectively. Another subrace, the Alpine, mainly inhabits the central and eastern regions of the continent. Some split our race further into Dinarics and Eastern Baltics. Many whites are a blend of these subraces or the strains within them, but many also stand apart. Ten or twenty millennia of evolutionary sifting in the harsh, unforgiving northern winters have given the Nordics a clear edge, though, again, we're speaking of averages, and there's plenty of overlap in genetic quality among all whites. This Nordic edge is evinced in the somewhat more orderly and industrious northern countries, and the fact that southerners go north to find better-paying jobs, while northerners go south to take vacations, as I've seen repeatedly in my travels all over Europe, while hardly ever seeing the opposite — and I say this as a man of southern Italian ancestry. Actually, the subject of racial differentiation in human beings fascinates me, but any open inquiry into it will be met by ear-splitting cries of "racism" from the army of academics, pundits and demagogues who would have to find an honest job if their dogma were to dry up.

As I've said, there's a very great number of whites in this country alone who don't take the dogma of racial equality seriously. But it's impossible to openly say so. If you're a teacher, you'll quickly find yourself on the unemployment line if you dare to discuss what Albert Schweitzer said about racial differences, even though he did more to alleviate the sufferings of blacks with his own hands than anyone who ever lived. It would be career and social suicide to talk about differences in brain structure when next year's high school S.A.T. scores come out. As happens every year, the scores in Hempstead, Roosevelt and other towns with high black populations are embarrassingly lower than in white towns, with the usual wringing of hands and evasion of essential facts. Anyone publicly stating that blacks on average are innately inferior to whites in cognitive intelligence and capacity for social order will be body-slammed by the media, ostracized by their colleagues, likely fired from his or her job, and possibly assaulted and threatened with death. In recent decades, some or all of these things have happened to many people, including scientists like Arthur Jensen and William Shockley (the co-inventor of the transistor), who studied the biology of race in painstaking detail and concluded that racial differentiation is primarily a matter of genetics and not environment. In a scene reminiscent of the Inquisition, Jensen, a professor at the University of California, was actually summoned to a videotaped tribunal before his liberal colleagues in 1969 and made to defend his position. Since then, the dogmatic strictures surrounding the race issue have become even more narrow.

❖

Now, let's take a look at the medical dogma of vaccination — the doctrine that vaccines have been a wonderful blessing for mankind, that they have greatly eliminated the scourge of disease, and their benefits far surpass the risks they carry. On the scale of dogmatic authority, vaccination occupies middle ground, that is to say, a majority of people unconditionally swallow these lies, but there have always been plenty of cracks in the edifice, and these cracks have become wider with the piling on of so many new childhood vaccines in the last thirty years. Still, out of fear, ignorance, or some other reason, even many people who are well aware of the dangers that vaccines pose and have been fighting the good fight to get the truth out, will not challenge the dogma. For example, in the introduction to his excellent and powerful book *Evidence of Harm*, which delved into the connection between mercury and autism, David Kirby wrote:

This is not an antivaccine book. Childhood immunization was perhaps one of the greatest public health achievements of the twentieth century, and vaccines will continue to play a crucial role in our lives as we enter an uncertain age of emerging diseases and potential bioterrorism. Some parents, fearing harmful effects, have been tempted not to vaccinate their children. Most people would agree that this is foolhardy and dangerous.... When vaccination rates fall, disease rates rise.

I assume he made these assertions to placate the editors at St. Martins Press, or because he really believes them. In either case, it's confusing and inconsistent in light of the book's thesis, a blemish on an otherwise superb piece of research.

Valeri Williams of WFAA-TV in Dallas, previously mentioned, who did a series of reports in 2002 that lifted the veil on government malfeasance regarding vaccine safety, and who was in turn fired by WFAA and had her videos and transcripts deleted from the station's website, began her narrative with these words:

From the outset it must be stressed that this report is not anti-vaccination. Every person interviewed for this story believes in the importance of having children immunized.

Opposite the dedication page of her gripping book *A Stolen Life*, Marge Grant, one of the unknown and uncountable parents in this country whose life was forever changed by vaccines when her son became a quadriplegic after a routine shot in infancy, wrote:

The author and publisher are not opposed to safe and effective vaccines. Our point is that parents should be accurately appraised (sic) *of the inherent vaccine risks so that they can be allowed to give informed consent or have the opportunity of withdrawing from the system without penalty if they do not wish to expose a child to those risks.*

I wholeheartedly agree with Mrs. Grant on freedom of choice, but as I've made clear in this work, the term "safe and effective vaccines" is an oxymoron. Unfortunately, many parents of vaccine-damaged children — really good people like Marge Grant and her husband, who reconstructed their lives to take care of their son — echo this line.

Even brilliant and sincere physicians who, through their work, are well aware of dangers associated with routine childhood vaccination, genuflect before the dogma. Dr. Andrew Wakefield is widely known, and has been widely attacked, for his discovery of a connection between the MMR vaccine and the onset of autism via the gastrointestinal tract. Yet he opened his testimony at a congressional hearing on the rising epidemic of autism, held on April 6, 2000, with these words:

Mr. Chairman and members of the committee: The purpose of this testimony is to report the results of the clinical and scientific investigation in a series of children with developmental disorders, principally autism. Nothing in this testimony should be construed as anti-vaccine; rather, the author advocates the safest vaccination strategies for the protection of children and the control of communicable disease.

Dr. Bruce West, whose newsletter *Health Alert* I have subscribed to for many years, and for whom I have the highest respect, also bends the knee. Dr. West, who frequently denounces vaccinations in his publication, and who once wrote, "You'd have to tie me down and beat me unconscious before I would allow my son to be shot up with any flu vaccine," came up with this *non sequitur* in the January 2007 issue: "There is little doubt that vaccines have done some good. But the true story of modern vaccination is a sordid one indeed." Little doubt that vaccines have done some good? And the story of only *modern* vaccination is sordid? Nothing from his pen that I have read elsewhere conforms to this statement. Was this a way of showing he's not *totally* opposed to the prevailing dogma?

❖

A striking feature of all major dogmas is that they have been administered by every variety of human scum, under whom rivers of innocent blood have flowed. Communism, its parallels with mass vaccination, and some of the monsters who pretended to put it into practice, is the subject of the next chapter; the psychopaths behind the vaccination dogma will be covered later in these pages. Taking second place in history's genocidal sweepstakes is Christianity, mainly, but by no means exclusively, while dominated by the Roman Catholic Church. One would have to search hard to find a more motley collection of freaks and assassins than among the popes of old. Wars, conquests, persecutions and burnings of heretics and witches aside, the most ghoulish episode of papal behavior was the Cadaver Synod of 897 A.D., when Pope Stephen VII had the body of his predecessor, Pope Formosus, dead for nine months, exhumed and "put on trial" for various transgressions after being propped up on a makeshift throne in the church of St. John Lateran in Rome. Screaming at the rotting corpse, Stephen eventually found Formosus guilty on all counts, nullified all his acts and ordinations, and had three fingers of his right hand that had been used to make blessings cut off, before ordering him reburied in a common grave. There then followed a vicious tug of war, lasting seven years, between Formosus's friends and enemies, including succeeding popes, in which his remains were dragged through the streets of Rome, dumped with weights in the Tiber River, fished out and reburied with full honors, re-exhumed, beheaded, dressed in papal vestments and given a decent burial yet again. All for the love of God.

The phantom gassed six million have always served the state of Israel as a bludgeon to stave off criticism of Zionist terrorism that has killed or dispossessed millions of Arabs. As far back as the World War Two era, before Holocaustianity became the *de facto* Western religion and even before it was called the Holocaust, the embryonic dogma, fueled by Old Testament vengeance, was the driving force behind the *real* Holocaust — death by fire — visited upon civilians in scores of German cities by British and American airplanes, and the unknown postwar war against the German people which amounted to history's greatest mass expulsion during which millions starved. These tragedies wiped out twenty percent of Germany's population. Every prime minister of Israel has invoked the Holocaust, each one has condoned killing innocent Arabs, some quite openly, and a few, like Menachem Begin and Ariel Sharon were truly bloodcurdling fanatics, compared to whom people like Elie Wiesel and Simon Wiesenthal seem normal.

In the same vein, wealthy race-agitating pimps like Al Sharpton and Jesse Jackson, both professed Baptist ministers, are small fry when viewing the dogma of racial equality as a whole. The crackbrained notion that all human beings of all races are innately equal has its origins in the French Revolution which, soon after it got underway in 1789, regressed into unrestrained mob violence and terror, symbolized by the guillotine and mass barge drownings in the Seine. The terror was supervised in large part by Georges Danton and Jean-Paul Marat, both hideously ugly creatures. The madness of this, the first revolution of its kind, whose slogan was "Liberty, Equality and Fraternity," made its way to the French colony of Haiti, where black slaves revolted and butchered every white man, woman and child unable to escape by boat, some five thousand of them — one of those tidbits of history they don't teach in school. The zeal inspired by this dogma has flared up sporadically throughout American history, from the gruesome Nat Turner slave rebellion of 1831, to the Civil War and shameful Reconstruction era — another important historical event kept out of the classroom — down to the race riots that have erupted in numerous American cities in the past century, reaching the peak of violence in the late 1960s. Since all races are inherently equal, and since there's such a clear disparity in the quality of life between white and black, it follows in many black minds that the white man, with his cruel institution of slavery in the past, and his discriminatory practices in the present, is the root cause of their problems. This is the message that has been drummed into them from an early age, so how can it be otherwise? Anyone who thinks that the genocidal mentality unleashed during the French Revolution (which, while not overtly racial, featured the lowest types of white rabble dragging down and killing whites of naturally aristocratic bearing) is a relic of ages past, should check out some of the more radical black websites and racially charged, profanity-laced exhortations to mass murder easily found on the Internet.

Just as liars have a way of getting along with each other, dogmas have a way of converging. It's no surprise that Sabin's compulsory polio vaccine experiment on all young Russians took place in a hardline communist dictatorship. Holocaust lies and communist lies, both Jewish at their source, interfaced easily, and racial equality was a staple of communist doctrine. The racial equalitarian dogma, so loudly championed by the abolitionists who played a key role in pitting North against South during the build-up to our Civil War, has meshed smoothly with recent variations of Christianity. Nor have any of the major Christian churches in America ever been ideologically opposed to the communist subversion of yesterday and the Holocaust hucksterism of today. As often as not they have found common ground. Earlier I quoted William "Parson" Brownlow, the preacher appointed governor of Tennessee during Reconstruction, who wanted to exterminate every man, woman and child south of Pennsylvania.

All in the name of brotherhood and the love of Jesus.

❖

Objective truth, the mortal enemy of dogma, packs a wallop. There is something about the calm way it expresses itself that is recognized even by those who fear and hate it. Those with the truth on their side do not shriek, assault or imprison their opponents. Evolutionists never tried to stop people from going to church; revisionist historians don't want to jail self-described Holocaust survivors for publishing wild stories; I'm not trying to interfere with parents who want their children vaccinated, even though I think they're making a big mistake. Truth has never hired enforcers to make people believe it. Above all, truth is timeless. No astronomer or biologist has shown Galileo or Darwin to be wrong; only two or three established historians have made a feeble attempt to refute the revisionists. On the other hand, the upholders of dogma squirm and scurry about or crack down when the light of truth shines too brightly on them. Christian dogma has undergone so many changes in two thousand years that an Augustine monk wouldn't recognize what passes for his religion today. Indeed, Christianity was never spread by peace, love or rational thought but by the sword. If the ancient Marcionites had beaten back the Albigensian Crusaders in the thirteenth century, there would be no Old Testament in the Bible today. In their brief reign thus far, Holocaust Pharisees have had to modify their dogma several times in response to the findings of honest historians. The number of the imaginary gasees at Auschwitz has been officially and drastically lowered, as mentioned, and the claptrap about Nazis making soap from Jewish fat and lampshades from Jewish skin, which I was solemnly taught as a schoolboy, has been quietly abandoned.

In the annals of history, dogmas come and go, popping up in different eras and among different cultures, then fading away. The Marxist dogma, brutally enforced and as false in its proclamations as racial equality, lasted not much more than a century, though like a 95-year-old man on life support, a handful of true believers keep it from expiring. Holocaustianity will continue to hold sway over the vast majority until the Jewish stranglehold on the news and entertainment media is broken, and perhaps long after that. The most ancient of dogmas, the Catholic branch of Christianity, admits of its errors only with great passage of time. It wasn't until 1835 that Galileo's works were dropped from the Church's *Index of Prohibited Books*; not until 1992 that Pope John Paul issued a declaration that the great scientist had been treated unjustly. Only after the last vestiges of Catholic political jurisdiction disappeared did a few of Bruno's admirers take action, in 1889 erecting a monument in his honor on the very spot where he was fed to the flames. And while in 2000, the four hundredth anniversary of his execution, a prominent cardinal declared it a "sad episode," he still defended Bruno's persecutors. The theory of evolution still gives Christians of all denominations fits, even though there have been lame attempts to integrate it with the Jesus legend, most notably by the French Jesuit priest Pierre Teilhard de Chardin, whose works were censured by the Vatican.

In the case of vaccination, the dogma has undergone countless revisions since Edward Jenner, when confronted with the failure of his cowpox vaccination to prevent smallpox, started making excuses. Since then the nomenclature games, the statistics juggling, the switching of ingredients, the revamping of doses and schedules, and the lies and cover-ups have been carried out past the point of absurdity. It's a dogma that has come under steadily increasing fire, but it's still an extremely lucrative one that spans the globe, most of all here in America, the fear of the crusader's sword having been replaced by the fear of the policeman's gun. Ugly threats and suspicious deaths have also entered the picture in recent years. Given the slothful nature of human minds, the lying media, the best public relations firms that money can buy, and the uncertain threat of force, the vaccine dogma may be around for a long time. That would be a disaster. It's up to a few of us, and especially new parents, to push back as hard as we can, and if it comes right down to it, to kill, the way the brave Hungarians killed in their failed revolt against the Communist beast in 1956. For the moment, in these still peaceful days in America, the dogma will continue to inflict pain, suffering and heartbreak on a daily basis, and will continue to reap millions of sad stories that will be known to only a few.

All for the love of our children.

The tragedy of Ignaz Semmelweis

ONE OF the most basic practices of hygiene, a practice most of us have never thought twice about, is washing one's hands with soap and water after touching anything that might be contaminated by germs. Yet, incredible as it seems, until 170 years ago no one had made the connection between dirty hands and the spread of disease.

The first man to do so — at least publicly — was Ignaz Semmelweis, a Hungarian doctor. Throughout the Europe of Semmelweis's day, the mid 1800s, countless women died of puerperal fever, a mysterious disease that occurred right after childbirth. Mortality rates were as high as twenty-five percent in some cities. It was common knowledge that this disease was rampant in hospitals, so many women chose to give birth at home, which was a far safer option. They knew nothing about medicine, but they were afraid of doctors and hospitals, and rightly so.

One thing struck Semmelweis, who was an intern at Vienna General Hospital: women seemed to come down with puerperal fever after their babies were delivered by doctors who had just performed autopsies, and had traces of blood and pus on their hands. In his mind there was a definite cause and effect, even though he couldn't explain it scientifically. The best he could do was attribute it to "humors," an ancient term for vital fluids which, he theorized, were carried from room to room by doctors who neither changed their stained smocks nor washed their hands.

To test his theory, he set up a separate delivery ward and required all doctors to wash their hands in chlorinated lime solution and change into fresh, clean clothes before entering. The results were instant and dramatic: the death rate from childbirth plummeted from twelve percent to two percent. His superiors, however, were not impressed. Everyone "knew" that puerperal fever was non-preventable. How dare this upstart come along and suggest they were not only unsanitary but ignorant? But Semmelweis stood firm. The connection between mortality following childbirth and doctors' neglect of hygiene was just so obvious that nothing could make him recant.

The years that followed were marked by endless acrimony between Semmelweis and the arrogant medical authorities of the era. It's a story as old as man: the powers that be, their puny minds cemented shut, persecuting the rebel who comes along and upsets the dogmatic applecart. Harassed and ridiculed, Semmelweis left Vienna and took up residence at St. Rochus Hospital in Budapest, Hungary, where he achieved the same remarkable results. This great doctor personally saved more than a thousand lives. His troubles with the big shots of organized medicine continued, however, and he eventually cracked under the strain. At 47, he was forcibly committed to a mental hospital and died shortly thereafter, possibly as a result of being severely beaten by guards. Years later, the authorities conceded that he had been right all along.

Today, you'll find a respectful column or two in just about every account of Semmelweis's tragic life. Think about him and what he went through the next time you wash your hands after handling raw meat or using the toilet. And incidentally, one reason that American hospitals, in this twenty-first century, have become such hotbeds of infection is because overworked doctors, rushing through their rounds, *still* don't take the time to wash their hands between examining patients.

But the reason I've written this brief tribute to Ignaz Semmelweis is to point out that even the most simple and obvious truths are belittled by those who stand to lose money or prestige or both by their proclamations — that human nature being what it is, there are powerful people who would rather see millions suffer or die than admit they are wrong.

Yesterday it was an imbalance of humors or a psychosomatic disorder that killed women after childbirth; today it's a genetic defect that causes autism, streptococcal cellulitis that's behind sudden infant death syndrome, and a hundred other flimflams. I understand how Semmelweis must have felt, banging his head against the wall, wondering how supposedly intelligent people cannot see a truth as plain as their faces in the mirror. How much more obvious can it get that vaccines are harming not helping our children? That they are a fraud and a disaster? What more does one need to show for proof? When will it all end?

Vaccination is medical communism

THE READER born from about 1970 onward might find my use of the word "communism" throughout this book out of place. Having grown up during the so-called Cold War, with the brooding menace of the Soviet Union and the quest of its leaders to subjugate every human being in the world under a communist dictatorship — a menace I now believe to be more contrived than real — that word will always be part of my vocabulary. I often use it as a synonym for tyranny, and to describe those horrible human specimens who lust equally for political power and blood.

Outwardly the world today is much different from the world of 1965. The misery and degradation of life under communism, which at its peak covered one-third of the earth's surface, has, with the exception of North Korea, all but disappeared. I had the singular experience, as a traveler, of briefly observing life under communism in six east European countries in the late seventies, and again in North Korea in 2013. If I had to describe the atmosphere in one word, it would be "bleak." You could see it and feel it, more so in some countries than in others, Romania and North Korea being the worst.

There is no universally accepted definition of communism because it means so many different things to so many different people. This stems in part from the fact that *Das Kapital*, the major work of Karl Marx, the founder of communism, is one of the most unreadable and indigestible piles of printed matter ever published. This did not prevent a hundred years' worth of "experts" from proclaiming it one of the most influential books of modern times. There is wide disagreement, however, even among our experts, as to whether *Das Kapital* is a philosophy of history, an expose of social injustice, a blueprint for revolution or something else. For many of us non-experts who have read it, or tried to read it, the book is just endless verbiage — in a word, trash.

Many people are familiar with the core of Marxist theory, that once the workers of all nations, the proletariat, overthrow their capitalist oppressors, the world will undergo a transition period of unspecified duration until the state withers away and Utopia, a permanent state of near perfection, is reached. Many adherents of Marxism who rose to power published their own views and had their own ideas on how to overcome obstacles on the road to Heaven on Earth. I've read a little bit of Enver Hoxha and Kim Il-Sung, and several pages of Chairman Mao's famous *Little Red Book*. It's all stuffy nonsense, occasionally spiced with rants against the bourgeoisie, the fascist thugs, the counter-revolutionary bandits and so forth. All this intellectual dust omits one ugly fact: communist theory, applied to the real world, killed off well over one hundred million people in the twentieth century. It is the most murderous ideology ever conceived.

A feature of every communist takeover has been the execution or imprisonment of all persons considered opponents. This was not always carried out on the same scale. The victims killed off in Vietnam and the former Yugoslavia, for example, do not compare percentage-wise to the number of people put to death in Russia, China and Cambodia. Not only political malcontents, but those whose background was deemed to be in the "enemy class" and often many ordinary people considered bothersome were systematically disposed of. Russia and China accounted for the great majority of victims. In the vast Russian penal system, Joseph Stalin regarded human beings as machine parts needed to build the Soviet empire, then discarded when they were worn out, to be replaced by new ones. During the 1959 to 1962 "Great Leap Forward" — an insane scheme hatched in the brain of Mao Tse-tung (or Zedong) to rapidly industrialize China and totally collectivize the country's agricultural sector — some thirty million died of starvation, both unplanned and deliberately inflicted by depriving people of food, a method also used in the engineered famine of the 1930s which claimed about seven million lives in the Ukraine, known there as the Holodomor.

Few, I think, would disagree that communist theory has been thoroughly discredited. Actually, nothing about communism has ever been theoretical, any more than you can have a theory based on the belief that men can grow wings. Alexander Solzhenitsyn, the great Russian writer who spent eight years as a prisoner in the Soviet forced labor camps, and whose book *The Gulag Archipelago*, published in 1973, revealed for the first time details of life under the Stalinist terror, compared the way a communist explains the nature of man to a surgeon who uses a meat cleaver to perform a delicate operation.

The train wreck of ideas called communism or Marxism or dialectical materialism or Marxist-Leninist theory was always a distraction employed by the fanatical savages known as Bolsheviks who clawed their way to power — abortions of our species for whom shooting, drowning, torturing or starving people to death in vast numbers was a source of accomplishment and amusement. It seems incredible that such monsters could hold the equivalent of a White House cabinet post in a country like Russia. Here is a description of one such creature, a Jew as most of the Bolsheviks were, named Goloshchekin as it appears in *The Last Days of the Romanovs* by Robert Wilton, foreign correspondent for *The Times* of London, stationed in Russia in 1917 during that country's convulsive revolution:

He had been selected for rulership of the Urals with an eye to other than political activities. He was bloodthirsty in an abnormal degree, even for a Red chieftain. People who knew him at Ekaterinburg describe Goloshchekin as a homicidal sadist. He never attended executions, but insisted upon hearing a detailed account of them. He huddled in bed shivering and quaking till the executioner came with his report, and would listen to his description of tortures with a frenzy of joy, begging for further details, gloating over the expressions, gestures and death throes of the victims as they passed before his diseased vision.

It's foolish to think that all the tragedies of the twentieth century were a perversion of Marxism as it was originally intended. Like all the Bolsheviks who later took over Russia, Marx was repulsive. While he did not aspire to political office and never killed anyone with his own hands, he contributed absolutely nothing of value to posterity. A miserable man who hated the world, he barely provided for his seven children, who lived in filth and hunger — four died in infancy or at an early age — and aside from writing newspaper articles, which earned a meager income, he never had a job, instead surviving on handouts from his longtime collaborator Friedrich Engels. A review of a book published in 1962, *Karl Marx, Master of Fraud* by S.M. Riis, who as a young man had interviewed a family that lived next door to Marx, states that "they had regarded Marx and the bizarre creatures who frequented his house as a gang of 'thieves and liars.'"

But it is Marx's own words that are the most revealing. This is what Marx wrote in an article titled "The Victory of the Counter-Revolution in Vienna," as it appeared in *Neue Rheinische Zeitung*, a German newspaper he edited, on November 7, 1848: "The very cannibalism of the counter-revolution will convince the nations that there is only one way in which the murderous death agonies of the old society and the bloody birth throes of the new society can be shortened, simplified and concentrated, and that way is revolutionary terrorism." Engels penned many more violent threats against the social order of his day, excerpts of which can easily be found on the Internet.

Sometimes I pause and question the authenticity of some of these accounts and quotations. Is it possible that they were embellished, or even entirely fabricated? The most shocking quote I have ever come across was attributed to Lenin and appeared in a short, obscure book, *The Crime of Moscow in Vynnytisia* (subtitled *Testimony on the Murder of 9,439 Ukrainians by the Soviet NKVD*): "Three-quarters of mankind must die if necessary, to ensure the other quarter for Communism." Did Lenin really say that, or did somebody make it up? Is it really possible that, if he had the opportunity, he would have murdered three or four billion human beings? I must say I don't know, but considering that no one, to my knowledge, disputes the numerous genocidal crimes carried out under the communist banner, all of which were grossly underreported or entirely ignored by our "free press," it behooves every nation to stand guard against the vestiges of this ideology in whatever form they may appear, and there are plenty of these in America today.

Communism as a belief system is pretty much extinct, but the subhumans who were once drawn to it are still with us — human nature does not change in one hundred or one thousand years — and only a fool would deny that a fair number are employed at various government agencies. Indeed, from the latter years of the Second World War into the 1960s, Washington, D.C. was a hub of the international communist conspiracy, as John Stormer documented in *None Dare Call It Treason*. In addition, there are still plenty of low-IQ thugs in law enforcement and the military — criminals with badges and weapons willing to do their dirty work, as the Waco holocaust of 1993 clearly demonstrated. Some facts about this atrocity will never be known, including the veracity of one account that some individuals at the Department of Justice celebrated with high-fives and jokes about a "Texas-sized barbecue" as seventy-six innocent people were incinerated. But again, to think that there aren't plenty of Goloshchekins on the federal payroll right now is to be very naïve.

❖

Compulsory vaccination, as it stands in America, is the kind of policy one would expect when a central government run by the sweepings of humanity cracks the whip and state governments fall in line. There is nothing "free" or "democratic" about it — democracy being one of those fuzzy words that sounds nice but eludes definition — but rather a system with many features we associate with communist tyranny. It's not as barbaric as the extreme forms of communism mentioned above, but it still exhibits the characteristics of communism in practice: a gross ignorance of man's place in the natural world; an indifference towards the wants and needs of ordinary people; legions of dimwitted, uncaring bureaucrats; resentment and disgust among the populace. Above all, mass vaccination destroys people: directly in the case of the hundreds of thousands who have already died or been made invalids, and in a greater sense in the millions who will never be born because of the fertility damage and the physical or emotional incapacities suffered by the current generation that will prevent many of them from having children. That's what the future probably holds.

Whether or not my premonition is correct, anyone who examines the words and deeds of vaccine pushers will come up against the communist mentality. During a seminar on infectious diseases in 1992, for example, Dr. George Peter, the chairman of the American Academy of Pediatrics, gave a reason for his recommendation that all infants be vaccinated against Hepatitis B: "Children are accessible." In other words, children are small, they can't resist, and there are enough laws mandating vaccination as a requirement for school attendance so that their parents are unlikely to object. Dr. Martin Smith, another former spokesman of the AAP, offered an even more sinister directive when he said, "Children of the nation are soldiers in the defense of this country against disease." He didn't mention that soldiers die when they defend their country, but others have come right out and said that some children must be sacrificed "for the welfare, safety and comfort of the nation." It would be interesting to find out how many pediatricians think like this, and of these, how many have not vaccinated their own children. Of course, this pseudo-patriotic ploy is just another example of the twisted logic of vaccine advocates, because if vaccines worked, all that sensible parents would have to do is bring their children in to get their shots, then they wouldn't have to worry about diseases being spread by unvaccinated children — children of parents like me.

Most vaccine advocates are not communist butchers at heart. Many of them are just a reincarnation of the fashionable dupes of the 1920s and 1930s who really thought that the Russian experiment would evolve into paradise on earth. One of these, a famous journalist named Lincoln Steffens who visited the Soviet Union in 1919, exulted "I have seen the future, and it works." He later changed his tune. Lenin called these people "useful idiots," and there have been many such idiots who have spoken out in favor of vaccines. Hollywood, long a hotbed of communist types, has provided more than its share. There is something about these people that makes them blind to reality, if not warped to begin with. And they are not rare — not a hundred years ago, not now. Just look at all the people who voted for Hillary Clinton, a communist at heart no matter what she calls herself. Sixty-four million voters could not detect the inherent evil in this woman that so many others could sense even without being aware of her notoriously horrible behavior in private and the numerous felonies she committed during her political career; or, considering all the freaks that now swarm over America, perhaps a few million of these were communist in character themselves.

Social injustice, exploitation, inequality all must be wiped out to achieve an ideal society, the communist proclaims. And how is that to be accomplished? By killing everyone who stands in the way of that lofty goal, that's how. Diseases must be eradicated, the vaccinators tells us, so that we can look forward one day to a world of optimal health. And how will this be achieved? With vaccines of all kinds, and newer and better ones on the horizon. What this has already achieved, of course, is millions of wrecked young lives — and considering that an unknown number of vaccine doses already administered are ticking time bombs, many more will eventually be destroyed. Actually, many of these vaccine advocates, like communist fanatics, are not motivated by any noble sentiment, but rather by a lust for power, by an urge to brutally dominate others. If allowed to run their course, the result will be the same: the end of human life on a great scale, as happened in Russia, China, and Cambodia.

Of course there has never been a perfect world, or anything close to it, and it is insane to even think about the possibility of creating one. Human beings have always been divided among themselves by tribe, religion, language, and especially race. As long as such differences exist, and as long as such different groups live near or among each other, or covet the same territory, there will be social friction. Since *Homo sapiens* first appeared on the scene, have there been ten years on this planet without conflict? And even within the most homogeneous and stable societies, there will always be natural differences and inequalities among individuals, both physical and psycho-emotional, that will always provoke jealousy and hostility, so we might as well get with the program. The best we can do is to be guided and led by the most superior, by those who are natural aristocrats in heart and mind — those precious few who can bring out the best in us and curb our destructive impulses.

By the same token, there never has been anything like perfect health. We all get sick at times on the journey through life, and in the end we all die. Naturally, most of us want to maintain excellent health, for ourselves and of course for our children, and one part of that is accepting that at times they will fall ill to certain pathogens, and temporary pain or discomfort needs to be endured without medical intervention. In fact, this is beneficial to the developing immune system. And another aspect of preserving health is understanding that while modern medicine has done some wonderful things, it has also spawned some totally crazy ideas, one of them being the notion, pushed by evil people and their ignorant fellow travelers, that vaccines can usher in a world virtually free of disease.

Vaccinations are medical communism, and like Marxist theory the whole filthy lie is going to fall apart some day.

For our soldiers who fought in Vietnam

MOST people reading this book were too young to remember the Vietnam War, or were born after the war ended. Very few who did not live through that event are emotionally affected by it. In fact, I've asked several people born after 1975 what they learned of the Vietnam War in school, and most told me that it was hardly ever mentioned. I find this incredible, though I suppose I shouldn't, because from kindergarten to the day I graduated college, as a history major no less, I never heard a single word about the Korean War, which ended in 1953, the year I was born. Is there a country on this planet whose citizens know less about their own history than America?

The Vietnam War was the war of my youth. I was eleven when the first American soldier died in combat in that country, nineteen when the last one died. A few more were killed during the frenzied evacuation of U.S. citizens as North Vietnamese troops closed in on Saigon — the capital of the former South Vietnam, renamed Ho Chi Minh City — some two years later, in April 1975. That was half a lifetime ago. And I have never been in the military. Yet the Vietnam War still has a hold on me, like no other war ever has. I've read more than twenty books on the subject, half of them written by men who fought there, and I've spent countless hours on the Internet watching old footage of nameless battles as well as battles whose names I remember well.

My earliest recollection of the war was some time in 1966 when I was in junior high school. One student took it upon herself to organize a letter writing campaign to "our boys" — that was the popular term back then — fighting in Vietnam, letting them know we were behind them all the way. Along with letters, she requested that we bring in small items like soap, toothbrushes and chewing gum to be sent to Vietnam to keep our soldiers' spirits up. I wondered why our government could not supply such cheap, simple items to its fighting men, but this seemed like a proper gesture and I made my contribution. I *did* support our boys in Vietnam; this was America, my country, forever in the right, and we were the good guys who would eventually prevail as we had prevailed in every war we had ever fought. In fact, all the indications were that we were winning and the war would be over soon.

Of course at this stage of my life, and right up to America's defeat in Vietnam nine years later, I was totally ignorant, steeped as I was in all the glorious myths about America and having seen almost nothing of the outside world. I would later learn, from reading accounts of our men who fought in Vietnam, that the television screen never conveyed the everyday realities on the ground. I've heard conservatives complain that the "liberal media" were biased against the Vietnam War, that they instilled a defeatist attitude in the American public that ultimately led to military defeat, but that's nonsense. What the news media did, at least in the early years, was preserve the status quo by leaving the public totally confused about what was happening in Vietnam. They did this by uncritically reporting as facts all the lies that the generals and military spokesmen were feeding them, leading us to believe that things were going well and predicting that an American victory was not far off. In *The Grunts*, Charles Anderson, a Marine Corps infantryman, mocked the kind of press statement that could be expected after thirteen of his buddies were killed when a North Vietnamese mortar shell landed squarely on a ton of high explosive ammunition sitting on a pallet: "The U.S. command in Saigon announced today that scattered actions with enemy forces accounted for three hundred fifty-eight communist dead. American losses were described as light." That's exactly the kind of thing I recall reading and hearing a hundred times, and it always sounded so encouraging, like we were well on our way to victory, and it was just a matter of time.

But we were not winning in Vietnam. As the years passed I became more and more mystified by the war, and more disturbed by the number of American soldiers dying halfway around the world. It didn't seem to make sense that we could not win a war against such a small and insignificant country, but at the same time, I had no reason to believe that the war was not being conducted properly. It just seemed to be the way of things that it kept grinding on with no end in sight. In the meantime, all I did was take away what I read in the papers and saw on television. Gradually, coverage of protests against the war increased as they became more vocal, and at times more violent. I didn't know what to make of all those young men burning their draft cards in public, and strange to say, fifty years later I still have mixed feelings. In one way, they were cowards for refusing to fight in Vietnam; in another way, they were bold for being so openly defiant and facing possible imprisonment, and wise for avoiding military service. But I don't want to give any undue credit here; you had to be a much better man to go to Vietnam than not to go.

Heavy fighting raged right through my high school years, 1967 to 1971. Some students wore bracelets inscribed with the names of American soldiers who were prisoners of war or missing in action. We cared — the war reached deep into our hearts. During the years of the most intense fighting, in the late sixties, the casualty rate was staggering, with 250 to 300 soldiers being killed every week. In no way am I downplaying the tragedy of those who have lost loved ones in Iraq and Afghanistan, but in Vietnam ten times more soldiers were dying each week, and it also must be remembered that young men were being drafted. (The military draft was abolished in 1973.)

Chaminade, my high school, has long had a reputation as one of the top academic institutions on Long Island. In addition, once a year, the entire student body attends a Gold Star Mass, with two uniformed Marines present, to honor those alumni, about fifty when I attended, who have fallen in foreign wars, including Vietnam. Yet, even though the war filled my entire four years there, and even though our teachers sometimes brought it up in class, they never said anything original or provocative: it was always the same old consensus reality.

❖

At my halfway point in high school — on June 27, 1969 to be exact — *Life*, for decades America's most popular magazine, came out with an issue that made a profound impression on me. The cover story was "The Faces of the American Dead in Vietnam: One Week's Toll." Across several pages were printed the photographs, names, ages, hometowns, service branches and ranks of nearly all the 242 men who, the article stated, were killed in action the week of May 28 to June 3. For the first time, the casualties we were taking in Vietnam were expressed not in hazy news reports and dry statistics, but in the faces of the young men in the prime of their lives who had just died in battle. So many wonderful young faces. Many were probably high school yearbook photos; half the men pictured in that magazine were no more than five years my senior; several were only three years older. I saw in those faces the seniors who walked the halls of my high school barely a year earlier; some of the older guys who had been in my boy scout troop; young men I'd worked with at various jobs. And there was something else that I'm certain I shared, even if on a subconscious level, with millions of other people who saw that issue of *Life*. These young men were America's best, the country's genetic treasure, now wasted in a war the purpose of which could not be explained to anyone's satisfaction. And these were the dead *in just one week* of a war that had been going on for more than four years. No one but their families and friends knew them as individuals, of course, but it was impossible to look at those faces without feeling a great sense of loss.

Sometime later, in a published statement, the editors of *Life* wrote that no other issue in the magazine's history had provoked such an outpouring of emotion among its readers. For once I think they were telling the truth. But looking back now, I don't for a moment believe what they had asserted in the introduction to that article, which made it sound like these men had been evenly divided between those who believed in the cause for which they were fighting, and those who were desperate to come home. Based on everything I've read, the few soldiers who arrived in Vietnam believing they were fighting for a noble cause had their illusions shattered in a matter of days. And then it came out that many, if not most of those pictured had not been killed in the time frame indicated, but earlier in May. When angry relatives contacted *Life* about this deliberate misrepresentation, the feeble reply, as I recall, was that they couldn't obtain all the details of those who had died in the designated week to make the June 27 issue, so they substituted other recent war dead. It seems that even in the rare instance when the media do something worthwhile, they still can't help lying.

By this time, dissent against the war had become so widespread, even among returning veterans, who had formed an organization called Vietnam Veterans Against the War, that it no longer seemed wrong to oppose it. The "domino theory" — the belief that the international communist movement would topple governments all over Southeast Asia, one by one, in its relentless drive to conquer the world, unless we stopped it — still made sense to me, as did the conservative appeal to support our boys and achieve a decisive military victory. There was tough talk among some over why we didn't just "drop the big one" on Hanoi, the capital of North Vietnam, and end the war immediately — the kind of bluster I've often heard from bigmouths who have never seen combat — as if instantly wiping out one or two hundred thousand people was no big deal. Such talk irritated me, but neither could I identify with the pacifist crowd, whining for peace and holding numerous demonstrations in support of an immediate end to the fighting. The pacifists did not offend me, however, and I made a clear distinction between them and the despicable creatures who marched and chanted for a communist victory while waving Viet Cong flags. These made my blood boil.

At this point in time, I was overwhelmed by everything that was going on. How could I support the war when so many men who had fought in Vietnam were speaking out against it? There were so many marches and rallies on college campuses and in the streets of the big cities — could all these people be wrong? It was incredible how the country was being torn apart by those who supported the war and those who opposed it. Then My Lai became a huge news story; nearly five hundred people in that village, including even babies, had been murdered in cold blood by our own army. Many refused to believe that Americans could do such a thing. Furthermore, we had taken the war into neighboring Laos and Cambodia, dropping a tremendous tonnage of bombs on those neutral countries through which the North Vietnamese Army was transporting supplies on jungle tracks and secondary roads into South Vietnam. When these facts became publicized there were more violent riots, which culminated in the gunning down of four students at Kent State University by the Ohio National Guard. Disgruntled combat vets were marching on Washington, throwing their ribbons and medals in the street like unwanted trash. I didn't know what to believe anymore. It was beyond my comprehension.

As the months passed, the war took on a new dimension for me. I would turn eighteen in late 1971, and would be eligible for the draft. I must say that American involvement in the land war was being phased out by the time I graduated high school, but for my birth year there was still a draft lottery, as there had been in previous years. The word was that the first thirty or forty birthdays picked would be told to report for a physical, though I later learned that 1953 was the first birth year in which no men were drafted. But at the time nothing was certain, and while the possibility of being drafted seemed remote, it was still there, and I felt especially vulnerable because I had dropped out of college after the first semester, and thus was ineligible for a deferment. With much apprehension I awaited the results of the draft lottery held on February 2, 1972. My birthday, November 24, came in at #180. Well up there. Safe.

Twenty-one years later I visited the Vietnam Wall Memorial in Washington D.C. Every American should go there and gaze at all those names to try to learn something — those, at least, who can deal with reality, by which I mean the treason of our elected and unelected officials. After all these years, one is still likely to see, as I did, war comrades or family members tracing the indented name of a loved one with pencil on a sheet of a paper held against the wall, or leaving a memento at its base. It is a moving experience. With mixed feelings of sadness and curiosity, I scanned the birth years of perhaps one thousand of our Vietnam dead in one of the registers near the Wall. It was no surprise that the great majority were born in the 1940s, particularly the latter half of that decade. Quite a few were born in 1950, not so many in 1951. I saw only one 1952 birth year, and none at all in 1953. But I was very close to the war in more ways than one.

❖

American combat operations ended in January 1973, and President Nixon started talking about "peace with honor." The slogan rang hollow with me, but I was glad the war was over. Of course, it wasn't over for the Vietnamese, but there was little news coverage of the continuing conflict until just before the fall of South Vietnam. Along with the rest of the country, I watched the TV news in grim silence as North Vietnamese tanks rolled triumphantly up to the presidential palace in Saigon, plowing right through the fence surrounding it. The South surrendered and the country was forcibly reunified under communist rule. And that's when the bitter reality set in: 58,191 American soldiers had died, and hundreds of thousands had been maimed in body or soul or both, for absolutely nothing. Their losses went hand in hand with the obliteration of much of the land through years of constant aerial bombing and use of defoliants, and the totally deserved hatred of America on the part of civilians not only in all of North Vietnam and among many in the South, but also in Laos and Cambodia. The loss of life from American air strikes in these countries ran close to one million with casualties mounting to the present day. (Few people are aware that Laos is the most heavily bombed country in history.) It was this secret air war, steadfastly denied by the Nixon administration, that generated popular support in Cambodia for the Khmer Rouge, which would soon carry out the most murderous communist program the world has ever seen — this a direct result of the American "good guys" fighting the communist threat.

When Vietnam went communist in 1975, I did not feel any of the hostility towards the American power structure that I do now, and as a self-styled conservative at the time, I would've written off the term "American power structure" as a delusion of the leftist mentality. Still, I could not begin to understand what had happened in Vietnam. None of the syndicated conservative columnists or conservative publications that I read back then could explain it to my satisfaction, other than to bemoan the fact that we lost a war that we could've won. It wasn't until three or four years after the war ended — around the time that I started reading dissident literature, and came across the writings of a few freethinkers who forcefully condemned the war while it was still being fought — that the elusive truth began to seep through. Around this time I read *A Rumor of War* by Philip Caputo, a Marine Corps second lieutenant who landed in Danang with the first American combat unit in March 1965. This wonderful book ripped the cover off all the pseudo-patriotic war propaganda and revealed, for the first time, the horrors of that conflict and what our soldiers had to endure. Indirectly, it also underscored what was beginning to dawn on me: that despite all the pep talk about freedom and democracy which characterized every administration since the end of World War Two — in fact, right up to the collapse of the Soviet Empire — no American in a position of power and influence was ever serious about vanquishing world communism.

❖

It was President John F. Kennedy who got us into the quagmire of Vietnam by beefing up the American military presence in that country and making pious noises about meeting the communist threat in Southeast Asia — this while increasing the leverage of communism in Laos, which shares a long border with Vietnam, and in other parts of the world. From the beginning, Kennedy's policies in Vietnam, which strengthened a despised ruling family hopelessly out of touch with its people, created all kinds of strife, portending the troubles that lay ahead. Some have speculated that Kennedy intended to clean up the mess that he himself created, and would have withdrawn most or all American military personnel from Vietnam had he survived and been re-elected. This might be true. While the Kennedy presidency was mainly a shabby one, his thinking is impossible to figure out. He was inconsistent in his words and actions, and took a lot of secrets to his early grave.

Lyndon Johnson inherited this festering problem, but in my view he bore the greatest responsibility for the Vietnam War — though no one should discount the influence of the "advisors" who swarm around every modern president. Johnson started the war using the false flag Gulf of Tonkin incident as a pretext — one of several naval incidents that have initiated American wars overseas, Pearl Harbor being the most notorious. He escalated the war to tremendous proportions. It was on his watch that the heaviest fighting took place and the most men died, though his successor, Richard Nixon, was also a bona fide war criminal, as was Nixon's national security adviser, the sinister Henry Kissinger.

And what was going on during this noble crusade against communism? Many things. For one, several large American corporations, in collaboration with Chase Manhattan Bank and certain federal agencies, constructed the Volgograd and Kama RIver truck plants in the Soviet Union, the latter of which became the largest truck factory in the world; trucks assembled here were used by the North Vietnamese Army to transport fuel and supplies for the purpose of killing American soldiers. This was in keeping with Johnson's decision in 1966 to relax export restrictions of numerous commodities and technology blueprints essential for weapons of all kinds, by declaring them "non-strategic." For another, earlier that same year, two Cuban cargo ship captains, who had left their vessels and sought political asylum in Spain, furnished our State Department with complete details of the routes and timetables of ships loaded with war materiel departing from Cuba with North Vietnam their final destination. Every one of these ships sailed through the Panama Canal, which then was under American control. Our Navy did nothing to intercept these ships at the Canal, in the waters off North Vietnam, or anywhere in between.

If that weren't enough, Johnson's biggest financial supporters of his entire political career, beginning as a Texas state senator, were the Brown brothers of Brown & Root, a firm that won contracts for huge military construction projects in South Vietnam. In other words, expanding the war was a good way for Johnson to pay off his political debts. That's my own conjecture, but I wouldn't put anything past LBJ, who had several people murdered while clawing his way to the top of the government heap, including his own sister, and had the interesting habit of briefing reporters while sitting on the toilet, and urinating on the White House lawn in the company of guests — or so I've read. I do believe these things, because many people close to him, in and out of politics, spoke and wrote about the Johnson they knew after his death. Apparently, they were too scared to spill the beans on this beast while he was still alive.

❖

I knew about ten men who fought in Vietnam, two of whom were wounded. Some I knew before they went there, some I got to know after they came back. All were blue collar types, what you would call regular guys — none were uncommonly good or uncommonly bad. I learned right away that men who have been in combat do not wish to talk to outsiders about their experiences. Instead, I learned of the day to day realities of the war from books written by veterans. The same themes came up again and again, even though they fought at different times in different regions: the personal principles that inspired them to enlist, or comply with the draft, which quickly disintegrated after they arrived in Vietnam; the growing realization of the futility of their mission; the incomprehension of guerilla warfare in an unfamiliar landscape, for which their combat training proved useless; the pointless patrols on dangerous jungle trails in the oppressive heat while heavily weighed down with combat gear, in which thousands of men were blown apart or hideously wounded by booby traps; the unsung casualties of heatstroke, malaria and other tropical hazards; the numerous casualties of friendly fire and various accidents and mishaps, affecting both soldiers and civilians, which were bound to occur in a war of such confused circumstances; the mental turmoil created by the objective of protecting innocent villagers from the communist Viet Cong, knowing that some of these same people sympathized with or even were members of this homegrown communist force, whom they were supposed to kill; the months on end of living in fear, filth and exhaustion, sleeping only three or four hours a day in mud; the grudging admiration for the determination of their enemies, which grew in direct proportion to the hatred of their own government; the mental state of caring about no one in the world except themselves and their comrades, and hoping for nothing other than surviving their tour of duty; the incompetence and indifference of their superiors who directed senseless operations from safe and comfortable locations far from the actual fighting; the contempt towards these same individuals, a contempt that often hardened into a burning hatred, and sometimes provoked acts of sabotage and even murder. (The Vietnam War contributed a new word to the English language: *frag*. To frag means to try to kill one's military commanders by throwing a hand grenade into their living quarters, or in some cases "accidentally" shooting them. There were nearly a thousand fragging incidents in Vietnam, about ten percent of which were successful.)

Nor did it take years for the madness to reveal itself. Philip Caputo related an almost surreal scene he witnessed during an early stage of the war. To the west, the Marines were fighting the Viet Cong; to the east, two factions of our ally, the South Vietnamese Army, were fighting each other — a civil war within a civil war. In disbelief he watched a South Vietnamese fighter plane roar downward and strafe a convoy of trucks packed with South Vietnamese soldiers. "I knew then that we could never win," he wrote. "With a government and an army like that in South Vietnam we could never hope to win the war. To go on with the war would be folly — worse than folly: it would be a crime, murder on a mass scale."

That was not the worst of it. In *Compromised*, the ultimate insider book about CIA drug trafficking and the making of the Clinton presidency, Terry Reed, an Air Force pilot, described how his upright midwestern values were cynically exploited by his government, and how he quickly became disenchanted with the way the Vietnam War was being conducted — a feeling that exploded into the open when he and other pilots were informed by an Air Force captain that there had been a policy change, and that American prisoner of war camps at strategic points along enemy supply lines, previously off limits, were now to be targeted: *he and his fellow pilots were being ordered to drop bombs on American POWs*. From page 22 of *Compromised*:

A small mutiny ensued and ended shortly thereafter with the stunned airmen standing at rigid attention while a full colonel read them the Uniform Code of Military Justice, specifically the section about the consequences of refusing to obey a direct order. They were guaranteed immediate lodging in the federal penitentiary at Fort Leavenworth, Kansas, if the orders were not obeyed immediately, and told that replacements from Hawaii would be there within 24 hours. The replacements would *follow the lawful orders being given. It was going to happen, with or without their cooperation.*

The bombs fell. That's the kind of government we have. That's the kind of person you'll find in our military brass. That was the reality — a small part of it — of the Vietnam War.

❖

Right through all this, our conservatives continued to fight the communist menace from behind their typewriters. No one is more anti-communist than I, but the greatest danger to America has always come from within, not from the Soviet colossus our government helped build, nor from anywhere else — least of all the backwater countries of Southeast Asia. In an earlier chapter I raked over the rubbish of communist dogma and its impossible designs for global domination. To think that communism could snowball into a monolithic force that would eventually swallow the whole planet was to think in very narrow terms, which is what conservatives do. Well before the Vietnam War began it had become obvious to any casual observer that Moscow could not keep a firm grip on the nations of Eastern Europe, not to mention its wavering influence in other parts of the world. Yugoslavia had left the Soviet camp, riots had broken out in Poland and East Germany, and for two momentous weeks in 1956, the people of Hungary had waged all out war with small arms against their oppressors in the streets of Budapest, with thousands of casualties on both sides, during which they pleaded in vain for military assistance from a supposedly anti-communist American government. When none came, the rebellion was crushed by Russian tanks.

So it's not just by benefit of hindsight to observe how mixed up conservatives were with their dominoes. But no one could have predicted how it would all play out. As it happened, the Vietnamese brand of communism, infused with a healthy nationalism that scorned American domination of the country, and French before it — another thing that conservatives can't understand — proved to be less murderous than anticipated, while next door in Cambodia the Khmer Rouge carried out a revolution horrifying even by communist standards, killing off nearly a quarter of the population. After four years of this nightmare, Vietnam invaded Cambodia and overthrew the regime, while renewing their ancient feud with their northern neighbor, China — where, of course, a communist government had been brought to power in 1949 with the indispensable help of Washington. China, which had supported the Khmer Rouge — as did some people in our own State Department — invaded Vietnam, launching a sporadic border war that dragged on for years — all these events exposing the absurdity of the domino scenario and the insanity of American foreign policy. Like everywhere else, this part of the world would have been spared a great deal of suffering if we had only stayed away from it. And even if these people had gone communist on their own and slaughtered each other in vast numbers, no American in his right mind should have cared, any more than a Vietnamese peasant living in 1863 should have cared about all the lives lost at Chickamauga and Gettysburg.

❖

Smedley Butler was a highly decorated Marine Corps major general who, over the course of thirty-four years, participated in American military actions all over the world, his last in France during World War One. He is one of the very few career military men to look back upon his years of duty with shame and disgust. He reflected on many things, most poignantly his visits to eighteen veterans hospitals around the country which housed 50,000 of what he called "the living dead" — forgotten casualties of the First World War. In a government hospital in Marion, Indiana he saw 1800 mentally destroyed men kept in pens like animals. Nothing ever changes in America, I thought as I read this. In 1935, after retiring from the service, he wrote a short book titled *War is a Racket*, in which he concluded that for his entire career he was a hireling for big business interests. Butler had particularly harsh words for those who profit from the human misery created by war. He got right to the point on page one:

At least 21,000 new millionaires and billionaires were made in the United States during the [First] World War. That many admitted their huge blood gains in their income tax returns. How many other war millionaires falsified their tax returns no one knows. How many of these war millionaires shouldered a rifle? How many of them dug a trench? How many of them knew what it meant to go hungry in a rat-infested dug-out? How many of them spent sleepless, frightened nights, ducking shells and shrapnel and machine gun bullets? How many of them parried the bayonet thrust of an enemy? How many of them were wounded or killed in battle?

Surely there were thousands of U.S. corporations that struck it rich during the Vietnam War, and it's not just the blatant merchants of death — the munitions manufacturers and the financiers — that rake in the wealth when war exacts its toll, but many others as well: the ship and aircraft builders along with the steel companies that supply their resources; the uniform companies; the chemical and petroleum firms; the boot makers and leather tanneries; the factories that make helmets, ponchos, mosquito nets, mess kits and all the other accoutrements needed by soldiers — not to forget the vaccine manufacturers since all the troops have to get their shots. It would take several pages to list all those who profit, directly and indirectly, from the hell of war.

In tandem with this is the military bureaucracy which, over the course of the last century, has swollen to leviathan proportions. The tax money sucked up by the federal agency known as the Department of Defense now totals more than the military expenditures of the next eight governments combined. On the eve of the Vietnam War this bureaucracy employed some 4.5 million people, one million of them civilians, and operated from a budget larger than the entire national budgets of nearly every country in the world. There were six to eight times as many American non-combat personnel in Vietnam as there were fighting men. War keeps a lot of people busy, and there were nice salaries and benefits for all the advisors, paper shufflers, supply clerks, truck drivers, doctors and nurses, pilots and crewmen shuttling cargo and troops, forklift operators routinely unloading soon-to-be-filled coffins — all the opportunists bred by bureaucracies which, in league with the private sector, feast on death and destruction. Like malignant tumors, bureaucracies take on a life of their own and their growth becomes impossible to control.

As mentioned earlier, the military is also a

sanctuary for all kinds of brutes and psychopaths incapable of productive work, men who find pleasure in killing and destroying for its own sake. War brings them out of the woodwork. In *A Rumor of War*, Philip Caputo recounts an incident in Okinawa while training for jungle warfare, shortly before being deployed to Vietnam. A beefy instructor walked into a classroom, buried a hatchet in the wall while screeching a war cry, and on a blackboard wrote the words "AMBUSHES ARE MURDER AND MURDER IS FUN." He then pulled out the hatchet, and brandishing it at his bewildered students, ordered them to recite out loud what he had written. George S. Patton, son of the famous World War Two general, was another one for whom killing provided endless entertainment. "I like to see the arms and legs fly," was one of his favorite lines. In 1968 he sent out Christmas cards reading, "From Colonel and Mrs. George S. Patton III — Peace on Earth." Inside the cards were color photographs of stacked, dismembered Viet Cong soldiers. At a farewell party shortly before he left Vietnam, Patton made the rounds wearing a peace medallion while showing off a polished enemy skull with a bullet hole in it. During a "pacification" campaign to wrest control of Kien Hoa province from the Viet Cong, thousands of noncombatant civilians were killed by American planes and helicopter gunships, all counted as enemy dead in a war for which the tally of bodies was the number one priority. Such atrocities took place all the time. "Death is our business and business is good" was painted on a structure used by some of these helicopter crews. These are just a few examples of the death cult mentality that one frequently encounters in literature on the Vietnam War. Many of those who were brainwashed into it, or who embraced it as a means of survival, had a great deal of trouble living with themselves after they returned home. Except for the comparatively few like those above, and the cold-blooded Chris Kyle types, killing people in a faraway country that poses no threat to one's own country is an unnatural act that can destroy a man's mind, even if it takes a while for the reality of it to set in. Some shrug it off, some get over it, some never do. Thousands have returned to Vietnam to meet their former enemies in a spirit of reconciliation, and heal the wounds of war, but many have never felt that way. In any case, the number of permanent casualties has been very high. God only knows how long the road of suicides, drug and alcohol addictions, broken homes and broken lives that began in Vietnam stretches.

❖

The official U.S. position of opposing communism in Southeast Asia was always a lie. From the Franklin Roosevelt administration onward, our government, aligned with the media, has aided and abetted nearly every communist power grab, even after the collapse of the Soviet empire, as in South Africa. War of all kinds declared by Washington D.C. has always been a farce. The war on communism, the war on terror, the war on this, the war on that. Back in 1964, Lyndon Johnson, while gearing up for the Vietnam War, launched his War on Poverty. Four riot-ravaged years later, after millions of tax dollars had quietly been siphoned to black revolutionaries, the people he claimed he set out to help were more impoverished than ever, owing to their own destructive behavior. In 1971 President Richard Nixon signed the National Cancer Act, officially declaring war on cancer. How's that war been coming along? That same year he inaugurated his War on Drugs, when there were less than a half million addicts in the entire country, and two federal agencies that enforced drug laws. Thirty years later the number of drug addicts had multiplied tenfold, while the anti-drug offensive had fanned out to include fifty-five agencies. Every American who has educated himself knows that rogue U.S. government officials, across a wide spectrum of agencies, constitute a major international drug trafficking network. These criminals make a fortune, illegal drugs flood the nation, more stringent anti-drug laws are passed expanding the police state enforcement apparatus at every level, and drug busts rise on a constant upward curve with America's "drug problem" getting worse all the time. It seems that whenever the government declares war on a problem, the problem becomes much worse by design.

Shortly after the events of September 11, 2001, the truth of which has always been smothered by the controlled media, President George Bush announced the War on Terror, which ushered in the boundless surveillance state that all of us have become accustomed to. The War *of* Terror would be a more fitting term. The immediate beneficiaries were companies like Boeing, Raytheon and Lockheed Martin, which make bombs used by the American military to kill people in foreign countries. This is terrorism. But Lockheed Martin also wants to keep our shores safe from terrorists, so they do a lot of business with the Department of Homeland Security, founded in 2002, and its many spinoff agencies. One of their swindles is the Transportation Worker Identification Credential, or TWIC. If, like me, you are one of two million Americans whose job requires that you enter a waterfront facility (barges deliver oil to harbor terminals, where I sometimes load my truck), you must have a TWIC card to gain entry. To get one, you have to go to a designated location to be photographed, fingerprinted and entered into Big Brother's database. A TWIC card costs $132, check made payable to Lockheed Martin. Two million times $132 equals $264,000,000 — and you have to renew it and repay it every five years. Not a bad cash cow. There are similar procedures and costs required of commercial drivers who transport hazardous materials, a category that includes all petroleum products. Plus, if you own an oil company, as I did for three years, you will receive literature in the mail from various hustlers cashing in on the anti-terror bonanza by offering seminars, security programs, manuals and such, and to try to scare you into buying their products, they cite recent federal dictates that include harsh penalties, including prison terms, for noncompliance. This is just the tiny bit I've seen of it. The War on Terror, it's called.

Out of the same mold — though an official term has not yet been coined — is the war on childhood disease. It's a scam to empower the burgeoning "health" bureaucracies and enrich the vaccine makers, but many Americans actually think it's driven by good intentions. In reality, politicians in every state are bought off by lobbyists employed by firms like Merck, Novartis and GlaxoSmithKline to enact laws mandating scores of vaccinations as a requirement for school attendance. And like Lockheed Martin, they work both sides of the street: if their vaccines bring on diabetes or asthma or attention deficit disorder in your child, they'll be happy to sell you Ritalin or Advair or Dulera to "solve" the problem they've created, which might lead to more problems for which they'll be happy to sell you still more drugs.

Like all wars, the war on childhood disease creates jobs. It accounts for the skyrocketing number of pediatricians, a type of doctor I never heard of when I was young. We just had a family doctor. According to Dr. Robert Mendelsohn, vaccines are the "bread and butter" of pediatric practice. As more and more children are damaged by shots, they are referred to more and more specialists, keeping more doctors' offices, laboratories and other medical sidelines busy. Those in special education probably have benefited the most. Special education is a growth industry with no end in sight. And I don't mean to denigrate the rank and file. There are many kind people, mostly women, in this field — I've known several who have worked with my children — just as I'm sure there were many caring, dedicated doctors and nurses who treated the traumatic wounds of our soldiers in Vietnam, perhaps even some true saints who acted purely from self-sacrifice, wanting nothing in return (not to gloss over the many disgraceful military doctors and VA hospital horror stories). But war casualties provided good pay and benefits for medical personnel, and without those perks I doubt most would have worked in South Vietnam. I never heard of one who spoke out against a meaningless war that was mangling the bodies of so many young American men. And my observation of people who administer programs for handicapped children convinces me that they would oppose any idea, such as abolishing vaccines, that might put them out of work.

❖

As someone who reads constantly, and with a million books on the market, I carefully select those that I think will be worth my time. One kind of book I don't often read is the memoirs of some president, or general, or other public figure because these books tend to be self-serving and omit a great deal of importance. Nevertheless, I made an exception for Robert McNamara's *In Retrospect*, subtitled *The Tragedy and Lessons of Vietnam*. I retain many images of the Vietnam War — images of things that took place in Vietnam as well as here at home. One indelible image is that of Robert McNamara, who served as defense secretary through the entire Kennedy and all but the last year of the Johnson administrations. Specifically, it was the news conferences that I saw on television, with McNamara waving his pointer at a map of Vietnam, and explaining, in his mechanical way, the latest war developments. I didn't know what he was talking about, and as it turned out, neither did he, but even back then, at the age of twelve, I sensed that there was something abnormal about this man whose middle name, believe it or not, was Strange.

McNamara was probably blamed more than anyone else, even the incompetent top Army general William Westmoreland, for everything that went wrong in Vietnam, even though he quit his post short of the war's halfway point. He deserved much of that blame, though it should have been spread among three presidents and thirty or so cabinet members and military advisors whose names frequently appear in his book. As the war ground on with no end in sight, McNamara came to be despised by all segments of the population. To the peace crowd he was a mass murderer who directed the aerial bombing of civilian areas in North Vietnam. To two-fisted Americans, whose sons and brothers were fighting in South Vietnam or were likely to be drafted soon, this Harvard egghead, once described as "an IBM machine on legs," was an alien geek whose bungled policies were getting a lot of young American men killed for unclear reasons.

The author describes some unpleasant confrontations with anti-war protestors, a few of which almost turned violent. The stresses and strains of the job led him to resign at the end of February 1968, and he became president of the World Bank. Four years later, while on a passenger ferry to Martha's Vineyard, a popular holiday destination for the Washington elite, someone recognized him and tried to throw him in the ocean. That's how high emotions ran during the Vietnam War. He chose not to mention this incident in his book.

For the most part, *In Retrospect* is a litany of meetings, conferences, debates, assessments, messages, fact-finding tours, arguments, discussions, memorandums, agreements, disagreements, re-assessments — all of which lead nowhere. The impression one gets is that of a bunch of cub scouts spending three years trying to fix a broken nuclear reactor, only to kill off the surrounding population with a massive radiation leak. In the last chapter he enumerates and discusses all the reasons why the war dragged on and ended the way it did. He could've compressed it into three sentences: "We didn't know what we were doing. We made a lot of mistakes. Sorry about that."

McNamara neglects to mention the sordid deeds of John and Robert Kennedy, Lyndon Johnson, and others he was close to. They weren't perfect, he tells us, they made mistakes like you and me, but deep down they were good people and devoted public servants of our great democracy. Apparently he really believed this. There also is a great deal that he doesn't mention about the war, including almost everything I've discussed above. This didn't surprise me, the author's claim in the preface that he is finally leveling with the American people and his many admissions of poor judgment notwithstanding. For example, McNamara, who had risen within the Ford Motor Company to become its president before he was tapped for his government post, makes no mention of Ford's longtime prominence in the industrial sector of the Soviet Union, North Vietnam's chief military supplier. (I would not be surprised if the tanks that rumbled into Saigon in April 1975, sealing a communist victory, were powered by Ford engines.) There isn't a word about the CIA's nefarious Phoenix Program, which gave free rein to sadists in American and South Vietnamese uniforms, who murdered thousands of innocent civilians. He offers a brief, vague account of his activities during World War Two, which I would have completely forgotten about had I not discovered what he left out.

While reading the book, I did some poking around on the Internet and learned that McNamara, who never saw combat, worked closely with General Curtis Lemay, the B-29 Hulagu, who in the closing months of the war flattened Japan. As a statistical analyst, McNamara's task was to determine how this could be done in the most cost-effective way. He was the original megadeath intellectual. It was he who recommended the firebombing of Tokyo on March 10, 1945, which left behind the ashes of more than 80,000 men, women and children. With McNamara's guidance, Lemay continued to do an effective job, incinerating a half million Japanese civilians and burning much of Japan to the ground — all this before the two atom bombs were dropped on Hiroshima and Nagasaki. Although he leaves all this out of his book, he was interviewed about it in the film *The Fog of War*, made in 2003, eight years after *In Retrospect* was published. Eighty-seven at the time, McNamara came across to me the same way he did forty years earlier — talking in circles, trying to sound important, saying nothing. There is no indication that he ever looked upon the countless innocent victims of his crimes as anything but "collateral damage," that sickening term regularly used by military mouthpieces. Considering McNamara's expertise in ending human life, I'm left wondering what he meant by "population planning," which in his book he says he implemented as World Bank president, but which, typically, he does not elaborate on. Did he mean sterilizing Third World populations under the guise of high-minded vaccination campaigns?

On the whole, *In Retrospect* is a dull, shallow book riddled with omissions — like any book, a reflection of its author. The bibliography contains eighty books and other publications about the Vietnam War, nearly all with sterile, scholarly-sounding titles. Not one book written by a combat veteran — the kind of book that will touch the emotions of any normal human being — is on McNamara's list. What are emotions to a robot? Robert McNamara was a nonentity addicted to abstract concepts, having no connection to America's working and middle class, having no real knowledge of history or world affairs, whose decisions caused hundreds of thousands of people who wanted nothing to do with war, including Americans, to die in agony. If this book is good for one thing, it's this: he has unwittingly given us a close look at soulless, bloodless, super-rich Corporate Man as he functions day to day in the highest councils of business and government. McNamara has reappeared in the likes of Dick Cheney, George Bush, Donald Rumsfeld, Rex Tillerson, Donald Trump and others — cowards and war pimps all who have never been anywhere near a battlefield and in one way or another weaseled out of being sent to Vietnam (in Rumsfeld's case, Korea). How many other McNamara clones are there, I wonder, in the executive offices of the big pharmaceutical firms and all the agencies and institutions that are behind the childhood vaccine racket? For how many more years will these people be allowed to destroy our kids?

❖

I've spent most of the last forty years, in between stretches of work, travel, marriage and divorce, living in various places in Massapequa and neighboring Amityville, on the south shore of Long Island. Massapequa was the hometown of Ron Kovic, now a California native who, having been raised from birth on all the overblown myths about America, enlisted in the Marines at 18 and volunteered for combat in Vietnam. On January 20, 1968, during his second tour, an enemy bullet smashed into his spine, paralyzing him for life from the chest down. Six years later he wrote a book, *Born on the Fourth of July*, which to date has sold more than a million copies and in 1989 was made into a movie, making him famous. He writes of many things, from his idyllic childhood in Massapequa to shouting down Richard Nixon from his wheelchair at the 1972 Republican convention in Miami, and everything in between. In the most recent edition, published in 2005, he penned an introduction condemning the war in Iraq that began two years earlier. Kovic has been an anti-war activist for many years, and while I don't agree with him on every point, his book, like every book written by one of the small fraternity of men who have experienced combat, is well worth reading. He has risen above his bitterness and aspired to a higher level of patriotism — a man who has never hated his country, but rather the lies he was fed while growing up and the people who still tell them.

Every Memorial Day we remember our war dead, but no such day is set aside for the greater number who survived as physical wrecks. Kovic's work is a rarity in this respect. Several scenes have stayed in my mind. He is lying helplessly in his filth in a rundown VA hospital in the Bronx, staffed by lazy, uncaring aides. For an hour he shouts and throws things through an open door, trying to get attention, until an orderly finally shows up. "I'm a Vietnam veteran," Kovic says. "I fought in Vietnam and I've got a right to be treated decently." The orderly replies, "Vietnam don't mean nothin' to me.... You can take your Vietnam and shove it up your ass." That's a fitting epitaph to all those fine young men who were severely wounded and became useless slabs of flesh, unknown, unwanted and unvisited in Army hospitals throughout America. While demonstrating against the war years later on several occasions, Kovic was spat on, called a communist traitor, dumped from his wheelchair into the street, and kicked and beaten. Undoubtedly his patriotic assailants had never been to war.

I found it hard to believe that people could do this to someone who had suffered such a tragic wound fighting for his country, even if they disagreed with his tactics, until I read about a scene that took place in the hallway of the Hart Office Building in Washington D.C., as described in *Evidence of Harm*. A group of parents from around the country, some with their autistic children in tow, were waiting to meet with a Republican health official on the issue of vaccine injury, when an autistic boy began screaming. A young woman emerged from a staff room and angrily told his mother to control her son, but the boy would not be consoled — a situation I know all too well. The staffer then shouted to the assembled parents, "You're all a bunch of freaks! Your son is a freak!"

Some of the women, shocked beyond words, began to cry. The bureaucrat called the Capitol police to have them all ejected, and as the cops stood there, unsure of what to do, she sneered, "You *people*, you're just using your kids to abuse the system to get rich!" One parent videotaped the entire incident. Such is the arrogance of our public servants — not all of them, of course, but their overwhelming mindset is the preservation of the status quo at all costs. America goes to war and vaccinates its kids for the good of society — and woe to those who fall through the cracks, or fight the System.

❖

Back in the early 1980s, when I still watched a little television, I saw a show that featured a panel discussion among four or five Vietnam vets who were reflecting on the war. One of them said something like this: "If there's one thing we learned from Vietnam, it's that we'll never again get involved in a war in which there's no clear objective, where we don't even know what we're fighting for." I thought, "Amen," and I'd bet that just about everyone else watching that show felt the same way. How wrong we all were.

Except for those who were personally affected by the Vietnam War, and a handful of others who have taken a keen interest in it, the lessons and events of the war have faded into oblivion. *Born on the Fourth of July* put Massapequa on the map, so to speak, but there are no reminders that Ron Kovic grew up here, on Hamilton Avenue — no signs or plaques, no parks or ballfields named in his honor, nothing at all mentioning the Vietnam War that I have seen, and I know this town well. I wouldn't be surprised if not a single student at Massapequa High School has heard of him. He is, after all, a symbol that all the pleasant myths about America that were drilled into our heads do not exemplify patriotism, but a grotesque caricature of patriotism. That makes a lot of people, especially the flag wavers, and the Fox News junkies, uncomfortable. After all, in Kovic's depth and courage, they see a reflection of their own emptiness and cowardice.

If you drive through Massapequa on the main thoroughfare, Merrick Road, you'll go right past John Burns Park, a large recreation complex, where you'll see the "Towers Of Freedom," two steel beams with a Liberty Bell replica between them that were erected shortly after the events of September 11, 2001. In the park, on a brick walkway leading to the towers, and along each side, there are many names of donors and sponsors, and some of police and firemen who lived nearby and perished when the twin towers collapsed. My lasting memory of 9/11 was delivering oil in the area for months afterward, as human remains were slowly recovered and identified, and several times coming across a barricaded street where a funeral mass was taking place, marked by huge crowds and an enormous American flag hanging between two fire engine ladders. In no way would I minimize the loss of those brave rescue workers. I do, however, discredit the bedtime stories engraved on large stones in that walkway, describing how terrorists belonging to al-Qaeda — an organization that, I am convinced, has never existed — crashed hijacked airplanes, one after another, into the World Trade Center, the Pentagon, and the Pennsylvania countryside. Almost directly across the street, on the front lawn of the high school, is a hunk of twisted metal, presumably from one of the fallen towers, and a plate reading "September 11, 2001." From time to time, I've noticed, more than a hundred full size American flags are planted in the lawn, then a week or two later they disappear. What this is all about, I have no idea, but obviously it's not education. How many kids in this school can find Vietnam — or for that matter, Iraq and Afghanistan — on a map, I wonder.

Over at the American Legion Hall by the Long Island Railroad station, both mentioned in Kovic's reminiscences, the military banners of each branch fly, along with the stars and stripes and the familiar black and white silhouetted POW-MIA flag below it, flying also at the Elks Club next door and at Burns Park, telling the world "You Are Not Forgotten." Words, words, words. Of course they're forgotten, except by their families. And not only in Vietnam, but in Korea and in Europe, where Stalin seized some 20,000 GIs from German prisoner of war camps who disappeared forever in the Gulag. For decades, numerous reports of sightings of American prisoners in Vietnam were dismissed by the military bureaucracy, which insisted they were all dead or accounted for. Read about it sometime, and unless you want to get on a list, don't make too much noise about it. Adjacent to the American Legion Hall is the fenced-in Veterans Memorial Garden, with several plaques. None mention Vietnam, but one honors our men and women in uniform who have served "wherever the cause of freedom and democracy has been threatened anywhere in the world." I'm amazed that even the members of a washed-up, brain-dead outfit like the American Legion can believe such a thing, if in fact they do, and I shudder to think there are plenty of Americans who still take it seriously.

In March 2003 I was living in Amityville, near the Massapequa line. The build-up to the second war to decimate Iraq, which the younger Bush called Operation Iraqi Freedom, was throbbing in the media airwaves, and a big "patriotic" rally was scheduled at John Burns Park. Never mind that Bush Senior's first war, Operation Desert Storm, launched in 1990, had left a million Iraqi civilians dead or destitute, and that thousands of our own soldiers were suffering from a strange illness called Gulf War Syndrome. Never mind that most Americans have never had the slightest idea of the actual situation in the Middle East. No, this time we were going to return and take care of unfinished business, destroying all those leftover weapons of mass destruction and getting rid of that evil dictator Saddam Hussein once and for all. We're Americans, we don't just talk about terrorism, we do something about it. I pulled into my driveway that night, unaware that the rally was in progress, and getting out of my car heard the thunderous chant "USA! USA!" carrying from Burns Park, a mile away. I suppose a case can be made that the people of Great Britain and France, countries with a much longer history than ours, are equally deficient in learning from the past, but to me there really seems to be something hopelessly insane about America.

❖

In 2013, during a six-week trip to the Far East, I finally made a pilgrimage to Vietnam, a country that impressed me as being on the right track, and if anything, less "communist" than America. Other than some museums and striking outdoor exhibits of captured or destroyed American weapons and planes in Hanoi and Saigon (most Vietnamese still refer to Ho Chi Minh City as Saigon) there were practically no reminders of the war, and certainly no hostility towards Americans. Vietnam is a pleasant country which attracts many Western tourists, and I met and chatted with several, but not one expressed anything more than a passing curiosity about the war. Most of the population, younger than forty, know nothing of what they and the rest of the world call "the American War,"and they know only the name Vietnam, no North and South. Everything about Le Loi, the main boulevard in the imperial city of Hue, was so normal I simply could not grasp that the very pavement beneath my feet had been trod by young American men locked in intense house-to-house fighting during the Tet offensive in 1968. It was only at the remote air base at Khe Sanh, relentlessly shelled that same year from the surrounding hills by the North Vietnamese Army, that I got a sense of "being there." There is now a small museum at the site, the past fighting explained from a Vietnamese perspective, and bunkers used by the Marines have been reconstructed. The air strip was torn up long ago and is covered by grass, but it's obvious where it was. An overcast sky and a strange calm hovered over the area. Having seen films of the siege of Khe Sanh on the Internet, and remembering that name so well from long ago, I could easily picture the whole area being pounded as Marines scrambled for their lives.

To honor, in a small way, my countrymen who fought and died there, I climbed Hill 937, better known as Hamburger Hill, a small peak in a remote jungle area barely a mile from the border with Laos. To do this I had to apply for a special permit, and I was required to go with a local guide, because unknown to me at the time, this region, and even more so large areas of Laos and Cambodia, is still littered with live bombs dropped by American planes long ago. Known as UXO, unexploded ordnance, many of them lie just beneath the soil surface and continue to pose a grave danger. Hamburger Hill, in a location of no signifi cance, was the scene of a bloody ten-day battle in May 1969. Our soldiers gave it that name because it was a human meat grinder. Sixteen days after taking the hill from the North Vietnamese Army, at a cost of more than 400 dead and wounded, they were ordered to abandon it. But it did give the rear echelon boys, who planned and directed it from their air-conditioned headquarters miles away, something to do.

I first learned about UXO in Cambodia, where I began my trip. There I saw the consequences of American bombing with my own eyes.

Near some of the temples at Angkor Wat, Cambodia's major attraction, and in nearby Siem Reap, where tourists stay, I saw several small groups of men, about forty in all, playing their traditional stringed instruments, hoping for a few coins from people walking by. Their legs had been blown off and they supported themselves on their stumps. The problem, from what I understand, is worse in Laos, but I didn't go there. In both countries farmers detonate these old bombs by accidentally striking them with their hoes. Children pick up small cluster bomblets, which look like tennis balls, and they sometimes go off as they throw them around. This happens a few times every week, now, as you read this. Google "UXO Laos" and it will take you two months to read everything that appears. The people of Cambodia and Laos are among the poorest and most docile in the world. They never bothered anyone. They had nothing to do with what was happening across the border in Vietnam. Nevertheless, in pursuing a North Vietnamese Army that was detouring through their territory, our fighter pilots killed them like bugs, and destroyed their land and their lives for generations to come. That's how we fought communism fifty years ago. That's how we fight terrorism today.

❖

This has been a rambling, subjective chapter. I began writing it with the intention of paying tribute to our men who fought in Vietnam, and that's how I'm going to end it. Some might find it out of place in this book, and there will be those who disagree with my ideas. Just as there are parents who are glad their kids got a lot of shots, I'm sure there are many veterans, even some who saw combat, who would take exception to what I've written, who look back with satisfaction at their military service for various reasons. I should say also that there's a small part of me that feels incomplete for not having been put to the ultimate test that any man can undergo — being thrown into mortal combat. (The much larger part of me is thankful that I missed out on that test.) I am not opposed to military service *per se*. I admire the Swiss system, in which all males are required to go through military training and all households required to have firearms. There is a world of difference between genuine national defense and pulverizing the planet for no sane reason. Switzerland, in the heart of Europe, has stayed out of all wars for more than three hundred years. Switzerland has no enemies and is universally respected. Switzerland is what America could have become.

I realize it's impossible for anyone who did not live through that era to fully relate to what I've written, just as I cannot fully relate to any other war, American or otherwise, that came before Vietnam. For that matter — and I don't want to sound unsympathetic to those who have experienced more recent heartache thanks to our war criminals in Washington D.C. — no war since Vietnam involving American soldiers has ever touched me as deeply. I feel what I feel. Somehow, the good fortune of being born just a little too late to be sent to Vietnam — if I would've been brave enough to go — and the misfortune of what happened to my children, are part of the tapestry of my life. Somehow it all goes together.

But it all goes together as well for those parents whom I hope will be enlightened by reading these lines. Just think of all those fine young men who went to Vietnam because they thought it was the right thing to do, because they didn't want to let their family and friends down, because they felt it was their patriotic obligation. Just think of all the psychopaths at the apex of power and wealth in America, whose deeds invariably have a negative impact on our lives. Just think of the millions of dolts in our military and public health bureaucracies, people incapable of empathy, people of little intelligence and no accountability, people who can justify anything that benefits their career, be it a child getting fifty vaccinations by age twelve, or a platoon taking heavy casualties fighting a make-believe enemy.

We live under a System that feeds on perpetual death and destruction. Sooner or later the best people in America are going to change that. Until then, don't get your children vaccinated. And urge them never to join the military. It all goes together.

The U.S. military: What is to be done?

THE OVERWHELMING volume of mainstream news reporting on vaccines focuses on children, with the subject of autism taking center stage, and journalists always fussing to remind us that study after study has proved that there's no connection. As if vaccines don't cripple children in other ways. As if every day adults don't become seriously ill after getting a shingles or flu shot in their doctor's office or at the local CVS or Walgreens.

Equally underreported is how vaccines have wrecked the lives of our military personnel and their family members. There are fewer options for avoiding dangerous shots in the military than there are for parents of school-aged children. Once those papers are signed, Uncle Sam owns you. Refusing an inoculation can result in anything from

dishonorable discharge to court martial and imprisonment. Those who suffer severe side effects are often told, predictably, that the shot had nothing to do with it, that it's stress, that they're just imagining things, and so on.

SPECIAL INVESTIGATION
LIFE
The Tiny Victims of Desert Storm
Has Our Country ABANDONED THEM?
Gulf War veteran U.S. Army Sgt. Paul Hanson and his son, Jayce, age three
NOVEMBER 1995/$3.95

Gulf War Syndrome has received only a smattering of media coverage over the last three decades, and the silent suffering of our veterans is drowned out by the perpetual Orwellian ruckus over the madman of the year and his nuclear weapons, or anything else in the endless stream of TV and front page pabulum. On very rare occasions the media draw our attention to something that really matters. One example was the heart-wrenching cover story of the November 1995 issue of *Life*, which focused on the plight of five young children born with a wide range of devastating birth defects to veterans of Operation Desert Storm. In some cases, both parents had been stricken with chronic health problems, even though only the father had been sent to the Middle East.

A photograph of Sergeant Paul Hanson holding his son Jayce, who was born with heart and blood abnormalities, and misshapen arms and legs that had to be amputated, appeared on the cover. Sergeant Hanson had fought in Iraq, and returned home with debilitating internal symptoms. Take a good look at the pain on this man's face, and ask yourself how many tragedies involving military service, never reported in the press, are played out every day in broken homes across all fifty states.

Because our troops were exposed to so many toxins during this conflict — among them insecticides, burning oil wells and garbage pits, radiation from shells lined with depleted uranium, nerve gas decontaminants and preventive drugs, as well as anthrax vaccine and in some an experimental anti-botulism vaccine — it's impossible to pin down any one agent, other than to say that their body fluids were saturated with poison. Needless to say, many were stonewalled by Army doctors. In the first Gulf War of 1990-91 alone, some 250,000 young Americans — more than one-third of the troops deployed to the Middle East — were so ravaged by American chemicals that they were forced to seek medical attention, their constellation of symptoms loosely defined as Gulf War Syndrome. Many more have been afflicted since 2003 while fighting abroad in the phony War on Terror, and will be sick until they die. This does not include all the psychological damage, substantiated in official figures of twenty suicides committed each day by those serving, past and present, in Iraq and Afghanistan. As to innocent victims of American terrorism in these two countries alone, the number is beyond calculation.

If the role of vaccines in Gulf War Syndrome suffered by U.S. infantry troops is murky, it's quite

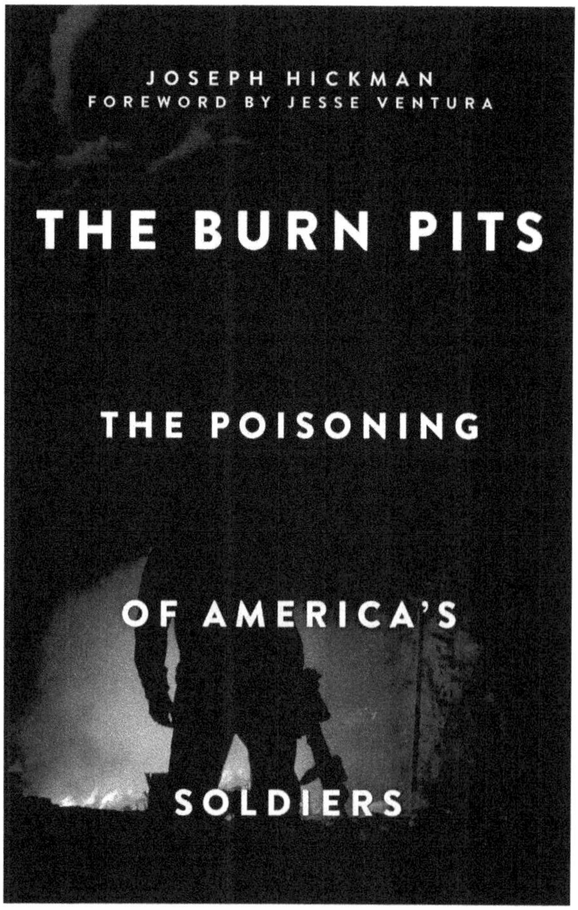

clear in the case of fighter pilots who did not come into contact with toxic substances prevalent at ground level. Lieutenant Colonel Thomas Heemstra, an Air Force fighter squadron commander who flew fifteen combat missions in the Middle East, wrote a book revealing how this unnecessary and very risky shot was forced on his fellow pilots, and the dissension it caused in the ranks. The evidence of its danger was undeniable, yet the attitude of the higher-ups continued to be "You're going to get this shot whether you like it or not." Not everyone submitted, but among those who did, many became so ill that they were never able to fly again. Some died.

This bombshell book, *Anthrax: A Deadly Shot in the Dark*, which you will not find in bookstores or libraries — not one of the 107 public libraries on Long Island has it — names names, and documents, at the highest levels, the unbelievable corruption and indifference to the lives of our men (and some women) in uniform. *The Burn Pits: The Poisoning of America's Soldiers* by Joseph Hickman, published in 2016, is another shocking book blackballed by the Establishment and unknown to the reading public. Either book has the power to inspire a coup d'état from within the military. After all, what kind of warmongering government does this sort of thing to its own warriors?

It's reassuring to know that there are men made of the right stuff like Thomas Heemstra in the military, though he was forced to retire for refusing the anthrax shot. On the other hand, in writing lines like "fighting terrorism around the world" and "the human rights we defend around the world," he apparently thinks it's patriotic to drop bombs on innocent people, and doesn't see himself as a terrorist, though much of the rest of the world does. Such spotty thinking is also typical of pediatricians who are blind to all the harm they cause by injecting children with scores of vaccines. It takes superior intelligence to fly an F-16 or get through medical school, but even some intelligent people lack the ability to think objectively, and can be conditioned to believe the nuttiest things.

Wars always generate suffering that lingers long after the fighting ends, like the radiation poisoning that continued to sicken and kill many thousands of Japanese years after the atomic bombs were dropped. The inhalation of toxic vapors for months or even years on end is a more recent by-product of the American war machine, a nexus of criminals in government, military, industry, finance and media.

The atrocities committed by these ghouls are so numerous and so dreadful that no one can possibly grasp the resulting scale of human misery. What hundreds of thousands of young Americans in uniform have gone through is terrible enough, but they are greatly outnumbered by innocent civilians in Vietnam, Iraq and other countries. Pictured below are children with severe birth defects whose parents were exposed to aerosolized toxins. In Vietnam, large areas of which were sprayed with the defoliant Agent Orange to deprive enemy forces of cover, close to a million victims live out their pitiful lives. Even today, in the third generation removed from the Vietnam War, deformities commonly occur in newborns, suggesting a permanent mutation of genetic codes. In Iraq the situation is tragic beyond description, with truly hideous birth defects seen in many babies born in the past twenty-five years, not to mention the astronomical rise in cancer rates in the general population. So far as I know, photographs like these have

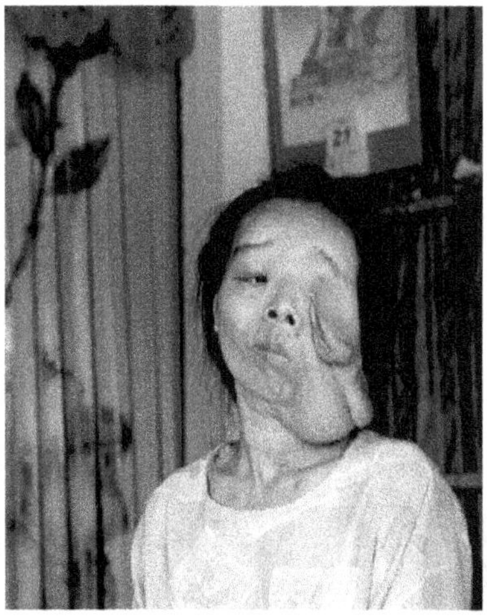

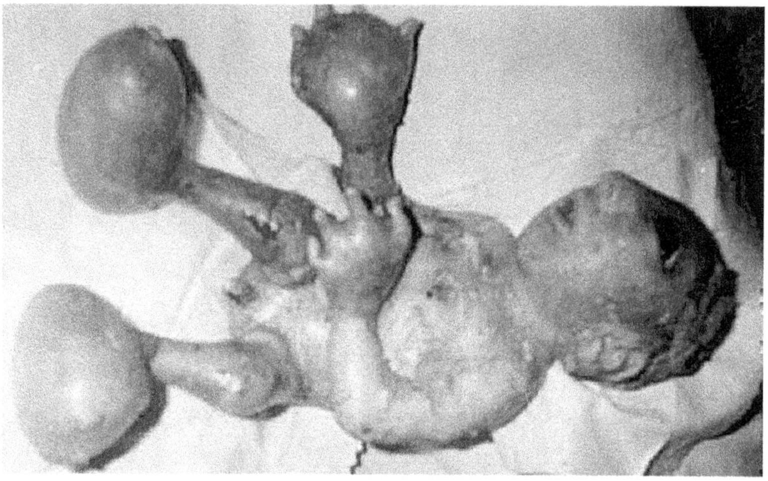

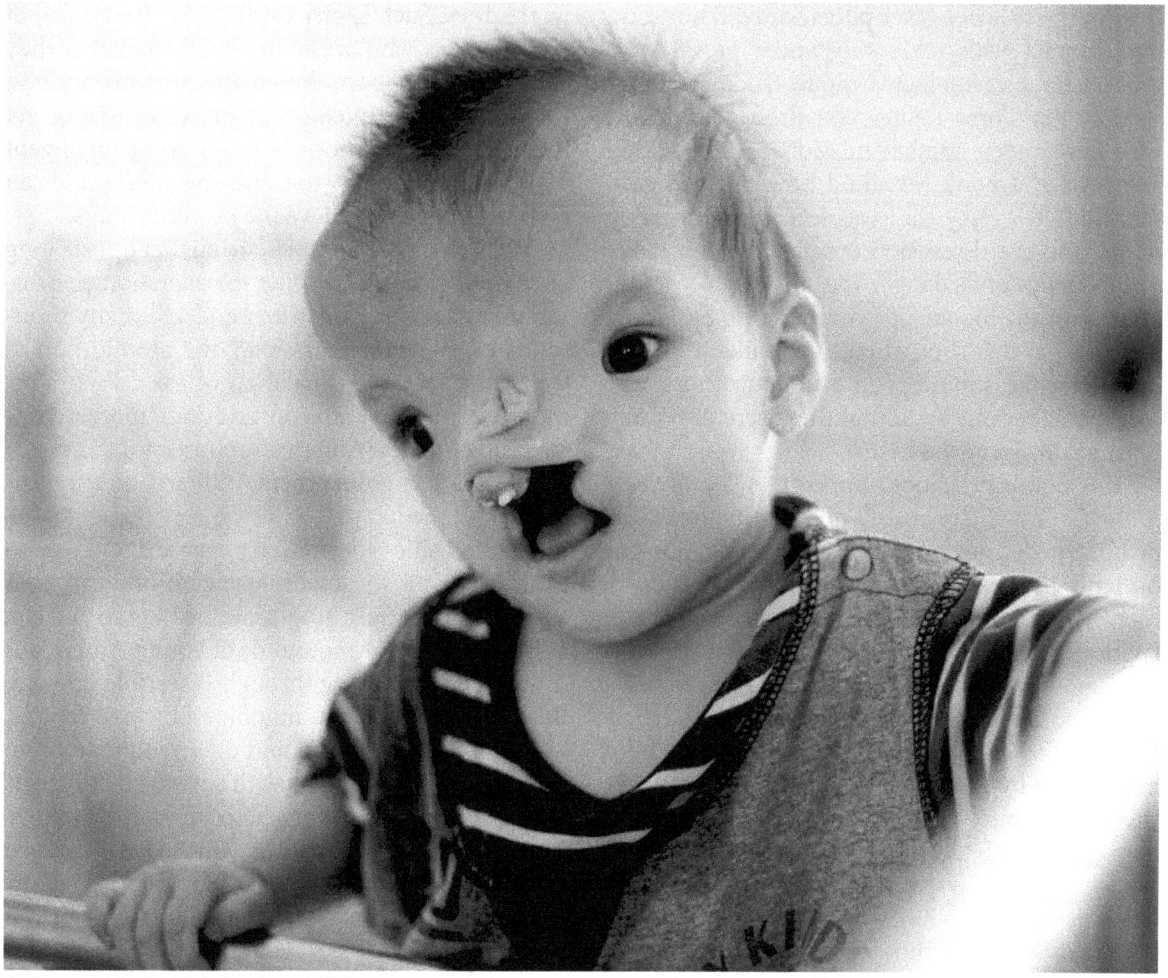

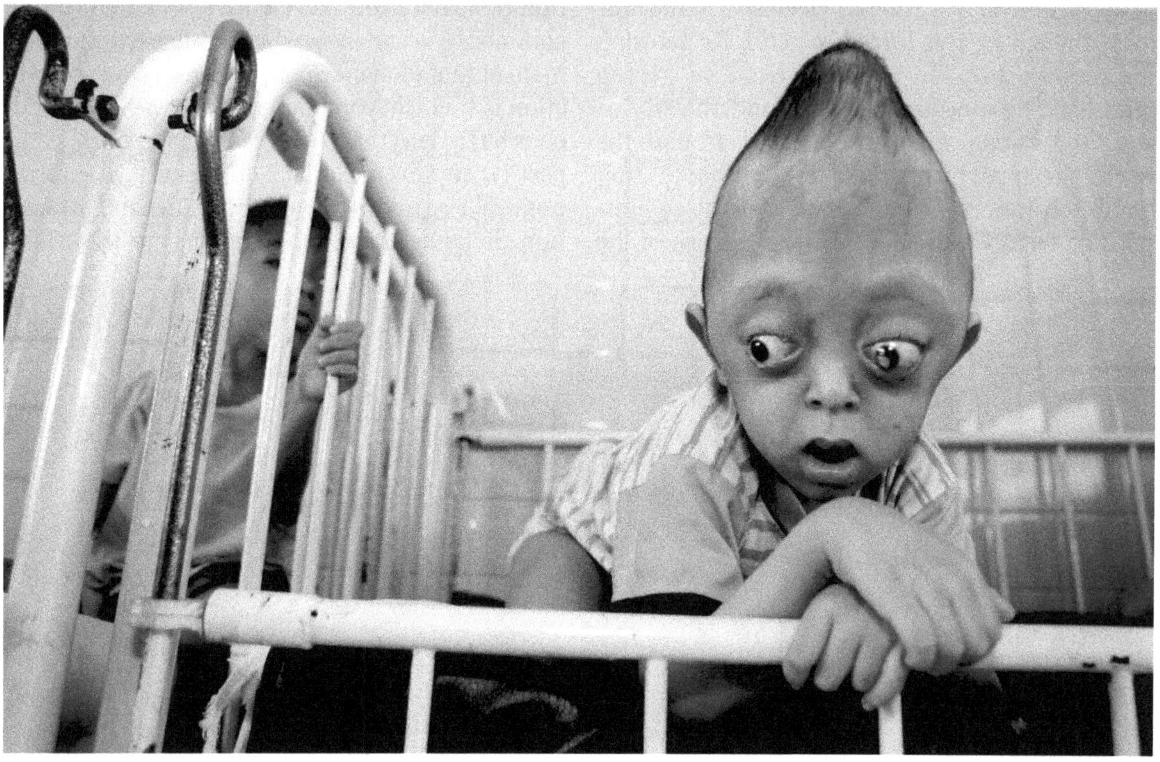

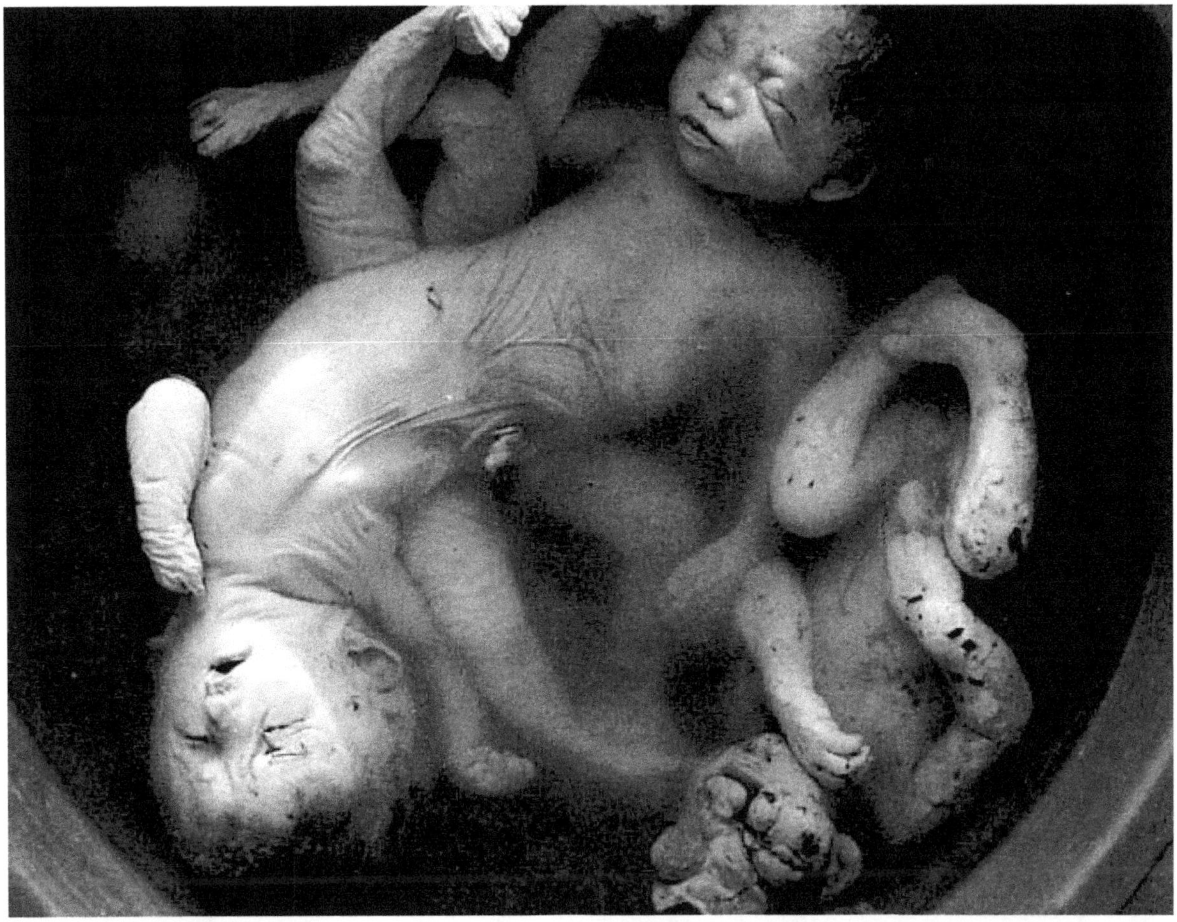

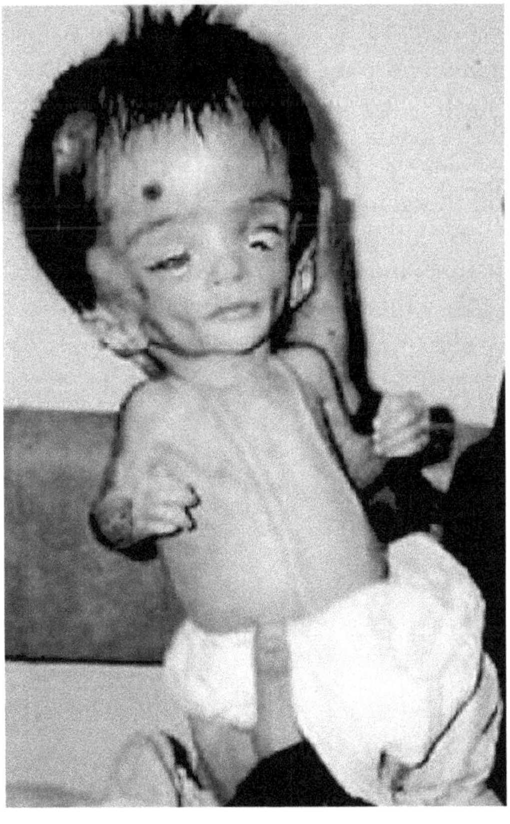

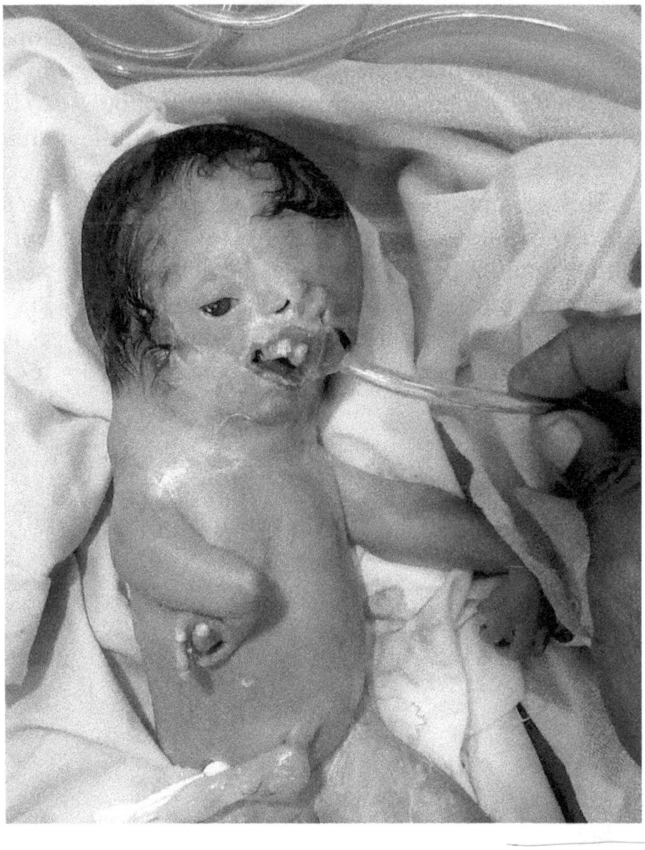

never been published in any mainstream American newspaper, magazine or news web site. In the rare stories written about this subject, any photos that appear show only slight defects that have little shock effect.

This prompts me to speculate on the future of American children yet to be born. How does the minute amount of biological and chemical sludge repeatedly injected into kids today compare in toxicity to what was breathed in — and perhaps ingested in trace amounts through contaminated food and water — by so many unfortunate people in Vietnam and Iraq? Will it damage their DNA to the point where their own children and grandchildren — assuming that they're born in the first place — will reap horrifying genetic consequences? Is all this going to lead to a large dying off of our species in this country? No one knows at this point in time, and while some dissenting researchers expressed their concerns about this as far back as the 1970s, far too few are asking these questions today.

Business is business

IF MONEY isn't the root of all evil, I can't imagine what is. It's an unpleasant fact that there are a lot of people in this world with no scruples whose main purpose in life is to acquire wealth without limit. "Corporate greed" defines most of them. To these people money is like a drug they can never get enough of, and to hell with the rest of the world. Lives wrecked every day by vaccines? New autism diagnoses, frantic trips to the emergency room, funerals? No problem. Business titans have bottomless purses, and there's an endless supply of politicians and journalists for sale.

I once got a firsthand lesson in this corporate mentality. On visitation weekends with my children I would drive a total of 480 miles between Long Island and northern Connecticut. I usually took my father's car, which sat in the driveway all the time, as he had stopped driving in his late eighties. Although I have long favored Japanese over American cars for their reliability, I figured his 2006 Lincoln Zephyr with 46,000 miles was less likely to break down than my 1997 Toyota Camry with 178,000 miles. For three years I'd had no problems with the Lincoln, even though it wasn't my style.

In April 2014 I was returning to Long Island with my son, and we were in stop and go traffic on the Northern State Parkway, only five miles from my dad's house. Suddenly the wrench icon, which I didn't even know existed, lit up on the dashboard, and the engine began to knock. As traffic picked up again, I stepped on the gas but the car continued to buck and would go no faster than 20 miles per hour even when I put the pedal to the floor. "What the hell is this?" I wondered. Fortunately, traffic slowed down again and I was able to get off at my exit. Unable to accelerate, I put on the four-way flashers and crawled to my father's mechanic, whom I knew to be honest and competent. I shut the car off, went inside, and told him what had happened. He came out and started up the car. The engine did not knock and the wrench icon did not light up. Everything seemed normal. He explained that it was a minor glitch and that the engine computer had now "recalibrated" itself, whatever that means. I then drove the short distance to my father's house and everything was fine. I shrugged it off and forgot about it. For the next six months I continued to make round trips to Connecticut with the Lincoln, clocking another five or six thousand miles without incident.

Then, in October, it happened again. I was driving north on the Merritt Parkway in Connecticut in a construction zone, just outside Milford, when the wrench light came on, the car started jerking, and I couldn't pick up speed. Once more, I was lucky to be in slow traffic, just a half mile from the Milford connector, which feeds into Route 1, the main business road that runs the length of the Connecticut shoreline. I exited with my flashers on, cars flying past me, anxious to stop at the first repair shop, which happened to be a Firestone Auto Care center. I parked the car, went inside, and explained to the manager what was going on. He had one of his mechanics bring the car into the garage to check it out. Ten minutes later he came in and said he could find nothing wrong, adding that I should've kept the car running the whole time, because that's the only way the diagnostic codes would've pinpointed the problem. Or something like that. All this newfangled computerized crap, I thought. They were very professional and didn't even charge me. Before I left, the manager printed an article from the Internet and gave it to me. I drove another forty miles north to pick up my son, then drove back to Long Island, the car running fine the whole time.

That night I read the article, which came from the web site carcomplaints.com. It was dated March 4, 2014 and headed "Ford Agrees to Fix 1,600,000 Cars That 'Limp Home.'" It began:

The National Highway Traffic Safety Administration (NHTSA) has closed their investigation into throttle body problems in Ford vehicles after Ford agreed to fix about 1,600,000 cars and SUVs. Apparently, Ford doesn't want to call it a recall, and NHTSA says it's not a recall. Instead, they are calling it a massive "customer satisfaction program" that extends the warranty of the electronic throttle body (ETB) for up to 10 years of service or 150,000 miles. NHTSA opened the investigation in February 2013, after thousands of complaints about ETB failures in model year 2009-2013 Ford Escape, Fusion, Mercury Mariner and Mercury Milan vehicles. The affected vehicles are equipped with 2.5 L and 3.0 L engines that seem to stall and lose power. Ford identified 59,807 warranty claims related to ETB replacements and traced the problem to electrical connectivity problems in the ETB. Owners reported vehicles that seemed to stall and completely lose power, but investigators found the vehicles entered a limited "limp home mode" that kept the cars moving but at drastically reduced speeds.

I dug up more information on the Internet, discovering that the problem extended beyond the models mentioned above, as in the case of my father's car. I read about forty stories written by people who, just like me, had been driving along when suddenly the car lost power but kept running. Some had been cruising on freeways in heavy traffic, and expressed the greatest relief at having been able to switch lanes and make it onto the shoulder without getting creamed. Many also were furious with Ford Motor Company. Some had been given the runaround, others had the ETB replaced only to find that the problem persisted.

I emailed the NHTSA on this and never received a reply. Whether they're in bed with the big auto makers like the FDA and CDC are with the big drug firms I can't say, but nothing in this country surprises me anymore. I brought the car to the Lincoln-Mercury dealership where my father had purchased it new. I explained the problem to the service manager, who played dumb. When I said that quite a few people had written about it online, he replied, "Oh, you can find anything on the Internet." He was polite and seemed sincere, but his loyalty to the company and maybe even holding his job required him to keep his mouth shut, I figured. What would I get out of him by being confrontational? Despite 59,807 warranty claims, and an agreement to fix 1,600,000 automobiles, nothing is wrong with the car. Nobody knows anything. He said if I could leave it for two days he'd have someone thoroughly check it. This I did. When I returned, he said they had found nothing wrong, and as a courtesy they had washed the car!

I mulled over having the computer replaced for $1200, but held off, uncertain that it would fix the problem. The car had been running fine for eight months after the second breakdown — I drove it around town now and then — but I wouldn't sell it in this condition. I almost felt like junking the damn thing. In the end my father gave it to his grandson, my nephew. I warned him that it could go into "limp mode" at any time, that there was simply no way of knowing if and when it would happen again. I also advised him to drive it only on local streets, where the speed limit was 30 to 40 mph, and never take it off Long Island. My brother was comfortable with that, and mentioned getting it fixed eventually, but in the back of my mind I wondered how I'd feel if my nephew were to disregard my advice, drive it on the parkway and have a serious accident. It became moot one night when he *did* drive the car on the parkway and had a serious accident, but not because it went into limp mode. In a juvenile fit of road rage, he lost control and totaled the Lincoln, walking away without a scratch. I think that straightened him out, and who knows, it might have canceled an appointment with fate. It certainly solved the problem. I suppose sometimes bad things really do happen for a good reason.

So far as I know, Ford never sent out a single notice warning those who had purchased their vehicles of this potentially deadly hazard. After all, it wasn't a recall. This is understandable, given that they have been swamped by so many complaints about unrelated defects that did result in recalls, as I learned, and even a company this huge probably doesn't have the resources to call back and fix, or pretend to fix, millions and millions of cars. And it seems the NHTSA is a party to the cover-up, as are the media. So, at this very moment, there are vast numbers of people driving Ford vehicles, unaware that lightning can strike at any moment. I'd like to know how many people have been killed or seriously injured already because of this electronic throttle body problem. It all smacks of the sordid silence wrapped around the issue of vaccine safety, which we're supposed to believe has been adequately addressed, but it hasn't at all because the harm continues. But as long as the money keeps rolling in, the corporate mentality of "who cares how many people die" keeps rolling along with it.

I write about this because, aside from all the vaccinations I got, it's the one episode I know of in which Big Business played fast and loose with my life. I'd put a lot of miles on that Lincoln driving on busy Interstate 95 in Connecticut, and I'd been tailgated countless times. I don't need to explain what would have happened had I been barreling along at 65 miles per hour, with the driver of an 18-wheeler twenty feet behind me, and the car suddenly lost power. Nor should Ford be singled out. One can easily find plenty of related horror stories involving the other auto makers. As with drug companies, the car companies furnish newspapers and television stations an enormous amount of advertising revenue, so the media go easy on them. One hand washes the other. That's just good business.

❖

There are any number of scurvy enterprises that come to mind when we think of unconscionable greed — war, human trafficking, prostitution, narcotic drugs to name a few. They've all been around for a long time, and they're not going anywhere. The scale of man's inhumanity to man in the pursuit of wealth is staggering. Among civilized nations, I suppose England would take first prize, though several others come in for dishonorable mention. I think here of the slave trade, tainted throughout Europe by Jewish involvement, which flourished for 245 years before it was abolished in 1807, and made fortunes for the owners of ships, factories and plantations, and as always, bankers. Records were not always kept, but historians are generally agreed on a figure of roughly 3.5 million slaves sold to British businessmen — by equally cruel African merchants who rounded up their fellow natives and transported them to embarkation points — with twenty percent of these dying on forced marches to the coast or while crossing the Atlantic.

In the early nineteenth century the British, who had colonized India, where opium poppy plants grew in profusion, flooded China with this lucrative drug, making addicts out of a large segment of the huge Chinese population. This prompted the emperor of China to take up arms in order to stem the supply, which became the Opium War of 1839-1842. More recently, in the late 1960s, the central government of Nigeria, another former British colony, crushed the independence of an oil-rich breakaway province known as Biafra, with the quiet blessing of Shell/British Petroleum. This was accomplished by blockading Biafra and systematically starving the population, resulting in more than a million deaths. Dozens of similar stories could be cited, especially among the old European colonial powers, but there's no need to go into all of it.

❖

Every now and then, a book that's actually worth reading appears on the *New York Times* bestseller list. *And the Band Played On*, published in 1987, is a study of the then nascent AIDS epidemic. The author, Randy Shilts, was an investigative journalist based in San Francisco who researched and wrote about AIDS full time. He was also an avowed homosexual, himself destined to die of AIDS in 1994. Against a backdrop of sad anecdotes of gay men slowly dying of this horrible disease — almost certainly man-made, and deliberately spread by way of vaccination, which Shilts apparently never realized and will be explored later — this book raises the curtain on all the infighting among federal public health agencies, politicians, gay activists and business interests as to the proper course to be taken in the face of such an unprecedented threat — a threat which, in the early 1980s, could not be quantified, and in the worst case scenario foreshadowed a pandemic equal to the history-changing Black Death of the fourteenth century.

To digress, younger readers might be amazed to learn that this business of accepting homosexuality as nothing more than an alternative lifestyle, so relentlessly pushed by the education system and the media, is fairly recent. We've come a long way since Thomas Jefferson could freely write that homosexual men should be castrated and lesbians have a hole punched in their nose cartilage. Right up to 1962, consensual sodomy was a felony in every state, and when I was young, words like "queer" and "fairy" could be tossed around without first looking over one's shoulder for the Thought Police. Back then, in the 1960s, the word "gay" had not yet been appropriated by pansies — it simply meant what it had always meant: "merry." To be sure, there were queer rumblings throughout that decade that I hadn't picked up, but it wasn't until the Stonewall riot in Greenwich Village on June 28, 1969 that the gay liberation movement burst into the open.

In reality there is nothing gay about homosexuality. Homos by definition are very troubled people. Rates of alcoholism, drug addiction, suicide and sexually transmitted diseases, among other woes, are far higher than in heterosexuals. An astonishing proportion of the most infamous mass murderers have been homosexuals. For obvious reasons, if everyone were queer *Homo sapiens* would go extinct in one generation. I'm certain that homosexuals are born, not made, although in these decadent times of gay pride parades and gay-straight high school clubs some weak natures are lured into making sexual choices that just fifty years ago they wouldn't have. It's unfortunate that due to a hormonal imbalance or quirky shuffle of genes or whatever the reason, a small minority end up being sexually attracted to those of the same gender. I pity those homosexuals who keep a low profile, leave children alone, and pass through life silently suffering from a distressing condition they have no control over. On the other hand, those who flaunt this perversion need to be dealt with radically, especially the worst of them, like a Texan named Robert Schwab who in 1983 suggested that his fellow AIDS-infected queers engage in "blood terrorism" — deliberately give blood if their demands for AIDS research money were not met by a certain date. But this will never happen in a country as far gone as America, where intelligent and forthright leadership does not exist, a point unintentionally made by Shilts.

So many names appear in his book that the violent clash of personalities splashed across its pages is not surprising. Nor was I surprised to learn that even among homosexuals there is a wide range of dispositions. There are those with a conscience who feel an obligation to protect not only others of their kind but also society at large from AIDS. Shilts was one. Then there are those filled with blind hatred, who harbor a death wish against the whole world, who think, "If I'm going to die, I'll take as many people with me as I can." Such sentiments are not uncommon, which is why, at last count, homosexuality is a crime in 74 countries; in 13 of these it rates the death penalty. One can only regret that it was not on the books in Florida in the late 1980s, when an AIDS-infected dentist named David Acer carried out his death wish. Knowing his days were numbered, he deliberately transmitted the virus — through routine dental procedures contaminated with his blood — to several of his patients, four of whom later died of AIDS. The first, a young woman named Kimberly Bergalis, received scattered news coverage, but Acer's heinous crime was passed off as accidental.

Much of *And the Band Played On* focuses on the bathhouse scene in San Francisco and New York, and on an institution most people regard with respect — the blood bank. For the uninitiated, I won't go into any detail on gay bathhouses, other than to say that they were dens of wildly promiscuous, biologically insane, "anything goes" sexual activity. Even among queers there were calls to shut them down in the early 1980s because they were such obvious breeding grounds for the AIDS virus. This became one of many bitterly divisive issues, according to Shilts. Naturally, the bathhouse owners adamantly opposed the measure, denying that their establishments were killing off their patrons — young men who lived for nothing but barnyard erotic pleasure. When they died, they were easily replaced by others who didn't care about smashing their lives and the lives of others on the rock of totally twisted sex. These bathhouses were big money makers, operating legally from coast to coast. The attitude of their feral owners was corporate America's business ethic: get rich any way you can, nothing else matters. (As the death toll from AIDS mounted, most of these places went out of business for lack of customers, or were padlocked by local health departments. Nationwide, fewer than 70 remain open.)

I would never have imagined, however, that those who run the nation's blood banks are cut from the same cloth. I knew nothing about this industry and its profitability in supplying blood to the nation's hospitals for use by hemophiliacs and surgery patients in need of transfusions. Rather than deal with the significant increase in labor and other costs involved in screening blood and refusing it from high-risk homosexual donors, even after thousands of innocents were tragically infected with AIDS, these businessmen simply went into denial. In fact, this book is a superb exposition of the psychology of denial — and not only denial but greed, incompetence, and the specter that, should a plague of history-changing magnitude ever threaten America, those in charge will do absolutely nothing but sit around and argue about it. Just as with the SV-40 contaminated polio vaccines, and the carelessness of putting so much mercury in other vaccines years later, the last thing they want to do is alarm the public. That would be very bad for business.

Rampant homosexuality itself is a plague, the carriers of which can never be tolerated in a healthy society. It's ironic that, had they stayed in the closet where they belong, a chain of vaccination crimes that led to the AIDS epidemic would never have happened. And as I've said, not all homosexuals are evil. There are some who deep down inside know that their sexual inclination is a curse, a dead end, and wish to do something good in their lifetime. Randy Shilts seems to have been somewhere in the middle. What he failed to grasp was that, even though one can empathize with the terrible suffering of a dying AIDS patient who did not knowingly give the disease to others, the wasting away of a person who has had deviant sexual encounters with more than five hundred different men — quite common among homos — is of no more concern to a monogamous heterosexual than the fate of a bum living under a bridge and drinking from a bottle of vodka morning, noon and night. Nevertheless, he left behind a timeless chronicle that exposed the sorrows of a self-destructive lifestyle that is anything but "gay," and of equal importance, what big business in America is all about.

And vaccines are big business.

Studies studies studies

RIGHT up to the time I first began to read about vaccines, that is, my late thirties, I never gave a thought to anything I heard or read about medical studies. We all know how it goes: "According to a recent study by..." "Several studies have confirmed that..." "A study just released by Cornell University has found that..." A study says this, a study says that, then another study comes along and contradicts a previous study. And without thinking about it, most of us just believe it for a few moments until something else occupies our attention. After all, these are doctors, scientists, highly intelligent and conscientious men and women, trained specialists in their fields — the experts. Looking back, I'm amazed at how naïve and ignorant I was about the lack of substance in so many studies. How can anyone put any credence in a study about vaccines — or any other drug for that matter — when that study is funded by the company that makes them? This is the rule, not the exception, I learned. It would be funny if it weren't so crooked.

A study is a careful analysis of data that have been compiled. Of course, there's nothing wrong with this if it's done the right way. In fact, studies conducted rigorously and honestly are the benchmark of true science and have contributed immensely to our knowledge of how the world works. For example, it was an unassuming Austrian friar, Gregor Mendel, who, in the mid-nineteenth century, crossing pea plants in his monastery garden and meticulously recording the characteristics of succeeding generations, established the law of genetic inheritance, even though the profound importance of his discovery did not sink in until after his death. Unfortunately, however, honesty and integrity have always been rare qualities among those who carry out vaccine studies, and their findings reflect this. In fact, what are called "studies" in the vaccine world are often sequels to criminal experiments.

After a while, it becomes redundant reading about the corruption inherent in studies related to human health, and not only those involving vaccines. So much has already been written about it that it seems pointless to write more. Nevertheless, vaccine studies merit a brief discussion, because they are so often cited by our opponents to bolster their case. Several years ago I made note of a few comments by honest doctors about medical studies in general. From Dr. Bruce West, in the July 2009 issue of his newsletter *Health Alert*: "No one — not you, me, your doctor, or anyone else — can glean the truth from most medical studies any more. The conflicts of interest and the slanting, omitting, and skewing of the data have become too pervasive to be able to make sense of most research. In the end, follow the money." And this from Dr. Russel Smith, who wrote copiously on the cholesterol-lowering statin drug racket: "These conflicts of interest are so pervasive in medical research that it is astonishing that anyone can take the results of most of that research seriously, particularly as it almost always 'proves' the great efficacy of each and every drug evaluated. I cannot accept as scientific evidence the results of any study in which the researchers receive pay, honoraria, 'consultation' fees, travel/hotel expenses, etc."

There's a saying, "Figures don't lie but liars figure" that seems to be apropos here. Just jerk the numbers and you can come up with any conclusion you like. Or employ any other underhanded method to arrive at a "finding" that suits you. Robert F. Kennedy Jr., a familiar face in this battle, has said, "One could design an epidemiological study to prove that sex does not make you pregnant — just remove all the pregnant people from the pool before you study it!" Actually, he was making a witty analogy to a 2007 CDC study that claimed to disprove an association between thimerosal exposure and autism. This particular study was widely cited by the vaccine pushers, including leading light Paul Offit, as definitive, as the last word on the subject. Its chief author, CDC researcher Dr. William Thompson, later came clean and, filled with remorse, explained how fraud was built into this study, and also into other studies on the connection between vaccines and autism that he co-authored in 2004 and 2010.

Interestingly enough, despite all the bogus studies that have put a positive spin on vaccines, a large number conducted by pro-vaccine researchers have indicated otherwise, even if this requires some reading between the lines. In Viera Schiebner's book *Vaccines*, there are well over four hundred references to such studies that have appeared in mainstream medical publications over the decades. Admittedly, I haven't read any of these studies. Life is short, and I don't have time to read everything. In fact, I've read a grand total of one study on vaccines. Medical studies are tedious reading anyway, and to my mind no vaccine study on either side of the issue is going to stand as the final word. As I wrote earlier, in this world of swirling lies I regard as absolute proof only that which I've seen or haven't seen with my own eyes. For that reason, there are very few references to studies in this book, even though I do believe that studies that point to the worthlessness or dangers of vaccines are honest, simply because these researchers have nothing to gain from publishing their conclusions other than having their names smeared.

I'll tell you about the one study that I took time to examine, because it led to one of the very few amusing anecdotes I came across in my thirty years of studying vaccines, though the study itself was anything but funny. I've read two versions of it, which is why I'm always wary of statistics, and why I believe as the unequivocal truth only that which I've seen. According to one source that I failed to record, in the late 1980s, WHO conducted a particularly fiendish experiment on some 15,000 babies in Haiti and the west African nations of Senegal and Guinea-Bissau. They were injected with a measles vaccine called Edmonston-Zagreb High Titer which WHO's field workers knew to be extremely potent. About 2000 of these babies died. The good doctors at WHO had ordered an additional *250 million* of these doses for further experiments, but when the results of this criminally insane project became widely known, further experimentation was called off. An article on this episode with a much different spin, which I read in all its labyrinthine language, appeared in the October 12, 1991 issue of *The Lancet.* This lame article mentions only Senegal, says nothing about 250 million ready-to-use doses, puts the number of subjects at only 1015, and the number of deaths — depending on how you interpret the murky data presented — at 123, and mentions that an equally super-concentrated measles vaccine from a different manufacturer was also used, as well as a standard vaccine for comparison purposes. Whichever source is the more accurate, both affirm that these field researchers *knew* that many infants would die. The *Lancet* article admits it, stating that "a mortality surveillance was set up both to check safety and to evaluate the vaccination strategy." (Note the euphemism. Check safety? Doesn't "mortality surveillance" mean check deadliness?) Despite the enormity of this crime, it was repeated a few years later with the same EZ-HT vaccine in a joint CDC-Los Angeles County Department of Health-Kaiser Permanente "study" using 1458 mostly black and Hispanic babies as guinea pigs after their parents had signed a fraudulent consent form. (Kaiser Permanente is one of those inscrutable "health care" consortiums.) Apparently, several were seriously harmed but only one died, almost certainly because they were much better nourished to begin with than the African babies. I don't recall if it was during the experiment or in the aftermath when the outrage came to light and word spread through the L.A. ghetto, but in response to a door-knocking campaign to get their children "immunized," a big black mama greeted the messenger holding a large kitchen knife, and he fled. The story may be apocryphal, but having observed several black folks in near violent mode, I can easily picture such a scene. I'm not in the habit of paying compliments to black welfare mothers, but how can one not side with such a woman for obeying her maternal instinct? Is she not more intelligent in a fundamental sense than a white soccer mom with a PhD who has her baby vaccinated to obey the code of social respectability?

I haven't kept pace with the latest vaccine studies because human nature does not need to be updated every ten or fifteen years. In any case, the admissions of insider Thompson would seem to obviate any need to keep track of these studies. For what they're worth, I've included five summaries of studies as I wrote them when I first began working on this book:

A 1988 study sought to explain the rising incidence of mumps in the U.S. The authors noted that acceptance of the live mumps vaccine, licensed in 1967, was only gradual, and despite low immunization rates, reported cases of mumps declined. They also acknowledged that recent immunization rates were very high — 80 million doses administered. Therefore, the vaccine doesn't work, and it might well have caused the disease it was supposed to prevent. But, resorting to mental gymnastics, the authors reached the opposite conclusion: that the resurgence of mumps was due to failure to vaccinate all susceptible persons.

Two types of Hong Kong flu vaccine were tested on 1200 students on October 30 and 31, 1968. Also participating in this study was an unvaccinated control group. Within a few days, confirmed cases of Hong Kong flu broke out among the vaccinated. But to "prove" that the vaccine was effective, only cases that occurred after November 16 were included. When *all* cases of the flu were taken into account, the incidence of flu in all three groups was found to be the same.

In a disgraceful Hib vaccine trial conducted from July 1988 to August 1990 on Navajo Indian babies, half of the 5200 infants were given Hib in combination with DPT and OPV (oral polio vaccine), the other half only DPT and OPV with placebo (2 mg of lactose). After most had received their second dose, the experiment was stopped. Eighty-three cases of invasive infection were reported with a somewhat higher incidence in the placebo group (2.3% vs. 1.7%). In each group, eight babies died. However, in summarizing their findings, these sixteen deaths were ignored by the trial committee because they considered them unrelated to the vaccinations.

A study undertaken in 1978-79 to downplay the risks of the pertussis vaccine reported that out of 15,752 shots that were given to children only 18 serious reactions (shock-collapse or convulsions) occurred. Three to five shots were administered to each child. Thus, what appears to be a severe reaction rate of 1 in 875 is really more like 1 in 200 children. The study was a sham in other ways, too: the number of children studied was not given; the groups were pre-screened to exclude high-risk children; reactions that occurred more than 48 hours following vaccination were not counted; and high-pitched screaming, which indicates cerebral irritation, was not classified as a serious reaction.

A famous 2002 Danish study, trumpeted by the CDC as well as the *New York Times* and the *New England Journal of Medicine*, conclusively disproved, they said, that thimerosal caused autism — the clincher being that autism rates spiked in Denmark after the preservative was removed from vaccines. Closer inspection revealed several irregularities, however. At various junctures of the study period, the guidelines were changed. For the first eleven years, for example, the authors had counted only children admitted to hospitals, where autism is rarely diagnosed, but from 1995 they included outpatients as well. In 1992, the year thimerosal was removed in Denmark, they began including cases diagnosed in a busy Copenhagen clinic which accounted for 20% of the country's cases; prior to this, the clinic's records had been ignored. The following year, new diagnostic codes for autistic spectrum disorders were introduced in Denmark which "may have stimulated reporting of autism cases."

Question: If vaccines really do what their proponents say they do, why do they so often resort to tricks like this?

The final word belongs to the parents. In the late 1990s, a group of angry, intelligent and aggressive parents of autistic children calling themselves Safe Minds joined forces and went after the medicrats. Their story is documented in David Kirby's *Evidence Of Harm*. They did fine work, causing some of the CDC's top people to break into a heavy sweat, but from what I gather, over the years their energy was squandered over internal conflicts and disagreements with other autism organizations. Still, the truth of the matter is to be found not in reams of dense and mystifying studies, but as is so often the case, in plain English which takes seconds to read, as in these barbs from Safe Minds parents: "Interesting how you can design a study to support your own hypothesis....What a piece of junk.... My jaw drops every time I see these guys go to work.... They are going to hype every piece-of-crap study they manage to get out. They inflate the inferences they draw beyond any reasonable limit, and suppress and dismiss any evidence to the contrary.... What kind of study is it where you change the data in response to peer-review comments? The data is what the data is. You don't go back and change it when people make suggestions. How can you change it?"

The absurd notion of "safe and effective" vaccines

SELDOM DO people ask critical questions about things they know little or nothing about. They leave that to the "experts." The pediatrician takes a clean, neatly labeled vial from the refrigerator, and very rarely do parents give a moment's thought as to how the vaccine inside that vial was manufactured. In this high-tech, super-efficient world in which we live, why should they? If they asked to read the label, which hardly any parents do, they would see the name of a pharmaceutical firm, the name of the vaccine, a lot number, an expiration date, and the words "Keep Refrigerated," and that would reassure them that everything looks just fine. But what went into the production of that vaccine? What, exactly, does that little glass vial contain?

No matter what your line of work — whether teacher, construction worker, dairy farmer, police officer, or a thousand other occupations — you have inside knowledge about which outsiders know practically nothing. In many cases, this includes a few little "secrets" that you usually don't share with those outside the workplace.

I have never been inside a vaccine production plant, and aside from what I've read on the subject, can't tell you anything about the manufacturing process. But I have worked at many different jobs in my life, and I know something about incompetence, indifference and bad attitudes. This does not describe everyone, of course; there are those with an admirable work ethic, who are caring and are outstanding at what they do. In my experience, however, not many people excel at their jobs, nor are they really dedicated; most are merely adequate. And there is no reason to believe that this rule does not apply to those employed in the production of vaccines.

The home heating oil business, my employment for most of my life, also has its high-tech side which must be a source of wonder to some of the millions of homeowners in the northeast who heat with fuel oil as well as propane. Most opt for automatic delivery, which simply means that, rather than call their company when they are running low on fuel, the truck comes when the computer calculates that they need another delivery. This is determined by their past usage rate, and the ideal number of gallons chosen by the company, which can be programmed into the computer. Most companies consider it ideal to deliver when the customer has roughly one-quarter of a tank; that's a profitable sale, but not cutting it too close to running out. For various reasons, automatic delivery is a more cost-effective way to run an oil or propane business, and it has a good selling point. "You don't have to worry about a thing," the customer is told. "The truck will be there when you need a delivery. The computer has it all figured out."

If you have a good oil company run by people who know what they're doing, this is true in most cases. There are homeowners who have been on automatic delivery for twenty years or more and have never run out of fuel. They come home from work, open their mailbox, and five or six times a year, mainly during the cold weather months, there's a streamlined delivery ticket with their name and address, words like "K Factor" and "Degree Day" in small print, and three or four rows of numbers generated by the computer. What does it all mean? Unless you're in the business, you probably don't have a clue. But it looks so neat and precise, you can't help but marvel at how things are done so efficiently in our advanced technological age.

Now let me share some inside information with you. Over the past thirty years, driving for seven different companies, I have seen well over a thousand homes and businesses on automatic delivery run out of oil (and I'm not talking about deliveries that were withheld because of credit problems, which is a separate issue). Needless to say, this does not make for happy customers, especially in frigid weather. This happens for so many different reasons that it would be impractical to list half of them here. But I'll fill you in on a few things I've seen.

The most common source of runouts, or "drys" as we call them, is the driver failing to fill the tank properly. This happens for various reasons, most of them inexcusable. When a tank is not filled, the driver is supposed to indicate this on the copy of the delivery ticket that he turns in to the office at the end of the day, but many, being less than stellar workers, fail to do so. On the office end, it's a big advantage if your people have a knack for numbers in this mathematically oriented business, but this is rarely the case. Most office workers do their job by rote, failing to "pick up" on things. So, for example, if a delivery to a standard tank holding

250 usable gallons is made right on schedule and should be about 180 gallons, but the driver puts in only 112 gallons, they don't question it, as they should; they just do what they were trained to do — post it in the computer as a full delivery. Based on this erroneous information, the computer now calculates that this home is burning oil at a much lower rate than in the past, and schedules the next delivery accordingly. I guarantee you that every winter day, at least fifteen or twenty homes on Long Island run out of oil for this reason alone.

Another common reason is an office worker posting a delivery to the wrong account. This happens fairly often when a customer owns three or four homes; the same name appears in consecutive columns on the computer screen, and if you click on the right name but the wrong address — an easy mistake to make — the delivery is posted to a house where no delivery was made, practically ensuring that the tank will run dry.

Delivery tickets fall behind desks, land in the wastebasket by mistake, end up beneath truck seats. It's not a frequent occurrence, but it does happen. To tell a story on myself, once I walked out of the office with a stack of thirty delivery tickets and put them on the rear bumper before fueling my truck. Of course I forgot about them, until I glanced in my mirror after pulling out on the road and saw them fluttering through the air. I think I found all of them; if not, one or two people ran out of oil a week or so later. The better companies keep a record of tickets in the driver's possession so they'll know something is amiss if the delivery is not made that day and he doesn't bring the ticket back; or, they periodically run "exception reports" on the computer to identify unseen irregularities in order to prevent runouts. Most, however, do not take these precautions.

There are several automatic delivery computer programs an oil company can choose from. Some are user friendly and some are a nightmare. Some have built-in glitches that make me shake my head at those managers — and there are many in the oil business, as there are in every field — who think that more technology always means more progress. One company I worked for installed a program which, unknown to them, would not print delivery tickets for customers on a monthly budget plan unless a separate payment for each month was made. For example, if the monthly payment was $200 and the customer made an advance payment of $400 to cover both January and February, the computer would flag this as a delinquent account because no check for February was received. The house would then run out of oil in February, because no ticket was ever printed. This is obviously ridiculous, and a great way to lose good customers. Several tanks ran dry before the company figured out the problem.

Trucks break down, drivers get sick or hurt, work gets backed up and routes have to be changed. If it's unseasonably cold, with plenty of winter ahead, panic can set in. It's a stressful, sometimes frantic business. No computer can replace a good dispatcher, the person who sorts out delivery tickets and makes up routes, but most dispatchers have never driven an oil truck and lack a "feel" for the job. When a company falls behind, the dispatcher has to start juggling tickets and routes, and a common mistake is to reschedule heavy users with average and light users, which always results in a few drys.

January 1994 was the roughest month I ever worked. In addition to extreme cold, we had plenty of snow with a few ice storms thrown in. It's hard enough to keep pace with a prolonged cold snap; frozen precipitation, especially ice, slows you down even more. Then we lost two of our twelve drivers; one slipped on the ice while pulling the hose and broke his collarbone, while the other quit to take care of his terminally ill wife. I worked twenty-nine days that month, twelve hours a day, and the rest of the crew put in an equally exhausting amount of time, but we could not keep our heads above water: people were burning oil faster than we could deliver it, and tanks began running dry. This was a good company with a smooth delivery operation, but we were simply overwhelmed by misfortune and Mother Nature. More than one hundred customers ran out of oil that month. I know of similar things that happened to other companies during long stretches of bitter cold, but due more to mismanagement and lack of foresight.

Sometimes even sabotage enters the picture. On my first day with a certain full service company I went out with the senior driver to familiarize myself with their procedures. He soon began to gripe about many things, one of them always being given the toughest routes. To show his displeasure, he pulled out a delivery ticket and ripped it in half in front of my eyes. One dry, on the way. There was also a serviceman there with driving experience, although he much preferred fixing oil burners to delivering. One day he was told to drive, using a spare truck. He complied, but he was furious. I happened to use the truck a few weeks later and found two crumpled delivery tickets on the floor; both houses probably ran out of oil. Few places I've worked for had so many terrible attitudes and such a lack of brains across the board. Things were done so poorly at this company of about three thousand accounts that nearly every winter morning, the answering service, which handled the phones from 10 PM to 6 AM, faxed the names and addresses of four or five customers who had run dry during the night.

Some oil customers, much less than half, keep an eye on their gauges or stick their tanks and call the office when they get noticeably low, in which case a truck can usually get there in time to prevent a runout. Those who run dry, however, learn the hard way — like losing heat at midnight when it's fifteen degrees outside — that this high-tech, computerized world of ours is not nearly as efficient as it's made out to be. And this discovery often comes with a line I must've heard 500 times: "I thought I was on automatic delivery."

❖

Now, do you really believe that incompetence, human error, computer problems, malicious acts of disgruntled employees, and events beyond human control play no part in vaccine production facilities? Even if, theoretically speaking, it were possible to make safe and effective vaccines, the highly complex and exacting process involved practically guarantees that things often will not go as planned. Nor is there any reason to believe that the people who work at these places are exceptionally bright.

And there are issues beyond production, one of which is brought up by the authors of *A Shot in the Dark*: refrigeration. Vaccines are supposed to be kept cold, but not frozen, at all times. But are they? Just like meat or dairy products that you buy at the supermarket, vaccines, which contain biological matter, degrade if not refrigerated, and they might become more toxic if the temperature soars. But who has ever tracked the journey of a vaccine vial from factory to distribution center to doctor's office to needle? Are these packed in sufficiently cold receptacles when they leave the factory or are they just shipped like ordinary merchandise then re-refrigerated at a distribution point — perhaps after being left to sit for hours at a loading dock on a warm day? Even if they're surrounded by ice packs, do they arrive at the doctor's office before the packs have thawed? Do you think the average UPS or FedEx driver knows, or even cares, about the possible consequences of a shipment of vaccines left to bake in his truck for six or seven hours when it's ninety degrees outside? Or how about a doctor or assistant who forgets to put the vaccines in the refrigerator and leaves them on the counter overnight, or over the weekend?

To return to the beginning of the manufacturing process, is the seed virus that is used free of biological debris or other impurities? Is it kept at the optimal cold temperature to prevent it from becoming stronger or weaker than desired? What if the alarm system that warns of a change in temperature during storage malfunctions? Are samples always withdrawn from the broth and examined microscopically, as they are supposed to be? Are certain viral strains that are considered too strong to be attenuated (weakened or killed) always separated from acceptable strains? Is equipment that has come into contact with bacteria or viruses always sterilized in an autoclave, according to safety protocols?

Just reading what goes into these vaccines makes me wonder how anyone came up with these recipes, and on what basis the ratio of ingredients was determined. For example, in addition to biological matter, the Hib vaccine, which is one of the simpler ones, contains aluminum hydroxphosphate sulfate, ethanol, enzymes, phenol, detergent and fermentation medium. Do factories ever run out of these or other ingredients, due to an oversight or a delayed shipment? If so, are other chemicals like glycine and formaldehyde used to replace them? Do workers just dump tap water in from time to time to fill a five gallon bucket or fifty-five gallon drum or whatever it is they use to store vaccines prior to packaging? Do they ever deliberately switch labels — for example, passing off a DPT lot as an MMR lot because someone screwed up and this saves time and effort? Do they ever package vaccines that they know were prepared improperly so they don't have to go through the trouble of making a new batch?

The questions I have posed do not even take into account the capacity for malfeasance. It would be naïve to think that there are no sociopaths employed by pharmaceutical firms, people who wreck vaccine production to spite their companies, or hate the world, or any other reason — who, for example, knowingly add an extra amount of one of the more toxic ingredients. Such an event would have much graver consequences than tearing up an oil delivery ticket. How often things like this happen it is impossible to say. We do know that, from time to time, an alarming cluster of vaccine deaths and injuries can be traced to a particular lot — a phenomenon known as "hot lots." Common sense tells us that something went wrong during the manufacturing process of a hot lot that made it highly toxic due to human error, shoddy workmanship or criminal behavior. I'm inclined to believe that, as in all occupations in the population at large, the number of inept or uncaring workers at vaccine factories far outnumbers the number who are genuinely evil, but this is not much of a consolation to those victims on the receiving end of a bad vaccine brew.

Nevertheless, criminality has always been a hallmark of this field, and anyone who researches the subject will come across episodes in which authorities knowingly put children at risk by giving the green light to suspect vaccine lots, a classic example being Jonas Salk during the Cutter incident. It would be sheer folly to think that the most powerful decision makers took this lesson to heart and revised vaccine testing procedures to protect the public. We're left wondering how many vaccine manufacturing plants are operating at this very moment under the filthy conditions that prevailed at Cutter Labs in the 1950s. And how many poorly trained or unfit lab technicians are employed by them — especially in this day and age of hiring Third Worlders at cut-rate wages.

That nothing has changed over a half century in the mindset of the vaccine pushers was made clear in an interview with a retired American vaccine researcher who worked for many years at major pharmaceutical houses and at the National Institute of Health. Fearing retribution from the FBI or IRS and possible loss of his pension for speaking out, he used the pseudonym Mark Randall. The interview was published in the February-March 2006 issue of the Australian magazine *Nexus*. The interviewer, Jon Rappoport, has written extensively on the vaccine racket and other taboo topics, but in this article he defers to the authority of an eyewitness insider who gives every indication of being honest — one formerly of the "inner circle" as he describes himself. Not only does Randall have an intimate knowledge of what goes on in the rooms where vaccines are made, he helped develop some of them. He speaks of his disillusionment and despair upon learning, after conducting a personal investigation, that his life's work was riddled with fraud and lies. He tells us that, if he had a child now and the situation required it, he would move out of state, change the family name, disappear — do *anything* to avoid getting his child vaccinated. He offers a sketch of his colleagues — on the whole, certainly not wicked people, but rather ordinary people who wouldn't think of rocking the boat, true believers and defenders of the dogma who plug their ears as soon as they hear something about vaccines being unsafe. Titled *Vaccine Dangers and Vested Interests*, this invaluable interview can be read in its entirety on nexusmagazine.com. The following excerpts spoken by Randall address the unworkable idea of creating safe and effective vaccines:

The public believes that these labs, these manufacturing facilities, are the cleanest places in the world. That is not true. Contamination occurs all the time. You get all sorts of debris introduced into vaccines. I'm talking about ...the actual lab conditions. The mistakes. The careless errors... I'll give you some of [the contaminants] *I came across, and I'll also give you what colleagues of mine found. Here's a partial list. In the Rimavex measles vaccine, we found various chicken viruses. In polio vaccine, we found acanthamoeba, which is a so-called "brain-eating" amoeba. Simian cytomegalovirus in polio vaccine. Simian foamy virus in the rotavirus vaccine. Bird-cancer viruses in the MMR vaccine. Various microorganisms in the anthrax vaccine. I've found potentially dangerous enzyme inhibitors in several vaccines. Duck, dog and rabbit viruses in the rubella vaccine. Avian leucosis virus in the flu vaccine. Pestivirus in the MMR vaccine... And if you try to calculate what damage these contaminants can cause, well, we don't really know because no testing has been done, or very little testing. It's a game of roulette. You take your chances... I have found what I believed were bacterial fragments and polio virus in these vaccines from time to time, which may have come from* [aborted human] *fetal tissue. When you look for contaminants in vaccines, you can come up with material that is puzzling. You know it shouldn't be there, but you don't know exactly what you've got. I have found what I believed was a very small fragment of human hair and also human mucus. I have found what can only be called "foreign protein," which could mean almost anything. It could mean protein from viruses... Remember, this material is going into the bloodstream without passing through some of the ordinary immune defences...Of course,*

I'm not even mentioning the standard chemicals... which are purposely put into vaccines [as preservatives].

There are decent and intelligent people, some of whom have done a great service in spreading the word about vaccine hazards, who nevertheless foolishly cling to the ideal of safe and effective vaccines. Proponents have had well over 200 years to come up with one, just one, vaccine with that guarantee and they have utterly failed. Given their ingredients, not to mention human fallibility and corruption, the notion of producing vaccines that are assuredly safe and effective is an absurdity and a dangerous delusion. The bottom line is this: there is no way that any doctor or nurse can know what's inside a vial of vaccine that he or she injects into a child, nor what will or will not happen as a consequence.

The NCVIA and VAERS debacle

ONE OF the worst bills ever passed and signed into law in this country, on November 14, 1986, was the National Childhood Vaccine Injury Act (NCVIA), which actually was part of a much larger health bill. Ironically, it was conceived in 1983 with the best of intentions by a small group of parents whose children were seriously injured by vaccines, one of whom eventually died. But little by little, over the course of three years, as politicians pawed over it, as vaccine makers and medical organizations stuck their noses under the tent, modifications, stipulations and compromises were worked in to the point where it became twisted beyond recognition.

The worst aspect of this law is that it cleared the pharmaceutical companies of all civil liability in connection with vaccine injuries and deaths, setting up a judicial monstrosity heavily stacked against parents and their handicapped children. Prior to the passage of the NCVIA, there had been so many lawsuits filed against pharmaceutical firms that some had dropped out of the vaccine market. Now they could no longer be sued, so they got back into it. Now parents had to take their case to what is informally called "vaccine court," and any awards would be paid from a fund subsidized by a small tax imposed on every childhood shot. This opened the floodgates to a whole new crop of compulsory vaccines and the millions of new casualties that have accumulated as a result.

In passing this legislation, it supposedly was the intent of Congress to create a swift, fair, nonadversarial payment program as a substitute for the grueling litigation involved in civil tort, which sometimes resulted in huge awards, and more often in nothing. A special master presides over the hearings, in which Justice Department lawyers represent the Secretary of Health and Human Services (HHS), the umbrella agency covering the vast bureaucratic empire of health-related federal agencies. The Secretary becomes a co-defendant in each case. In such an atmosphere, created by a government committed to the belief that childhood vaccines are very safe and effective, it was terribly naïve of anyone to expect things to run smoothly. And they didn't. As cases piled up, there was frequent bitterness over unjust decisions. Awards were granted or denied based on hair-splitting definitions of vaccine injuries as related in the medical literature. There was a modicum of justice for about half the families who had filed petitions. Most of these involved children maimed by the DPT vaccine, who would need professional care for the rest of their lives.

What little fairness there was in the system fizzled out with the incoming Clinton administration in 1993, and the appointment of Donna Shalala as the new HHS Secretary. Irate over the official recognition of vaccination hazards, Shalala, a Hillary crony, and one very mean and militant woman, changed the rules. Shalala had arrived after serving five years as chancellor at the University of Wisconsin, where she had been a sort of dominatrix of politically correct speech codes. These rules had riled and intimidated students and teachers alike, and were so outrageous that even the American Civil Liberties Union opposed them, challenging her in federal court and winning, which forced her to strike them down. Now, in her new post, the chunky lesbian was going to show America she meant business. Her first move was to narrow the medical criteria for compensation, already restricted and unreasonable, to the point of absurdity. Reactions to the pertussis vaccine in particular were radically redefined. One lawyer who represented children in vaccine court remarked, "What you now have can be compared to a federal program that will compensate anyone who is in a plane crash in a snowstorm within ten miles of Tahiti. Nobody will ever qualify." And as if to drive her point home, shortly after publishing the final rules in the Federal Register (Vol. 60, No. 26, February 8, 1995) Shalala fought the parents of 19-year-old Maggie Whitecotton — crippled earlier in life by a DPT shot and now confined to a wheelchair — all the way to the Supreme Court. Later that year, this beastly woman confiscated the Social Security numbers assigned to newborn babies as a first step in establishing — through the linking of separate state vaccine registers — a national electronic tracking system, to ensure universal compliance with federal vaccine policies. That project ran aground in several states but remains a utopian dream of public health Stalinists in Washington D.C.

Needless to say, with very few exceptions all of this has been swept under the rug by our news media. One exception was an article headed "Vaccine Injury Claims Face Grueling Fight," which appeared in the November 29, 2004 issue of the *Los Angeles Times*. Read this long article on the Internet. Read the names and stories of real people with real handicapped children, and the hell they go through with government lawyers who eat off your tax dollars. Read about how these heartless shysters nickel and dime grieving parents over petty items like wheelchair maintenance and special bedsheets. Read about the witness tampering, the disqualifications on legal technicalities, the cases that drag on for ten years or more. Read all this and then ask yourself how long America can go on like this.

❖

The Vaccine Adverse Event Reporting System (VAERS) was part of the NCVIA. Its stated purpose was for the government to keep track of serious reactions to vaccines by requiring doctors to report them, but due to widespread indifference on the part of both doctors and bureaucrats, this turned out to be another flop. I have read estimates ranging from 82% to 99% of adverse vaccine reactions going unreported, the latter figure coming from former FDA commissioner David Kessler in 1993. Few Americans know about VAERS. Many doctors are in denial about vaccine injuries and refuse to cooperate. Other requirements, like keeping records of dates and lot numbers of vaccines given, the signature of the person administering them, and more, are similarly ignored. No disciplinary action is ever taken. The NCVIA is a joke, a national disgrace.

VAERS seems to have been a bone thrown by the federal government to anti-vax activists. It was clear from the start that the medicrats had no intention of changing their policies despite countless reports of adverse vaccine reactions. It is equally obvious that they knew they were sitting on a powder keg and had no desire to share this information with the public. Stonewalling and monkey business are the stock in trade of these people. The first successful attempt to pry out the data was made by the National Vaccine Information Center (NVIC) whose directors had propelled the fruitless NVCIA legislation. I have some reservations about this group and the character of its co-founder, Barbara Loe Fisher, a prominent voice in this ongoing battle, but there's no question that the NVIC has done a good job in educating parents.

The NVIC was able to procure the earliest VAERS records, stored on computer disks, after filing a request under the Freedom of Information Act (FOIA). The data, however, was indecipherable, and they had to get outside help to unscramble it. In the report they issued, the pages of computerized and tabulated lists, covering VAERS records from July 1990 through July 1995, are preceded by a detailed account of their efforts and findings. It's a sad but not surprising record of bureaucratic sloppiness: data entry errors, duplicate reports, filing delays, omission of details, etc. They mention, for example, that in some instances it took eleven months for reports to be entered into the system, and that some lot numbers are wrong, containing too many characters, or having numbers missing, or a letter where a number should be. They also mention that the FDA does not follow up on reports to monitor the condition of injured children, as they're supposed to do, nor do they recall hot lots, vaccine batches with very high reports of severe reactions, even though it's easy to do so. The "hottest" lot resulted in 234 events, eleven of which were fatal. Think about that for a moment: *the government knows there are exceptionally dangerous vaccines out there that can be recalled without difficulty, yet they do nothing.* Some details regarding hot lots are not accessible, as the FDA considers this confidential information shared only between them and vaccine manufacturers. The phenomenon of hot lots shows how botched-up vaccine production can be; clearly mistakes are sometimes made in the lab or factory that result in some lots being much more dangerous than others.

After examining the VAERS reports compiled by the NVIC, I decided to do some research of my own, and purchased seven disks copied from disks turned over by the FDA in compliance with the FOIA. These covered the years 1998 through 2002. The first disk I put in the computer came up blank (the rest worked). This is a common occurrence which reflects either incompetence or chicanery on the part of FDA employees. The seminal book *Evidence of Harm* relates many examples of this. There are more. The November 2017 issue of the newsletter *Health Alert* asserts that, for a long time, some browsers, namely Firefox and Internet Explorer, have blocked access to the VAERS website vaers.hhs.gov, and an error page pops up. Using the computer at my local library, I usually had no problem getting to VAERS data, but sometimes these words appeared: "This operation has been canceled due to restrictions in effect on this computer. Please contact your system administrator." It would seem that, since this information is so explosive, since it blows the received wisdom about vaccination to smithereens, there have been attempts to shield it from public view, but that's only speculation on my part.

Hopefully this data will always be accessible on the world wide web, because it's an open admission of the serious and commonplace injuries caused by vaccines, an admission made by the federal public health bureaucracy itself — the same bureaucracy that's always telling us out of the other side of its mouth how safe vaccines are. I don't know how any thinking American, and especially parents and grandparents of young children, can examine this data without being shell-shocked. The volume of these reports is simply staggering. I scanned only about one hundred pages from the VAERS website, and if these pages represent the average of daily reports, by my count we're looking at roughly 18,000 reports per year, keeping in mind that these might well represent less than one-tenth of alarming reactions that occur nationwide.

The aforementioned NVIC report, published on October 10, 1995, at the dawn of the Internet age, was the first of its kind. I have a bound copy, and in thumbing through it, the importance of all this data instantly becomes clear, as does the painstaking work it took to arrange it. The format, however, is outdated and the print is so tiny as to be nearly illegible. Furthermore, it covers only adverse reactions to the DTP (diphtheria, tetanus and whole cell pertussis) vaccine, the DTaP (DT and acellular pertussis) vaccine, and the DTPH (DTaP plus Hib) vaccine; only DTaP is still in use in the U.S. (As an aside, the high risk of the DTP [or DPT] shot due to the whole cell pertussis component was known for fifty years, and no one ever did a damn thing about it. DTP was phased out in the U.S. in 1996. However, some countries still use it, and there has even been some discussion in official circles of resuming its use here, because with whooping cough on the rise it has become obvious that the DTaP vaccine doesn't work. The madness never ends.)

The NVIC researchers pointed out that many times other vaccines are given simultaneously, which can result in statistical inaccuracy. They stressed that their list contained what they believed were current lots as of July 31, 1995, meaning that many more adverse events from those lots surely occurred after that date. To further complicate things, they had no way of knowing how many doses from each lot had been administered. For what it's worth, these are the official totals from the DTP, DTaP and DTPH vaccines as compiled by VAERS from July 1990 through July 1995. Events: 18,657, of which all but 39 involved children aged from infancy to six years. Emergency room visits: 8,425. Life threatening conditions: 293. Hospitalizations: 2,387. Permanent disabilities: 201. Unknown recovery status: 2,947. No recovery (excluding deaths): 587. Deaths: 638. Incidentally, while the VAERS database was established in 1990, these figures are not available on their web site. Also, if the megabyte count given for each year on the VAERS site is any indication, the number of reports filed in recent years has multiplied about tenfold from the early 1990s, and in all likelihood so have the above figures.

Reproduced below are pages I randomly chose to reprint from vaers.hhs.gov covering the years 2001, 2012 and 2016. The data on each report takes up more space than can be reprinted on a single page, so in order to print a page for this book showing the state of residence, age and sex of an individual, as I have done in the 2001 and 2016 samples, I had to leave out detailed symptoms, which appear by clicking on a different tab. I show these details in the 2012 sample, though they too sometimes run off the page. I have reproduced 25 pages to impress two things on the reader:

first, the enormous number of reports that pour into VAERS, and second, the futility of the whole system, which appears to be just another federal boondoggle.

In analyzing the data, I noticed the same kind of mistakes that the NVIC investigators ran into. In many cases ID numbers are out of sequence. The dates they were received often do not correspond to the dates they were recorded, and in some cases appear to have been sitting around for months until someone noticed them and entered them into the system. On some pages there are numerous blank spaces where state and age should appear. Also, on some pages, the sex is indicated on all or nearly all entries, while a "U" for unknown appears in profusion on other pages.

The first page, beginning with ID# 172499 on July 2, 2001, displays passable work, with just one state and two report date omissions, plus four sequence number gaps. The next four pages containing 208 entries (not shown) are also satisfactory. Beginning on July 9, however, there's a considerable number of omissions, though this page is not nearly as bad as some of the 2016 pages that follow.

Several issues arise from the early January 2016 sample, over and above the numerous "sex unknowns" and blank spaces where states and ages should appear. (I've excluded several pages in this time slot to economize space.) Why, compared to earlier years, do middle-aged and older people now outnumber children? The astronomical number of reports, 67, from New Jersey, filed on one date, January 6 — all except one showing adults aged 27 to 64 (twenty-two show no age) — strongly suggests injection with a flu vaccine hot lot. The same suspicion, though of a different vaccine, hangs over forty-nine reports of infants aged one month to one year from California recorded on January 6 and 7. Beginning with ID# 618161, 191 consecutive reports appear that show no recorded date, and of these, 160 show no reported date either. (The page beginning with ID# 618269 hints at yet another hot lot, in Florida.) When a date is not recorded, the symbol ######## appears, or occasionally the space is left blank. Actually, there's a high number of ########s in the 2012 sample as well, but it's much worse in 2016. If you log on to the VAERS site and scroll down through the calendar year of 2016, you'll find many more of these symbols than you'll find dates. In fact, the entire VAERS database is strewn with them. There are other anomalies that I won't go into. The whole thing is a record-keeping nightmare. Anyone exhibiting such laziness or incompetence or both in the private sector would be fired, but of course the rules are different for bureaucrats. From the little I've discussed here, it should be obvious that VAERS is a travesty that serves no useful purpose whatever, other than unwittingly to document the barbarity of vaccination for all the world to see.

VAERS_ID	S_DATE	STATE	AGE_YRS	SEX	RPT_DATE	SYMPTOM	SYM_CNT	SYM01
172499	7/2/2001	DE	0.5	F	6/19/2001	Had vaccin	4	DYSPHAG
172500	7/2/2001	CA	0.2	F	5/19/2001	Erythema r	2	ERYTHEM
172501	7/2/2001	CA	1	M	6/19/2001	#10; 3-4mr	2	NODULE S
172502	7/2/2001	CT	1	F	6/20/2001	Persistent	1	SCREAMIN
172503	7/2/2001	MN	0.5	F	6/19/2001	Right acute	4	AGITATIOI
172504	7/2/2001	NC	0.3	F	6/26/2001	No data pr	1	REACT UN
172505	7/2/2001	VA	5	M	6/21/2001	Got chicke	1	INFECT
172506	7/2/2001	VA	1.3	F	6/21/2001	Got chicke	1	INFECT
172507	7/2/2001	CT	0.3	M	6/20/2001	About 12 h	2	AGITATIOI
172508	7/2/2001	MA	0.2	M	6/20/2001	3 hours, pc	2	AGITATIOI
172509	7/2/2001	IL	1	F	6/21/2001	Red knot a	2	HYSN INJE
172510	7/2/2001	IN	0.5	F	6/20/2001	Developed	4	AGITATIOI
172511	7/2/2001	HI	1.5	M	6/20/2001	Received D	4	EDEMA IN
172512	7/2/2001	CA	3.3	M	6/20/2001	Left deltoid	3	EDEMA IN
172513	7/2/2001	FL	0.3	M	5/13/2001	Pt had a se	2	ANEMIA
172514	7/2/2001	CO	4	M	6/17/2001	Onset of ce	5	CELLULITI
172515	7/2/2001	NH	1	M	6/19/2001	Fever of 10	1	FEVER
172530	7/2/2001	OH	1	F	6/6/2001	The aftern	9	AGITATIOI
172531	7/2/2001	MA	6	M		Patient dev	2	RASH MAC
172533	7/2/2001	KS	23	F	3/13/2001	Patient aw	6	ASTHENIA
172534	7/2/2001	WA	5	M	3/26/2001	Patient dev	4	EDEMA IN
172535	7/2/2001	WA	34	F	4/18/2001	One hour a	3	ASTHENIA
172536	7/2/2001	NE	5	F	6/9/2001	The evenin	3	EDEMA IN
172537	7/2/2001	TX	4	M	6/13/2001	Patient dev	2	HYSN INJE
172538	7/2/2001	CO	1	M	6/20/2001	Patient dev	3	EDEMA IN
172539	7/2/2001	CA	4.5	F	6/18/2001	Patient wa	5	EDEMA IN
172541	7/2/2001	WA	1.5	F	6/20/2001	Patient dev	3	EDEMA
172542	7/2/2001	CO	1.4	F	6/17/2001	Patient dev	4	FEVER
172543	7/2/2001	OR	1.9	M	6/20/2001	Patient dev	2	EDEMA IN
172544	7/2/2001	FL	23	F	6/20/2001	Patient did	1	REACT UN
172545	7/2/2001	FL	2.9	F	6/20/2001	Patient dev	3	CELLULITI
172546	7/2/2001	MN	1.1	F	6/21/2001	Patient dev	5	AGITATIOI
172547	7/2/2001	NV	22	F	6/20/2001	Patient dev	5	ARTHRAL
172548	7/2/2001	FL	30	F	6/11/2001	Patient rep	5	ARTHRAL
172549	7/2/2001	MD	15	M	6/22/2001	Patient dev	1	ASTHMA
172550	7/2/2001	WI	61	F	6/15/2001	On 6/14/01	5	ARTHRAL
172551	7/2/2001	WI	19	M	6/20/2001	Patient cor	3	PRURITUS
172552	7/2/2001	TX	28	F		Patient dev	3	EDEMA IN
172553	7/2/2001	FL	1.2	F	6/14/2001	Patient dev	1	RASH MAC
172554	7/2/2001	NY	11	F	6/21/2001	Patient dev	1	URTICARI
172555	7/2/2001	CA	72	F	6/15/2001	Approxima	1	HYSN INJE
172556	7/2/2001	AZ	4	M	6/5/2001	Patient dev	4	EDEMA IN
172573	7/2/2001	IL	1	F	6/27/2001	Varicella - I	1	INFECT
172574	7/2/2001	IA	5	M	6/27/2001	Erythema,	3	EDEMA IN
172575	7/2/2001	MI	4.8	M	5/21/2001	One wk aft	6	AGITATIOI
172576	7/2/2001	CA	1.3	F	5/31/2001	Local indur	3	FEVER
172577	7/2/2001		23	M	6/2/2001	Erythema v	2	HYSN INJE
172578	7/2/2001	MI	1	M	5/17/2001	Child had h	2	RHINITIS
172579	7/2/2001	NC	1.5	M	6/18/2001	Cellulitis ar	2	CELLULITI
172580	7/2/2001	WA	5	M	6/20/2001	24 hrs post	4	EDEMA IN
172581	7/2/2001	GA	38	F	6/21/2001	Painful knc	3	ASTHENIA

172800	7/6/2001	ME	2.3	F	6/22/2001	Pt develop	2	RASH
172985	7/6/2001	WI	0.3	F	6/18/2001	Found dea	3	HEM LUNG
170495	7/9/2001	FL	3.3	U		Arthritic typ	3	ARTHRITIS
170748	7/9/2001	TX	44	F	5/30/2001	A 44 yr old	5	ALLERG R
170749	7/9/2001			M	5/30/2001	Informatior	1	FEVER
170750	7/9/2001	PA	18	F	5/30/2001	Info has be	2	DIZZINESS
170751	7/9/2001	MI	1	U	5/30/2001	Info has be	1	HYSN INJE
170752	7/9/2001		4	M	5/30/2001	Informatior	2	HYSN INJE
172898	7/9/2001	WA		M	7/2/2001	Medical his	5	GUILLAIN
172899	7/9/2001	OH	21	M	6/27/2001	I began the	1	ILEITIS
172944	7/9/2001	IL	0.3	M	1/26/2000	"A sales re	3	AGITATIOI
173039	7/9/2001			F	7/5/2001	A physiciar	2	SIDS
170753	7/10/2001	NY		U	5/30/2001	Informatior	2	FEVER
170754	7/10/2001		18	M	5/30/2001	Informatior	4	FEVER
170755	7/10/2001			F	5/30/2001	Informatior	1	NO DRUG
170756	7/10/2001	PA	1	M	5/30/2001	Informatior	1	ALOPECIA
170757	7/10/2001	NJ	4	M	6/30/2001	Informatior	1	HYSN INJE
170758	7/10/2001		1.4	F	5/30/2001	Informatior	2	AGITATIOI
170759	7/10/2001			U	5/30/2001	Musculosk	2	FEVER
170760	7/10/2001			U	5/30/2001	Rash - Info	1	RASH
173006	7/10/2001		1.2	M	6/25/2001	In June 19	19	AGITATIOI
173007	7/10/2001	FL	1.4	F	6/25/2001	The week a	5	ARTHRAL
173008	7/10/2001		1.1	M	6/25/2001	In March 1	12	AUTISM
173009	7/10/2001	MO	68	M	3/28/2002	On 6/8/01,	4	CELLULITI
173012	7/10/2001	VA	0.4	M	6/25/2001	On 7/12/99	1	ASTHMA
173013	7/10/2001	VA	0.3	M	6/25/2001	On 4/28/00	4	ANOMALY
173040	7/10/2001		58	M	7/3/2001	Informatior	17	ACIDOSIS
173042	7/10/2001	OK	0.7	F	6/6/2001	15-20 minu	3	DYSPNEA
173044	7/10/2001	AK	7.8	M	12/7/2001	On 5/25/01	3	HEMATUR
173047	7/10/2001	NJ	1	M	6/29/2001	From maile	6	APNEA
170761	7/11/2001	CA		U	5/30/2001	Informatior	1	REACT UN
170762	7/11/2001	IL	1	M	5/30/2001	Informatior	1	FEVER
170763	7/11/2001	PA	1.1	M	5/30/2001	Informatior	5	EAR DIS
170764	7/11/2001		1	M	5/30/2001	Fever, Ano	6	ANOREXI/
170765	7/11/2001	GA	5	F	5/30/2001	Injection Si	4	EDEMA IN
170766	7/11/2001	NY	5	F	5/30/2001	Rash Inforı	1	RASH
170767	7/11/2001	NE	1.3	F	5/30/2001	Eryhema M	5	ERYTHEM
170768	7/11/2001			U	5/30/2001	Rash - Info	1	RASH
170769	7/11/2001		30	M	5/30/2001	Parotitis - I	1	SIALADEN
170770	7/11/2001	MO	5	M	5/30/2001	Injection sit	3	EDEMA IN
170771	7/11/2001	PA	16	M	5/3/2001	Info has be	3	DIZZINESS
170772	7/11/2001	OH	1.3	M	5/30/2001	Informatior	4	HERNIA
170773	7/11/2001			U	5/30/2001	Parotitis; Ir	1	SIALADEN
170774	7/11/2001	WA	25	F	5/30/2001	Headache,	6	HEADACH
170775	7/11/2001		10	M	5/30/2001	Rash - Info	1	RASH
172843	7/11/2001	AK	9	M	2/27/2001	Within 4 or	5	NAUSEA
172844	7/11/2001	AK	6	F	3/15/2001	Pt had sore	7	FEVER
173048	7/11/2001	GA	1.1	M	6/11/2001	Taken to E	12	AGITATIOI
173072	7/11/2001	NV	1	M	6/6/2001	4 hours, pc	7	AGITATIOI
173074	7/11/2001	CA	1.2	M		After pt's 2	2	DEAF
173075	7/11/2001	WA	1	M	7/2/2001	Febrile sei	1	FEBRILE S
173076	7/11/2001	IL	1.8	M	6/26/2001	Client had	4	CONVULS

617755	1/5/2016	FL				U	1/5/2016	This spontaneous repc
617757	1/5/2016	FL	46	46		M	########	Patient reported that l
617758	1/5/2016	FL	46	46		M	########	Patient received PPSV
617765	1/5/2016	MA				U	1/5/2016	Information has been
617767	1/5/2016		71			F	1/5/2016	This spontaneous repc
617886	1/5/2016	FL	78	78		M	########	Patient presented witl
617887	1/5/2016	FL	60	60		F	########	Rash and urticaria pru
617888	1/5/2016	FL	36	36		F	########	Skin rash. Loratadine 1
617597	1/6/2016	CA	0.5	0	0.5	F	1/6/2016	Prolonged, Atypical Fe
617604	1/6/2016	NJ	3	3		M	1/6/2016	On 12/4/15 around 8 I
617619	1/6/2016					M	1/6/2016	This case was reportec
617628	1/6/2016	CA	69	69		F	########	Mid-lower LBP radiatir
617629	1/6/2016					F	########	Initial unsolicited repo
617631	1/6/2016	AZ	56	56		F	########	Received injection ear
617632	1/6/2016	GA	38	38		F	########	This is a spontaneous ı
617633	1/6/2016	NJ	61	61		U	1/6/2016	On 9/30, DOH was not
617634	1/6/2016		70	70		F	########	This is a spontaneous ı
617637	1/6/2016	NE	62	62		M	########	Swollen arm.
617638	1/6/2016	NJ	35	35		U	1/6/2016	On 9/30, DOH was not
617639	1/6/2016	NJ	44	44		U	1/6/2016	On 9/30, DOH was not
617640	1/6/2016	MI	0.34	0	0.3	M	1/6/2016	Vaccine was administe
617641	1/6/2016	TN	61	61		F	1/6/2016	Patient received a pne
617642	1/6/2016	NJ				U		On 9/30, DOH was not
617644	1/6/2016	NJ	38	38		U	1/6/2016	On 9/30, DOH was not
617645	1/6/2016	NJ	45	45		U	1/6/2016	On 9/30, DOH was not
617646	1/6/2016	NJ	31	31		U	1/6/2016	On 9/30, DOH was not
617647	1/6/2016	NJ	43	43		U	1/6/2016	On 9/30, DOH was not
617648	1/6/2016	NJ	33	33		U	1/6/2016	On 9/30, DOH was not
617649	1/6/2016	NJ	39	39		U	1/6/2016	On 9/30, DOH was not
617650	1/6/2016	NJ	56	56		U	1/6/2016	On 9/30, DOH was not
617651	1/6/2016	NJ	41	41		U	1/6/2016	On 9/30, DOH was not
617652	1/6/2016	NJ	33	33		U	1/6/2016	On 9/30, DOH was not
617654	1/6/2016	NJ	44	44		U	1/6/2016	On 9/30, DOH was not
617655	1/6/2016	NJ				U	1/6/2016	On 9/30, agency was r
617656	1/6/2016	NJ				U	1/6/2016	On 9/30, DOH was not
617657	1/6/2016	NJ				U	1/6/2016	On 9/30, DOH was not
617658	1/6/2016	NJ				U	1/6/2016	On 9/30, DOH was not
617665	1/6/2016	NJ				U	1/6/2016	On 9/30, agency was r
617668	1/6/2016	NJ				U	1/6/2016	On 9/30, DOH was not
617672	1/6/2016	NJ				U	1/6/2016	On 9/30, DOH was not
617673	1/6/2016	NJ	39	39		U	1/5/2016	On 9/30, DOH was not
617674	1/6/2016	NJ				U	1/6/2016	On 9/30, DOH was not
617676	1/6/2016	NJ				U	1/6/2016	On 9/30, agency was r
617677	1/6/2016	NJ				U	1/6/2016	On 9/30, DOH was not
617678	1/6/2016	NJ				U	1/6/2016	On 9/30, DOH was not
617679	1/6/2016	NJ	64	64		U	1/5/2016	On 9/30, DOH was not
617680	1/6/2016	NJ				U	1/6/2016	On 9/30, DOH was not

617681	1/6/2016	NJ				U	1/6/2016 On 9/30, DOH was not
617682	1/6/2016	NJ	54	54		U	1/5/2016 On 9/30, DOH was not
617684	1/6/2016	NJ				U	1/6/2016 On 9/30, DOH was not
617685	1/6/2016	NJ				U	1/6/2016 On 9/30, DOH was not
617686	1/6/2016	NJ	41	41		U	1/5/2016 On 9/30, agency was r
617687	1/6/2016	NJ				M	1/6/2016 On 9/30, DOH was not
617689	1/6/2016	NJ	48	48		U	1/6/2016 On 9/30, DOH was not
617690	1/6/2016	NJ				U	1/6/2016 On 9/30, DOH was not
617691	1/6/2016	NJ				M	1/6/2016 On 9/30, DOH was not
617693	1/6/2016	NJ	30	30		U	1/5/2016 On 9/30, DOH was not
617694	1/6/2016	NJ	58	58		M	1/5/2016 On 9/30, DOH was not
617695	1/6/2016	NJ	52	52		U	1/5/2016 On 9/30, DOH was not
617696	1/6/2016	NJ	29	29		U	1/5/2016 On 9/30, DOH was not
617697	1/6/2016	NJ	52	52		F	1/5/2016 On 9/30, DOH was not
617698	1/6/2016	NJ	58	58		U	1/5/2016 On 9/30, DOH was not
617699	1/6/2016	NJ	47	47		U	1/5/2016 On 9/30, DOH was not
617700	1/6/2016	NJ	36	36		F	1/5/2016 On 9/30, DOH was not
617701	1/6/2016	NJ	50	50		F	1/6/2016 On 9/30, DOH was not
617703	1/6/2016	NJ	56	56		U	1/5/2016 On 9/30, DOH was not
617704	1/6/2016	VA	61	61		M	1/6/2016 Headache, flushed hot
617705	1/6/2016	CA	0.51	0	0.5	M	1/6/2016 Right thigh swollen.
617706	1/6/2016	AZ	47	47		F	1/6/2016 Pt reports that she is s
617707	1/6/2016	IA	0.18	0	0.2	M	1/6/2016 Patient was given .25r
617708	1/6/2016	NJ	62	62		M	1/5/2016 On 9/30, DOH was not
617709	1/6/2016	NJ				U	1/5/2016 On 9/30, DOH was not
617710	1/6/2016	NJ	51	51		U	1/5/2016 On 9/30, DOH was not
617711	1/6/2016	NJ	44	44		U	1/5/2016 On 9/30, DOH was not
617712	1/6/2016	NJ	62	62		U	1/5/2016 On 9/30, DOH was not
617713	1/6/2016	NJ	60	60		U	1/5/2016 On 9/30, DOH was not
617714	1/6/2016	NJ	35	35		M	1/5/2016 On 9/30, DOH was not
617715	1/6/2016	NJ	49	49		U	1/5/2016 On 9/30, DOH was not
617716	1/6/2016	NJ	27	27		U	1/5/2016 On 9/30, DOH was not
617717	1/6/2016	NJ	56	56		U	1/6/2016 On 9/30, DOH was not
617718	1/6/2016	NJ	30	30		U	1/5/2016 On 9/30, DOH was not
617719	1/6/2016	NJ	37	37		U	1/5/2016 On 9/30, DOH was not
617720	1/6/2016	NJ	35	35		U	1/5/2016 On 9/30, DOH was not
617721	1/6/2016	NJ	47	47		M	1/5/2016 On 9/30, DOH was not
617722	1/6/2016	NJ	39	39		U	1/5/2016 On 9/30, DOH was not
617723	1/6/2016	NJ	45	45		U	1/5/2016 On 9/30, DOH was not
617724	1/6/2016	NJ	37	37		M	1/5/2016 On 9/30, DOH was not
617725	1/6/2016	NJ	30	30		U	1/5/2016 On 9/30, DOH was not
617729	1/6/2016	NJ	52	52		U	1/5/2016 On 9/30, DOH was not
617730	1/6/2016	PA	81	81		M	Severe local reaction v
617731	1/6/2016	MN	31	31		F	1/6/2016 The Tdap injections we
617732	1/6/2016					U	1/6/2016 This spontaneous repc
617733	1/6/2016	MN	26	26		F	1/6/2016 The Tdap injections we
617734	1/6/2016	MN	29	29		F	1/6/2016 The Tdap injections we

617805	1/6/2016	TN	50	50		F	1/6/2016	SHE HAD SORE ARM F
617806	1/6/2016	OR	0.37	0	0.4	F	1/6/2016	No adverse effect in m
617807	1/6/2016	MI	17	17		F	1/6/2016	Hives, redness, and sw
617808	1/6/2016	LA	19	19		F	1/6/2016	JOINT PAIN IN NON-V
617809	1/6/2016	WA	26	26		M	1/6/2016	I first began having pr
617810	1/6/2016	SD	67	67		F	1/6/2016	Patient presented to p
617811	1/6/2016	CA	4	4		F	1/6/2016	1" X 1/2" welt above i
617812	1/6/2016	CA	66	66		M	1/6/2016	Fever, chills, sore left
617813	1/6/2016	CA	59	59		M	1/6/2016	Herpetic rash along de
617817	1/6/2016		18			F	1/6/2016	This spontaneous repc
617818	1/6/2016	OH	0.35	0	0.3	M	1/6/2016	Information has been
617819	1/6/2016	CA	1.03	1	0	F	########	During/after vaccinati
617820	1/6/2016	CA	1.09	1	0.1	F	1/6/2016	Pt has a reaction to th
617822	1/6/2016	WA	0.38	0	0.4	M	1/6/2016	Developed hives to le
617872	1/6/2016		63			F	1/6/2016	Information has been
617874	1/6/2016	OH				F	1/6/2016	Information has been
617889	1/6/2016					U	1/6/2016	Information has been
617890	1/6/2016					F	1/6/2016	Information has been
617893	1/6/2016					U	1/6/2016	This spontaneous repc
617914	1/6/2016		16	16		F	1/6/2016	Case number PHEH20
617915	1/6/2016		16	16		M	1/6/2016	Case number PHEH20
617924	1/6/2016	CA	0.37	0	0.4	F	1/6/2016	This initial spontaneou
617925	1/6/2016	CA	0.43	0	0.4	F	1/6/2016	This initial spontaneou
617926	1/6/2016		16	16		M	1/6/2016	Case number PHEH20
617927	1/6/2016	CA	0.16	0	0.2	M	1/6/2016	This initial spontaneou
617928	1/6/2016	CA	0.15	0	0.2	F	1/6/2016	This initial spontaneou
617929	1/6/2016					U	1/6/2016	This spontaneous repc
617930	1/6/2016	CA	0.35	0	0.3	M	1/6/2016	This initial spontaneou
617931	1/6/2016	CA	0.12	0	0.1	M	1/6/2016	This initial spontaneou
617932	1/6/2016	CA	0.36	0	0.4	M	1/6/2016	This initial spontaneou
617933	1/6/2016	CA	0.15	0	0.1	F	1/6/2016	This initial spontaneou
617934	1/6/2016		14	14		M	1/6/2016	Case number PHEH20
617935	1/6/2016		16	16		M	1/6/2016	Case number PHEH20
617936	1/6/2016	CA	0.11	0	0.1	M	1/6/2016	This initial spontaneou
617937	1/6/2016	CA	0.32	0	0.3	M	1/6/2016	This initial spontaneou
617938	1/6/2016	CA	0.16	0	0.2	M	1/6/2016	This initial spontaneou
617939	1/6/2016	CA	0.31	0	0.3	F	1/6/2016	This initial spontaneou
617940	1/6/2016		16	16		M	1/6/2016	Case number PHEH20
617941	1/6/2016	CA	0.23	0	0.2	F	1/6/2016	This initial spontaneou
617942	1/6/2016	CA	0.59	0	0.6	M	1/6/2016	This initial spontaneou
617948	1/6/2016	CA	0.21	0	0.2	M	1/6/2016	This initial spontaneou
617949	1/6/2016	CA	0.54	0	0.5	M	1/6/2016	This initial spontaneou
617950	1/6/2016	CA	0.33	0	0.3	M	1/6/2016	This initial spontaneou
617951	1/6/2016	CA	0.35	0	0.4	U	1/6/2016	This initial spontaneou
617952	1/6/2016					U	1/6/2016	This spontaneous repc
617953	1/6/2016	CA	0.52	0	0.5	M	1/6/2016	This initial spontaneou
617954	1/6/2016		16	16		M	1/6/2016	Case number PHEH20

617955	1/6/2016		11	11		F	1/6/2016	Case number PHEH20:
617957	1/6/2016	WI	81	81		M	########	Fever and rigors, night
617958	1/6/2016	CA	0.39	0	0.4	F	1/6/2016	This initial spontaneou
617959	1/6/2016	CA	0.51	0	0.5	M	1/6/2016	This initial spontaneou
617960	1/6/2016	CA	0.44	0	0.4	F	1/6/2016	This initial spontaneou
617961	1/6/2016	CA	0.18	0	0.2	M	1/6/2016	This initial spontaneou
617962	1/6/2016	CA	0.2	0	0.2	F	1/6/2016	This spontaneous repc
617963	1/6/2016	CA	0.65	0	0.6	M	1/6/2016	This initial spontaneou
617984	1/6/2016	CA	0.16	0	0.2	F	1/6/2016	This initial spontaneou
617989	1/6/2016					U	1/6/2016	Information has been
617991	1/6/2016	CA	0.51	0	0.5	M	1/6/2016	This initial spontaneou
617993	1/6/2016	CA				U	1/6/2016	This initial spontaneou
617994	1/6/2016	CA	0.32	0	0.3	F	1/6/2016	This initial spontaneou
617997	1/6/2016	CA	0.32	0	0.3	F	1/6/2016	This initial spontaneou
617999	1/6/2016	CA	0.35	0	0.4	M	1/6/2016	This initial spontaneou
618001	1/6/2016	CA	0.57	0	0.6	F	1/6/2016	This initial spontaneou
618004	1/6/2016	CA	0.17	0	0.2	F	1/6/2016	This initial spontaneou
618005	1/6/2016					U	1/6/2016	This spontaneous repc
618006	1/6/2016					U	1/6/2016	This spontaneous repc
618007	1/6/2016					U	1/6/2016	Information has been
618008	1/6/2016	NY				F	1/6/2016	Information has been
617815	1/7/2016	OH	18	18		M	1/7/2016	About 5 minutes after
617816	1/7/2016	VI	11	11		F	1/7/2016	Syncope, extreme pale
617821	1/7/2016	TN	68	68		M	1/5/2016	Shooting left arm pain
617826	1/7/2016	FL	11	11		M	1/7/2016	Cellulitis on injection s
617846	1/7/2016	VA	36	36		F	1/7/2016	Stiffening and sorenes
617847	1/7/2016	PA	1.19	1	0.2	M	1/7/2016	Began with on going h
617848	1/7/2016	TN	67	67		F	1/7/2016	Antibiotics, Z pack, Ste
617849	1/7/2016	FL				U	1/7/2016	This spontaneous repc
617850	1/7/2016					U	1/7/2016	This spontaneous repc
617851	1/7/2016	IN				U	1/7/2016	Thus spontaneous rep
617852	1/7/2016					U	1/7/2016	This spontaneous repc
617853	1/7/2016	KS	4	4		F	1/6/2016	4/15/15 Mother prese
617854	1/7/2016	FL				U	1/7/2016	This spontaneous repc
617855	1/7/2016					U	1/7/2016	This spontaneous repc
617856	1/7/2016					M	1/7/2016	This spontaneous repc
617857	1/7/2016	FL	19	19		M	########	Patient experienced pa
617858	1/7/2016					F	1/7/2016	This spontaneous repc
617859	1/7/2016	GA	60	60		M	1/4/2016	Severe vertigo began :
617860	1/7/2016	MA	6	6		M	1/2/2016	Hives 1 day post influe
617861	1/7/2016					U	1/7/2016	This spontaneous repc
617862	1/7/2016	PA	20	20		F	########	Had symptoms of ulce
617863	1/7/2016	NC	15	15		F	########	Chronic abd pain, bloa
617864	1/7/2016	NJ	1.25	1	0.3	M	########	Hives at site of MMR v
617865	1/7/2016	PA	78	78		F	1/7/2016	Admitted to Hospital 1
617866	1/7/2016	CO	31	31		F	########	Patient is the mother (
617867	1/7/2016	CA	71	71		F	########	Patient got Pneumonia

617967	1/7/2016	PA	56	56		F	1/7/2016	Extreme pain in left ar
617968	1/7/2016	FL	47	47		M	1/7/2016	Sweating, Dizzy, blurry
617969	1/7/2016	OH	43	43		F	1/7/2016	Pt reports lymphadeni
617970	1/7/2016					F	1/7/2016	This case was reportec
618013	1/7/2016					U	1/7/2016	This spontaneous repc
618014	1/7/2016	GA	78	78		F	1/7/2016	Lethargy, nausea/vom
618015	1/7/2016	CA	0.28	0	0.3	M	1/7/2016	This initial spontaneou
618016	1/7/2016	CA	0.15	0	0.1	M	1/7/2016	This initial spontaneou
618019	1/7/2016	CA	0.39	0	0.4	F	1/7/2016	This initial spontaneou
618092	1/7/2016					U	1/7/2016	This spontaneous repc
618093	1/7/2016					U	1/7/2016	This spontaneous repc
618094	1/7/2016	CA	0.17	0	0.2	M	1/7/2016	This initial spontaneou
618095	1/7/2016	CA	0.36	0	0.4	M	1/7/2016	This initial spontaneou
618096	1/7/2016					U	1/7/2016	Information has been
618097	1/7/2016					U	1/7/2016	This spontaneous repc
618098	1/7/2016	CA	0.25	0	0.3	M	1/7/2016	This initial spontaneou
618099	1/7/2016	CA	0.13	0	0.1	F	1/7/2016	This initial spontaneou
618100	1/7/2016					U	1/7/2016	This spontaneous repc
618101	1/7/2016					U	1/7/2016	This spontaneous repc
618102	1/7/2016					U	1/7/2016	This spontaneous repc
618103	1/7/2016	FL				U	1/7/2016	This spontaneous repc
618104	1/7/2016	FL				M	1/7/2016	This spontaneous repc
618105	1/7/2016	NJ				U	1/7/2016	This spontaneous repc
618106	1/7/2016	CA	0.18	0	0.2	M	1/7/2016	This initial spontaneou
618107	1/7/2016	CA	0.58	0	0.6	F	1/7/2016	This initial spontaneou
618108	1/7/2016					U	1/7/2016	Information has been
618109	1/7/2016					U	1/7/2016	Information has been
618110	1/7/2016	NJ				U	1/7/2016	This spontaneous repc
618111	1/7/2016					U	1/7/2016	Information has been
618112	1/7/2016					U	1/7/2016	This spontaneous repc
618113	1/7/2016	CA	0.17	0	0.2	F	1/7/2016	This initial spontaneou
618114	1/7/2016	CA	0.57	0	0.6	F	1/7/2016	This initial spontaneou
618115	1/7/2016	CA	0.28	0	0.3	M	1/7/2016	This initial spontaneou
618116	1/7/2016	NY	65	65		M		ZOSTAVAX at pharmac
618117	1/7/2016	TX	0.33	0	0.3	F	1/7/2016	Sterile abscesses form
618118	1/7/2016	CA	0.12	0	0.1	M	1/7/2016	This initial spontaneou
618119	1/7/2016	NJ				U	1/7/2016	This spontaneous repc
618120	1/7/2016	NJ				U	1/7/2016	This spontaneous repc
618121	1/7/2016					U	1/7/2016	This spontaneous repc
618122	1/7/2016					U	1/7/2016	This spontaneous repc
618123	1/7/2016		53			U	1/7/2016	This spontaneous repc
618124	1/7/2016	CA	28	28		M	1/7/2016	Pain at injection site. F
618125	1/7/2016					U	1/7/2016	This spontaneous repc
618126	1/7/2016					U	1/7/2016	This spontaneous repc
618127	1/7/2016					U	1/7/2016	This spontaneous repc
618128	1/7/2016					U	1/7/2016	This spontaneous repc
618134	1/7/2016	CA	0.17	0	0.2	F	1/7/2016	This initial spontaneou

618215	1/8/2016	MA				U	1/8/2016	This spontaneous repc
618217	1/8/2016	TX				U	1/8/2016	This spontaneous repc
618218	1/8/2016					U	1/8/2016	This spontaneous repc
618219	1/8/2016	FL				U	1/8/2016	This spontaneous repc
618220	1/8/2016					U	1/8/2016	This spontaneous repc
618221	1/8/2016					U	1/8/2016	This spontaneous repc
618222	1/8/2016					U	1/8/2016	This spontaneous repc
618223	1/8/2016	IN				M	1/8/2016	This spontaneous repc
618224	1/8/2016	FL				U	1/8/2016	This spontaneous repc
618225	1/8/2016	FL	16	16		F	1/8/2016	This spontaneous pros
618226	1/8/2016					U	1/8/2016	This spontaneous repc
618227	1/8/2016	FL				U	1/8/2016	This spontaneous repc
618228	1/8/2016	FL				U	1/8/2016	This spontaneous repc
618240	1/8/2016	NH	1.01	1	0	M	1/8/2016	20-50 papular lesion 1
618253	1/8/2016		12			U	1/8/2016	This spontaneous repc
618256	1/8/2016					U	1/8/2016	This spontaneous repc
618257	1/8/2016	FL				U	1/8/2016	This spontaneous repc
618258	1/8/2016	FL				U	1/8/2016	This spontaneous repc
618259	1/8/2016	FL				U	1/8/2016	This spontaneous repc
618260	1/8/2016					U	1/8/2016	This spontaneous repc
618261	1/8/2016					U	1/8/2016	This spontaneous repc
618262	1/8/2016	FL				U	1/8/2016	This spontaneous repc
618264	1/8/2016					U	1/8/2016	This spontaneous repc
618280	1/8/2016					U	1/8/2016	This spontaneous repc
618281	1/8/2016		14	14		F	1/8/2016	This spontaneous repc
618288	1/8/2016					U	1/8/2016	This initial spontaneou
618296	1/8/2016		0.8			U	1/8/2016	This spontaneous repc
618308	1/8/2016					U	1/8/2016	This initial spontaneou
618310	1/8/2016		11			F	1/8/2016	This spontaneous repc
618312	1/8/2016					U	1/8/2016	This initial spontaneou
618321	1/8/2016					U	1/8/2016	This initial spontaneou
618327	1/8/2016					U	1/8/2016	This initial spontaneou
618332	1/8/2016					U	1/8/2016	This spontaneous repc
618333	1/8/2016					U	1/8/2016	This spontaneous repc
618154	1/9/2016	NY	47	47		M	1/9/2016	Anaphylactic reaction.
618155	1/9/2016	NY	22	22		F	1/9/2016	Patient felt burning in
618156	1/9/2016	UT	63	63		F	1/9/2016	Moderate pain at injec
618157	1/9/2016	GA	57	57		F	1/9/2016	(Left) Arm/muscle pai
618158	1/9/2016	WI	65	65		F	1/9/2016	Patient reported that a
618159	1/9/2016	FL	67	67		M	1/9/2016	PATIENT SAYS SHOULI
618160	1/9/2016	OR	63	63		M	1/9/2016	Extreme loss of streng
618188	1/9/2016	SC	60	60		F	1/9/2016	After vaccine the patie
618191	1/9/2016	PR	65	65		F	1/9/2016	2 day fever, itching, sv
618193	1/9/2016	WA	46	46		F	1/9/2016	Pt not treated for rxn I
618161	########	MA	74	74		F	########	Patient came to the ph
618162	########	MO	34	34		M	########	The patient reported r
618163	########	FL	65	65		F	########	Very sore at injection a

618164	########	NY	60	60		F	########	PATIENT HAS UNUSU/
618165	########	CO	70	70		F	########	Arm achy (arthritic-like
618166	########	NY	63	63		F	########	Patient has severe infl
618167	########	VA	50	50		F	########	The location of my arn
618168	########	CA	19	19		M	########	Fever, Dizziness, Naus
618270	########	NY	91	91		M	########	Patient's left upper arr
618334	########					M	########	This spontaneous repc
618169	########	CA	35	35		F	########	Neck pain, head pain,
618170	########	IL	0.08	0	0.1	M	########	Raw, open, reddish bli
618171	########	HI	72	72		M	########	Severe pain, swelling f
618172	########	NY	60	60		M	########	Hives, 4 or 5 on my ch
618173	########	MA	1.16	1	0.2	F	########	Red bumps/rash cover
618174	########					U	########	This sponta Y
618179	########	GA	65	65		F	########	Swelling, induration ar
618189	########	DC	25	25		F		Not complete. Already
618190	########	CA	75	75		F	1/9/2016	Patient came in to pha
618207	########	PA	46	46		F	########	Marked pain of left arr
618216	########	LA	63	63		M	########	Swelling at injection si
618229	########	MD	68	68		F	########	Started the evening of
618230	########	IL	54	54		F	########	Very thirsty, sore thro
618231	########	FL	70	70		F	########	Began with mild itchin
618232	########	LA	0.17	0	0.2	M	########	Inconsolable crying for
618234	########	IL	66	66		F	1/8/2016	This is a spontaneous r
618235	########					M	1/8/2016	Initial unsolicited repo
618236	########					M	1/8/2016	Initial unsolicited repo
618237	########	NY	44	44		F	1/1/2016	Pain/burning in (L) arn
618238	########	MA	48	48		F	########	Itching, redness, swell
618239	########	CA	58	58		F	########	Vaccine given 10/8/15
618241	########	CA	64	64		M	########	Pt received ZOSTAVAX
618242	########	SC	73	73		F	1/8/2016	1 deltoid red and warr
618243	########	MA	60	60		F	########	Started "w/ hot, itchy
618244	########					U	########	This spontaneous repc
618245	########	MI	23	23		F	1/8/2016	None stated.
618246	########	OH	56	56		F	1/4/2016	Area around where ZC
618247	########	MT	1.02	1	0	F	########	Per parent child had h
618248	########	NJ	0.32	0	0.3	M	########	Hacking Cough, Whee
618249	########	AZ	23	23		F	########	Hard lump injection si
618250	########	MI	57	57		F	########	Patient was supposed
618251	########	MI	76	76		F	########	4 days after getting flu
618252	########	WA	69	69		F	1/4/2016	Patient reported pain
618254	########	CO	11	11		F	########	After vaccines and blo
618255	########	CO	13	13		F	1/8/2016	Patient received MEN/
618263	########	OR	40	40		M	########	Flu like sxs, generalize
618265	########	IL	61	61		F	1/4/2016	Rapid heartbeat, dizzy
618266	########	SD	0.52	0	0.5	M	1/6/2016	45 minutes after recei
618267	########	CA	72			M	1/5/2016	Pt reported 3 weeks o
618268	########		71	71		F	########	This spontaneous repc

618269	########	WA	13	13		M	1/4/2016	Patient had wheal and
618271	########	AL	76	76		F	1/6/2016	Patient went next day
618272	########		63			F	########	This spontaneous repc
618273	########	NC	0.6			M	########	Date 1-6-16. Fever. Ra
618274	########	CA	53	53		F	########	Itching and hives on 1:
618275	########	FL				U	########	This spontaneous repc
618276	########	AZ	38	38		F	########	Large welt at site of in
618277	########	AZ	29	29		F	1/2/2016	Possible reaction to Va
618278	########	MA	0.17	0	0.2	M		Redness and swelling a
618279	########	NC	65	65		M	1/8/2016	In March, right side leg
618282	########	CA	52	52		F	1/5/2016	Pt became dizzy imme
618283	########	FL				U	########	This spontaneous repc
618284	########	FL				U	########	This spontaneous repc
618285	########	AZ	50	50		F	########	Patient vaccinated wit
618286	########					U	########	This spontaneous repc
618287	########					F	########	Information has been
618289	########	FL				U	########	This spontaneous repc
618291	########					M	########	This spontaneous repc
618292	########		60			F	########	This spontaneous repc
618293	########	FL				U	########	This spontaneous repc
618294	########	MA	32	32		F	########	Area warm to touch, r
618297	########	FL				U	########	This spontaneous repc
618298	########	FL				U	########	This spontaneous repc
618299	########	FL				U	########	This spontaneous repc
618300	########	FL				M	########	This spontaneous repc
618301	########	FL				U	########	This spontaneous repc
618302	########	FL				U	########	This spontaneous repc
618303	########	FL				U	########	This spontaneous repc
618304	########	FL				U	########	This spontaneous repc
618306	########	FL				U	########	This spontaneous repc
618307	########	RI				U		After administration o
618309	########	FL				U	########	This spontaneous repc
618311	########					F	########	Information has been
618313	########	NC	36	36		M	########	Left shoulder pain afte
618314	########	TN	0.27	0	0.3	M	########	Small rash began to br
618315	########	ME	22	22		F	########	Almost immediate for
618316	########	ND	0.78	0	0.8	F	########	Mother phones public
618317	########	GA	72	72		F	########	Low BP (85/60), increa
618318	########	MN	64	64		F	########	Started with arm pain
618319	########	NC	50	50		M	########	Fatigue, fever (measu
618320	########	IL	85	85		F	########	Redness and swelling a
618322	########	NV	17	17		F	########	Daughter passed out c
618323	########	PA	71	71		F	########	Patient is stated nerve
618324	########		68	68		M	########	Patient seen in the em
618325	########	CA	58	58		F	########	Approximately two da
618326	########	NV	23	23		F	########	Pt returned to clinic 2
618328	########	KS	73	73		F	########	Pruritic red rash left d

618329	########	GA	72	72		F	########	Upset ston Y
618330	########	TX	32	32		M	########	Felt feverish, noticed s
618331	########	IA	1.27	1	0.3	F	########	Seizure like activity, la
618335	########					U	########	This spontaneous repc
618336	########					U	########	This spontaneous repc
618337	########	FL				U	########	This spontaneous repc
618338	########	FL				U	########	This spontaneous repc
618339	########	TX				U	########	This spontaneous repc
618340	########					U	########	This spontaneous repc
618341	########					U	########	This spontaneous repc
618342	########	FL				M	########	This spontaneous repc
618343	########	FL				U	########	This spontaneous repc
618344	########	FL				U	########	This spontaneous repc
618345	########	CA	62	62		F	########	Next morning noticed
618346	########	NC	4	4		M	1/8/2016	Patient came to office
618353	########	CA	0.54	0	0.5	M	########	Baby soon developed
618354	########	MT	17	17		M	########	No adverse reaction. F
618355	########	WA	12	12		F	########	Pt reported lump to th
618356	########	CA	74	74		M	########	Severe flu symptoms.
618357	########	OR	53	53		F	1/8/2016	This case was reportec
618358	########	WA	0.35	0	0.3	F	########	Bloody stools.
618390	########	WI	15	15		M	########	Pt's mother states pt b
618391	########	NC	40	40		F	########	Red rash on arm wher
618392	########	TX	66	66		F	########	Patient had pain after
618393	########	DC	11	11		F	########	Approximately 1-2 mir
618452	########	FL				U	########	Information has been
648439	########	CA	74	74		U		Numbness of area - an
618363	########	KS	0.01	0	0	F	########	Fatigue, weakness, de
618364	########					U	########	This spont Y
618365	########					F	########	This spontaneous repc
618373	########	SC	14	14		F	########	Guillain Barre. 10/25 N
618382	########	FL				U	########	This spontaneous repc
618383	########	MI				U	########	This spontaneous repc
618384	########		15	15		F	########	Case number PHEH20
618385	########		16	16		M	########	Case number PHEH20
618386	########	FL	66	66		F	########	Numbness starting/wc
618387	########	FL				U	########	This spontaneous repc
618388	########		8			U	########	This spontaneous repc
618389	########	FL				U	########	This spontaneous repc
618394	########	OH				U	########	This spontaneous repc
618395	########	FL		0	0	M	########	Lethargy weakness sle
618397	########	NC	12	12		F	########	Type 1 Diabetes.
618398	########					U	########	This spontaneous repc
618399	########					U	########	This spontaneous repc
618400	########		81	81		F	1/9/2016	ZOSTAVAX immunizati
618401	########	FL				U	########	This spontaneous repc
618402	########		62	62		F	1/8/2016	This patient received a

618403	########					U	########	This spontaneous repc
618404	########		10			F	########	This spontaneous repc
618405	########	KS	47	47		F	########	Pain noted in left deltc
618406	########	MT	65	65		M	1/5/2016	Developed a red rash t
618407	########	FL				U	########	This spontaneous repc
618408	########	MI				U	########	This spontaneous repc
618409	########	FL				U	########	This spontaneous repc
618410	########	MI				U	########	This spontaneous repc
618411	########					F	########	This sponta Y
618412	########					U	########	This spontaneous repc
618413	########					U	########	This spontaneous repc
618414	########					U	########	This spontaneous repc
618415	########		61	61		M	1/9/2016	Shoulder pain from flu
618416	########	FL				U	########	This spontaneous repc
618417	########	FL				U	########	This spontaneous repc
618418	########	IN	22	22		U		Per health departmen
618419	########	CA	67	67		F	########	24hrs later pt develop
618420	########	TX				U	########	This spontaneous repc
618421	########	FL				U	########	This spontaneous repc
618422	########	TX	36	36		F	########	In the middle of the ni
618423	########	NC	1.25	1	0.2	M	########	From ED H and P: HPI
618424	########	VA	8	8		F	########	Swelling, redness, and
618426	########		11	11		F	1/8/2016	Patient in for well che
618427	########					U	########	This spontaneous repc
618428	########	TX				U	########	This spontaneous repc
618429	########	FL				U	########	This spontaneous repc
618430	########	PA	52	52		F	1/9/2016	Needed to take flu shc
618431	########	NY	82	82		M	1/5/2016	(L) upper arm pain, sw
618432	########	FL				U	########	This spontaneous repc
618433	########		67	67		F	########	This case, manufactur
618434	########	OH	20	20		F	1/7/2016	Patient immediately c
618435	########	IL	21	21		F	1/7/2016	Lightheaded, dizzy, cl
618436	########	VA	70	70		F	1/7/2016	Severe swelling from s
618437	########	NY	69	69		U		Patient stated he start
618439	########					U	1/8/2016	This case was reportec
618441	########	FL				U	########	This spontaneous repc
618443	########	TX	81	81		M	########	This spontaneous repc
618444	########					U	########	This spontaneous repc
618445	########	MI				U	########	This spontaneous repc
618447	########	FL				U	########	This spontaneous repc
618448	########					U	########	This spontaneous repc
618449	########		67	67		F	########	Case number PHEH20
618450	########	FL				U	########	This spontaneous repc
618451	########	FL				U	########	This spontaneous repc
618454	########					U	########	Information has been
618456	########					U	########	Information has been
618457	########	OH	56	56		F	########	Patient experienced re

1/1/2012	PATIENT EXPERIENCED FELT FLUID SURGE A˙	Y			Y
1/1/2012	Bloodshot eyes, hoarse voice, scratchy throat, extreme fatigue, fever, achy, soreness at injection site,				
1/1/2012	Rash (6 inches by 3 inches) severe itching, pain, swelling. I am still having these symptoms (6:00 PM,				
1/1/2012	Immediately following shot had lots of pain. Has lots of redness, swelling, warm to touch. Following				
1/1/2012	Facial swelling, headache, body aches, tiredness. Patient self medicated with Advil, Sudafed, Xyzal. P				
1/2/2012	Ongoing, wide spread muscle twitches. Injec	Y			
1/2/2012	Fever of 105 degrees and unconsolable.				
1/2/2012	Repeated projectile vomiting on administration of vaccine, attempted three times in office with same				
1/2/2012	Shingles type sensitivity on left side (lower back and upper leg), continuing for several days, followed				
1/2/2012	Red rash at site of injection and surrounding 3-4 inches of arm. Small amount of swelling at site of in				
1/2/2012	Swollen, red, warm posterior left arm, imprc	Y			
1/2/2012	It should be noted up front that before this i	Y	Y	2	
1/2/2012	Eye roll, affect change, stopped smiling, laughing etc. Sa	Y		5	Y
	Bright red rash immediately after injection (at site) with burning.				
1/2/2012	Patient had VARIVAX (#2) immunization on 12-30-2011 at Right arm. The Mother brought child back t				
########	Patient had low grade fever of 99 degrees or	Y			
1/2/2012	Fever 99-101.1. Redness at injection site.				
########	Soreness and weakness/pain, difficulty movi	Y			
	Nonspecific rash (L) hand on arm of vaccine that should be self-limiting.				
########	Mother reports that approximately six hours	Y			
1/3/2012	Seen hospital ER on 1/1/12 Dx with pericardi	Y			
########	(R) leg -> 6 x 6 cm erythematous reaction on ant/lateral aspect of leg. (L) leg -> 10 x 12 cm erythemat				
########	Swelling, nodule immediately. No treatment. Encouraged use of warm compress.				
########	Patient had nodule noted on left arm after vaccine was administered.				
1/3/2012	24yo male trainee. On 12/23 he awakened w	Y	Y	6	
1/3/2012	20y male who presents with left facial asymmetry shortly after receiving vaccinations. Over the cours				
########	It was reported in a published article that a 4	Y			
########	Information has been received via letter from the father of a 10 year old female who on a				Y
########	Information has been received from a physic	Y			
########	Information has been received from a registered nurse, for GARDASIL, a Pregnancy Registry product,				
########	Information has been received from a nurse	Y	Y	6 Y	
########	Information has been received from a nurse	Y			
1/3/2012	Hives with itching started morning 1-2-12 sig	Y			
	12/21/11 son reports patient with temp 101	Y			
1/3/2012	Patient developed arm pain where injection was given one day after receiving. Area continues to be s				
########	Information has been received from a registered nurse concerning herself, a 6			Y	
########	Information has been received from a 31 yea	Y			
########	Information has been received from a registe	Y			
1/3/2012	Client received flu vaccine at 12:47 PM. Patient stated she was feeling s/e from injection that was rai				
1/3/2012	Received vaccine at 1:08 PM. At approximate	Y			
1/3/2012	Nurse administered flu vaccine at 2:30 PM. P	Y			
1/3/2012	This 18 year old female who was otherwise always healthy and active received a 3rd Gard				Y
1/3/2012	Chicken pox like rash on torso, confirmed as	Y			
1/3/2012	Employee received vaccine at 11:38 AM. At 12:48 PM she returned complaining of feeling flushed, he				
1/3/2012	11:56 AM Patient stated "I am afraid of needles." Nurse used distraction and slow breathing techniqu				
1/3/2012	Joint pain was reported to pharmacist on January 2, 2012.				
########	Information has been received from a license	Y			

1/3/2012 Rash, supportive care.
######## Patient claims arm became weak on Saturday, Dec. 17, 2011, but not sore, just weakness.
######## Cellulitis, itching. Linezolid 600mg po ordered to treat cellulitis. BENADRYL 50mg TID prn - received 4
######## Patient reported swelling, pain, tenderness, & warm to the touch 2 days after receiving FLUVIRIN & P
1/3/2012 Rash, supportive care. Y
1/3/2012 KINRIX dose was given before age 4-6 yrs. Client called on 01/03/12 no adverse reactions noted.
######## Pt fainted after flu & HPV vaccine.
None stated.
######## Patient got a flu shot at 12:33 PM said he started throwing up at 5 PM (12/19/11) and continued thro
######## Atypical delayed onset (3 days after shot) of Y
######## (R) thigh became swollen and pt had a lump Y
1/3/2012 Hives - mostly around eyes & face but also oı Y
1/3/2012 1 hour post nasal spray developed headache, pallor, mild fever and night time hallucinations. This las
######## Rash, felt very bad and progressively got better after first day, had mild problems breathing, no treat
######## Patient's daughter reported to myself (RN ar Y
######## Swelling, erythema. Y
######## Vomiting, fussy behavior, diarrhea.
######## Developed red, blistering rash at injection site 1-2 days after injection, spread to trunk, other arm.
1/3/2012 Rash on lower part of trunk, fever. Resolved with Tylenol and Benadryl oral.
1/3/2012 Papular/pustular rash developed on right arr Y Y 14
1/3/2012 At 10:37 AM client complained of tingling on right side of arm, face, and whole right side of body. 10:
1/3/2012 Participant returned one hour after receiving injection stating feeling lightheaded, queasy, "like I am
1/3/2012 High-pitched cry for 1.5 hours (never again before or since). Stopped babbling for about 6 Y
1/3/2012 Erythema, edema, itching, pain at injection site.
1/3/2012 Guillain Barre syndrome. (L) esotı Y Y Y
1/3/2012 Patient complained of being tired and weak Y Y 15
1/3/2012 Immediately following IM injection of Flulaval 0.5cc to left arm the participants injection site had loca
1/3/2012 Child received duplicate dose of IPV. Given individual vaccine along with combination vaccine, which
1/3/2012 Left shoulder pain, near the site of the vaccine, but not overlying vaccine site. Pain has persisted for 2
1/3/2012 S: Patient stated, "I don't like seeing the needle." Within 5 minutes of the injection patient reports fe
1/3/2012 Flu shot (Fluvirin) given high up in upper arm rather than all previously received shots in lower half o
1/3/2012 Day after injections, tenderness and erythem Y Y 1
1/3/2012 She developed weakness in the right face an Y
1/3/2012 Irritable, inconsolable, fever, loose stool, walk off balance, exhaustion. Contacted pediatri Y
1/3/2012 Constant pain in left shoulder and down arm Y
1/3/2012 Pt's mother called 12-21-11 relates pt had fe Y
1/3/2012 Child rec'd KINRIX vaccine as booster for DTAP/Polio series and he is < 4 yrs of age. Vaccines were #4
1/2/2012 Thursday, December 29, 2011 9:40a.m. 15 m Y
1/3/2012 21 yo developed LE weakness on Y Y Y 11
1/4/2012 Uncontrollable crying and cringing, baby turned beet red.
######## Initial report was received from a consumer Y
1/3/2012 Received influenza injection on 12/9/11. On Y
######## Skin rash and Urticaria, Pruritus. Treatment: Diphenhydramine 50 MG PRN, Triamcinolone 40 MG On
1/4/2012 Received immunization on 12/20/11. States Y
1/3/2012 Initial and follow up information has been received from a nurse practitioner, for VARIVAX (Merck), a
######## Red area on thigh about 24 hr. after shot. No Y
Patient seen on 12/21/11 for new onset righ Y

1/4/2012	Achiness, Congestion, which got			Y					Y
1/4/2012	Client presented 20 minutes after receiving vaccines to RN with hives noted on R eyebrow, below R e								
1/4/2012	Rash at injection site, generalized hives, breath holding					Y	1		
1/4/2012	Soreness in shoulder that has continued for t				Y				
1/4/2012	"Paralyzed" left shoulder-site of injection.				Y				
########	3 weeks after tetanus booster, pa			Y	Y	Y			Y
1/3/2012	Information has been received from a certified medical assistant concerning a 12 year old female pat								
1/4/2012	Within about 5 days post-vaccine receipt, pa				Y				
########	Hospitalized due to "unable to walk" one morning when					Y	11		
1/3/2012	14 yo. male with (L) 7th cranial nerve palsy onset 12-25-11. S/P FLUMIST 12-7-11.								
1/4/2012	Pt rec'd 4 vaccines at PE appt with PCP. He knew he passes out with vaccine administration so was lyi								
########	Case number PHEH2011US04447, is an initial spontaneous report received from a nurse on 15 Sep 2C								
1/4/2012	1-3-12 Received vaccines and approx 8 to 10				Y				
1/4/2012	C/o raised, hot, swollen, sore left arm that began at the start of her 11PM-7AM shift and worsened d								
1/4/2012	Mother states pt had redness and swelling a				Y				
1/4/2012	Right arm very sore, unable to pick things up				Y				
1/4/2012	Low grade fever. Skin rashes.								
1/4/2012	Received FLUMIST Nasal on 12/30/11 - was s				Y				
	Arm especially sore at shot site after flu for 3 days.								
########	This is a 63-year-old male veteran arriving vi				Y				
########	Patient was diagnosed at community hospital with Guill					Y	9		
1/4/2012	Left upper thigh with 3 cm induration & erythema with visible puncture warmth & pain to touch.								
########	Red, swollen Lt. arm. Measured 1 1/2 inch diam. Patient was given ibuprofen (400 mg) for the pain ai								
########	No adverse event. (TWINRIX vaccine had been expired when given to pt.)								
1/4/2012	Client stated approximately 2 hours after rec				Y				
1/4/2012	Parasthesias in bilateral lower ext, progressi				Y	Y	3		
1/4/2012	Fever and "diffuse macular eruption over whole body" that lasted 3 weeks. Start Triamcinolone Acetc								
1/4/2012	This was very first flu vaccination recieved. C				Y				
1/4/2012	Patient began to have auditory hallucination				Y	Y	14	Y	Y
1/4/2012	Excess thirst, frequent urination. By mid-Nov				Y				
1/4/2012	Rash, started 2 wks ago, burning, painful left lower side and rash broke out 2 days ago. Dx - Shingles.								
1/4/2012	Fever that started shortly after receiving immunizations up to 102 degrees. Rash present along entire								
########	Day of immunization, pt awoke with visual hallucinations. Pt stated those things are flying all around.								
1/4/2012	Nasal congestion, runny nose, facial pressure, slight temp since 12-28-11. Given AMOXIL 875mg BID								
1/4/2012	Pt received injection on 8-24-2011. Pt went i				Y				
########	Received VARIVAX on 12/21/11. On 12/22/11 at 12:00 pm post. Lt. forearm sore and redden. Feet itc								
1/4/2012	Patient came in with concern after receiving				Y				
1/4/2012	Within 12 hours of vaccine, infant had increase in work of breathing and required increased respirato								
1/4/2012	Within 12 hours of vaccine, infant had increase in work of breathing and required increased respirato								
1/5/2012	Large raised, hard, red welt on leg below inj				Y				
1/4/2012	1/4/12 10:35A Pt received HPV shot in (R) deltoid without problem. Talking & responding appropriat								
########	Initial report was received from a health care professional on 12 December 2011. A 40-year-old male								
########	High fever 105 on 12-18, 19 2 days later red papular rash on trunk went away 2 days came back more								
1/5/2012	Tickly in throat about 3 hrs post injection. Ra				Y				
########	Infant dev	Y	########	Y	Y	Y			
########	Patient called pharmacy 12-10-11 to report muscle and joint pain that started on day after vaccinatio								
########	Dizziness, blurred vision, coughing - 20 min.				Y				

########	5 days after Tdap right deltoid site swollen tenderness about golf ball size swelling, right arm sore pai			
########	Cutaneous nodularity with transient inflammation (R) thigh.			
########	Skin flushing - diffuse, splotchy with pruritis, Y			
########	Developed sx of wheezing, headache, nasal (Y			
########	Aggravation of pre-existing peripheral neuropathy - states dx of Charcot-Marie-Tooth - but had taken			
########	Pt reports aggravation of fibromyalgia symptoms.			
########	FLUZONE received 10/04/2011 - developed (Y			
########	Diffuse itching 30 min - 1 hour; swelling of lo Y			
########	Received FLUZONE at 0700, developed flush Y			
########	32 y.o. female, pregnant, 30 weeks gestation, rc'd inactivates flu vac 1/6/12 IM, (L) deltoid. C/O rash			
########	Within 1 hr had site tenderness, erythema, s Y			
1/6/2012	Edema & red ring around injection site. Tender to touch. Ice to area prn. Also ADVIL prn.			
########	After 2 injection pt. stated she felt like she was going to pass out. St. her vision was going dark. Face (			
########	Information has been received from a physic Y	Y		2
########	Information has been received from a licens(Y			
########	3 cm x 3 cm circular wheal of erythema, war Y			
########	1-4-2012 started with red papules on rt- shoulder. Seen in office 1-9-12 with papules over torso head			
1/5/2012	Within 1 to 2 minutes after receiving this shot my daughter fell back into a chair and had convulsions			
	Bruising at and below injection site. Some swelling. Instructed to take BENADRYL and call family MD i			
########	Lg area approximately 2" below injection site that had fluid, redness and warmth to touch. Client stat			
########	Serum sickness enlarged (small apple size kn Y	Y		12
########	CLIENT WOKE UP WITH LOWER BACK PAIN A 5 ON THE SCALE OF 1:10. ALSO HAD SOME SLIGHT STIFI			
########	RED RAISED RASH TO CHEEKS AND TORSO, FRONT AND BACK.			
########	My back started aching and I also got a mild headache. The next morning, I woke up with a worse hea			
########	I noticed patient was covered in petechiae a Y			
	None stated.			
1/6/2012	Pt had redness, swelling, low grade 99.5 deg Y			
########	Vaccinated 4 Dec 11. Discovered 19 Dec 11. She was 7 wks pregnant.			
1/8/2012	Fevers & chills started at 2300. Took (1) one 200 mg Ibuprofen at 0200 and 0600.			
1/5/2012	Started swelling 6 hours post chicken pox vaccine. 48 hours after vax wheal & induration 60 mm x 70			
1/8/2012	Member describes SOB, having to take shallow breaths. Along with periods of feeling flush.			
########	Patient received MENACTRA vaccine on 12-16-11 in right deltoid muscle. Mother noticed swelling of			
########	None stated.			
########	Right upper arm swollen, red, warm to touch, and painful. Mom states "cool compress was applied a			
########	Terrible head and neck pain.			
########	Fever of 101, followed a few days later by rash on trunk.			
########	Hives occurring almost daily lasting 30 min to 1 hour. Exacerbated by cold temperature. Diffuse, itch			
########	Numbness, pain in limbs loss of vision & bal Y		Y	
########	2 - 3 facial hives. Swollen lower lip. Approximately 90 minutes after injection. Took BENADRYL 25 mg.			
########	Symptoms included tingling in both extremities, along w/numbness both arms, "itchy sensations", m			
########	1-11-12 Pt calls approx 300pm talks to nurse Y			
########	Became very red-faced and had trouble brea Y			
########	This retrospective pregnancy case was reported by a consumer (also the subject) and described the o			
########	Felt sharp pain and burning in (R) deltoid inj Y			
########	Within 12 hours of receiving vaccine pt was feeling ill: nausea, vomiting, severe headache. On 01/07/			
########	Mother calls facility reports child might have had a reaction. Reports fever. 01/10/2012 0900 Mother			
########	"Painful spasms located on the lateral side of my left leg for approx. 20 -30 minutes", says patient. Nc			

	Pt got "hot flashes" and felt faint and had to sit down for about half an hour.			
1/3/2012	Nausea, fatigue, blurred vision, motion sickness, moody, heartburn, leg & feet pain (neuropathy). Mi			
	Pt got "hot flashes" and felt faint and had to sit down for about half an hour.			
########	Pt returned to clinic 3 days after receiving ZOSTAVAX. c/o redness, swelling, & mild itching at injectio			
########	Pt returned to clinic 3 days after receiving ZC	Y		
1/3/2012	Patient stated developed painful red spot on forehead that was oozing and draining day after ZOSTA\			
########	Pt at 2 & 4 when vaccine administered.			
1/3/2012	Beginning at 7 pm shaking, heart palpitation:	Y		
1/6/2012	Redness at injection site that went away in 2 weeks after administration after adding ice pack to the r			
########	Feverish, chest pain, treated in ER.	Y		
1/6/2012	Rt deltoid swelling, redness and fever > 48 hrs.			
1/5/2012	(L) deltoid swelling per parents. Parent reported 01/05/12. Vaccine given 9/27/11.			
########	Developed noticeable diminished hearing or	Y		
1/4/2012	Pain on vaccine administration. Ache and pain 1 week severe. Ice, ibuprofen. Pain decreasing slowly (			
########	Swelling, started size of baseball then covere	Y		
########	#NAME?			
1/5/2012	None stated.			
########	Injection 11/29, arm sore, fever up to 102 or	Y		
1/6/2012	Redness to Rt. deltoid - rash.			
1/5/2012	Swelling in arm, heat in arm. Swelling in face, redness in face & arm.			
1/3/2012	Fever 104-105 x2 days.			
########	Mother gave child acetaminophen prior to se	Y		
########	Site red, swollen and warm to the touch and painful.			
########	Onset of clinical symptoms of infantile spasn	Y	Y	3
1/5/2012	30 minutes after receiving vaccine, I experiei	Y		
########	Wild type varicella rash.			
########	This case was reported by a healthcare profe	Y		
########	Localized edema.			
########	Patient's injection site from varicella vaccine (upper right arm) swollen and red. <24 hours later, red t			
########	Rash similar to hives. Tuesday very severe muscle stiff, could not move leg, Tuesday evening very stif			
########	No adverse event at this time for patient symptoms. Patient presented to stations at military mass ir			
########	Vaso-vagal episode due to bearing down and hyperventilation in the setting of receiving three vaccin			
########	They gave her the shot and about 30 second:	Y		
########	Grandmother called 1-9-2012 to report child developed swelling of tongue and lips on 1-5-2012. She			
########	He contracted up his arms head went down and went unresponsive for about 15 seconds before we c			
########	Migraine headache and ordinary headachy on next three days. Fatigue. Chills. Mild fleeting dizzy.			
########	Raised bump, bruise, swelling, stiff arm, war	Y		
########	Cellulitis with erythema; mass approximately 3cm diam	Y		2
########	Left arm is still sore after approximately 6 weeks from receiving injection. He reported this 1-11-12.			
########	Flu-like symptoms for 2 weeks. Initially fever to 102. Pe	Y		8
1/9/2012	Immunizations given and infant broke out wi	Y		
1/9/2012	Nausea, vomiting, fever, body aches.			
########	Redness, swelling at site on left thigh, sore to touch. No fever.			
########	Patient received a ZOSTAVAX vaccine on 1/9/12 in her right arm and on 1/11/12 she noticed her inje(			
########	Shoulder pain in the arm which was injected with the vaccine. The patient is still in pain.			
########	Left shoulder which was the location of the s	Y		
########	1-10-12 4 cm induration, tender, red, hot (R) arm. 1-11-12 2 cm induration, no redness, no tendernes			

########	Hives. BENADRYL and Methylprednisolone.			
########	Shot in left arm. After about 1 wk pain. Pain size of a dime near shot. Pain spot got larger to size of a s			
	Red hard spot after inject for at least 12 hrs - MD gave pt. KEFLEX; severe muscle pain in arm to point			
1/4/2012	In 1 hour, patient had episode of limpness, a	Y		
########	My son got vaccinated with the combo flu/H	Y		
########	About a month after I was vaccinated - I wok	Y		
########	On 10/26/11 at 10:30 AM my daughter received the flu/H1N1 combo mist. On 10/27/11 in the AM sh			
########	Sudden crash of the nervous system - shakin	Y		
########	Pain since 24 hours after injection, generalized aches, now recovered from this. Currently has localize			
	High fever of 103+ on 1/2 dose. Can't take full dose so my doctor gave 1/2 + 1/2.			
1/5/2012	"Large Lump to right upper extremity".			
1/5/2012	"Redness, Swelling, Soreness x 3 days, no further problems".			
1/5/2012	High Fever 104 degrees, very stuffy nose, muscle aches, cough. Ibuprofen every 5 hours for 5 days.			
1/5/2012	Localized swelling. BENADRYL given.			
########	12/27/2011 9:00 PM severe pain in right arm & headache patient took TYLENOL then oxycodone. 12/			
########	Pt describes a soreness in injection arm which has persisted since 12/16/11.			
########	Mother called the doctor and reported that the baby cried in a "high pitch" for a few hours post vacci			
########	On 12-22-11 client & mom came to WIC offic	Y		
########	Body aches, flushed, fatigue.			
########	Fever, headache, fatigue, chest pressure wheeze.			
########	None			
########	Pt received DTaP & Hep A. That evening, vomited 6 times then had 2 episodes where he had a "blank			
1/5/2012	Muscle myalgia within 24 hours of administration of vaccine.			
1/5/2012	Mother state patient started twitching on the right side of his torso. It only lasted seconds. Patient ha			
1/5/2012	Severe bruising on entire front side of upper Right arm (arm she received Zostavax vaccine.) Slight inf			
1/5/2012	PT IMMEDIATLY HAD BLOODY NOSE FOLLOWING INTRANASAL FLU ADMINISTRATION RESOLVED AFT			
1/5/2012	Symptoms of narcolepsy diagnosed and treated over the next year. Stimulants prescribed (and event			
1/5/2012	90 min seizure. 100 degree F fever. Acute re:	Y	Y	3
1/5/2012	Bowel obstruction; hospitalization.			
1/5/2012	Within 24 hours of the influenza injection, pt	Y		
1/5/2012	Contact Dermatitis worsened, with hives. Se	Y		
1/5/2012	Extremely high heart rate, uncontrollable br	Y		
1/5/2012	This case was reported by a healthcare professional (nu	Y		4
1/5/2012	Pt noticed next day at injection site was sore and hard. On 1/3/12 pt noticed that site was hard, red, a			
1/5/2012	None.			
1/5/2012	Guillain Barre Syndrome, acute Inflammator	Y	Y	11
1/5/2012	My son woke up with a bad rash on face and	Y		
1/5/2012	Patient had swelling and redness from injection site (RAT) to right ankle. Observed patient for 30 min			
1/5/2012	Numbness, twitching and pain involving all 4	Y		
1/5/2012	Maximum conductive hearing loss on the lef	Y		
1/5/2012	Migraine-type headache day following immunization with ZOSTAVAX. Self treatment with ibuprofen.			
1/5/2012	(L) arm swelling, tenderness, redness at site of injection.			
1/5/2012	Rash on top of thighs. Itchy and very red.			
########	Patient was given influenza vaccine 11/12/2011 returned to pharmacy 12/31/11 complaining of pain			
1/5/2012	Inconsolable crying x 7 days without fever, n	Y		
1/5/2012	Itching, then hives, treated with Benadryl, prednisone, Zantac, as of 20 Dec still had some hives.			
1/6/2012	He developed the classic case of Rubella-mild rash from head to toe, mild fever, malaise, stuffy nose,			

1/5/2012	Received vaccine 2 days ago - 1/3/12 - left arm pain at site. Complaining of right eye twitching & occa				
1/6/2012	Vaccine administered 1/3/2012 at 1450. Initial symptoms pain, soreness and unable to raise arm, the				
1/5/2012	Child had acute onset of significant stuttering 2 days after receiving the KINRIX vaccine. Child had hig				
1/5/2012	Patient seen on 1/5/12. States the evening of the vaccine administration she noted pain no swelling i				
1/6/2012	Patient became extremely lethargic, staring into space for more than five hours at a time..no respons				
1/6/2012	Left leg swelling and redness.				
1/6/2012	Hives on trunk.				
1/6/2012	Redness, heat, pain at site. Also "startled" awake numerous times throughout the night and was mor				
########	Pt. was vaccinated by student nurse who was under clinical supervision by nursing instructor. Pt state				
1/5/2012	This literature report (initial receipt 26-Dec-2011) concerns a 52-year old male patient, who had a his				
1/3/2012	Vomiting the day after rotavirus #3 - full recc	Y			
	12/5/2011 C/O swelling, itching, redness (L) upper arm since immunization 11/28/2011. Fever - impr				
1/3/2012	Rash on chest X 3-4 days.				
1/2/2012	Pt received Tdap in ER on 12/3/11 following a laceration. She called our risk manager on 12/13/11 to				
1/3/2012	11-2-11 developed "shingles" (blisters on top of left hip) lasted through November & went away very				
1/6/2012	Felt flu like symptoms, very exhausted, could not stand for more than a few minutes at a time. This w				
1/6/2012	Local site reaction. Redness and slightly elevated skin - warm to the touch. Patient denies pain at site				
1/6/2012	He awoke from falling asleep after the shot and was inconsolable. He wouldn't eat or nurs				Y
1/6/2012	States hair started falling out, morning after vaccine given. More every day, till call yesterday. Hair th				
1/6/2012	Swollen itchy feet, lethargy, feeling "sick". Next day - terrible pain in right ankle and left wrist, feeling				
1/6/2012	Uncontrollable crying immediately. Within a few hours his thigh turned very red with hard spot. His t				
########	Child had seizure approx 8 hours after immu	Y			
1/6/2012	Mom called, child received IMZ yesterday at 10:15am, by 1pm child would not bear wt. on either leg.				
1/5/2012	Injection site swelling & pain with radiation to axilla, subjective fevers & myalgias x 10 days. Pt. recei				
1/6/2012	Rash to face, leg, chest, back.	Y			
1/6/2012	This case was reported by a healthcare profe	Y	Y		
1/6/2012	Patient was give flu vaccine 1/4/12 approx 4	Y			
1/5/2012	Hives/rash starting on inner arms & upper back on Wed, Jan 4th. Spread to lower torso, thighs and se				
1/6/2012	Redness, and pain at injection site of subcuta	Y			
1/6/2012	Red spots all over since this AM. No fever. Positive pruritus.				
1/6/2012	Progressive weakness in lower extremities, a	Y	Y	32	
1/6/2012	Had a low grade fever (100.5) and was not acting himself. Crying almost all the time, wouldn't eat, ar				
1/6/2012	Small rash on left thigh noted 12/20/2011. Similar rash noted 12/30/11 on left wrist. As of 01/06/12,				
1/6/2012	Erythema approx 50mm in Length by 30mm width, area warm but not hot, slightly firm.				
1/6/2012	Sandpaper like rash over entire body for the following 3 days. Mother did not bring child for a provide				
1/6/2012	Within an hour of the vaccine my arm began	Y			
1/6/2012	Mother called 01/05/2012 to report that client has had a sore arm since vaccinations in Nov 2011. Ha				
1/6/2012	Tremors in upper body, lasting <1 minute (occurred on 01/05/2012). No treatment. Decreased appet				
1/6/2012	Acutely could not hear out of the right ear. E	Y			Y
1/6/2012	2 M old previously healthy boy who was adm	Y	Y	2	
1/6/2012	Soreness and Numbness starting approx 2:30 pm at injection site and continuing to full arm. Still un-c				
1/6/2012	CHILD HAD LOW GRADE FEVER FOR 2 DAYS,	Y			
1/6/2012	Extreme soreness in arm and shoulder. Dizziness.				
1/6/2012	Pt states that she has itchiness and redness all over her front trunk area starting since 01/05/12 night				
1/6/2012	Swelling, itchy red rash at injection site. Rash grew in size. Fever 101 F.				
1/6/2012	Fever, body aches, joint pain, stomach cramps, diarrhea, fatigue.				
1/6/2012	6.5 cm induration redness, swelling, itching.	Y			

1/6/2012 Child had seizure approximately 30 minutes Y
1/6/2012 Patient became lightheaded after receiving vaccine. She also experienced tinnitus. She became pale.
1/6/2012 Aching sore arm about 5 hours after vaccine: Y
######## 3 month 4 week female presents Y Y Y 4
1/7/2012 C/O SWELLING FROM MID UPPER ARM DOWN TO ELBOW.
1/7/2012 Pain, swelling, and red, itchy rash spreading on arm from the site of the vaccination. This continued t
1/7/2012 Severe allergic reaction. Symtoms - rash, itch Y
1/7/2012 Guillain Barre. Y Y 7
1/7/2012 Severe headache, sinus pain, dizzy right after shot & mild/mod. continue next day.
1/7/2012 Symptoms: SkinRash & UrticariaPruritus. Tre Y
1/7/2012 Red, inflamed arm & pain to touch on left deltoid. (Right arm was and still is in sling from another cau
1/8/2012 Unable to defecate. Unable to speak clearly. Intense fear. Nausea. Headache. Runny nose. Earaches.
1/8/2012 Crying, Pain, Unable to Defecate. Prior to vac Y
1/8/2012 Large amount of swelling very red and itchy Y
1/8/2012 Red bumps covering legs - began at back of t Y
1/8/2012 Swollen right deltoid, red ring on surface of the skin about the size of a quarter, tingling of fingers.
1/8/2012 Red splotchy rash around where the vaccine was admistered. It was hot and looked like a whelp.
1/8/2012 Headache, itchy/burning eyes, especially left eye, joint aches, tired, sneezing, runny nose, large welt
1/8/2012 Patient admitted 11/7/11 and discharge 11/8/11 readmitted 11/10/11 and dis Y
1/8/2012 Pain in arm at site of injection when lifting arm to shoulder level for three days straight.
Redness, hot to the touch, hard.
######## Patient complains of hives, chills, vomiting that began within a few hours after receiving the vaccine,
1/9/2012 Runny nose, cong for 2 weeks general malais Y
1/9/2012 Temp. started at 101.2 took Tylenol that was Y
1/9/2012 Call from pt's mother (visit 12/28/11) 01/06/12. Concern: 6 yr old "bump and redness" at the injectio
1/7/2012 Macular papular rash & pruritis to body - throat fullness. Do + MEDROL pak & Hydroxyzine PRN.
1/6/2012 Information has been received from a physic Y
1/6/2012 Information has been received from a 65 yea Y
######## Injection site became very indurated warm, tender.
With erythema at site of vaccine and ongoin Y
1/3/2012 12/20/11 Evening - affected arm became red, swollen, hot & painful. Sat 12/31/11 the redness starte
######## Patient complained of pain upon administration of shot & came back into pharmacy complaining on 1
######## None stated.
1/3/2012 None.
1/2/2012 None.
1/2/2012 Mother called 1 hour after patient received second influenza vaccine: Infant's abdomen covered in re
None stated.
1/2/2012 I got five rabies shots between Friday Decem Y
1/2/2012 Paresthesias - numbness, tingling involving both upper and lower extremities - Sx come and go but st
######## Pt. received a Tdap & PNEUMOVAX & the night after the vaccines she developed fever, malaise, dren
1/3/2012 Headache, redness, hot at injection site. Low Y
Nausea & vomit.
1/3/2012 Inflammatory reaction (+) swollen.
1/3/2012 Swelling after shot same day went to ER. Giv Y
######## Injection given early.
######## 4 hospital visits in a two week period. Primar Y Y 6
######## Patient received pneumococcal vaccine Dec 12 as an inpatient. He was seen today in outpatient clinic

########	Profuse vomiting for 3 days following ROTARIX vaccine.				
########	Right after injection. Child has seizure. Sent t	Y			
########	Shivering of jaw, legs shaking, trouble walking which resulted in being unable to walk until the next n				
########	PT. FELL TO GROUND AFTER EXITING CLINIC,	Y			
########	Sxs began as a muscle cramp in arm followed by swelling that extended to the elbow and arm pit lym				
########	High fever, bad headache, dizziness, leg muscles hurt and feet hurt, sharp pain throughout and heavy				
########	Itching and burning of left corner of right ey	Y			
########	Cold, ear infecation, lack of: Motor skills, not social, lost words, does not talk any more.				
########	Presented to ED with continued right should	Y			
########	Aches and pains. I attibuted it to getting bot	Y			
########	Red, raised and swollen lump at the site of the injection. Painful to the touch.				
########	5 minutes after vaccination patient stated she had a 4/10 h/a,pain in the back of her head, and feelin				
########	Employee reported experiencing (L) facial nu	Y			
########	Mouth/tongue painful.				
1/5/2012	Initial report was received on 08 November	Y			
1/6/2012	Initial report was received on 03 January 201	Y			
########	Patient had a large area red and swollen on right arm. (Deltoid region).				
1/9/2012	Pt developed shingles on face (below (R) eye	Y			
########	This case was reported by a physician and de	Y	Y	15	Y
########	Shot given too high in shoulder - extreme pa	Y	Y	1	
########	Vaccine admin on 12/29/11. On 1/07/12 dev	Y			
########	While playing or just sitting up her head wou	Y			
########	Malaise, chilling "global headache" states "most severe headache I've ever had" gradual onset that in				
########	Right arm weakness, tingling, numbness one	Y			
########	Approximately 3 hours after receiving vaccine patient started to wheeze and have a sore throat and h				
########	Nonpruritic papular rash on face and neck.				
########	Patient underwent both colonoscopy with pc	Y	Y	2	
########	Fever 102.				
########	Muscle fatigue and tiredness.				
########	Swelling, itching, slight redness, warm to touch and induration. Redness and induration 2 inches in di				
########	Red mark about the size of a 50 cent piece with raised bumps with yellow tops slight ooze, like poisor				
########	Left arm was sore.				
########	Vaccine 1100 on 01-10-2012; began to flare with swelling & inflammation at 1600. 9 cm x 7.5 cm (RT)				
########	Patient presents today with a rxn to vaccine, her (L) arm has a red area the diameter of a baseball tha				
1/9/2012	At injection site (Right outer thigh) welt measured 2cm x 2.5cm with surrounding area red measuring				
########	Appears to be generalized vaccinia; diffuse n	Y			
########	Swollen tender muscle at injection site - wor	Y			
########	9-27-09, 6-3-10, 3-4-11] shingles outbreak; ab & pain treatment given.				
########	Pain and swelling in right foot and ankle, eve	Y			
########	Patient was given outdated vaccine. Flu Mist was outdated on 12/25/2011.				
	Raised site of erythema and tenderness approx. 2-3 inches in diameter. Pruritic rash - day 3.				
	None stated.				
########	Big red skin rash (3" in diameter) around injection site. The rash is itchy, her entire upper arm achy ar				
1/5/2012	Weakness requiring medical treatment.	Y	Y	16	
########	"Rash" arms/feet headaches. Vomiting/fatigue (erythema multiforme).				
1/5/2012	HPV #1 received 10-17-11 at well check. Headache within 24 hours. Seen 11-1-11 ? atypical migraine.				
1/5/2012	Patient received PNEUMOVAX & FLUVIRIN in same arm, arm is sore, swollen, red, hard to move, & w				

########	Intense burning pain from injection site shoulder joint - > elbow. Lasted x 4 days also reported malais			
########	Pt. given pneumonia vaccine and had an imn	Y		
########	Fever, stomachache, head ache, neck ache, general soreness, sore throat, congestion.			
########	Seen at this office for consult 1/6/2012. Doc	Y		Y
########	Disoriented, dizzy, headache, body aches, ex	Y		
########	Stomachache, sore throat, fever, headache, fatigue.			
########	Headaches; Legs going numb; Nausea; Chest pains-right side; Joint pains; Lower right abdominal pain			
########	Patient received injection and had a little pin	Y		
########	Fever, red streaks running down arm around	Y		
########	Received 2 vaccines at school clinic Tdap & MENACTRA 1/11/12 (R) arm very sore unable to move arr			
########	None stated.			
########	Patient reported inflammation at site of inje	Y		
########	Swollen tender nodules 3 weeks after shots.			
########	Approximately 20 minutes after receiving FLUMIST patient reported itching to right forearm. Quarter			
########	Pt received Varicella vaccine in right thigh or	Y		
########	After being vaccinated, the patient wasn't able to stand and hold her own weight. It seemed like her			
	Patient experienced redness, swelling. MD prescribed Cephalexin.			
########	Bloody, mucousy loose stools began 1-11-12	Y	Y	2
########	Arm which vaccine given became very painful - joints became painful & stiff - mobility/movement be			
########	Right leg - 1 inch red, warm but not tender nonfluctuant nodule on Rt. thigh at injection site of flu an			
########	Information has been received from a physic	Y		
########	Information has been received from a Registered Pharmacist (R.Ph.) concerning a 68 year	Y		
########	1/12/12 12 x 8 cm area of erythema with warmth and tenderness. No mass, no discharge (L) deltoid a			
########	SWELLING AND PAIN LEFT DELTOID.			
########	Small blisters on same arm as the injection site that ruptured and itched. Applied OTC meds. Very littl			
########	Acute onset of vomiting and diarrhea. Has be	Y		
########	Pale and woozy/dizzy afer MCV4 followed by sleepiness. Instant migraine-intensity headache that re			
########	(1/9/12) Patient was seen in MD office for Pt	Y		
########	Pruritus, internal tremors, began 2 hours aft	Y		
########	(R) UA - had a rash site of FLUZONE. Now rash resolved & pin head size erythema, site of injection.			
1/6/2012	Patient present at clinic requesting to report	Y		
########	On 10/10/11 patient was diagnosed with shingles (10/8/11 onset). She is on immune-suppressing me			
########	This spontaneous consumer report (initial receipt: 03-Jan-2012) concerns a 73-year old male patient,			
########	Initial report was received on 05 January 201	Y		
########	Initial report was received from another mar	Y		
########	Initial report was received from a consumer,	Y		
########	Initial report was received on 19 October 20:	Y		
########	Initial report was received from a consumer,	Y		
########	Itching at injection site that lasted for 2 to 3 days. Injection site sensitive to being touched (feels bruis			
########	Initial report received on 04 January 2012 frc	Y		
########	Initial report received on 04 January 2012 from the Inve	Y		
########	Redness, swelling and itching of affected right arm. Mother reports redness and swelling decreased s			
########	Symptoms began approx. 2 weeks after vacc	Y		
########	Very intense seizures unlike those that she has normally. They lasted for about 2 minutes at a time ar			
########	Left arm pain and swelling. Treatment of ice	Y		
########	Patient stated she received INFLUENZA vaccine at that evening started feeling itchy. When she got up			
########	Leg, shoulders, arms joint pain; felt very weak. Received cortisone injections from Dr. Followed up w/			

######## WOKE UP FOLLOWING AM AFTER IMMUNIZATION AND C/O SORENESS TO TOUCH, REDNESS, ITCHIN
######## Patient experienced extensive inflammation and redness a the site of injection.
######## Patient stated discovered to have red, blotchy and itchy skin everywhere.
######## Shoulder and Arm Pain reported by pharmac Y Y
######## Approximately 24 hours following flu injectic Y
######## Approx 12:30 PM same day..excrutiating pain in left shoulder..unable to move arm at all...used ice ar
######## Fever, bone aches, stiff painful arm, chills- N Y
######## Pt. received flu shot, within 24 hrs. her left a Y
######## Patient received flu shots on 12/1/2011. Client states that she experienced redness pain & swelling a
######## 49 y.o. healthy female received Typhim-Vi 12/16/11 c/o fatigue 12/19/11, vertigo 12/20/11 and diarr
######## Pt received shot on 11/28/11. Pt came back on 1/13/12 and has a visible bruise and a small rash arou
######## Generalized malaise weakness the day after vaccination. Reactive arthritis, roaming from ankles, kne
######## This case w Y Y
######## This case was reported in a literature article and describ Y 8
######## Mother states child was jerking the day after Y
######## Patient was given the ZOSTAVAX on 9/19/11. On 10/10/11 his wife was diagnosed with shingles (10/
######## Nickel sized circular red mark over the injection site along with itching.
######## Routine fevers started, odd skin rashes, and Y
######## Severe pain, hard to breathe, vomiting, fevei Y
######## Fussy, crying, decreased appetite, symptoms Y
######## Swelling, warmth. Area is 4cm redness, 2 cm Y
######## High fever (104.0).
######## Severe fatigue and pain in back near right should blade. Pain is intermittent and occurs in throbbing
######## Infant began uncontrollably screaming and crying for 5 hours without stopping. Fever and swelling or
4 inch wide redness circle around injection site rash on back of neck.
######## Patient noticed a lump at vaccine site 2 weeks after the vaccine was administered.
######## Hives on arms, legs, and torso. Red, warm, s Y
######## Mother states fever 100.5 for 1 1/2 days (since injection) Rt side shaking and staring for 30 seconds tl
######## Patient developed chills, body aches, mild fever (didn't take actual temp) and local redness & swellin
######## See attached. Patient was vaccinated on Jan Y
######## Child had 4hr period of lethargy/non respon: Y
######## Started as red lump. 1-14-12 Dx with cellulitis & abscess. Treated with AUGMENTIN.
######## Local maculopapular rash at upper arm 48 hours after HepA injection given, pruritus.
######## Decrease appetite / Black stools / Y Y Y 2
######## Immediately following vaccination, pt had pallor, c/o dizziness, and "lights seeming too bright".
######## Limbic Seizures later diagnosed a Y Y Y 25
######## Localized allergic reaction: redness, swelling, pain around area of injection that lastest for 7 days. Sav
######## Red, warm induration at injection site.
######## Patient received vaccination on 9 Y Y Y 6
######## Immediate pain in left arm, which was unusu Y
######## Nacrolepsy. Y
######## Arm pain.
######## Abscess to left thigh. Y
######## Redness, swelling, numbness at site of HPV and MCV4.
######## Abscess on left thigh. Y
######## R arm swollen and hot elbow to shoulder; couldn't lift arm more than 1" without pain; fever/chills; 1/
######## Red rashy heat present tender to touch and hardened area approx 4 inches.

########	Pt reported feeling light-headed/dizziness, vision changes (tunnel vision) and shaky, also reported pa			
########	This case was reported by a physician (also the subject's husband) and described the occurrence of ly			
########	Patient showed signs of going into a catatoni	Y		
########	Tachycardic, difficulty breathing, nauseated.	Y		
########	Received vaccine at 1:50pm. Began to have site soreness and swelling that radiated to right shoulder			
########	Pt. presented to office 3 days after vaccine. Large swollen area, warm to touch. Has "improved" since			
########	Redness, mild induration (R) thigh. No tenderness.			
########	This case was reported by a healthcare professional and described the occurrence of passing out in a			
########	Information has been received from a regist(	Y		
########	Information has been received from a physician concerning a 15 year old male who was vaccinated w			
########	Information has been received from a 72 year old female patient with allergy to sulfa medications an			
1/8/2012	Bruise at injection site and lump under armpit noticed.			
1/5/2012	11/23/2011 - Pt presented for flu vaccine. Pt	Y		
########	Arm was sore and continued to stay sore for over 8-10 weeks. Had limited mobility in arm.			
1/9/2012	#1 9-20-07. #2 5-27-11. Has had 2 varicella vac - diagnosed varicella.			
########	Fever 100.1 F for 24 hr then 99.6, very fussy for 4-5 days, sudden episode of crying, abd. pain, cramps			
########	Has dime sized, slightly raised area at injection site. Area affected has diminished slightly over month			
########	Redness of (L) arm x 1 day & today. Arm is sore x 7 days. No fevers. No treatment.			
########	Arm pain, fever, erythema of (R) arm 5cm x :	Y		
1/6/2012	Client states she felt "funny" immediately after vaccination. Returned to clinic to have PPD read, and			
1/4/2012	In the evening after receiving vaccines - pt started to have hives (approx 6) on sides of face & neck sit			
########	Hep B vaccine given on 1/11/12. Call from Mom on 1/13/11 to report fever 100-102 degrees, fussine:			
########	12/17/11 - very stiff arm & tricep, decreased movement, dizziness during AM hours. Dizziness ended			
1/9/2012	Edema (L) thigh with mild ecchymosis.			
########	Uncontrolled screaming x 3-4 days.			
########	Redness, swelling, severe itching, pain, cellulitis. BENADRYL 50 mg Q 6hrs.			
########	Pt telephoned approximately 24 hours after	Y		
########	- Rash generalized, worse on buttocks since Sunday, U shaped hive on (L) cheek. Fever 101 - 102 Thur			
########	Slight swelling and redness at site of injection. Arm soreness radiating to shoulder - pc patient.			
########	6 mo vaccine given at noon on 1/9/2012 baby was more "sleepy" per mom that evening woke at 013(			
########	2 hours after immunizations, groggy, letharg	Y		
1/5/2012	Flu shot 12-13-11. Nasal congestion, upper respiratory infection cough, sore throat, laryngitis, fatigue			
1/9/2012	Pt had redness/watering of eyes. Reaction similar to pt history of allergic reaction to thimerosal (in e\			
1/5/2012	On 10/8/11 employee started feeling achy al	Y		
########	12/6 - 3 pm - received vaccines. 5 pm - (1) Burning throat; "itchy esophagus." (2) Felt like throat was "			
########	Bruising and petechia on limbs, torso and or(	Y	Y	4
1/6/2012	Within 1 degrees of administration of vaccine pt's Mom states dime sized hives appeared on abdome			
1/9/2012	Started tunnel-vision with muffled hearing; lost feeling from neck down; got lightheaded with heada(			
########	Reported to be irritable evening vaccine rec(	Y		
########	Redness, heat, soreness at injection site - red area under shot location. Patient reacted to PNEUMOV			
########	Swelling with redness from deltoid to elbow, accompanied by fever.			
1/9/2012	(L) arm with 2 x 1 1/2 cm area of erythema & hyperpigmentation.			
########	Difficulty breathing (wheezing); chest pain; c	Y	Y	4
1/6/2012	Seizure/status epilepticus. Received multiple medicatio(	Y		2
1/9/2012	Approximately 5 hrs after vaccination, child \	Y		
########	Sudden onset of high fever (104 degrees F), f	Y		
########	Warm/Red/Non-tender induration measurin	Y		

1/6/2012	Within 24 hours after receiving immunizatioı	Y			
########	Mother reports redness at site on injection. Pictures reveal improvement. No fever. Child does not aį				
########	Redness, pain, swelling 3 days after vaccine administered on (R) deltoid 6 days after vaccine administ				
########	Severe swelling and induration of upper arm at site of injection, requiring use of prednisone to preve				
1/3/2012	1-27-11 given ZOTAVAX IM (L) deltoid in error. 6cm x 6cm red, tender and swollen area developed in				
########	Initial stick did not hurt but on injection it felt like fire in my shoulder and to elbow. Later pain was so				
########	Could not reach vaccination report page so am reporting here. After receiving the flu shot I began itcl				
1/9/2012	Loss of subcutaneous tissue at site of injection, 1 inch in diameter of (L) arm.				
1/9/2012	Patient presented back to the clir Y	Y			
1/9/2012	"High fever, dizziness, paleness and mild difficult breathing as well as nausea. Was seen at urgent ca				
1/9/2012	HERPES ZOSTER IN A CHILD VACCINATED WI	Y			
1/9/2012	MOM STATES PT BROKE OUT IN HIVES ALL OVER BODY, GAVE BENADRYL WITH SOME RELIEF.				
1/9/2012	Fever day after vaccination, st, slight nausea. Pt states higher fever 1-8-2012 T103. No treatment, m				
1/9/2012	Stinging in arm, followed by dizziness, then light convulsions. Still on table so laid her down. She can				
1/9/2012	Since 12/17/2011 symptoms include: Fatigue, malaise, dizziness, abdominal pain, pain in the back of				
1/1/2012	Swelling, red, irritation at sign of injection. 3	Y			
1/9/2012	Feeling warm/heat, unable to sleep, nervous	Y			
1/6/2012	Syncope approximately 30 sec duration.				
1/9/2012	The patient developed hives approximately !	Y			
1/9/2012	Felt nauseous immediately after inj. which continued through the night and off/on until present (1/9,				
1/9/2012	Headache in back of head and pain in neck. !	Y			
1/9/2012	Tongue turned a reddish purple and the lips	Y			
1/9/2012	VOMITING 10 DAYS. WENT TO EMERGENCY	Y	Y	10	Y
1/9/2012	Swelling, redness, pain and induration-cellul	Y			
1/9/2012	Red flat rash on upper chest. Given Ibuprofen 200 mg. Advised to take Benadryl 25 mg 1-2 as neede‹				
1/9/2012	Right deltoid discoloration, swelling, tenderness 7/10 pain, right bicep twitching intermittently every				
1/9/2012	Patient began feeling tired and weak. Body a	Y			
1/9/2012	Swelling and itchiness.	Y			
1/9/2012	Swelling, redness & discomfort (R) deltoid were pneumococcal polysaccharide vaccine was administe				
1/9/2012	Felt ill 2 days post administration - site extremely swollen & hot took BENADRYL 12/17/11 & felt bett				
1/9/2012	3 immunizations given 1/4/11. On 1/6 developed diffuse urticarial appearing rash without symptoms				
1/9/2012	Several hours after injection of PENTACEL (R) thigh became warm, red & swollen. Mom says patient I				
1/9/2012	Perioral numbness, weakness, bilateral rhytł	Y	Y	4	
1/9/2012	Patient reports the day after vaccine given h	Y			
	None stated.				
1/9/2012	Child given PENTACEL in (L) leg at end of moı	Y			
1/9/2012	Had immunization at 3pm (L) leg -> PENTACEL. Call from mom at 715 pm. Crying, inconsolable, left tł				
1/4/2012	Temp of 105 degrees night of vaccine (he also had concurrent BOM).				
1/9/2012	Flu vaccine administered to 4 month old. No adverse event..				
1/3/2012	Pt reported (L) arm pain and swelling at and around injection site. Red, hot and painful to touch. Use‹				
########	* Patient had erythema & site was hot, this happened two days later. Advised by our PAC to put ice o				
1/2/2012	Child had flu vaccine on 12/16/11 from there	Y	Y	8	
1/9/2012	Day after vacc. fist size mark showed up at inj. site. Site is hard, hot, lump. Pt doesn't have fever or dy				
1/6/2012	Injection site redness & tenderness for two days.				
########	Pt. developed hives - given diphenhydramin‹	Y			
1/9/2012	Information has been received from a registered nurse and a family nurse practitioner, for GARDASIL				
########	Pt "felt dizzy throat and tongue tingling" epir	Y			

Now that you've taken a look at a few sobering pages from the VAERS database, consider this: How safe are vaccines? *So safe that, if you're on Medicare or Medicaid and have a serious reaction, a code number is pre-assigned to each and every one of them.* Look at the codes from the Medicaid and Medicare Web site on the following pages.

CMS.gov
Centers for Medicare & Medicaid Services

ICD-10 Code Lookup

Please Note: Enter a Code or keyword to conduct a search for ICD-10 Codes. To populate the ICD-10 Code Field on the Advanced Search page, click on the code link in the display list. The results page will close and your selection will display on the Advanced Search page.

Enter ICD-10 description keyword(s):

vaccine Search

ICD-10 CODE	ICD-10 CODE DESCRIPTION
A80.0	Acute paralytic poliomyelitis, vaccine-associated
B08.011	Vaccinia not from vaccine
T50.A11A	Poisoning by pertussis vaccine, including combinations with a pertussis component, accidental (unintentional), initial encounter
T50.A11D	Poisoning by pertussis vaccine, including combinations with a pertussis component, accidental (unintentional), subsequent encounter
T50.A11S	Poisoning by pertussis vaccine, including combinations with a pertussis component, accidental (unintentional), sequela
T50.A12A	Poisoning by pertussis vaccine, including combinations with a pertussis component, intentional self-harm, initial encounter
T50.A12D	Poisoning by pertussis vaccine, including combinations with a pertussis component, intentional self-harm, subsequent encounter
T50.A12S	Poisoning by pertussis vaccine, including combinations with a pertussis component, intentional self-harm, sequela
T50.A13A	Poisoning by pertussis vaccine, including combinations with a pertussis component, assault, initial encounter
T50.A13D	Poisoning by pertussis vaccine, including combinations with a pertussis component, assault, subsequent encounter
T50.A13S	Poisoning by pertussis vaccine, including combinations with a pertussis component, assault, sequela
T50.A14A	Poisoning by pertussis vaccine, including combinations with a pertussis component, undetermined, initial encounter
T50.A14D	Poisoning by pertussis vaccine, including combinations with a pertussis component, undetermined, subsequent encounter
T50.A14S	Poisoning by pertussis vaccine, including combinations with a pertussis component, undetermined, sequela
T50.A15A	Adverse effect of pertussis vaccine, including combinations with a pertussis component, initial encounter
T50.A15D	Adverse effect of pertussis vaccine, including combinations with a pertussis component, subsequent encounter
T50.A15S	Adverse effect of pertussis vaccine, including combinations with a pertussis component, sequela
T50.A16A	Underdosing of pertussis vaccine, including combinations with a pertussis component, initial encounter
T50.A16D	Underdosing of pertussis vaccine, including combinations with a pertussis component, subsequent encounter
T50.A16S	Underdosing of pertussis vaccine, including combinations with a pertussis component, sequela
T50.A21A	Poisoning by mixed bacterial vaccines without a pertussis component, accidental (unintentional), initial encounter
T50.A21D	Poisoning by mixed bacterial vaccines without a pertussis component, accidental (unintentional), subsequent encounter
T50.A21S	Poisoning by mixed bacterial vaccines without a pertussis component, accidental (unintentional), sequela
T50.A22A	Poisoning by mixed bacterial vaccines without a pertussis component, intentional self-harm, initial encounter
T50.A22D	Poisoning by mixed bacterial vaccines without a pertussis component, intentional self-harm, subsequent encounter
T50.A22S	Poisoning by mixed bacterial vaccines without a pertussis component, intentional self-harm, sequela
T50.A23A	Poisoning by mixed bacterial vaccines without a pertussis component, assault, initial encounter
T50.A23D	Poisoning by mixed bacterial vaccines without a pertussis component, assault, subsequent encounter
T50.A23S	Poisoning by mixed bacterial vaccines without a pertussis component, assault, sequela
T50.A24A	Poisoning by mixed bacterial vaccines without a pertussis component, undetermined, initial encounter
T50.A24D	Poisoning by mixed bacterial vaccines without a pertussis component, undetermined, subsequent encounter
T50.A24S	Poisoning by mixed bacterial vaccines without a pertussis component, undetermined, sequela
T50.A25A	Adverse effect of mixed bacterial vaccines without a pertussis component, initial encounter
T50.A25D	Adverse effect of mixed bacterial vaccines without a pertussis component, subsequent encounter
T50.A25S	Adverse effect of mixed bacterial vaccines without a pertussis component, sequela
T50.A26A	Underdosing of mixed bacterial vaccines without a pertussis component, initial encounter
T50.A26D	Underdosing of mixed bacterial vaccines without a pertussis component, subsequent encounter
T50.A26S	Underdosing of mixed bacterial vaccines without a pertussis component, sequela
T50.A91A	Poisoning by other bacterial vaccines, accidental (unintentional), initial encounter
T50.A91D	Poisoning by other bacterial vaccines, accidental (unintentional), subsequent encounter
T50.A91S	Poisoning by other bacterial vaccines, accidental (unintentional), sequela
T50.A92A	Poisoning by other bacterial vaccines, intentional self-harm, initial encounter
T50.A92D	Poisoning by other bacterial vaccines, intentional self-harm, subsequent encounter
T50.A92S	Poisoning by other bacterial vaccines, intentional self-harm, sequela
T50.A93A	Poisoning by other bacterial vaccines, assault, initial encounter
T50.A93D	Poisoning by other bacterial vaccines, assault, subsequent encounter
T50.A93S	Poisoning by other bacterial vaccines, assault, sequela
T50.A94A	Poisoning by other bacterial vaccines, undetermined, initial encounter
T50.A94D	Poisoning by other bacterial vaccines, undetermined, subsequent encounter
T50.A94S	Poisoning by other bacterial vaccines, undetermined, sequela

T50.A95A	Adverse effect of other bacterial vaccines, initial encounter
T50.A95D	Adverse effect of other bacterial vaccines, subsequent encounter
T50.A95S	Adverse effect of other bacterial vaccines, sequela
T50.A96A	Underdosing of other bacterial vaccines, initial encounter
T50.A96D	Underdosing of other bacterial vaccines, subsequent encounter
T50.A96S	Underdosing of other bacterial vaccines, sequela
T50.B11A	Poisoning by smallpox vaccines, accidental (unintentional), initial encounter
T50.B11D	Poisoning by smallpox vaccines, accidental (unintentional), subsequent encounter
T50.B11S	Poisoning by smallpox vaccines, accidental (unintentional), sequela
T50.B12A	Poisoning by smallpox vaccines, intentional self-harm, initial encounter
T50.B12D	Poisoning by smallpox vaccines, intentional self-harm, subsequent encounter
T50.B12S	Poisoning by smallpox vaccines, intentional self-harm, sequela
T50.B13A	Poisoning by smallpox vaccines, assault, initial encounter
T50.B13D	Poisoning by smallpox vaccines, assault, subsequent encounter
T50.B13S	Poisoning by smallpox vaccines, assault, sequela
T50.B14A	Poisoning by smallpox vaccines, undetermined, initial encounter
T50.B14D	Poisoning by smallpox vaccines, undetermined, subsequent encounter
T50.B14S	Poisoning by smallpox vaccines, undetermined, sequela
T50.B15A	Adverse effect of smallpox vaccines, initial encounter
T50.B15D	Adverse effect of smallpox vaccines, subsequent encounter
T50.B15S	Adverse effect of smallpox vaccines, sequela
T50.B16A	Underdosing of smallpox vaccines, initial encounter
T50.B16D	Underdosing of smallpox vaccines, subsequent encounter
T50.B16S	Underdosing of smallpox vaccines, sequela
T50.B91A	Poisoning by other viral vaccines, accidental (unintentional), initial encounter
T50.B91D	Poisoning by other viral vaccines, accidental (unintentional), subsequent encounter
T50.B91S	Poisoning by other viral vaccines, accidental (unintentional), sequela
T50.B92A	Poisoning by other viral vaccines, intentional self-harm, initial encounter
T50.B92D	Poisoning by other viral vaccines, intentional self-harm, subsequent encounter
T50.B92S	Poisoning by other viral vaccines, intentional self-harm, sequela
T50.B93A	Poisoning by other viral vaccines, assault, initial encounter
T50.B93D	Poisoning by other viral vaccines, assault, subsequent encounter
T50.B93S	Poisoning by other viral vaccines, assault, sequela
T50.B94A	Poisoning by other viral vaccines, undetermined, initial encounter
T50.B94D	Poisoning by other viral vaccines, undetermined, subsequent encounter
T50.B94S	Poisoning by other viral vaccines, undetermined, sequela
T50.B95A	Adverse effect of other viral vaccines, initial encounter
T50.B95D	Adverse effect of other viral vaccines, subsequent encounter
T50.B95S	Adverse effect of other viral vaccines, sequela
T50.B96A	Underdosing of other viral vaccines, initial encounter
T50.B96D	Underdosing of other viral vaccines, subsequent encounter
T50.B96S	Underdosing of other viral vaccines, sequela
T50.Z91A	Poisoning by other vaccines and biological substances, accidental (unintentional), initial encounter
T50.Z91D	Poisoning by other vaccines and biological substances, accidental (unintentional), subsequent encounter
T50.Z91S	Poisoning by other vaccines and biological substances, accidental (unintentional), sequela
T50.Z92A	Poisoning by other vaccines and biological substances, intentional self-harm, initial encounter
T50.Z92D	Poisoning by other vaccines and biological substances, intentional self-harm, subsequent encounter
T50.Z92S	Poisoning by other vaccines and biological substances, intentional self-harm, sequela
T50.Z93A	Poisoning by other vaccines and biological substances, assault, initial encounter
T50.Z93D	Poisoning by other vaccines and biological substances, assault, subsequent encounter
T50.Z93S	Poisoning by other vaccines and biological substances, assault, sequela
T50.Z94A	Poisoning by other vaccines and biological substances, undetermined, initial encounter
T50.Z94D	Poisoning by other vaccines and biological substances, undetermined, subsequent encounter
T50.Z94S	Poisoning by other vaccines and biological substances, undetermined, sequela
T50.Z95A	Adverse effect of other vaccines and biological substances, initial encounter
T50.Z95D	Adverse effect of other vaccines and biological substances, subsequent encounter
T50.Z95S	Adverse effect of other vaccines and biological substances, sequela
T50.Z96A	Underdosing of other vaccines and biological substances, initial encounter
T50.Z96D	Underdosing of other vaccines and biological substances, subsequent encounter
T50.Z96S	Underdosing of other vaccines and biological substances, sequela
Z28.04	Immunization not carried out because of patient allergy to vaccine or component
Z88.7	Allergy status to serum and vaccine status

Home CMS.gov

A federal government website managed and paid for by the U.S. Centers for Medicare & Medicaid Services. 7500 Security Boulevard, Baltimore, MD 21244

Some thoughts on rabies and Down's Syndrome

"SHOULD I get a rabies shot?" That's the first question most people ask after they are bitten by a strange dog or wild animal. The rabies "shot" is actually a series of shots, and at one time the most painful and deadly of all vaccinations. Its extreme danger was concealed by the fact that it was only rarely given.

The problem with this vaccine goes way back to when it was first conceived and used by the French biologist Louis Pasteur in 1885. Pasteur was as much a fraud, a popularizer, and a thief of other scientists' ideas as was Jonas Salk, a fact much bandied about even before his death in 1895. It was confirmed in 1971, seven years after his grandson released his notebooks. Yet his legend lives on, as do his dubious concepts of health and sickness, specifically his version of the germ theory, from which the questionable process of pasteurization evolved. His opponents convincingly argue that this practice is unnecessary and strips food of its nutrition. Perhaps in rare instances it does prevent serious illness by killing harmful bacteria, but it also kills beneficial bacteria.

As crazy as it might sound, it appears that there is no such thing as rabies — at least in humans. Certainly people have died of septicemia (blood poisoning) after being bitten, as virulent microorganisms from an animal's mouth can enter the bloodstream, but this is not rabies, and no rabies vaccine is going to prevent it. What the rabies vaccine *does* is provoke a variety of delirious behavior patterns (such as a horror of drinking water — hence the scientific term for rabies, "hydrophobia") in a category known as "Korsakoff's Psychosis," which can be permanent. "Monsieur Pasteur does not cure hydrophobia — he gives it!" said Professor Michel Peter, a contemporary of Pasteur.

Throughout the animal kingdom, viruses that are deadly to certain species are harmless to others, and rabies appears to be one of these. People cannot get rabies because of this species barrier. For the last 130 years, what has been called rabies is the result of being injected by the rabies vaccine. In his booklet *Lethal Injections*, Dr. William Douglass wrote: "I challenge you to find proof that the shots work rather than kill. The medical literature is replete with cases of people who refused the shots and lived. In fact, I could not find a *single case* of a victim who died of rabies as a result of a dog bite. As there were no antibiotics, many people died of blood poisoning, but *not* rabies."

Dr. Peter Cole, a former medical officer of health in Ontario, agrees. He was featured in a news article in the January 28, 1988 Toronto *Globe and Mail.* Not only did he refuse to support tighter controls of vicious dogs, but he also opposed routine rabies vaccination of household dogs and cats. "In twenty years, there have been thousands of confirmed rabid animals," said Dr. Cole. "Hundreds of people have been exposed to bites, and not all of them have been able to get the vaccine. Yet no one got rabies. That should tell you something."

Ghislaine Lanctot, another renegade Canadian doctor who broadsided the Quebec medical establishment in the 1990s with her book *The Medical Mafia*, goes even further. She wrote: "Rabies, it seems, does not exist. It appears to be an imaginary illness." She refers to an obscure booklet *The Fraud of Rabies*, a compilation of statements by veterinarians and kennel owners distributed by the California Animal Defense and Anti-Vivisection League. One point they make is that simply being told, after an animal bite, "You have rabies," even when no symptoms are present, is enough to make some persons become hysterical. Here we are reminded of a phenomenon in Africa discussed earlier, where a native, just slightly ill, is told by his witchdoctor, "There is nothing I can do for you. You are going to die." Whereupon the devastated native returns to his hut, lies down, and does exactly that. The minds of most white people are not much more developed.

Another point made in *The Fraud of Rabies*, according to Lanctot, is that animals that act "strange" — supposedly a telltale sign of rabies — might only be hungry and miserable. This seems plausible. From time to time most of us have encountered a few human castoffs on the street muttering gibberish or otherwise acting weird, who might only have been hungry and desperate. Whatever the reason for their odd behavior, they certainly weren't rabid.

I'll leave it to the reader to verify the claim that there's no such thing as rabies. I have no strong opinion on it, but given mankind's eternal addiction to hogwash, that's where my own thinking leans. One thing is certain: the Pasteur vaccine in use for over a century was a medical monstrosity. Over the years, several freethinking researchers have stated that Pasteur hijacked medicine and rode it down the wrong track — a runaway train still going in the wrong direction. Yet schoolchildren still learn of his supposedly great achievements, and in Paris an enormous statue honoring him still stands. Myths die hard.

The terrible affliction of Korsakoff's Psychosis resulting from Pasteur's rabies vaccine was taken up by the French Academy of Medicine in the 1950s. The January 14, 1956 issue of *JAMA* featured an in-depth article about it. And the vaccinations went on — and go on. The good news is that this useless vaccine has been diluted almost to the composition of tap water, which is an altogether different but still excellent reason to avoid it.

❖

Down's Syndrome was named after a British doctor, John Langdon Down, who first studied it in 1862, and published his findings four years later. Back then, and even in recent times, the term "Mongoloid idiot" was used when referring to people with this unfortunate condition. Most everyone has seen people with Down's Syndrome, and the physical traits are well-known: small stature, small hands, feet, ears and neck, and distinctive, slightly Mongoloid facial features. All have varying degrees of mental retardation, and many have internal organ disorders as well. Average life expectancy is well below that of the general population.

Human beings have 46 chromosomes, the building blocks of reproduction — 23 from each parent. But for a reason that no one has ever been able to explain, sometimes a mutation in cell development results in a forty-seventh chromosome. To get a bit more technical, there is an extra copy of genetic material on chromosome number 21. This defines Down's Syndrome. About 1 in 800 American babies are born with it.

There are good reasons to believe that vaccines spawned Down's Syndrome. It's highly significant that it first appeared just two generations after the birth of vaccination, and in the very country where they were first made and used on a wide scale. It seems there is no reference to it anywhere prior to the 1860s. During her 1996 trial in Quebec, the ruling of which barred her from practicing medicine, rebel physician Ghislaine Lanctot spoke at length about the dangers of vaccines. At one point, she flatly stated that Down's Syndrome was caused by Jenner's vaccine, and also that in Germany the first child ever seen with this condition was reported in 1922. If true, this is dynamite. Another noteworthy fact is that Down's Syndrome occurs in every human race, and needless to say, members of all races were heavily vaccinated through the twentieth century. So far as I know, peculiar biological abnormalities are nearly always confined to one race or subrace, but in this case it is universal.

I never got past high school biology, so I don't pretend to be an authority here, but it would seem possible for that strange forty-seventh chromosome to be a result of reckless engineering in the laboratory. Every vaccination is a roll of the genetic dice, even if most doctors are totally unaware of it. Remember that vaccine cultures are grown on the tissue of ducks, pigs, dogs and other animals that harbor viruses. Viruses are reservoirs of genetic material that, when pumped directly into the bloodstream, transfer DNA. They can invade the cells of the new host and imprint them with the genetic codes of a different species. This was known as early as the 1950s, and it's a frightening thought. Who's to say what kind of biological havoc this can lead to? We already know that vaccines can cause the immune system to go haywire. Is it not conceivable that they might have the same effect on the life process itself?

Earlier in her career, Dr. Lanctot dramatized this by visualizing human beings born with mouse tails. She was employing hyperbole to draw attention to what was going on at the microscopic level, but it brought her an avalanche of ridicule from her many enemies in the media and medical establishment. Those with Down's Syndrome don't have tails, of course, but they do have slight deformities that defy explanation. Honest research can answer the question. Someone should scour the medical literature prior to the early 1800s. If the characteristics of Down's Syndrome existed that far back and were recorded, that would clearly disprove a vaccine connection. If, on the other hand, John Langdon Down was indeed the first to write about it, that would constitute powerful evidence of yet another horrible disaster brought about by vaccines.

Vaccines for foreign travel (don't get any) and other lessons learned

WRITING CAN be a pleasure, but not when the subject is vaccination, at least not for me. Much of it has been drudgery, and much I've written with a heavy heart. In this chapter I'm going to lighten up, reminisce, and have some fun. Having seen the world, I'm going to share with you some entertaining stories and opinions. I know that most Americans do not have passports, and of those who do travel abroad or have in the past, most have not visited countries where health standards fall far short of ours. Therefore, what I'm about to discuss is irrelevant to the great majority of readers. But you can still learn a few things. And if you *do* happen to be an intrepid traveler or intend to become one, you already know how important it is to safeguard your health. That diseases, both serious and transitory, are far more prevalent in less developed countries, accentuates the fact that advancements in hygiene and sanitation, *not* vaccines, led to their virtual disappearance in the West.

Before I begin, let me say that I did most of my traveling in the 1980s, and during that time I kept abreast of the health situation, or better said the disease situation, throughout the world. Since 2007 I've done only a little research in connection with two trips I made to Asia and one to Central America. I'm pretty sure the information in the pages that follow is accurate, but I can't vouch for it. It's *your* responsibility, before leaving home, to consult the best sources and get the most reliable current information. With the Internet, which did not exist in the 1980s, all the information you need to know is a few keystrokes away.

I feel I owe it to myself, as much as to the adventurous reader, to include this chapter, even though it obviously has nothing to do with childhood vaccinations. After all, had I never traveled far and wide, I never would have gotten all the shots that began a chain of events that led to my children's lifelong handicaps. If not for them, I'm sure I never would have felt so intense about this issue, and never would have had the incentive to write this book. Yet, if as a young man I could have plotted my future, I would have done nothing differently regarding foreign travel except to avoid all vaccines. I can't imagine going through life without seeing the world, and I feel truly blessed in having done so. Most people, of course, don't feel this way, and can't relate to someone who feels this way, but for those who might make just *one* trip involving a health risk, I'll do my best to point you in the right direction.

The issues of safety and effectiveness surrounding "travel" vaccines are no different to those of childhood vaccines, though needless to say they affect far fewer people. Still, it's worth noting that the advisories published on the travel pages of the CDC website duplicate their stance on childhood vaccines. There's the same urgency to vaccinate for no reason, the same scare tactics, the same misinformation. I would never rely on *any* government "health" agency, in *any* country, for vital facts.

My position on travel vaccines also reflects that of childhood vaccines. It's based on firsthand observation and common sense, magnified by what I've read. Since my enlightenment in Nepal in 1987, which I wrote about at the beginning of this book, I've traveled to about 20 countries where health matters were a definite concern, but never considered getting a shot, nor under any possible circumstances in the future will I give it a moment's thought. The mere suggestion raises my blood pressure.

So obsessed is the CDC with vaccination that they recommend booster shots for several diseases on the U.S. childhood schedule list — for every country on earth! Take Norway, for example, by any standard one of the world's healthiest nations. Here's what you'll find on their Norway page, and everywhere else for that matter, and this goes out to *all* travelers: "Make sure you are up-to-date on routine vaccines before every trip. These vaccines include measles-mumps-rubella (MMR) vaccine, diphtheria-tetanus-pertussis vaccine, varicella (chickenpox) vaccine, polio vaccine, and your yearly flu shot." Also: "You can get hepatitis A through contaminated food or water in Norway, so talk to your doctor to see if the hepatitis A vaccine is right for you." This makes as much sense as considering a hep A shot before taking the train from Providence to Boston.

Of the 20 countries I mentioned, Cambodia was the one that most concerned me, health-wise, so I typed it in to see what the CDC had to say. Specifically, I sought out malaria, a very serious disease which I will return to. According to the CDC website, as it appeared in February 2019 – unchanged from when I first read it years earlier – malaria is "Present throughout the country, including Siem Reap city. None in the city of Phnom Penh, and the temple complex of Angkor Wat." The huge temple of Angkor Wat and many smaller temples nearby are scattered throughout a jungle area just outside the small city of Siem Reap. Angkor Wat is Cambodia's top tourist attraction, and nearly all travelers who visit it stay in Siem Reap, with its many hotels, restaurants and other amenities. Malaria is transmitted by the bite of infected mosquitoes. Who doesn't know that there are more mosquitoes in a jungle than in a city? Some people who work for the CDC, apparently. And the same people haven't figured out that if there's malaria at Point A, it will also exist three miles away, at Point B. Quite a few people leave their hotels in the morning darkness to photograph Angkor Wat at sunrise, a beautiful sight. The anopheles mosquito, the species that carries the malaria virus, is active from dusk to dawn. Anyone foolish enough to believe the claims posted by the CDC is flirting with a potentially fatal disease.

Unlike the thriving culture of corruption in the CDC's childhood vaccine division, this is obviously a matter of ignorance, of typical bureaucratic incompetence. Probably very few CDC employees have ever heard of Angkor Wat, or can even find Cambodia on a map. As a rule, people with government jobs, other than those whose work takes them overseas, travel very little. In fact, of the thousands of travelers I've met over forty years, I can't recall meeting a single bureaucrat from *any* country. And I'm not trying to disparage all people employed at various levels of government. It's just that, in the nature of things, very few who opt for the security of a government job are adventurous, know much about the outside world, and are qualified to give advice on avoiding diseases. And the unfortunate flip side of this is that people in general, including many travelers, respect the authority of government, take it for granted that, on such an arcane subject as health risks in foreign countries, these bureaucrats know what they're talking about. Incidentally, I'm not the only one who feels this way. While researching, in 2016, a trip to Central America that I took the following year, I came across the website costaricaexpeditions.com, which criticized the CDC for their stale and inaccurate health information about Costa Rica.

The people whose advice is most sound on this matter are those who have lived in a particular country for a long time. Western doctors who have worked in clinics or hospitals of impoverished countries are another reliable source. Through the Internet, many of them share with the rest of the world the things they've seen, and haven't seen. Another place to go is the travel blogs which abound on the net. You can ask questions which will be answered by travelers who have been where you're thinking about going — people who will discuss their own experiences as well as what they've picked up from others. The Thorn Tree, a division of Lonely Planet, the world's largest publisher of travel guide books, is a good place to connect, as is Trip Advisor, the world's largest independent travel forum.

While surfing the Internet on the subject of Cambodia, I discovered aboutasiatravel.com, the website of a British-based travel company specializing in Southeast Asia. This is what they write about malaria in that country:

Cambodia malaria risk should be considered carefully if traveling to the border regions of Cambodia but is not a significant problem for the visitor around the main tourist areas of Angkor Wat, Siem Reap, Tonle Sap and Phnom Penh. It is worth noting that our team of expatriates and locals have lived in Siem Reap for a number of years now and have no incidence of malaria. We do not take anti-malarial medication but are careful in the way we dress and in our use of insect repellent, especially at dawn and dusk.

They then cited "four important principles of protection" from malaria, according to WHO (like the CDC, a gangster outfit in large part, though I do agree with what they wrote about malaria prevention), and followed up with "UK travel advice in regard to the various anti-malaria medications available," adding: "No anti-malarial drug is 100% effective, however, or free of side effects, and we strongly recommend avoiding bites as the best defence — wear long sleeves and trousers (especially in the evenings), use an insect repellent, and sleep under a mosquito net (or in a screened room)."

I agree with everything these people have written. There's both wisdom and kindness in their advice. Yes, kindness. We explorers are a worldwide family and we look out for each other, taking newcomers under our wing. Like parents of vaccine-damaged children who sound the alarm for new and prospective parents to hear, many of us wish to warn others of travel risks of any kind when and where they exist, and reassure them when and where they're negligible.

Health risks vary in the lagging (or if you prefer, "developing") countries of the world, by which I mean most of them outside North America, Europe, some scattered Western outposts and Japan. Some others come fairly close to our standards. Risks can vary in different areas of the same country. If you intend to travel, even once, to one or more of the many nations that fall into the "Third World" category — and you really should if there's adventure in your DNA — you need to look at it this way, keeping in mind that the numbers I'm about to throw out are purely hypothetical. Let's say the odds of contracting a debilitating or potentially fatal disease here at home are 4000 to 1. Now you've arrived somewhere in Asia, Africa or South America, and the odds have instantly become 40 to 1. The odds are still very much in your favor, but you have to accept the fact that they've just increased a hundredfold — and more than that, I assure you, if you're in tropical Africa. An analogy can be made with white-water rafting and its different levels of danger. You can control the risk factor, but even in Class 1, the safest level, there are no guarantees. You can still be thrown into the water (as once happened to me in Class 1), be injured (I wasn't), or even drown if your foot becomes wedged between rocks (very unlikely, but still possible). The only way to avoid all risk is to stay home.

You also need to know, even if you're fastidious about what you eat and drink, that there's a fair chance, especially if you've never been exposed to our dear planet's many nasty bugs, that you'll suffer a brief bout of what the Australians cheerfully call "V & D" (vomiting and diarrhea), or just the D. In either case, it usually clears up on its own in a few days, and is nothing to worry about. But if you're squeamish about this, you might want to limit your foreign travel to Europe, where you need not worry about getting sick anywhere. You can eat, drink and contemplate mosquitoes with nothing more than the annoyance you do here at home. Don't bring any medicine for the occasion, and — perish the thought — don't get vaccinated, not even if going to Norway. I've been almost everywhere on the continent, and even in what is probably the poorest country, Albania, I ate and drank with abandon and never had a problem.

But there are a few caveats. I became quite ill on the Trans-Siberian railway trip I made in 1980, when Russia was still under the lash of communism. The meals on that train were awful — the worst food I've ever eaten. The train made brief stops at many small towns, and on the platforms peasant women sold pine nuts, mushrooms, pickles, cole slaw, the latter spooned into cones of newspaper — the small measure of free enterprise allowed them. I bought and ate these snacks indiscriminately, since they tasted so much better than the glop on the train. Actually, this was in Asia, not Europe; I'd begun the trip with a small group of Australians in the Siberian port city of Nahodka, after a two day ferry ride from Japan. The trip was broken by stopping off a few days in the Siberian cities of Irkutsk and Novosibirsk, where the hotel food was almost as bad as on the train. There was a slight improvement when we got to Moscow, but only slight.

I paid the price for indulging my comestible curiosity, as I occasionally would in the future. Sometimes I got away with it. But not in Irkutsk. I clearly remember sitting on the train there, waiting to depart, and looking at a portrait of Soviet leader Leonid Brezhnev on the platform — and coming an eyelash away from falling to the floor, though I had never fainted in my life. That didn't happen again, but I felt miserable for ten days, in the process becoming an expert on the various models of Soviet-era toilets. I can only guess that the cole slaw was to blame, because nobody else ate it and nobody else got as sick as I did; possibly it was the mayonnaise, the newspaper ink, the water used to wash the cabbage — who knows. Years later I thought about this. There's no disease I've read about that's typified by lightheadedness and persistent diarrhea, yet also an absence of fever and vomiting. I can only guess that it was the way some strange bug manifested itself in my system. The mysteries of the organism. On several occasions while traveling I've experienced vague intestinal and respiratory discomforts that came and went quickly enough but always remained a riddle. This does, however, demonstrate two things: first — and I'll keep repeating it — we who take a dim view of vaccines stand on solid ground when we point out that advanced sanitation and hygiene means less disease; second, our bodies are well-equipped to conquer illness on their own, without medical intervention.

I learned this the hard way during an epic rail journey that should have been most enjoyable, but wasn't. I'd never been this sick for this long before, was totally ignorant about my condition, and didn't know what to do, though I was determined not to seek medical help in a country where everything was so shabby. One of the Australians gave me some charcoal pills which, she said, were supposed to absorb the bad stuff circulating in my system, but they did nothing. Someone else gave me Lomotil, a prescription drug that cakes watery stools and gives you the false impression that you're getting better. In fact it cures nothing, carries the possibility of side effects, and after learning these things I never considered using it again. In the end, it was my immune system that did the trick.

Actually, my body had already taught me this after my first trip to Europe in 1978. Shortly after returning home I realized I had an internal issue, the symptoms of which were a chronic bloated feeling, sulfur-tasting burps and extremely foul-smelling gas. (You know something's wrong when your own farts make you gag!) There was no pain, just mild discomfort, but it was constant. Since I've always mistrusted doctors, I simply lived with it — for eight months. Then one day I woke up and it was gone — just like that! I had no idea what had ailed me until years later when I read about the symptoms of giardia, a common amoebic parasite that contaminates drinking water. That was what I had. Antibiotics can kill it, I read.

Although it was all very educational, had I known about giardiasis at the time I probably would have gone to a doctor to get a prescription for the proper antibiotic. I'm appalled at the gross abuse of antibiotics in this country, especially by doctors and parents who think nothing of giving them to children, as if they're as harmless as breath mints. But I feel they *do* have their place, and can even save lives *sometimes*, unlike vaccines, which are an open and shut case. Here too I could not make a definite connection, but I suspect I got it from drinking at public water fountains — dumb, dumb, dumb — most likely in the old Yugoslavia. Interestingly there was a cholera outbreak in Naples in 1973, I also learned, which does not dissuade me from feeling that you need not be a health vigilante in southern Europe, with one exception. I was astounded to read, a few years ago, that malaria has made a comeback in Greece. Reported cases are still few and far between, and the invasion of illegal immigrants from Africa might be the cause, but I don't know. With malaria, it behooves the traveler to seek the latest information and act on it.

One footnote here. In 2001 I spoke to an American doctor who had just returned from a medical symposium in St. Petersburg, and he told me that the food was wretched. This really surprised me, because it was ten years after the collapse of communism, and St. Petersburg is the most European of all Russian cities; I assumed that things had really changed for the better in that country. I write these words sixteen years later, and thirty-seven years since I was in Russia myself, so I make no judgment here. I will say that in 2014 I traveled to the Baltic states, Estonia, Latvia and Lithuania, nations that became independent of the Soviet Union when it cracked, and the food there was very good, the eating establishments as spotless as anywhere as in Scandinavia.

In the spirit of offering friendly advice, a great way to enjoy delicious food and local specialties at low cost is to go to supermarkets, and sometimes even department stores, in the Baltic countries and in fact all over Europe, where a wide variety is prepared daily and sold by weight from display cases, as is fairly common in our own country. But it's much more fun when the person serving you doesn't speak a word of English.

❖

As much as I've traveled, I've never lived in a foreign country. Those who intend to take up long term residence in an impoverished part of the world, and particularly those who will be living in primitive conditions, will need to acquire more information than I provide in this chapter. There is, for example, some risk of picking up an exotic disease from daily contact with soil, and in bathing in bodies of water, especially stagnant ones. Also, I'm not going to get into sexually transmitted diseases; if having sex of any kind with foreign natives is your thing, then quite frankly I have no interest in advising you about anything. Here I will focus only on insect-borne diseases, and disease spread by contaminated food and water — and only those that I regard as significant risks. In my opinion there are only two, malaria and hepatitis A, that belong in this category, though a few others deserve passing mention.

It would be most unfortunate, and possibly tragic, to contract yellow fever or Japanese encephalitis, both transmitted by mosquitoes, but they seem to be rare enough, and the risk so tiny, that I've never worried about them. Another disease caused by mosquitoes, dengue fever, is not rare at all, and has been on the rise in the tropics worldwide in recent years. From what I've read dengue fever is not life-threatening, and runs its course in one or two weeks, but during the most acute phase you get an absolutely excruciating headache. Notable also is that, unlike the malaria-transmitting female anopheles mosquito, which bites only from dusk to dawn (especially around dusk), the species responsible for dengue bites day and night, and furthermore is as much a threat in urban as in rural areas. As for the storied tsetse fly and sleeping sickness, which is supposed to be the scourge of Africa, or so I've read, in a cumulative five months of travel south of the Sahara I never saw one. Maybe I was just lucky.

No vaccines were ever made for dengue fever and sleeping sickness. There is a vaccine for Japanese encephalitis, which occurs in much of Asia, but only the ignorant would want to be injected with it — people as ignorant as I was when I got my yellow fever vaccination and booster in 1981, thinking that it would protect me and never giving a thought that it could harm me. In fact, the yellow fever vaccine appears to be the most dangerous of the travel vaccines, in a league with the whole cell pertussis shot that maimed and killed so many infants before it was taken off the market. But I was told nothing by the nurse who administered it at the Kennedy Airport medical office. I was only asked if I was allergic to eggs, and after I said no, she prepared the shot and stuck the needle into my arm.

Would I be remiss here if I failed to mention yet another gift of the mosquito, West Nile virus, of which we've heard a great deal here in the New York area over the past seven or eight summers? Not at all. Although I've been west of the Nile, east of the Nile, and on the Nile, I never heard it mentioned there, nor read about it in the health sections of guidebooks. With the exception of malaria, there are so many microbes carried by biting insects in the world's tropical regions that do not seem to me to pose any danger, and thus do not merit further discussion, aside from two sentences on Zika virus. In my not so humble opinion, the Zika virus is about as credible as Piltdown Man. If you're worried about getting Zika, you might as well be worried about getting the cooties.

❖

Food and drink are the usual culprits when travelers get sick. It's not exactly a savory subject, but the fact is that human waste that is not properly disposed of is the source of a lot of disease in the Third World, or as a wag once called it, the Turd World. In many countries it's not unusual to see people defecate in public, or the results of what they leave behind. No one ever cleans it up. Where plumbing exists, it's of the most rudimentary kind, usually culverts of foul, grayish water running alongside the road. Also, for many, toilet paper is an unknown luxury; instead they use a small can of water, or sometimes just their fingers to wipe themselves. As disgusting as it sounds, most travelers get sick from ingesting human excrement. It takes only a microscopic speck on the leg of a fly that's landed on your omelette, on the finger of someone preparing food, or in waste water that somehow gets mixed with the tap water supply.

Some diseases spread this way are potential killers, like cholera and typhoid. Amoebic dysentery is a bad one too. But to be honest, I've never met nor even heard of a traveler who came down with a life-threatening disease from contaminated food or water, which is not to say it doesn't happen. But most travelers deal with short-lived bugs that resolve quickly enough on their own, though it's always a bummer to get sick while traveling. Then again, there are exceptions, such as my own bout with giardiasis. In my countless "health chats" with fellow travelers, I met only one who shared this particular misfortune with me.

The one gut disease that's a definite risk is hepatitis A. In fact, malaria and hep A are the *only* serious diseases contracted by a fair number of travelers, ten or so in each case — in addition to two who came down with dengue fever — whom I've personally met, as opposed to hearsay. If you like to "go native" in your dining habits, or otherwise take chances with dubious hygiene, you have a good chance of returning home with a case of hep A. It's very common in the Third World. It won't kill you, but it *will* knock you out for a month or two, during which time you'll feel miserable and require plenty of rest and total abstention from alcohol while your liver recuperates. Once you recover you'll never get it again, but obviously it's something to avoid in the first place.

While hepatitis A is not much of a concern here

at home, it's not as rare as you might think. In fact, small outbreaks, likely the result of immigrants employed in restaurants or elsewhere in the food service industry, have become common nationwide. There's also a small risk in eating raw clams and oysters. I know a few people here on Long Island who became ill this way, and one woman who got hepatitis. The risk is highest right after a heavy rain which washes animal waste into the shallow bays where clams are harvested, and before they have a chance to cleanse themselves. This is what I once read, anyway, and it makes sense.

An injection of gamma globulin, which I discussed at the beginning of this book, and which is a human blood product loaded with antibodies and technically not a vaccine, supposedly provides protection against hep A. And as the line goes, if you get the shot and you still get hepatitis, it will "probably" be a milder case than if you didn't get the shot. I remain skeptical. Although I've never heard of anyone contracting AIDS or hepatitis B from gamma globulin serum, which is supposedly purified, the fact is that it's taken from other people's blood, I don't know where that blood came from, and I want no part of it. Also, I've traveled to the Third World more often than not without the presumed benefit of this injection, and I've never had hep A, even though, in my earlier traveling days at least, I took more chances than I should have. Parenthetically, gamma globulin has been pushed aside by the recently concocted hepatitis A vaccine, which is now the shot of choice among travelers who don't know better.

Mexico is one country which a considerable number of Americans have visited, and gotten sick. In fact, Mexico is as bad as it gets, at least with stomach problems. Everyone who's gone there has heard the warning "Don't drink the water," advice which is as simple as it is priceless. It goes without saying that you should never drink tap water in countries like Mexico, but it's much trickier to avoid traces of water used to wash food and cutlery. Salads are probably the biggest danger in this regard, and should be avoided, though I've given in to temptation now and then without getting sick. On the other hand, I do believe in the benefits of eating fresh fruits and vegetables while traveling, and shopping and haggling at local markets is part of the experience. Fruit you can peel, like bananas and oranges, is always safe, and I believe that most fruits and vegetables, other than the leafy variety which may have been washed in tap water, are generally okay. In any case I've often eaten tomatoes, apples and the like and never gotten sick. On every trip I take along a small knife and peeler, which always come in handy.

There are those, mostly throwbacks from the flower power generation, who believe in going native. Back in the 1970s, hepatitis A was so common among the long-haired set, especially on the well-worn overland route from Europe to India, that it was jokingly called "hippietitis." I recall seeing a couple of these types eating at a kind of field kitchen that was part of a large outdoor market in LaPaz, Bolivia. Actually, it was quite colorful, the Indian women in their frilly dresses and bowler hats, a legacy of the Spanish conquest, doing a brisk business dishing out meals to locals sitting at long makeshift tables, and washing everything — food, pots, pans and cutlery, all of it left dripping wet — from water burbling from a rusted pipe. I can't imagine a better way to come down with hep. I remember a similar scene in India — fashionable young people eating with their hands alongside villagers at a roadside concession. All I can say is, if you can eat like this, even a few times, without becoming seriously ill, God bless you.

I've often wondered who came up with the term "Third World." What about the Second World? Quite a few countries fall into an intermediate category in terms of health standards. While they're below ours, they're well above many others, even though in isolated areas conditions might be primitive. I'm thinking here of China and some countries in Southeast Asia, and in particular a five-week trip to Singapore, Malaysia and Thailand in 1983. During this time I ate a lot of street food, particularly in Singapore, a culinary paradise. The general rule is that food served piping hot is safe. I followed this rule and had no problems in any of these countries, stoking my belly with chicken satays, noodle dishes and heavenly soups as I worked my way north. I even trekked for three days in the tribal region of northern Thailand and stayed perfectly healthy. Then, shortly before I was due to fly to Australia, I made a terrible mistake. Wandering around the Thai capital of Bangkok the last few days, I came across an English-language flier advertising an all-you-can-eat lunch buffet for eight dollars at the Bangkok Peninsula Hotel, the most luxurious hotel in Thailand. Here was an opportunity to dine like a business tycoon, something I'd never done, at an absurdly low price, so I had to try it. What I remember is walking a mile in the sultry heat, arriving at the hotel unshaven, dripping sweat, and feeling uncomfortable as servants fluttered around me, calling me "sir," and offering to cater to my every whim. It was not my kind of place even though, as you can imagine, the spread was sumptuous. Food connoisseur that I am, I wanted to sample a little of everything. This included two raw oysters.

Twelve hours later, around midnight, I woke up with agonizing stomach cramps. For the next half day I spent as much time in the bathroom as I did in bed, moaning out loud. Never in my life had I experienced pain like this, and never since. The cramps lasted well into the afternoon, before they began to subside. Fortunately, I had one more night in Bangkok before I was scheduled to fly to Perth; I would never have made it onto an airplane in that condition. I did manage to fly out the next day.

For an entire week there was a bonfire in the pit of my stomach, and although the unbearable cramps never returned, I obviously wasn't getting better. I don't like to admit defeat, but this time I went to a clinic in Australia. The doctor there prescribed the antibiotic metronizadole, which I began taking immediately. Within 48 hours I was back to normal — proof that antibiotics, when used judiciously, can be a godsend. That was the only time I sought medical assistance while traveling; I remain a firm believer in giving the body a chance to heal itself when sickness strikes. And the other lesson here is that it's much safer to eat hot food on the street than to eat raw shellfish where the rich and famous dine.

❖

If you're willing to take risks — and the longer the trip, the more likely you are to let your guard down — it's a matter of hit or miss. Why did I get so sick in Bangkok, and on the Trans-Siberian, and in Niger when I ate what I think were fried intestines from a guy selling them out of a metal pail, and not when I ate sparrows on a stick in Burma, sheep pancreas in Morocco (disgusting!), mare's milk on the Kyrgyzstan steppes, and shrimp that tasted faintly of ammonia at a restaurant in Ecuador? It's as much the luck of the draw as it is with vaccine side effects, I suppose. I don't have all the answers.

In that connection I cannot estimate the risk of dying from cholera or typhoid; I can only tell you that I believe it's so small that it's silly to worry about it. Back in the day when I knew nothing about vaccines, I got several of them before leaving home. Once I got a cholera shot from our family doctor, who was not keen on it, telling me that it had a low efficacy rate, and that the disease is rare and occurs only in well-publicized epidemics. I have no idea if this is true, though cholera does seem to crop up only during times of total social breakdown, such as wars and devastating earthquakes. At the time, I just wanted the shot, seeing it as a cheap insurance policy, so he reluctantly gave it to me. I've written about him earlier. As honorable a doctor as he was, he did not mention side effects, possibly because he didn't know of any. Come to think of it, it seems strange that he even had a supply on hand. Over the years I was injected with more cholera vaccines than any other, and I've often wondered if that was the one that damaged my fertility.

I first visited Africa in 1981, returning the following year, and again in 1988. Back then, yellow fever and cholera vaccinations were required — on paper anyway — for entry to several countries on that continent, and *only* on that continent. Since that time, the cholera vaccination requirement has been dropped everywhere, but the yellow fever vaccination is still obligatory, and now some countries in South America require it, which was *not* the case when I went there in 1985. So anyone considering a journey to that region needs to look into this.

The World Health Organization, as I've said, is a global version of the CDC, that is, an international crime syndicate of vaccination zealots. WHO publishes an official, bilingual (English and French) booklet in which vaccinations are recorded. The current certificate has been slightly revised from the one issued to me in 1981. It's available online from WHO headquarters in Switzerland at a cost of $30 for a minimum quantity of fifty. If you go to any medical establishment that offers travel vaccines you'll get one free, with required shots documented.

As I write these words, the yellow fever situation is in a state of flux, not to mention total confusion, and I'm not sure why I'm writing this at all as it's quite possible that not a single person who reads this book will care. For what it's worth, WHO currently has a list of about twenty African and South American countries in the yellow fever doghouse, by which I mean they have declared that unvaccinated people who live in or visit these countries can spread it elsewhere. These countries, under the spell of WHO, require a vaccination certificate for entry — but only some of them, and only if arriving from another country on the WHO blacklist. In the rest of the world, many countries require proof of vaccination upon entry from an infected country, but many don't. On top of all this, a few years ago, in a most untypical move, which illustrates the "flip a coin" pseudo-science of vaccinology, the learned doctors at WHO reversed the longstanding policy that the yellow fever shot was effective for only ten years, after which travelers would have to get another one, and declared that it conferred immunity for life. Since I still have the certificate in which my 1981 shot was recorded, this would be great news if I intend to return to Africa or South America, which I don't. In the same pronouncement, however, they cautioned that many immigration officials the world over are unaware of the new regulation, so travelers should expect problems.

Knowing Africa, I can fill you in on a few things. And let me preface my remarks by saying that we all know that forgery is very naughty, and far be it from me to entice you to commit a crime — though I have no idea under what jurisdiction. It makes much more sense to be a good little law-abiding dumbbell and become infertile or get rheumatoid arthritis, doesn't it? I'll simply tell you what I would *consider* doing if I could turn the clock back, based on what I have learned, but what I will *never* do, because it doesn't matter to me anymore. I should point out, first of all, that even if proof of vaccination is required, that doesn't mean you'll be asked to show documentation. Nothing in Africa makes sense, and few officials are even aware of any vaccination requirements for foreign travelers. In the dozen or so countries I visited where yellow fever vaccination was supposedly required for entry, only twice was I asked to show my WHO certificate — in the *middle* of the Congo, on a ramshackle ferry of all places, by a meek man wearing a white smock, on which a red cross was stitched, who called himself a doctor, and at the airport in Kano, Nigeria as I was about to *leave* the country on a flight to London. In neither case did they really seem to care.

If I could go back in time, the one thing I would never do is get the shot. And having learned about Africa, I wouldn't waste thirty dollars to obtain an official WHO vaccination certificate, even if I knew of a printer who would make a fake stamp or a corrupt doctor who would just stamp and sign off on it, though these are options. Here's what I'd do. I'd print a fictitious letterhead in English or French, or both, depending on my destination — one or the other is an official language in the great majority of African countries, and you might have to pay for the services of a translator — reading something like OFFICE OF THE SUPREME DIRECTOR OF NATIONAL HEALTH OF THE UNITED STATES OF AMERICA. I would then type in a sentence or two solemnly asserting that I was duly vaccinated on such and such a date, and stamp it. I'm told that a coin pressed on an ink pad, then applied to paper, makes a fine imitation of an official seal. Am I writing these words in jest? Not at all. Africans are awed by grandiose titles, especially those bestowed in faraway white countries, and would never question something like this.

I crossed less than half as many borders in South America as I did in Africa, but I'd follow the same procedure, with the difference that you'll want to get a letter printed in Spanish. If you show up without anything, my belief, based on my experience crossing many Third World borders, is that if you do run into a difficult official on this matter, who demands the certificate, in one way or another you can get around it. This includes bribery, and be aware that in much of the world, a bribe is not considered a crime but rather a compulsory "tip." Sometimes it's even demanded in a roundabout way: "you must pay special entry tax." At worst, you'll probably anger an honest official; it's nothing like trying to bribe a cop here to get out of a speeding ticket, which is very stupid and will get you locked up. Instead of being self-righteous, look at it this way: many immigration officials in poor countries are poorly paid, and a few extra dollars will buy their kids a solid meal. If the outright offer of a bribe rubs you the wrong way, be creative. Feign astonishment, become indignant, and say something like "But your embassy in my country told me that you don't require a certificate, that I can just make a ten dollar donation to fight yellow fever." My hunch is that the only place you might have a problem with this is at the airport check-in counter if your first destination is a country that requires the official WHO certificate. A by-the-book ticket agent might refuse you a boarding pass if you don't have it. The whole thing is just so ridiculous, but this is the nutcake world in which we live.

❖

Throughout this chapter, I have touched upon the one disease that concerns me, or rather scares me, more than all other diseases combined: malaria. Malaria is and always has been a major killer, and there is a shockingly low awareness among many adventure travelers of just how widespread and lethal this disease is in the impoverished, tropical regions of the world. I had known this for a long time, but it really hit me in 2013 during my two nights in the aforementioned Siem Reap as I ate dinner at a crowded outdoor restaurant. The majority of tourists here were from Australia and New Zealand. Most of them were wearing shorts, and I saw insect bites on many bare legs. This, to me, is pure insanity in a country like Cambodia, claims by some that Siem Reap is malaria-free notwithstanding (to say nothing of the risk of dengue fever).

I am unaware of reliable statistics, but my best guess, based on everything I've read, is that two to three thousand Western travelers contract malaria each year, and of these, around one hundred die of it. Africa has the highest incidence by far. Except at high altitudes, and in the far south, where the disease tends to be spotty and seasonal, malaria is absolutely rife throughout sub-Saharan Africa. Every thirty seconds an African native, usually a child, dies of malaria. It's Nature's ruthless way of keeping the human population in check.

As far back as twenty years ago I remember reading a few glowing news stories of a malaria vaccine on the horizon. Nothing ever came of it, but as I write, GlaxoSmithKline has come up with a vaccine which was scheduled to be tested in Africa in 2018. The fact that it's supported by the Bill and Melinda Gates Foundation means that, in case the vaccine is actually released, its purpose will almost certainly be to kill Africans or make them infertile. Those who think I'm just shooting my pen off here should withhold judgment until they read this book through.

The malaria risk factor varies greatly, in some cases within the same country. Ecuador, Peru and Bolivia, for example, encompass both the Andes mountain chain and the Amazon rainforest; a day's travel can take you from an area of no risk (mosquitoes cannot live above seven thousand feet, roughly) to another where there's definitely malaria. It is my opinion that in some countries designated as low risk, Turkey for example, the disease occurs so infrequently that there's nothing at all to worry about. Where the risk is high, however — and in most of Africa and some Pacific islands it's *very* high — it's imperative that you take all reasonable precautions, and even unreasonable ones. But short of wrapping yourself up like a mummy to avoid mosquito bites, there are no unreasonable precautions when it comes to malaria prevention.

Malaria has long been my bugaboo. One reason I'm so obsessed with it is that, despite going to extreme lengths to keep mosquitoes away, they want you in the worst way, and given the tiniest chance they'll get you. I learned this firsthand at the beginning of a two-month overland camping trip on the coast of Kenya, where the malaria risk is way up there. The evening was so hot and humid that I threw open the front flap of my tent for ventilation, stretching out on top of, not inside, my sleeping bag. I had covered every square inch of skin, wrapping my neck and head in a towel, putting on gloves that I had brought for the purpose, and placing rubber bands around the bottom of my pants to hold them against my socks. I gave no thought, however, to the open slits in my long-sleeve shirt. When I woke up I discovered that I'd been bitten about fifteen times, and pondered my lumpy wrists with much anxiety. Three weeks later, somewhere in the Congo, the trip leader and a young British fellow came down with fever and chills, malaria's telltale symptoms, later confirmed by a blood test. Like me, they were taking preventive medication, which obviously does not guarantee immunity. It's a certainty they were infected in Kenya or Tanzania, likely on the sweltering Indian Ocean coast. At various times in my travels I've dodged bullets, but never have I felt more fortunate than that night when malaria took a shot at me and missed.

No effort should be spared in keeping mosquitoes away from your skin in malaria zones, especially around dusk, when they start raising hell. Always wear long pants, shoes and socks, never shorts and sandals. Personally speaking, I would *never* wear shorts, even during daylight hours, and as long as it's not very hot, I prefer to wear a long-sleeve shirt at night. Apply insect repellent to all exposed skin before the sun goes down. Be especially wary around areas of standing water, including hotel bathrooms; this is where mosquitoes breed. If your room has a fan, use it and put it near you while sleeping; mosquitoes don't like air currents. Mosquito nets work great, but make sure there are no holes, however tiny, because these buzzing bastards will find it and eat you alive. Mosquito coils, which burn for several hours and give off a fragrant smoke similar to incense, also keep the little devils at bay. In Africa I found them to be very effective while sleeping on bare floors at missions, though to my chagrin, after I bought a few packs here at home in 2013 for use in Cambodia and Vietnam, they burned out after ten minutes. When I read the fine print on the package, I discovered it was typical junk made in China.

What's also perplexing about malaria is that there are so many variables and uncertainties. We do know, or think we know, a few things, for example that where there are seasonal changes; the risk is lower in the dry season than in the wet. But how much lower? There are different strains, the most common of which are *vivax* and *falciparum*, the latter far deadlier, and said to be the prevalent strain throughout Africa and Southeast Asia, but not Central and South America. But if you're bitten, how can you possibly know not only whether or not the mosquito is infected, and if it is, which strain it carries? What percentage of mosquitoes are carriers? Are some people naturally resistant to the malaria virus? In addition, there are numerous brands of insect repellents on the market, ranging from 20% to 100% DEET, and newer ones made from natural ingredients, like picaridin, that are free of DEET, which can eat through plastic. Are you better off using something with DEET, which despite its toxicity, has been known for decades to be highly effective and long-lasting? Generally speaking, in countries with malaria, urban areas carry a much lower risk — and sometimes, reportedly, zero risk — compared to rural areas, due to various control measures (though I don't believe this is true of any African city). But are you *really* safe if you're bitten in a city? And what of the various prophylactic pills available? As we have seen, there is no guarantee that they will prevent malaria (though they might), and additionally, some are supposedly more effective against specific strains, but in some areas resistance to these drugs — which are also used as treatment in higher doses in the event of infection — has been reported. Furthermore, all of them carry a wide variety of possible side effects, some of them quite horrible. It's a real quandary. Nevertheless, if you're headed to malaria country, particularly a high-risk area, you *must* do your homework and make an informed decision. Here, I will offer a little advice based on my own experience and reading, reminding you that every word in this chapter is my personal opinion and might differ from that of someone else who also writes from knowledge and observation.

Malaria is the *only* disease on planet Earth for which I would seriously consider taking preventive medication after assessing the benefit versus risk equation. I took a two-month course of Chloroquine on my first trip to Africa, and a three-month course of Fansidar on my second; naturally, my friendly neighborhood pharmacist never informed me about risks, small as they were, like retina damage and kidney failure, nor today, I'm sure, with all the new drugs, are travelers apprised of reported side effects, a fairly common one which seems to be hallucinations. On my more recent travels to malarial regions I brought pills with me but did not take any, even though I did get bitten on a few occasions due to my own negligence — in Cambodia and in Honduras. Both times I seriously considered taking them, but chose not to. There's always that nagging question: "Am I doing the right thing?" Only the passage of time proved that I did. Actually, I am so goddamn sick and tired of worrying about malaria, which is more life-threatening the older you get, that now, at age 65, I have vowed never to set foot in malaria country again. (We'll see about that.)

My final word on the subject is this: at the first sign of fever in a malaria zone, or for about three months afterward — and many if not most people experience the first symptoms after returning home — you need to see a doctor *immediately* to get a blood test. And *never* go to a general practitioner, but rather to a specialist in tropical medicine. (This holds true for all diseases possibly acquired abroad, which most doctors know nothing about.) Left untreated, *falciparum* malaria can kill within days of the onset of symptoms.

In a book about vaccines, I think I've written enough about a disease that has no vaccine relevance. To sum up, in addition to having acquired a lifelong pathological hatred of all mosquitoes, and going out of my way to kill every one I see, I will summarize the cardinal rules of staying healthy while traveling beyond the Western world and Japan, which I always abide by.

NEVER GET ANY VACCINATIONS. Any urgent travel notices put out by the CDC or any other organization about the danger of being unvaccinated against meningitis, polio or anything else should be laughed at. As with childhood vaccines, every travel vaccine is a profit-driven fraud. Undoubtedly many well-intentioned people would be aghast at this advice: the health sections of every website and travel guidebook to undeveloped countries routinely recommend them. I'm simply telling you what I believe and what I live by. If you intend to visit countries caught in the web of yellow fever vaccination, you'll have to give some serious thought on how to handle it.

NEVER, EVER, DRINK TAP WATER, OR WATER FROM AN UNKNOWN SOURCE. This holds true even if you've been assured by seemingly responsible people that it's safe. They're right most of the time, but if you keep believing it you're going to pay the piper eventually. (I don't drink tap water anywhere, *period*, if not for the risk of getting sick, the chemicals.) It's really very simple: why would you take a chance, even a small chance, when so many different kinds of microbes have been known to contaminate public water supplies? Except in the most isolated or backward areas of the world, bottled water is readily available. Buy it. Drink it. Brush your teeth with it. (And if you order water in a restaurant, make sure the seal is unbroken when you screw the cap off.) Boiled water, of course, is safe, so hot tea or coffee is never a problem, and the same goes for beer and bottled soft drinks. I'm inclined to believe that ice is not as risky, and a margarita or pina colada on the rocks can sure be tempting on a hot evening in the tropics. The better places purify their water before freezing it, but as a rule I skip the ice cubes.

Especially in countries where water has a bad reputation (India must be the worst), never eat salads or any other kind of raw food that is customarily washed. Mayonnaise, creams and sauces should also be avoided. Although cheese is generally safe and yogurt, I believe, always so, and beneficial as well, use discretion with dairy products in impoverished areas, especially ice cream, even if it's factory-made. In markets you'll sometimes see ice cream scooped and sold from pots. Avoid this like the plague, as you might get it.

When eating out, a good rule of thumb is to go to restaurants where patrons and staff *look* healthy, as opposed to "hole in the wall" establishments. In countries like India, where dire poverty and affluence exist side by side, you'll see this discrepancy everywhere, and eating well is not at all expensive. Steaming hot food brought to your table at a clean restaurant is as close as you'll get to a guarantee against nasty stomach bugs. Of course, from time to time you might want to take a chance by trying something exotic, or you might find yourself in a situation where you're offending someone's hospitality by refusing questionable food or drink offered to you; or again, if you're with a group, peer pressure might compel you to eat something you normally wouldn't. This is all part of adventurous travel, and you'll just have to deal with it.

ALWAYS BUY INSURANCE BEFORE LEAVING HOME. Several companies, easily found on the Internet, sell comprehensive, short-term travel insurance policies that cover all contingencies that the policy you might have at home might not. It's cheap, and without it you can lose your shirt if, for example, you have to be hospitalized or evacuated. In fact, you might not be hospitalized or evacuated unless you pay up first, or show proof that the money is forthcoming. I wouldn't think of going abroad, even to Europe, without travel insurance. I'm sorry I can't tell you how it works because, thank God, I've never had to use it. On my last five or six trips I've purchased from World Nomads, even though I don't know anything about them (I just like the name). Bad things can happen to good people, and when you least expect it. In Vietnam, for example, I stepped on a tree root the wrong way and sprained my ankle. A little worse and I might have required a cast. Be prudent, not parsimonious.

On the subject of unforeseen accident or illness, it's not a bad idea to bring along a broad spectrum antibiotic (and of course malaria pills, if applicable), keeping in mind that these drugs can have rare but life-wrecking side effects and should be used only as a last resort. They might not be available abroad, or unscrupulous pharmacists may sell you expired medication, or even substitute something entirely different.

Lastly, if going to Thailand, DO NOT EAT RAW OYSTERS AT THE BANGKOK PENINSULA HOTEL.

❖

It's a pity that most Americans go through life without ever visiting a single foreign country, and of those who do have passports, few stray from the well-trod tourist paths of popular destinations in Europe and the Caribbean. I can understand why most people will not or cannot explore the world, but I have a hard time relating to those who are not even curious about distant lands and peoples.

If nothing else, I hope this chapter will benefit those born with adventure in their blood. For the more timid I would say go to Europe at least, or at the very least travel America. Other than having and raising children, I can't imagine anything more enriching than travel, and I feel truly blessed in having seen the world.

Most of my travels have been solo, getting around on public transport, such as it is, with no set plans. You don't have to be well off to travel this way. Traveling in Third World countries, in fact, is much cheaper than living in the U.S. Although I prefer privacy, I've stayed in youth hostels now and then, most recently in China when I was 59, where even now you can get a bed for ten bucks. Alternatively, you can throw your sleeping bag down in a secluded spot for free, as I did many times when I hitchhiked around America in the 1970s, and years later on my first trip to Europe. If you can forgive me one more boast, I'm the very rare traveler who has spent less than $1 for a "hotel" room not once but twice — in wretched Khulna, Bangladesh and the remote frontier town of Boyuibe, Bolivia. This was out of desperation not choice, but writing about it now puts a smile on my face. And might I add that I've yet to meet anyone else who spent less than $100 a week in Japan, the formidable costs in that country being offset by the ease of hitchhiking and the wonderful hospitality of the Japanese.

Admittedly, my thirst for adventure and willingness to put up with hassles has faded with age, and in recent years I've opted for offbeat travel with small groups. For those uncomfortable with the idea of "winging it" on their own, there are outfits that offer unusual trips spanning the globe. Even if Europe is your limit, they have intriguing itineraries on that continent that are so much better than the sappy motorcoach tours that most Americans opt for. I've done several trips with British-based Explore (exploreworldwide.com) and give them the highest marks. Exodus is also very good, I've heard. There are others, more expensive, out there. Just remember, no matter who you book with, always ignore advice on vaccinations because sorry to say, they have it all wrong.

❖

In the passport photograph on the next page, taken when I was 31, I'm poring over a world atlas and contemplating my next journey, which turned out to be one of my longest, in South America. At my side, dozing off, is my terrier mix Straw. A more lovable dog never lived. Although Straw was a senior citizen at the time, I really wanted to take him with me. After making a few inquiries, only half seriously, I was astounded to learn that this was indeed possible. The only requirement was proof of a rabies vaccination, which I still had from Straw's puppyhood. I was further amazed when the U.S. passport office in New York City told me they could issue me a new passport showing both of us. Who would've thought?

Straw had to go in an approved kennel in the cargo hold on the flight from New York to Quito, Ecuador, our starting point. Although it's legal to travel with a canine companion everywhere in South America — I'd done my homework on that — I often ran into problems. Only the cheapest fleabag hotels would let us stay, and these only after pleading in my muddled Spanish. Likewise, many bus company and railway station clerks refused to sell me a ticket, but with persistence, I always found someone willing. In fact, the difficulties were balanced by the many people who were as accommodating as they were amused.

The biggest obstacle was trying to cross the border from Bolivia into Paraguay. The Paraguay immigration official was a real bastard, flatly refusing us entry. I remained holed up on the Bolivia side for two days, trying and failing four times to get across, before giving up and rerouting myself into Paraguay by way of Argentina. At that border post the official stamped me in with a smile and a handshake, which goes to show what a difference different personalities can make.

To this day, people don't believe me when I tell them I traveled all over South America with my dog. And Straw really enjoyed himself — he even loved the food! If he had been able to talk, he could've told you some great stories towards the end of his fourteen years.[1]

1 Actually, Straw never left Long Island. If you believe what you just read without any skepticism, you'll probably want to take the CDC's advice and get nine booster shots before heading to Norway. (Oh my goodness gracious, I wrote this as a spoof, and a year later, just prior to completing the final draft, I discovered that it *is* possible to take your dog with you when traveling abroad! To be sure, there's red tape involved, but you can do it. And since people in the more indigent countries often transport a goat, sheep, or chicken or two from here to there, why not a dog? See pettravel.com for details.)

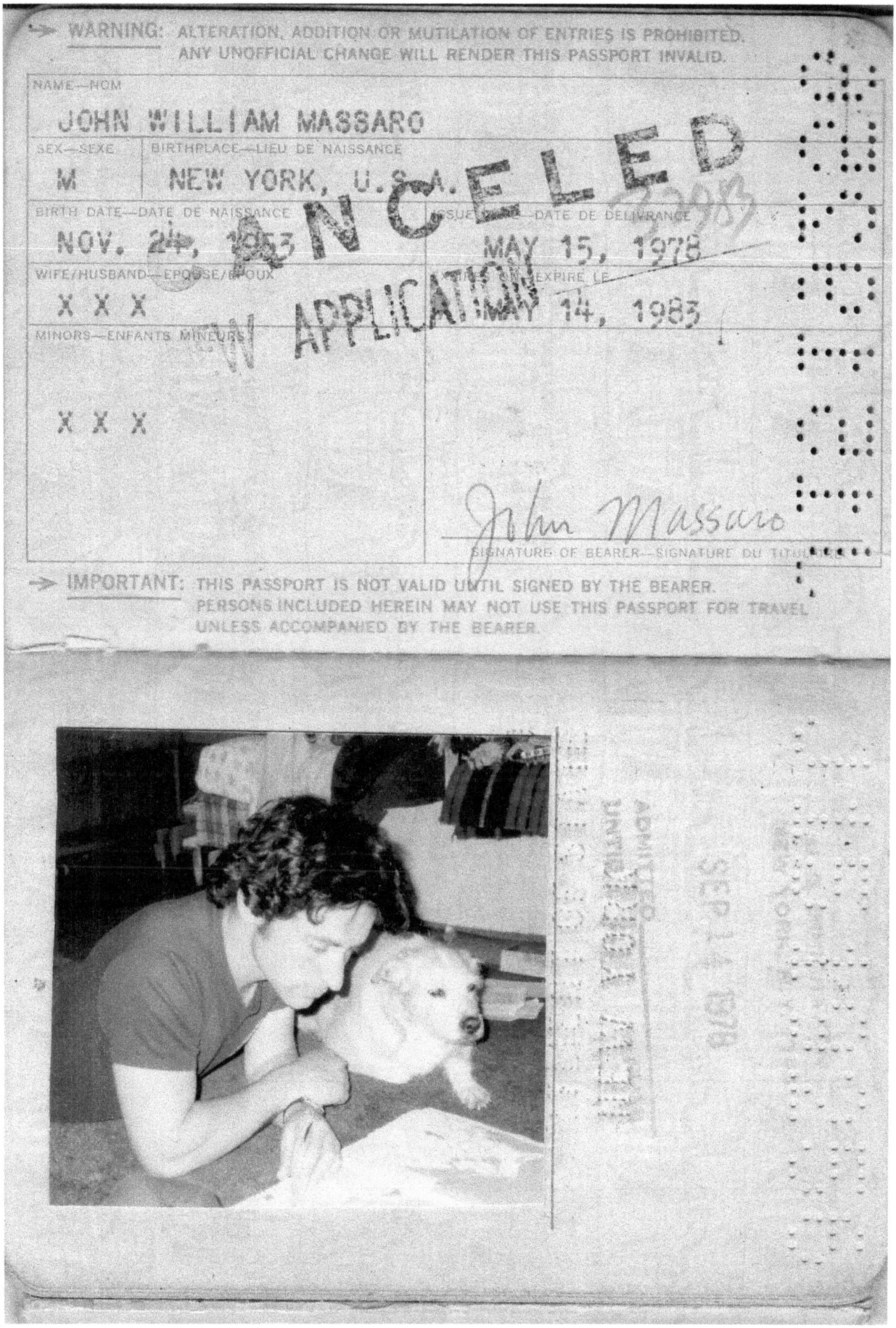
WARNING: ALTERATION, ADDITION OR MUTILATION OF ENTRIES IS PROHIBITED.
ANY UNOFFICIAL CHANGE WILL RENDER THIS PASSPORT INVALID.

NAME—NOM
JOHN WILLIAM MASSARO

SEX—SEXE
M

BIRTHPLACE—LIEU DE NAISSANCE
NEW YORK, U.S.A.

BIRTH DATE—DATE DE NAISSANCE
NOV. 24, 1953

ISSUE DATE—DATE DE DELIVRANCE
MAY 15, 1978

WIFE/HUSBAND—EPOUSE/EPOUX
X X X

EXPIRES ON—EXPIRE LE
MAY 14, 1983

MINORS—ENFANTS MINEURS
X X X

CANCELED
NEW APPLICATION

John Massaro
SIGNATURE OF BEARER—SIGNATURE DU TITULAIRE

IMPORTANT: THIS PASSPORT IS NOT VALID UNTIL SIGNED BY THE BEARER.
PERSONS INCLUDED HEREIN MAY NOT USE THIS PASSPORT FOR TRAVEL
UNLESS ACCOMPANIED BY THE BEARER.

ADMITTED
SEP 14 1978

INTERNATIONAL CERTIFICATE OF VACCINATION OR REVACCINATION AGAINST CHOLERA
CERTIFICAT INTERNATIONAL DE VACCINATION OU DE REVACCINATION CONTRE LE CHOLÉRA

This is to certify that / Je soussigné(e) certifie que JOHN MASSARO — sex / sexe MALE

whose signature follows / dont la signature suit John Massaro — date of birth / né(e) le 24NOV1953

has on the date indicated been vaccinated or revaccinated against cholera.
a été vacciné(e) ou revacciné(e) contre le choléra à date indiquée.

Date	Signature, professional status, and address of vaccinator / Signature, titre, et adresse du vaccinateur	Approved stamp / Cachet autorisé
1. 6-2-81	LEON D. STAR, M.D. Medical Office Kennedy Int'l Airport Jamaica, N.Y. 11430 — Cholera I Wyeth 106206 — Leon D. Star MD.	OFFICIAL VACCINATION NEW YORK 31-040674 U.S.A. 1.
2. 18 June 1981.	LEON D. STAR, M.D. Medical Office Kennedy Int'l Airport Jamaica, N.Y. 11430 — Cholera II Wyeth 106209 — Leon D Star MD	OFFICIAL VACCINATION NEW YORK 31-040674 U. S. A. 2.
3. 23 April 1982	LEON D. STAR, M.D. Medical Office Kennedy Int'l Airport Jamaica, N.Y. 11430 — Cholera (B) Wyeth 106604 — Leon D. Star MD	OFFICIAL VACCINATION NEW YORK 31-040674 U. S. A. 3.
4. MAY 12 1983	LEON D. STAR, M.D. Medical Office Kennedy Int'l Airport Jamaica, N.Y. — Wyeth 106904 0.5 cc — Dr. L.D. Star MD	OFFICIAL VACCINATION NEW YORK 31-040674 U. S. A. 4.
5. 6/13/84	Leon D Star MD — LEON D. STAR, M.D. Medical Office Kennedy Int'l Airport Jamaica, N.Y. 11430 — Wyeth 12000D2A	OFFICIAL VACCINATION NEW YORK 31-040674 U. S. A 5.
6. NOV 14 1986	CORPORATE HEALTH EXAMINERS 437 MADISON AVE. New York, N.Y. 10022 — [signature]	OFFICIAL VACCINATION NEW YORK 31-040674 U. S. A. 6.
7.		7.

Vaccines ruined my life

SANDRA NEWMAN

writes about the suffering she has experienced since receiving the typhoid and polio vaccinations.

In January 1991 I decided to have a tetanus booster, mainly because I hadn't had one since my school days. (I'm now 43.) I called my doctor's office to enquire about having vaccinations carried out, and made an appointment.

I asked the nurse if there were any side effects, particularly in light of the length of time since my last tetanus vaccination, but was told there weren't. Eventually, I got the shot without any problems.

In early July 1991, I returned to my doctor's office to ask what vaccinations I'd need for Morocco. They suggested typhoid and polio.

A few days later, I went along to the doctor's, where another nurse gave me an injection for the typhoid and the live oral polio vaccine. I'd never had either vaccination before. In fact, the only time that I remember having vaccinations was in school, apart from tetanus in my teens.

The nurse said I might feel achy afterwards, but said there were no other side effects.

That evening, I developed a temperature, with aches and pains in my arms and legs. As the nurse had told me to expect some aching, I didn't report any of this, although at the time I thought that it was rather severe. I also suffered an enlarged, painful gland in my neck, which came on a short while after the vaccination.

I also had a general feeling of malaise. I couldn't define what it was except to say that I thought that it was a bug. I tried to shake it off. About two weeks later, while I was out walking, one of my legs "gave out." It felt as though my legs were both weak and they were numb.

Some time after that, my legs started to feel as though they were burning. This, coupled with the pain, forced me to make an appointment at the doctor's again. I saw my doctor and reported to her my enlarged gland and flu-like symptoms.

Three months later, I returned to her office to complain that I kept suffering weakness and pains in my legs that I'd never felt before. My doctor just stared at me. When I asked her if she thought that I should rest my legs or keep on walking, she replied, "keep on walking."

Since that summer when I got the vaccines, my condition has steadily deteriorated over the years, and I am now at the stage of being able to take only a few steps before I experience the pains and a horrible numbness in my legs, which forces me to sit down. Any kind of movement gives me the same pain, even if I travel in a car.

The crises came in January 1992, when my legs were so painful and weak that I could hardly stand. Again I returned to my doctor. I also had large glands and a sore throat. I could hardly hold my head up because the right side of my neck hurt. A month later she requested a blood test.

'I have been resting solidly for nearly five months to try and ease the burning pain.'

By then, I was off work and feeling quite ill. I looked pale and still had the pains in both legs. My hands were affected, too. They burn when I have done too much, and there is a weakness there.

Now, three years later, I suffer constant pains in both legs with numbness and burning. I suffer pains and weakness in my hands, and pains in my arms, but I don't dare lift anything heavy as that also hurts my legs. Writing or repetitive movements hurts my hands.

I also suffer from an abnormality in the way my thought processes work. I have memory loss and I often stop in mid-sentence.

Besides the limb problems, I suffer ear aches and a kind of deafness, plus frequent infected neck glands, which only clear up with antibiotics. I also have serious problems with balance, unsteady walking and falling.

With each step I take, the pain increases. I will seize up with pain and have to stop, after which it takes many hours for the pain to unravel.

These effects have all had a devastating effect on my life. I am now totally house bound. I have been resting solidly for nearly five months to try to get the burning pain to ease. Although it has eased somewhat, the pain and numbness are constant when I attempt to walk.

Eighteen months ago, an intern at my local hospital concluded that I'd suffered possible vaccine damage to the central nervous system, disconnecting various nerves at the base of the spine. These nerves are giving the wrong signals and sensations, causing constant severe pain. That same day I was diagnosed as having CFS.

Ever since I have had these problems I have had considerable difficulties with doctors. I've made a claim of vaccine damage to confirm that my injuries are at least partly caused by vaccination. But my medical notes indicate both that I had a pre-existing condition, and in any case, it's all in my mind.

However, doctors now have diagnosed the problem as Guillain-Barre Syndrome. When I contacted someone from the Guillain-Barre Society, he told me that mine was the worse case he's ever seen.

The side effects that I've suffered from the vaccine are consistent with known side effects of the polio vaccine. That and the close connection in time between receiving the vaccine and the onset of my illness makes it a virtual certainty that the vaccine caused my illness.

My doctor has now admitted that I have nerve damage, caused by vaccine damage.

By the time I had the vaccination I was 38 years old and had been alive for nearly 14,000 days. If a condition such as the one I suffer from was going to happen spontaneously, it would have happened on any one of the other days of my life. There is something like a one in 14,000 chance that I would have "caught" it immediately after the vaccination.

Sandra Newman

Sandra Newman (not her real name) is suing the manufacturers of the drugs for compensation.

Passport to infertility

ALONG WITH my first travel shots, I was given an official international vaccination certificate in which they were recorded, shown on page 184.

This certificate is still used. As the years went by, and the spaces filled in with more and more shots, I came to look upon it as a unique keepsake, as exotic as many of the stamps in my passport. Little did I know how much those shots would change the course of my life.

Vaccines for foreign travel are a small part of the industry, since only a tiny segment of the population travels to those regions of the world where tropical diseases are a concern. For that reason, accurate information and statistics on adverse reactions to vaccinations are hard to come by.

Also, these shots are given to adults acting on their own free will. There's no legal pressure to get them, and they have the option of simply refusing to travel to those countries where proof of inoculation, mainly yellow fever, is supposed to be shown upon entry. Therefore, few people care about this.

I never had a bad reaction to a travel shot, and for a long time I never discussed the matter with other travelers. Recently, however, I learned that my cousin's daughter passed out after getting a hepatitis A shot for a trip to Tanzania, and I spoke to a middle-aged British woman I met in Costa Rica who told me that, a few years earlier, a yellow fever jab had caused her white blood cell count to fly off the chart. She was exhausted for seven months, she said, her life turned into a dreary routine of doctor appointments and blood tests, though she did make a full recovery.

She was lucky. The same cannot be said of the woman whose story, reprinted on the previous page, appeared in the publication *What Doctors Don't Tell You.*

What the odds are of a travel jab temporarily or permanently shattering a life I can't begin to guess. I can only advise the new generation of adventurers to never submit to a single vaccine of any kind, period.

R.I.P. Ingwe

I'VE READ a great deal on each of the vaccinations currently given to children, but not much on those given to the military and "travel shots" like the ones I got in the 1980s. I've touched upon these, however, to expand the big picture. And to expand it a little more I want to write a few words about vaccines that are routinely given to our pets. It should be self-evident that the same biological principles apply, which means that vaccines are as much an abomination for our dogs, cats and other domesticated animals as they are for our children. To take it further, you need to be careful with veterinarians who have swallowed the dogma, and I'm sure most of them have.

I've had two dogs and three cats in my life, and I don't have to convince anyone of the joy that living with a pet, or maybe I should say the right pet, can bring. With dogs you really can't go wrong, except to have a dumb one, which never happened to me, but even then they'll always love you. Cats are more dicey, as everyone knows. We had a great family cat, Henley, when I was young, but later, with my wife, a lifelong cat lover, I wasn't as lucky. I did not miss Fleur, the last pet I knew, after I got divorced and never saw her again. She was a bitchy little thing who would run away when I tried to pet her, and hiss if I tried to pick her up. There wasn't an ounce of affection in her. Who wants a cat like that? I'd rather have a pet salamander.

But it was the cat we had before Fleur that gave me the idea to write this chapter. My wife and I had just returned from a two-month trip to five countries in southern Africa, and after finding an apartment she decided to get a kitten, which was fine with our landlady because she loved cats and had three of her own. She ended up adopting one from a litter in a private home, and boy, did she pick a winner. We named him Ingwe, the word for "leopard" in one of the African languages. Before he was a year old, we knew we picked a good name because we had a miniature leopard on our hands. I would never have imagined that an ordinary house cat could be so vicious. I wouldn't wish a pet like this on you any more than I'd wish your daughter to go on a blind date with Charles Manson. Ingwe was a cat from hell.

Typical story: One night I was ready to go to sleep and Ingwe was on the bed. I pushed him off, a bit more roughly than I should have, so I reached down to give him a reassuring pat on the head. He went berserk on my arm with teeth and claws, just like a leopard does when killing its prey. I shouted something unprintable and Ingwe ran off. I got out of bed, turned the light on, and saw blood trickling down the inside of my arm, down the side of my pinky, and dripping on the floor. And yet in his own way Ingwe was close to us and only us. Once we went away for four or five days, our landlady looking after him, and when we returned he was outside. As I fumbled in my pocket for my keys to get into our apartment, he wouldn't stop rubbing against my leg — cat talk for "I'm so glad you're back."

One day my wife took him to the vet to get him tested for the feline leukemia virus, even though he had shown no signs of illness. The results came back: positive. Our cat would probably die well before his time. She was beside herself; I was saddened. But a few years went by and Ingwe showed no signs of disease. Those were the years my wife was becoming increasingly despondent over her failure to become pregnant, and when we started seeing fertility doctors. They weren't easy years. And now we had a cat with a likely death sentence.

A few months before his fourth birthday, amid all the infertility frustrations and doctor appointments, Ingwe began to look and act sick, and he quickly got worse. We knew the end was near. Then one day he was lying on the floor, howling in agony, a piercing, unnatural sound. There was only one thing to do — the hardest thing any pet owner has to do.

The vet came out of the examination room and asked me if I wanted to see my cat. Yes, I said. Ingwe's lifeless body was stretched out on a metal table, his green eyes wide open. I didn't think I'd get choked up, but I did. Funny how you can get attached to an animal you live with, even if it's a little monster.

❖

I've written this to pose a few questions about feline leukemia, the main one being, did this disease come about from an excess of vaccinations? It was first identified in the 1960s, which suggests that it very well might have. Vaccines have been implicated in human leukemia, so it follows that it might be the causative agent of feline leukemia in cats. Another potentially fatal disease which made its initial appearance in the late 1970s is canine parvovirus, closely related to another virus that attacks cats, which goes by the common name feline distemper. In fact, from the information on cat and dog diseases that I've skimmed over, there seems to be a connective web of viruses and vaccines involving both species, a web woven by recent human intervention. Everything about pet vaccines mirrors what has been done to our children. The principal fact is that there are now many more vaccines on the market than there used to be, a widely expanded vaccination schedule driven, as always, by the profit motive, and a concomitant increase in veterinarian appointments for sick animals. One clear consequence of vaccines is the growth of cancerous tumors at the injection site, which seems to happen a lot. Like autism, most of these conditions were unheard of decades ago — at least I never heard of them. And not only is there plenty of discussion on the net about the issue of cats and dogs being over-vaccinated, but both species, more so cats, have long been used in a wide range of bizarre Frankenstein experiments that cause immense pain and suffering before the animal is euthanized — and for what benefit, other than to satisfy the whims of robots who call themselves scientists? Is it not probable that foreign viruses have been injected into these unfortunate animals for observation purposes? On the other hand, they've now come out with a vaccine to *prevent* the same diseases that vaccines likely caused. It would appear that the field of veterinary science is as corrupt as everything else in America. It's also interesting to see that those who love their pets have plenty to say about the risks of all these shots, while websites run by the usual "experts" only tout the supposed benefits.

I'm not chastising all veterinarians, just as I would never rip all pediatricians. The vet we had for Ingwe was very nice, as was his staff. I was so impressed with the compassionate way they handled my last visit with Ingwe that I wrote them a thank you letter a few days later. But not all vets are like that. When I was a teen, we had a beautiful German shepherd named Cavi. In the summer of 1969 we made the mistake of boarding him when we went on a family vacation. The kennel was part of the same business practice of the vet we had, and was in the same building. A week or two after picking up Cavi, he began to develop suppurating sores and losing his fur. My father and I brought him back to be examined. The vet told us he had pyoderma, a skin disease not fatal, he said, but incurable, and we would just have to get used to it. He clipped away some fur that was about to fall out, then told us we could do the rest when we got home. Nearly fifty years later I still remember the uncaring attitude of this man, not to mention his complete lack of compunction over the near certainty that our dog had become infected in his kennel. Cavi got worse and worse, constantly scratching and biting at his fur. It didn't seem to be painful, but more of a constant tormenting itch. It kept spreading, and reached the point where half his body looked like pizza. This went on for four or five months until one day I came home from school and he wasn't there. My father, who had taken him in for the final ride, didn't have to give me the details. Cavi wasn't even two.

I'll tell you another sad story about a magnificent shepherd, a dog owned by one of my oil customers, an older gentleman. Usually he was in the back yard, and as he had no love for strangers, I always rang the front doorbell to have him brought inside before opening the gate to fill the oil tank in the back. One day the man came to the door, and nearly in tears, told me I wouldn't have to ring the bell anymore. A few weeks earlier, he said, he'd taken his dog to the vet to have his teeth cleaned — a procedure I'd never heard of, but which apparently is fairly common. It was a fatal mistake. I don't remember if it was the effects of the anesthesia or an infection that set in, but his dog rapidly went downhill and had to be put down. I told him what had happened to my own shepherd, and we chatted a while about this wonderful breed and the danger of trusting veterinarians.

❖

I recently hit upon the idea of adopting a cat, something I've never done on my own. There are plenty of shelters within twenty miles of my home, and it's so easy these days to just go on the computer and look at photos of hundreds of cats and dogs waiting to be placed in a home. I found a shelter, a small storefront actually, quite close to my apartment, and dropped in on a Saturday to check it out. There were about fifty cats there, most in small cages, some walking around. The place was clean, orderly and seemed to be an efficient volunteer-run operation. Two women were there, and I had a lot of questions which they answered patiently and pleasantly. I wasn't ready to make a decision, I wanted to think things over, but there was one cat named Jake that caught my eye and seemed to have a sweet disposition.

A few days later I called back to see if Jake was still available. He was. The woman I spoke to introduced herself as Elaine. She hadn't been there Saturday so I chatted her up a bit, disregarding one of life's cardinal rules that you should never tell people more than they need to know. I casually mentioned that one of the stipulations in my apartment rental lease, which I had signed fifteen years earlier, was that pets were not allowed, but my landlord had since mellowed and told me, when I asked him, that he had no objection to my getting a cat. All of this was true. I also told her that I was opposed to shooting up animals with vaccines, that I intended to feed my cat the same meat I ate, and that I preferred that Jake not be microchipped. I had brought up these subjects with the other two women, and while they were not in full agreement with me, my point of view did not seem to alarm them. Their policy was that, before adoption, every kitten or cat was vaccinated, microchipped, and tested for the feline leukemia virus. Of course I had no objection to the last part, and after my experience with Ingwe would never get a cat without that test being done and proving negative.

A few hours later I decided to drive down and have another look at Jake, and perhaps take him home. I walked in and asked the woman there if she was the one I'd spoken to on the phone. She immediately became brisk and told me she was uncomfortable with everything I had said. She didn't turn me away, but she did say that the first thing I'd have to do is submit a copy of my landlord's drivers license, his phone number, a copy of our original lease agreement, and a statement from him affirming that he had changed his mind and now allowed me to have a cat. That ended it. To me this was extremely unreasonable, and I told her so in a quiet way before walking out the door. The next day yet another woman from the same place called me to tell me again that Elaine, who apparently made all final decisions, was uncomfortable with my attitude. This volunteer was a little more polite but had the same mindset. I tried to explain why I felt the way I did about food, vaccinations and microchips, but I could tell she wasn't listening. The way I left it was for them to call me within a week if they had a change of heart. Of course they never did. I had no idea how some of the people who run these shelters go to great lengths in screening potential pet owners.

For those who are ignorant about this microchip business, as I was until recently, a microchip is a metal chip, the size of a grain of rice, that is surgically inserted beneath the skin of an animal, usually between the shoulder blades. It identifies the owner and can be read by a scanner. The rationale behind it is that if a dog or cat were to become lost, the owner can be contacted. It truly amazes me how people so readily buy into the newest high-tech gimmick without thinking about it, how it just takes a little salesmanship. May I ask how this is an improvement over a pet wearing a collar and tag that shows the owner's phone number? Is it not much easier for someone finding a stray dog or cat to call the phone number on a tag, rather than going through the rigmarole of getting a microchip scanned — and might such a person not want to be bothered? I wouldn't know the first step to take myself. And there are questions that most people never stop to consider: What detrimental health consequences might this foreign object cause? What about reports that this chip has been known to migrate to different parts of the body? And how effective are these scanners? Apparently some of them don't work very well and are incapable of reading the information on the chip.

As for pet food, I've done less than an hour of research on the subject, but some of what I've read validates what I've always felt. What I'll say here might offend some readers but I'll say it anyway. How can anyone believe that it's better for a cat or dog to eat the industrialized crud that's sold in supermarkets and pet supply stores, rather than to feed it fresh beef, poultry or fish and perhaps a few chopped vegetables — the same food we eat? Ever check out the ingredients on those cans, or worse, those big bags of dry food? Some of them are scary enough to read, and you're in for more of a scare if you analyze them individually. Yet so many people who love their pets, and want them to eat the healthiest kind of food, uncritically believe all the bilge they've lapped up, for the most part, from decades of television commercials, the gist of which is that all the top brands have been specially formulated to meet the dietary requirements of cats and dogs. Sure they have, just like the makers of Wonder Bread, Lucky Charms, and a thousand other abominations have fortified them with vitamins and minerals because they love children. As if they have no idea that their products have little nutrition and can lead to serious health problems down the road. As if the almighty dollar isn't the bottom line.

After giving up on Jake I searched the Internet for another shelter and found one not too far away with a pretty, young tuxedo cat — the coloring I prefer — named Dolly. I can't explain why, but I had a good feeling about this particular cat. I called the shelter and was sorry to hear that Dolly had a thyroid condition, and was doing okay, but was on medication. Why would a one-year-old cat have a thyroid condition? I think I'm within my rights to ask if it was caused by a terrible diet, or more likely, a vaccination. I do love animals, but I'm sorry, I'm not Mother Teresa and I would not knowingly adopt a sick one.

❖

Have I written more on this subject than called for? Some may think so but if, armed with the information herein, I spare just one pet owner the grief stemming from a bad decision, it's worth it. There are probably more than a billion human beings on this planet who share a bond of love with animals, a bond second in importance only to people whom they love. There are about two hundred million in America alone. The lessons I learned as a pet owner encompass more than just vaccination. I'll summarize them in one long paragraph, sprinkled with my usual contrariness.

If at all possible, adopt an animal that has never been vaccinated, and if you own one now, never allow it to have more vaccinations. The rabies vaccination can be dismissed as just another money-making scam, and like all vaccinations carries risks, but in some circumstances you may have to show proof of vaccination, in which case you'll just have to figure out what to do. Unless your pet has an obvious medical problem, keep him away from veterinarians who, like regular doctors, show every sign of having been "educated out of their brains." Don't fall for scare tactics, and don't be talked into unnecessary procedures which might seem harmless. I mentioned what happened to a fine dog from a teeth cleaning. In fact, I suspect this might be as unnecessary with people as it is with pets. Oral hygiene is definitely important, but beyond daily flossing and brushing, are there really any benefits to a professional cleaning, or is it just a scheme to pad the bank accounts of dentists? If you go away, it is highly inadvisable to put your dog or cat in a kennel, especially sight unseen. Just as children left in day care centers get sick much more often than children who stay home, animals in proximity to the secretions of others of their species, or kept in improperly cleaned cages, are much more likely to become infected. Having a friend or relative look after your pet is obviously the best way to go. Finally, in regard to your children or grandchildren, beware of the cat lady Elaines out there, employed as school or hospital nurses, public health officials, social workers, what have you — the "experts" who think they know it all, whose minds are cemented shut, who feel threatened by any viewpoint that rocks their secure little world. I'm sure that woman knows a lot more about cats than I do, but if she had done any investigation, she might have felt more uncomfortable with her own beliefs than with mine. She would have discovered that quite a few pet owners have questioned the supposed benefits of microchips and multiple vaccinations, and even more have condemned the top selling brands of cat and dog food, especially the dry variety, which the cats in her shelter were fed. I can easily picture someone like her making life difficult for parents submitting a perfectly legal philosophical or religious exemption as a prerequisite for their child attending school, or even contacting some swine at Child Protective Services. As in so many important matters, it often comes down to the personality of one individual and the luck of the draw.

Postscript: All about a cat

One year has passed after I put the finishing touches on this chapter. In that year I continued to live alone, as I have for many years, daydreaming about getting a four-legged companion. My landlord had given me the okay for a cat or small dog, and I looked into buying a Jack Russell terrier, but took no action, held back in part by the thought that no dog in the world could ever measure up to Straw, and I didn't want to be disappointed, especially since, at 65, any pet I were to adopt might be my last.

One day I stopped at an outdoor mall five miles from my home to buy something. I'd never been there before. In the mall was a PetSmart store, one in a nationwide chain of large pet supply stores. I figured I'd have a look. In a small area there were seven or eight cats and kittens looking out from their tiny cages. Most of them were strays that had been brought in by a local cat rescue organization. Handwritten signs on the windows outside their cages told their stories. One black cat, eleven months old, lay on his back and followed me as I slowly walked past him. For ten seconds there was intensive eye contact. What's in a look? Well, something. I couldn't leave the store without asking a worker a few questions about this cat, which had been given the awful name Zeus. She suggested I fill out a form and said that Ann, who handled cat adoptions, would be in tomorrow and would contact me.

Ann called me the next day and asked me to come in. We talked a bit. She asked me if I might be interested in adopting two cats. I said no. Zeus and I clicked somehow, I told her. Why don't you spend a few minutes with him and see how you get along, she said. She unlocked the little cat room, opened the cage, and we got acquainted. The first thing he did was bite my fingers, though in a playful way. Then he sparred with my shoelaces. Not being an authority on cat behavior, I was a bit uneasy about the biting, wondering if this was a sign of worse behavior to come, as in Son of Ingwe, but there was just something about him that I couldn't walk away from. It was a Wednesday. Ann told me I could pick him up on Friday, after the vet performed a final check-up. He had been neutered, examined and vaccinated seven months earlier, and given a clean bill of health. This time I didn't air my opinions on vaccinations and commercial cat food, but I did quietly mention that I preferred he not be microchipped. All cats that leave the store have to be chipped, Ann told me — company policy. I nodded and said no more.

I showed up on Friday with a cat carrier, a signed contract, and $125. I asked Ann if she had called any of the references I'd listed. No, she said, she just had a good feeling about me from our conversation and that was enough. What did I say about individual personalities and the luck of the draw? I'd already picked out another name for my new companion, after looking up the word for cat in nearly every language on the Internet. The one I liked best was Kochka — "cat" in the Czech tongue. Kochka and I have been together for seven months now, and even though I'm partial to dogs, and had been leaning towards getting one, I'm happy with the choice I made. I never imagined that a cat could complement my personality. This is a cat that, like all cats, wants his own space but at times also wants affection. Just like me. The biting business is interesting. He enjoys rough play, which I always instigate, and sinks his teeth and claws into my skin, but never viciously, never drawing blood, and if he gets too rambunctious, I raise my voice and he knows to knock it off, and starts licking my hand. The only time he really got angry and threatening was the one time I tried to clip his claws, so I gave up on that. What do I care if he scratches my furniture? And like the little wise guy I once was, he'll playfully swipe at me when I walk by, sometimes attacking my leg then running off, which always gets a chuckle out of me. Such a delightful little imp. No one could have told me that I would one day own a cat that, like Straw, I'd consider irreplaceable. But he's just that.

I've learned a few things along the way. I got him a snap and release collar with a laser engraved ID tag, and he promptly lost it. I got him another and one day he showed up at the doorstep without this one. The third time I bought an old-fashioned buckle collar and adjusted it tightly, taking care not to put pressure on his throat. He slipped out of that one too — no mean feat. How did he do it? I went on the net and learned that this is a gripe of many cat owners, but none seem to have figured out how they do it. A collar and tag costs 22 bucks, so after shelling out $66 in five weeks, I decided no more collars. So am I glad they put the microchip in? Yes, it's better than nothing if he gets lost. But it's clearly a scam, because the agreement I signed required that I would always keep him inside. This seems to be a stipulation at most cat adoption centers. I had no intention of abiding by it, of keeping Kochka cooped up in my small apartment — exaggerated outdoor hazards be damned — and he comes and goes indoors and outdoors as he pleases.

As for food, Kochka is no gourmand. I've given him morsels of chicken, steak and fish from my plate, and it's always the same story: he sniffs it and walks away. I've bought premium wet cat food in luscious flavors and that doesn't interest him either. What the hell? Is he holding out for lobster or caviar? I don't think so. Dry food is all he ever knew and all he will ever eat, so dry food it has to be. Actually, I discovered that PetSmart sells a much higher quality dry food than the bag of gravel that Purina calls cat chow, so even though it costs a lot more, that's what I buy. Only the best for my kitty.

Have you ever stopped to think of how it came to be that, in terms of household pet ownership, cats and dogs outdistance all other animals by a million miles? What is it about them that they easily and naturally form strong bonds with human beings, even though there have always been those who mistreat them and in some countries even kill and eat them? With dogs, the answer is easy because we see it everywhere, except perhaps in the toy breeds. We evolved together in mutually dependent roles of protection, and as hunting partners. But cats are different. We know their close connection with us goes back at least 5000 years, as evidenced by sacred statues of them found in the tombs of Egyptian pharaohs — and interestingly, as I've seen myself, they are much more esteemed as pets than are dogs in Arab countries — but how to explain it? Obviously, even though they tend to be far more aloof and independent than dogs, they see something in us, and we see something in them, but what it is is intangible. Even the way they purr is something of a mystery.

So it is that Kochka and I connect on some transcendental level. At this very moment my little buddy is napping on a rug eight feet away. When I finish writing this I'll walk over to stroke his fur. I know that he'll open his eyes for a moment, then close them again and smile in pure feline bliss, and the purring motor will kick right in. In addition to unfailing companionship, a cat can provide a wonderful diversion from the toil of writing and revising a book. But as I said, with cats it's hit or miss. This time I got lucky.

Reason and emotion: vaccines and guns

AT THE beginning of this book I took note of the explosive nature of the vaccine issue and the reason for it: our children, whose very lives hang in the balance of who is right and who is wrong. We. the anti-vaxxers, say that children are being brutalized and killed by vaccines; they, the pro-vaxxers, say that children are suffering that fate by *not* being vaccinated. A calm deliberation of fact and opinion seems to be impossible. Indeed, the two sides will never agree upon what is true and what is false. This is an issue that will always be fueled by raw emotion.

Of the two faculties, reason and emotion, the latter rules over the vast majority, a fact known to every demagogue, campaigning politician and newspaper editor. Nevertheless, it is reason — a calm and open-minded deliberation of what we perceive in the world around us — that has brought about every form of social and scientific progress. And it is an appeal to reason rather than emotion that invariably wins over those with a higher than average intelligence.

I once read a remarkable exchange, far too long to reproduce here, on the anti-vaccine website thinktwice.com that illustrates my point. It began with this email from a true believer:

You fucking bastards! Our town in Ontario, Canada is in the process of inoculation for meningitis after several recent deaths. Your web pages are being waved all over town as reasons not to immunize. I hope you sell a lot of books. I hope the proceeds assuage your conscience. It's statistically likely you'll kill a few kids a year with this horseshit. I truly hope you rot in hell.

In a civil reply, the thinktwice respondent criticized the tone of this unsolicited email, and cited a study that concluded that the meningitis vaccine itself was found to be the cause of meningitis in many cases. This prompted a lengthy response from the sender, obnoxious in places, but this time he did not use profanity and did not seem so sure of himself. Thinktwice then hit back with an equally long, incisive fourteen-point rebuttal of the case for vaccines that this man had built up. In it they offered to send him medical literature that he had never heard of, to which he emailed a final reply: "Please mail [copies of the studies and the academic vaccine publication]. I was operating from the viewpoint that you were deliberately being coy so you could sell books. Frankly, that isn't supported by your responses, so I'm operating on the thesis that one of us is profoundly misguided."

Despite his opening salvo, this man is to be commended for having an open mind, which is far more than can be said for most vaccine champions. I have yet to hear of someone who went through the reverse process — who at one time strongly opposed vaccines, then came to doubt his or her position. If somewhere there *is* an exchange like the one reproduced above that I've missed, one attacking the pro-vaxxers and being answered by logic and sound ideas, I'd like to know about it. But all I've ever encountered in my investigation has been shoddy arguments and name-calling.

In writing this book I have tried to stimulate both reason and emotion in the reader. Visual images can instantly arouse emotions, which is why I have placed several provocative photographs in these pages. If you would call this propaganda, I would reply that this word has a broad meaning. Though propaganda by definition is always emotionally charged, it can be based on truth as well as falsehood. Every picture and caption in this book is a reflection of the truth as I see it. Propaganda in the negative sense is never so rife as during wartime — especially wars in which those who control the mass media have an emotional stake — when populations are bombarded with torrents of lying words and photographs with brazenly falsified descriptions. But it's not only in times of war that the emotions of a nation are aroused. On a smaller scale, this has taken place and continues to take place on a wide variety of issues, vaccination being one of them.

Although I was born at the tail end of the nationwide crusade to wipe out polio — fifteen years after the National Foundation for Infantile Paralysis was formed, and seventeen months before Jonas Salk came along with his dangerous vaccine — some of the propaganda of that era washed over into my elementary school years. It was in a science book in the fifth grade that I first saw photographs of children crippled by polio. Juxtaposed with the photo of Salk holding his test tubes, this provoked a flood of emotion — on the one hand sorrow for the afflicted children, on the other veneration of Salk the savior. The images of these young polio victims were, I'm sure, real enough. But the objective truth of the matter is that, far from saving children from leg braces and iron lungs, Salk was responsible for putting more children *in* them. The number he *saved* from that cruel fate is very likely zero, though this will never be known with absolute certainty.

Nothing is easier than wielding a heart-wrenching photograph of a child crippled by polio and screaming at a person who opposes vaccines, "Is this what you want to return to!!??" When someone becomes enraged and seizes the moral high ground like this, his opponent is instantly beaten down because it becomes awkward to even try to offer a rational reply. "You're against the vaccine, you miserable bastard, therefore you must love to see children suffer and die" is the message conveyed. There are plenty of peabrains who crawl all over the Internet with this kind of argument. In the case of polio, I doubt that one of these people has ever taken the trouble to examine Salk's career. One encounters the same indignation when there is an outbreak, especially among children, of what is deceptively referred to as a "vaccine-preventable disease" — say, measles or whooping cough. The foregone conclusion is that the unvaccinated, or their parents, are to blame. That there are *always* individuals who have been vaccinated against the "vaccine-preventable" disease that strikes them during an outbreak, demonstrates for all to see that vaccines can and do fail. But the true believers have an answer for everything.

Even rational people can become irrational while discussing this contentious subject. Years ago I brought it up with the older of my two sisters, who like me was born too early to have received the measles vaccine. I never came down with measles as a child, but my sister did. She doesn't have a strong opinion about vaccines, but she does lean favorably towards them, as most people do. Wanting to start trouble, I began talking about the uselessness of the measles vaccine. She became indignant, telling me that our family doctor had said that she'd had a particularly bad case of this disease. She said she was five at the time, which means I was nine. "Well," I thought, "that's interesting because I don't even remember you getting the measles, so it couldn't have been *that* bad." But look at the faulty reasoning in someone who happens to be very bright: first, as "bad" as her case might have been, she made a full recovery with no lasting effects; second, she tacitly assumed that the vaccine, had it been available, would've prevented her illness; third, she seemed to be unaware that there are major risks associated with the MMR vaccine, which are spelled out in the product insert.

❖

Now let us examine a subject which at first glance seems to be entirely unrelated, but is in fact part of the same culture war being waged in America. The people who want to force vaccines on our children are very often the same individuals who want to disarm the citizenry in the name of gun control. And in both cases, it's for our own good, of course.

On December 14, 2012, a deranged and heavily armed young man named Adam Lanza walked into the Sandy Hook Elementary School in Newtown, Connecticut, and going from room to room, shot six teachers and twenty first-graders dead before taking his own life. Every mother and father could feel the pain of the parents of these little children, and was sickened, as I was, to learn of this tragedy. In the wake of this shooting, the nationwide media screech that "Something must be done about guns!" was as predictable as it was ear-splitting. It was all emotion and no reason. How can there be a rational debate when twenty little children lay dead? Not only was the country's long-maligned gun lobby, the National Rifle Association, thrown on the defensive, from some quarters came the cry that NRA should be designated as a terrorist organization! Nine months later, following another mass shooting at a naval installation in Washington D.C., a journalism professor at the University of Kansas publicly expressed his wish that the next victims should be the children of NRA members.

As an aside here, I would like to demystify the NRA, one of the media's favorite punching bags. I've been an NRA member for more than thirty years, only because I enjoy the articles in *American Hunter*, their monthly magazine. I've read hundreds of their editorials over the years, and am thoroughly familiar with their mindset. The head honchos at the NRA live in a fantasy world of an America reflected in wholesome television programs of the early 1960s like *The Andy Griffith Show* and *Leave it to Beaver*. They're always crowing about our freedoms and the need to stand up and fight for them — freedoms that exist only in their collective imagination. I always ignore their begging letters from the pen of longtime executive vice president Wayne LaPierre who, I learned only recently, earns about one million dollars a year, and who — it seems superfluous to mention it — ducked the draft during the Vietnam War. I do applaud the NRA's efforts on behalf of the Second Amendment, but in every other way this outfit — its board members, most of its rank and file, and its major supporters, several of them famous politicians and entertainers — epitomizes the cowardice and dead-end creed of conservatism. In their pathetic attempt to maintain a respectable appearance, they remain silent when the media go on the offensive against them — I think here in particular of some screaming front-page agitprop printed by the New York *Daily News*, which was so crazy it was funny — when they should be fighting back on the media's own mud-wrestling terms. What they should do is point out that the media as well as the gun control movement is heavily Jewish, print names and photographs to prove it, then inform their readers that most of the Bolsheviks who overthrew the Russian czar in 1917 were Jews, and they had a much easier time exterminating tens of millions of Russians once they had confiscated their guns. But the NRA leadership is far too "respectable" to do anything like that. If the NRA is a terrorist organization, so is the 4-H club.

❖

In the days after the Newtown shooting, parent advocacy groups sprang up, pleading with the public and their lawmakers to "do something." Connecticut, which already had highly restrictive gun laws on the books, promptly passed legislation tightening the screws even more. Grandstanding in front of the television cameras, our own tinhorn liberal governor, Andrew Cuomo, an outspoken NRA hater, rammed through a law that made it illegal to load more than seven bullets in any gun, making instant felons of many New York hunters who are not in the habit of reading legislative fine print. In an act that defied any kind of logic, one Cary McBride, editor at the *Journal Register*, a suburban newspaper, published the names and addresses, along with an interactive map, of all pistol permit holders in Rockland and Westchester counties, just north of New York City and west of the Connecticut state line. Unfortunately for her, the paper was inundated with furious letters and phone calls, and along with other staffers she lost her job, but not before hiring an armed guard for protection, even though the local police determined that there had been no unlawful threats. This revealed her rank hypocrisy, on the same level as that of doctors who preach vaccination but don't vaccinate their own children or themselves.

The right to possess guns and the right to vaccine choice are among the most passionate and polarizing issues in America. And as with vaccines, there are incontrovertible facts that allow us to accurately assess the subject of gun violence.

1968 was a watershed year for those of us who believe in the privilege as American citizens to purchase and own firearms without government encumbrance. Prior to the federal Gun Control Act of 1968, gun laws nationwide, for the most part, were true to the spirit of the Second Amendment, namely that "the right to keep and bear arms shall not be infringed." With the passage of the Gun Control Act, buyers and sellers of firearms now had to contend with numerous regulations and limitations. Among them, gun dealers had to obtain a license from the federal government, mail order purchases came to an end, and military surplus rifles could no longer be imported. Persons in certain categories, not all of them criminal, were excluded from buying guns and even receiving them as gifts. Interstate sales became highly restricted. Every firearm manufactured after the passage of the Gun Control Act had to have an engraved serial number. All these measures were passed into law for the ostensible purpose of curbing gun violence and saving lives.

In a majority of states it's still pretty easy to walk into a store, show proof of residence, and walk out with a rifle or shotgun of the type commonly used for hunting. In some states it's just as easy to buy a pistol or a military-style "assault rifle," the media definition of which is always purposely vague. Since 1968, however, every purchase involves more paperwork to keep bureaucrats busy, supplemented in recent decades by background checks conducted by the FBI. From 1998 to 2013, 167 million such checks were made. In addition, there are local ordinances in many of our big cities that make it difficult to legally own any kind of gun, while some, handguns in particular, are banned outright. Before 1968, in the nearly 200 years of American nationhood, when guns were much more accessible than they are today, not a single mass murder on elementary or high school property took place. To be sure, there have been numerous shootings on school grounds going back to colonial days, but all of these involved personal grudges with one or two individuals being targeted. In a handful of cases, which apparently never involved a fatality, drunks or miscreants took potshots at school buildings while classes were in session. But not once were students or teachers randomly shot at in or out of a classroom.

Since the slaughter in Newtown — I write these words in July 2016, one month after the killing of 49, a record number, by a lone gunman at a homosexual nightclub in Orlando, Florida — there have been many more mass shootings not involving schools, and by the time this book is published, others will be fresh in the public mind. There have been so many incidents of this kind in the past thirty years or so, in which groups of people unknown to a demented loner are sprayed by gunfire, that we have almost come to expect them as a periodic phenomenon, like deadly tornadoes or flash floods. Yet in the entire history of America before passage of the Gun Control Act — that is, in a period covering 192 years — there were only two random mass murders in which guns were used: the killing of thirteen on September 6, 1949 by Howard Unruh, who casually picked off passersby with his Luger as he strolled down a sidewalk in

Camden, New Jersey, and the sixteen who died on August 1, 1966 when Charles Whitman climbed to the top of the University of Texas tower in Austin and began shooting at people walking below. (I was twelve at the time and clearly remember this.)

In this age of constant lies, there are certain facts about these mass shootings that are hardly ever discussed. Virtually all of them involve one or more of the following elements: the killer was taking psychotropic drugs, which are acknowledged by their manufacturers to have adverse side effects, including suicidal and homicidal inclinations; the influence of violent and bizarre video games and movies that trivialize murder; the racially and culturally alien background of the killers; hatred and resentment of the murderous American military machine abroad. None of these were factors before 1968, when guns were more easily acquired by anyone: in the sixties there were no dangerous drugs like Cymbalta and Paxil prescribed to troubled teenagers; no dehumanizing films and fantasy games easily accessed on a computer screen, the newest genre of which is "torture porn" as labeled by its degenerate Hollywood creator (Lanza was hooked on these games, as well as the drug Celexa); the tidal wave of legal and illegal immigrants from all corners of the world, initiated by the Immigration and Nationality Act of 1965, had barely begun; and the U.S. military had not yet commenced its slaughter of civilians in the Islamic nations of the Middle East at the behest of that region's most lawless state, Israel.

I'd like to make a few more observations. Not many people are aware that homosexuals have committed a highly disproportionate number of mass murders, though many of these did not involve firearms. Quite a few of these shooters have been military veterans, their minds twisted by indoctrination in basic training and in some cases their combat experience — not to mention the vast, unknown number, probably well into the millions historically, of American vets who took their own lives, died in other violent circumstances, or committed murder on a small scale. (Whitman was a Marine Corps veteran, while Unruh, in addition to having fought in World War Two, was a homosexual. Omar Mateen, the Orlando nightclub shooter, of Afghan background, was widely reported to have been a closet queer himself, despite being married.)

Another glaring factor connected with gun crime in general, one the NRA is too frightened to mention and the media have always censored, is race. Vermont is the whitest state in the country, and one of the most lax in regard to the purchase and ownership of guns, yet murder and armed robbery are extremely rare. The same is true of rural western states like Wyoming, Montana and Idaho, where guns are as fixed an institution as marriage, and many pick-up trucks you see have a rifle in the rack. By contrast, Washington D.C., with a huge black population and stringent laws governing the sale and possession of firearms, is always a top contender in all categories of violent crime, as are other black hellholes like Newark and Baltimore, both in states where firearms statutes are among the most prohibitive in the country, New Jersey in particular.

Yet another component of the recent spate of mass shootings might well be mass vaccination. *Vaccination, Social Violence and Criminality: The Medical Assault on the Human Brain*, by Harris Coulter, co-author of an earlier stellar work, *A Shot in the Dark*, argues that there is a correlation between the high number of shots children receive today and the wide variety of cerebral and neurological disturbances, unseen a few generations ago. Since 1990, when this book was published, several more vaccines have been added to the CDC schedule. Coulter's thesis is convincing, though I believe that the penchant for violence among blacks and Hispanics, followed by the side effects of mood-altering drugs so commonly prescribed today, are paramount in understanding anti-social behavior. Coulter does mention these factors, but in my opinion does not place enough emphasis on them.

The facts are indisputable. Children born two generations ago, my generation, were on average much healthier in body and mind than young people today who have received, thirty, forty, fifty more vaccines than we did; gun violence was less common, and mass shootings almost unheard of, when guns were much easier to obtain, America was much whiter, and both psychotropic drugs and deviant entertainment were unknown. The hidden agenda, the ultimate aim, of those who push most forcibly for vaccination and gun restrictions — and again, by and large they're the same self-appointed elites — is domination, subjugation, total control over our lives. Although I've said some unkind things about the NRA, they're dead right when they say that the end game is complete disarmament of the population.

Unfortunately, we can't reset the clock to when

America was a better country; we can't turn back the tide of history. We *can* understand what's going on in America right now, and we *can* learn from the past in laying the groundwork for a brighter future.

Postscript

Some months after completing this chapter I came across a book titled *Nobody Died at Sandy Hook*. Eleven people contributed to it, in addition to the editors, Jim Fetzer and Mike Polacek. They claim that the mass shooting of December 14, 2012 was a hoax pulled off by the Federal Emergency Management Agency (FEMA), a division of the Department of Homeland Security, in conjunction with the news media, that it actually was a drill planned well in advance to promote gun control. Now while the federal government and the media often work in partnership to dupe Americans, this went too far; it irritates me that the truth movement in this country unfailingly attracts some real screwballs. Nevertheless, my curiosity won out and I ordered the book online, though not from Amazon, which had taken the unprecedented step, at the time, of banning it.

There is nothing nutty about this work. It raises some very serious issues, and most of it is quite convincing, indeed, astonishing. Some of it, though, is quite unconvincing. For example, I found the authors' analyses and explanations of many photographs very weak. The two main pillars of evidence put forth are first, that Sandy Hook Elementary had been closed in 2008 because of its dilapidated and hazardous condition: the school district used the building strictly for storage purposes, and the property had been selected as the site of an evacuation drill, which is what actually took place; second, the editors claim that they downloaded a 20-page manual published by FEMA confirming this, which they reproduce in full. Printed on the cover are the words, "The FEMA Manual For Sandy Hook," further down, "Site Activation Call-Down Drill Exercise Plan," and in brackets, "MASS CASUALTY DRILL." Below that, "Exercise Date 12/14/12" and "Publishing Date 10/08/12," and at the bottom, "For Official Use Only." The whole thing looks and reads authentic, though I realize it might be a forgery. My one major objection to the book's thesis is a cry of common sense: How can all these children and their parents (if they existed), the Connecticut state police, the medical examiner, and many others besides — how can all these people play their roles in such a gigantic hoax without a single person being brave enough to come forward and expose it? This is a classic example of how cunning minds on either side of a controversy can obfuscate the truth by piling layer upon layer of deception. I would put nothing past the media bosses, but I really can't believe they would broadcast a lie of this magnitude for fear of being discovered.

As it turns out, there's a great deal of discussion about this book on the Internet. Some of it supports the book's claims, though most of the reactions are hostile. Methinks, however, that the book's detractors doth protest too much. As for me, I'm undecided about what really happened in Newtown, Connecticut on December 14, 2012 and, after much thought, am open to any possibility. I'm only interested in the truth.

The truth of the matter would be easy to establish, as I see it. It all comes down to whether the school was operational in late 2012 or had been closed down years earlier, as the authors of this book claim. Any Newtown resident living on or near Dickinson Drive, where Sandy Hook Elementary stood, would know that. They would have seen plenty of activity on school days: school buses rolling in and out, parents dropping their kids off and picking them up, children playing outside and so forth — or they would have seen nothing at all taking place outside a deserted building. *That's why, in a country of constantly swirling lies and fake news, what one sees and hears with one's own eyes and ears is the only thing guaranteed to be true — and the most compelling evidence I cite for my position on vaccines.*

Postscript to a postscript

The October 1, 2017 shooting massacre at an outdoor country music concert in Las Vegas, and another bloodbath by gunfire at the Marjory Stoneman Douglas High School in Parkland, Florida on February 14, 2018 necessitate additional comment. Separating clear facts from mistaken but sincere beliefs from mainstream media malarkey from assertions of disinformation agents now seems an impossible task. What I will say here is that, while I do believe that these shootings did happen and many people died, something fishy is going on in America. I believe that there were multiple shooters involved in Las Vegas and that some survivors who spoke publicly about this were murdered — all this naturally hushed up or derided as conspiracy mongering by the controlled media. In the case of the high school in Parkland, no sooner did the smoke clear than an effeminate freak named David Hogg — who claimed to be a senior at the school, which I doubt, and the son of a retired FBI agent, which is probably true — emerged as a self-appointed, righteous spokesman against gun violence and the NRA. Often at his side sounding off was buzzcut bisexual Emma Gonzalez, another student. Hogg became an instant TV star, doing the news show gab circuit while the media as a whole went into overdrive with their anti-gun propaganda, in particular their doting coverage of militant students marching on the Tallahassee state capitol, and weeks later on the White House, demanding tighter gun laws and an outright ban on assault rifles, as if they knew what they were talking about. Hogg was flayed so widely and mercilessly on the net that he felt the need to publicly deny that he was a crisis actor, and did some verbal jousting of his own against what he called conspiracy theorists. How original. This event prompted James Fetzer to write another book, *The Parkland Puzzle*, which I found even more persuasive than his work on Sandy Hook. All I will say here is that something stinks to high heaven, and while I don't claim to know the real story, I firmly believe that Hogg is a disturbed attention seeker or professional troublemaker, funded by God knows who. As for all those gun-hating, marching students, and the supposed nationwide high school walkout on March 14, there was too much anger in those mostly ugly and alien faces to persuade me that this was a cross section of sincere though misguided American high school students. It was all too well-scripted, and I wouldn't be surprised if a hidden hand rounded up these students, if in fact most of them *were* students and not a rented mob.

The upshot of the Parkland episode — whatever happened or didn't happen there — and another illustration of how we have government by media — was a surge in anti-gun tyranny, namely a new flood of "red flag" state gun legislation, whereby a judge, upon petition, allows police to simply raid the home of a person deemed to be a possible threat, and seize all firearms — without a warrant, with no arrest being made, with no criminal charges filed. This obviously opens the door to all kinds of abuse, and sets the stage for incremental gun confiscation — all with the blessing of journalistic whores doing the dirty work of their bosses. After all, just think, we're nipping more of theses planned mass shootings in the bud, we're saving the lives of so many innocent children, so how can any reasonable person oppose taking guns away from dangerous people? And let's start by banning AR-15 military-style rifles. As of this writing, there is additional stealth legislation that has not been clarified.

There have been more mass shootings since I have written the above, and there will be more, followed by more televised demonstrations and marches, more anti-gun shrieks, and more aggressive gun-grabbing tactics. I am already inclined to believe that these killings or staged killings, as the case may be, are part of a cunning design to disarm law-abiding Americans, in which the media bosses are conspirators. It's all quite nebulous, but there are two things I know with certainty: the media bosses hate and fear the fact that there are so many privately owned firearms in America, and the people who want to take our guns away are the same ones who want to vaccinate our children to death.

What kind of people are these?

THERE ARE certain jobs that, by their very nature, attract only people who are unprincipled. Most of those who make a living in sales are not evil, of course, but all of them are at least a little dishonest; I've never met a salesman who wasn't. Anyone who owns or works in an adult bookstore is slime, at best. If you're a high roller in the real estate market, surely you don't care about local residents whose lives will be thrown into disarray by your construction projects. Ask our president.

Most lines of work, though, employ people whose traits run the gamut. The medical field is so vast and encompasses so may different personalities that it would be senseless to make generalizations. Many a black mark stains the history of medicine, but there are many bright spots as well. Think of Florence Nightingale, the angelic "Lady with the Lamp" as wounded British troops affectionately called her, ministering to them in Turkey during the Crimean War of 1853 to 1856. Not only did her exacting standards of hygiene save many lives and revolutionize the emergency treatment of battlefield casualties, on her own initiative she reformed all facets of the nursing profession, and is so remembered to this day. Joseph Lister, her contemporary, was another English pioneer who quietly transformed medicine by improving surgical techniques, and demonstrated that deadly infections could be prevented by sterilizing surgical instruments and applying antiseptic solution to wounds and post-operation incisions. Listerine, the famous mouthwash, was patented by an American chemist in 1879 and named in his honor. Then there's the Scottish doctor and microbiologist Alexander Fleming, who in 1928 made a landmark discovery in the annals of medical history — a mold called penicillin. Penicillin, used judiciously, killed the germs of some of mankind's worst scourges. His discovery ushered in the age of antibiotics. But the modest Fleming was not the kind to cash in on his achievement. Rather, he was the kind to repeatedly warn, with great foresight, that the indiscriminate use of antibiotics — indiscriminate as in "there's a lot of money to be made from these wonder drugs" — would lead to the widespread propagation of bacteria resistant to them, which sure enough has become a twenty-first century crisis.

Although I've made only a quick study of the lives of these three, not once did I encounter a bad word about them. Nothing suggests they were out for fame or fortune. Nightingale, Lister and Fleming were saints. They saved countless lives, and they paved the way for further medical progress.

Now we're constantly reminded how wonderful vaccines are, how they eliminated so many terrible diseases, and how grateful all of us should be for this. So it would seem to follow that, at the very least, two or three vaccine developers should stand tall as undisputed medical heroes, as individuals like the three I've cited who deserve our veneration, and who incidentally lived and worked in the same time frame as those who worked on vaccines. But this is not the case. Beginning with Edward Jenner, nearly every individual who has produced and tested vaccines has been a snake in the grass.

Vaccinology, by its very nature, draws people of questionable character at best. Look at it this way. It's inconceivable that any doctor who routinely performs abortions will go down in history as a great man or woman, because it's widely recognized that there's something dark about a person who makes a living by killing healthy human fetuses. Well, on the scale of amorality, the coerced vaccination of children is not far behind. The essence of vaccinology is experimentation with children, that is, gambling with their lives, and if most pediatricians don't realize this, those who make and test vaccines certainly do. There is indeed something "special" about vaccinologists as there is about abortionists.

When you peel away the myths and begin to examine the personalities and careers of these vaccine trailblazers, not to mention their most vocal supporters, it's amazing how easily they reveal themselves. There's nothing mysterious about them. They're just plain despicable people. Some are worse than others, of course. Salk and Sabin, the polio vaccine boys, were bad enough, but further on we'll meet a few who were more than a match for Jim Jones or Pol Pot. By comparison, Jenner and Pasteur were pikers. The number of children they experimented on with their vaccine concoctions was minuscule compared to what would follow in the twentieth century, but it should not be forgotten that Jenner, the father of vaccination, was a swindler who set the pattern for everything that came later. It also should not be forgotten that from the outset, many people were shocked and outraged by Jenner's vaccine road shows, and his steady stream of lies, evasions and alibis, right up to his death, only brought him more opprobrium.

❖

Looking back on twenty-five years of sporadic research, which gave me the opportunity to appraise the character of so many prominent figures in the vaccine movement, only two struck me as ethical. One was George Dick. The other was John Enders, one of the leading polio vaccine researchers, whom I'd read about, and whose photograph appeared, in David Oshinsky's *Polio: An American Story*. Why did he stand out? What was exceptional about him? Several things. Enders was one of the very few in that mad scramble who was above all the feuding, who seemed to be liked and respected by all his colleagues. He was also the only one of this group to be awarded the Nobel Prize, for his success in growing the polio virus in a test tube culture of animal tissue which provided a safe and ample supply for use in vaccines — and he accepted this prestigious award only on the condition that his two lab assistants share the honor with him. Not one of his peers seemed capable of such a noble gesture. Furthermore — and this might sound a bit silly — he *looked* dignified and intelligent; there was something about him that reminded me of the glowing image the word "scientist" held for me when I was young.

Well, I was in for a rude awakening. Upon further reading I learned that Enders had become disenchanted with the whole polio vaccine scene and went on to tinker with a measles vaccine which was, in fact, used in the U.S. for five years before being junked — in effect, shuffling from one failure to another. Not a sign of a brilliant mind. More seriously, prior to licensure in 1963 he tested his measles vaccine on thousands of children in Nigeria and on institutionalized children in New York City and Waltham, Massachusetts. He used them like laboratory mice. He willfully put them at risk, knowing that if they suffered or died there would be little or no recrimination. There's no indication that he tested the vaccine on his own grandchildren. As I see it, anyone who does things like this is not so different from those involved in child trafficking. Maybe not quite as depraved, but in the same league. John Enders was disconnected from normal morality and emotions like the rest of them.

One of the many ploys of the vaccinators is telling us that mass vaccination works for the greater good of society, that sacrificing a few children ensures that many more who would have died for lack of vaccines will live. Some have stated this openly. If the evidence supported them, I would go along with this, though very reluctantly and with a troubled conscience. There are times in life when tough choices have to be made. But this is not one of them. The greater good of society? What societal good has come from all the vaccine-wrecked kids and young adults living among us today?

The entire history of vaccination is smeared with failure, delusion, deceit and misconduct. Some good people in key positions have, over the years, stepped out of the shadows to tell the world what's going on. A few are named in this book. As for the rest who are drawn to the business of vaccines, every last one of them is drenched in corruption. There's not a single one who stands out as a medical hero, or anything close to it. And that makes perfect sense because, being the criminal enterprise that it is, the fake science of vaccinology attracts the worst kind of people.

The trouble with people

THE MAIN dilemma faced by parents who, for the love of their children, want to cut back on mandated vaccines or opt out altogether, is the issue of school attendance. Once a year I used to get a sheet in my mailbox spelling out the kindergarten and pre-K registration rules for the school district in Amityville, a town on Long Island where I lived for a long time. Here's what we read for the school year 2019-2020 pertaining to vaccines:

All registrants must provide written evidence, signed and stamped for your PHYSICIAN or HEALTH CLINIC that your child has been immunized against seven (7) diseases:

- *4 doses of DTaP for PK students, 4-5 doses of DTaP (last dose after 4 years old and 6 months after previous dose) for Kindergarten students.*
- *3 doses of Polio for PK students, 3-4 doses of Polio (last dose after 4 years old and 6 months after previous dose) for Kindergarten students.*
- *1 dose of MMR for PK students, 2 doses of MMR for Kindergarten students.*
- *3 doses of Hepatitis B for PK students, 3 doses of Hepatitis B for Kindergarten students (last dose after 6 months old).*
- *1 dose of Varicella for PK students and 2 doses of Varicella for Kindergarten students.*
- *1 to 4 doses of Haemophilus Influenza (HIB) for PK students only.*
- *1 to 4 doses of Prevnar 13 (PVC) for PK students only.*

I've tried to maintain an intellectual level in writing this book, and I think I've succeeded. But there are times when I read something that brings out the truck driver in me, and this is one of them. When I read this, I put myself in the place of parents sending a young child to school, and this was my reaction: ARE YOU OUT OF YOUR FREAKIN' MIND? THERE'S NO WAY I'M GONNA COMPLY WITH THIS CRIMINALLY INSANE HORSESHIT! I'm sure there are at least three or four million Americans who feel the same way, though I've only met a few in person. I attribute it to a heightened sensitivity. A keen sixth sense. A gut feeling that Paul Offit and his ilk are evil, and that the corruption among those responsible for vaccination policy in this country screams for a settling of accounts. But of course the vast majority don't feel this strongly. They might be a bit leery about all the vaccines their kids have to get to go to school, but they still go along with it. This is evidenced by the fact that school registration has not fallen off much as a result of stringent requirements like the above, which I believe are nearly identical in most states, though I should note that New York is one of only five, at this writing, that does not allow religious exemptions.

No one really knows how many American children are in compliance with these requirements. Statistics, which are always suspect, vary. I've reviewed several sources, and in averaging them out, my best approximation is as follows: 88% are in full compliance; 7% have an incomplete or delayed vaccination status; 2% are mostly or totally unvaccinated, their parents claiming a religious or philosophical exemption (as do some parents of the previous group); 3% are home-schooled, and virtually all of these are completely unvaccinated, or nearly so.

Even if these numbers are off, there's no doubt that the great majority of Americans would vigorously disagree with my contention that vaccines are an absolute fraud and all of them should be scrapped. But what do they base their beliefs on? I've answered that question in previous chapters, but maybe it's worth taking a closer look. Here's an interesting observation I've made of the roughly fifty people to whom I've expressed my anti-vaccine sentiments — family, extended family, friends, co-workers and other acquaintances, past and present. *Only one*, many years ago — my elderly landlady's daughter who, as I recall, was expecting her first grandchild — felt that it was important enough to look into, and followed through on my recommendation to read Neil Miller's *Vaccines: Are They Really Safe and Effective?* The book made an impression. "It makes you stop and think," she told me when I saw her next, and she ordered it for the West Islip public library, where she worked. But what about all the others? I'm talking about basically good people of average or superior intelligence who, I'm sure, are a cross section of at least a hundred million American adults. As far as I'm concerned, anyone among them who refuses to examine the evidence that torpedoes the vaccine dogma, is no more entitled to an opinion on the subject, and no more qualified to criticize this book of mine, than he is to play on the U.S. Olympic hockey team or sit behind the controls of a 747 jumbo jet at 30,000 feet.

Let me tell you about my immediate family — my parents and my three younger siblings. There are plenty of smart genes in the Massaro clan. My sisters ranked second and fourth, academically, in their high school graduating classes of more than 200, and my brother, who became a high school math teacher, was up there too, though not that close to the top. Although my mother was of average intelligence, my father was brilliant in mathematics and also at fixing things. He was an electrical engineer employed by Grumman Aerospace who in 1969 worked on the circuitry of the Apollo 11 lunar module, which took the first men to the moon. He was also savvy in the stock market and made some prudent decisions that helped all his children, especially me, after he left us at age 90. But no one in my family, except me, had a real thirst for knowledge. This led to opinions on politics and almost everything else that made me the black sheep of the family, a distinction I've always enjoyed. Over the years, I've learned not to intrude on the comfort zones of others, that the best way to keep the peace is to keep my mouth shut most of the time, because I'm not going to change anyone's mind anyway. We've always been mature enough to get along, with an occasional flare-up to keep the pot boiling. But the thing is, I never knew my parents nor my siblings to ever read a worthwhile book on a serious subject, and certainly never a "banned" one. It hurts me to say it, but my dear father had no mind of his own. He never felt strongly about anything, and was incapable of believing anything he hadn't seen on television or read in that wretched daily Long Island newspaper, *Newsday*. As much as I loved him, his mental sluggishness drove me crazy and made me lose my temper with him a few times. Getting his flu shot became an annual ritual that he followed for at least fifteen years, well into his eighties, and I reached the point where I gave up on trying to talk him out of it, as I had given up on discussing many other things. Fortunately it never did him any harm. Nor did the Lipitor he took for years, having bought into the cholesterol scam like so many others. When she was 68, my mother ended up in the hospital with pneumonia five weeks after getting a flu shot, and she stopped getting it after that, but not my dad. Of course I can't prove that there was a connection, just as I can't prove that Lipitor gave her the pancreatic cancer that ended her life at age 82. I was left with only a strong suspicion, but I never ran any of this by my father, because his doctor had the first and last word always, and how many doctors talk their patients out of statins and flu shots?

I wrote earlier about the resistance I encountered with the older of my two sisters, who does not have children, regarding measles and the measles vaccine. My brother and younger sister each have three children. I know they've all been vaccinated, but I don't know how thoroughly, and I don't want to know, and they don't want me to know. It's a sensitive issue. They know that I'm totally opposed to vaccines, and that I have read a great deal on the subject, but they've never asked to borrow one of my books, and so far as I know, never looked into it. It suffices that only those who have bats in their belfry, as my father used to pleasantly say about me, don't get their kids vaccinated, though it's possible that my sister, who is more health conscious than my brother — though also more liberal and more hostile to my views — spaced apart her daughters' shots and held off on a few of them. When I urged her to read the Miller book many years ago, she said, "Is that put out by one of your wacko publishers?" I will say, though, that a few years ago she emailed me asking my opinion about certain travel shots, as my niece enrolled in a college study program that would take her to three countries, one of which, Brazil, required a yellow fever vaccination. I advised her to avoid all vaccines. If it were me, I said, I'd just cut out Brazil, and if you think I'm being weird, I added, check out vaers.hhs.gov. I directed her to the pages documenting severe reactions, 25 of which I've reproduced in this book. Apparently, this got her to thinking, *finally*, because she sent me a terse reply: "Thanks for the information." But after that she never brought it up and I don't know if she ever looked at the VAERS site. My niece went ahead and got the yellow fever shot anyway, and my other niece got a typhoid fever shot for travel to Tonga, which is not required — a totally unnecessary risk. Neither had an adverse reaction, but obviously my advice didn't sink in. Then, just a few weeks ago — this is February 2020 — I learned from my sister, when I gently pressed her, that she and her husband, who also is very bright, got hepatitis A vaccinations prior to a trip to southeast Asia. Most curiously, my sister adopted a mixed breed puppy from a shelter, and after her pet became very ill after getting a rabies shot, she swore off all additional vaccinations and never took her to the vet again. I was surprised she even told me about that. Moxy is eleven and still going strong. I will never understand people, by which I mean people who nominally are quite intelligent. When I see someone as smart as my sister, ignoring my advice to read of all the severe reactions listed by VAERS, or reading them and still getting vaccinated, I start to wonder if I've wasted all my time and effort writing this book. Is it that something bad or tragic has to happen before people wake up, before dogma loses its grip on even gifted minds? That would appear to be the case. It seems to me that people generally don't become cynical or embittered towards the vaccine propagandists until they see the damage done with their own eyes, until tragedy hits home — and even then some are uncertain or apathetic about it.

❖

What's true of my family and social circle is undoubtedly true of the responsible and reasonably intelligent segment of our society. As I've said more than once in this book, the need to conform and to avoid unconventional ideas runs much deeper and stronger than the ability to think straight. Taking it to the next level, this phenomenon is even more pronounced, I've noticed, in people with some medical qualifications — people who seem more sure of themselves while being no less ignorant, although there are some dissenters. I think here, first of all, of the usually polite but ditzy young women who work as assistants in doctor offices, or at the local CVS or supermarket pharmacy, which frequently advertise their vaccination services — and are ready to cheerfully give you a flu or shingles shot, no appointment needed. This attitude is entrenched throughout the kingdom of modern medicine.

Sometimes I'm not only perplexed by all this, I'm speechless. By way of example, I've kept in touch with a British couple since meeting them in 1982 while traveling across Africa. When you think like me, you try to hold on to the few friends you have, even if they're separated by an ocean and you've seen them only once in the last 38 years. Colin and Trish are really fine people, grandparents, and bold adventurers who have been to 132 (!) countries. We have a tradition of exchanging Christmas cards and long, handwritten letters — Trish always writes them — filling each other in on our travels and what life has dealt us in the year gone by. Last year, I mentioned that I was writing this book, that I was adamantly anti-vaccine, and that I hadn't gotten a travel shot since 1986. I asked Trish, who is exactly my age, how she felt about this. Here is the relevant excerpt from her letter of December 14, 2018:

Having a healthcare background, I'm all for vaccinations and make sure we're up to date on ev-

erything before we travel. A couple of years back I was scratched by a monkey in Thailand, so was very grateful I'd had the pre-exposure rabies shots, or that could have been fatal. I made sure all the kids had their vaccinations when they were little — there are figures about the percentage uptake of vaccinations protecting those who don't have them but I wasn't prepared to take any risks. When I was little kids used to die from measles, mumps etc. and I wouldn't want to see a return to those days. None of us have had any adverse reactions (other than some localised tenderness) and I don't know anyone who has. There were theories some years ago about a link between the measles/mumps/rubella vaccine (MMR) and autism, which was subsequently discredited but not before some children had experienced damaging consequences from measles etc. So, as you can see, I beg to differ from you on this one!

Yikes! Not only in favor of vaccines, not only the usual worn out clichés, but a preventative shot against rabies?? Which could have prevented death by monkeyscratch??? I'm at a loss for words here, other than to say that Trish is a sweet and seemingly intelligent lady, and I'd feel awful if she ever reads this.

There wasn't a trace of arrogance in Trish's statement — that's just not who she is — but I've come across many displays of arrogant certainty, in varying degrees, about the benefits of vaccines, within the medical community and outside of it. People who think they know so much yet know nothing. People who sneer, who ridicule, who insult, who lecture us about what so many studies have proven, who compare us anti-vaxxers to flat-earthers. I've read it all, much of it in comment sections on the Internet. These know-it-alls have probably been with us since men learned to talk. Then there are those, usually bureaucrats, for whom there isn't even a controversy, who simply see anti-vax parents as lunatics or common criminals, and are all for locking up such parents and putting their children in foster homes where they can be forcibly vaccinated. To give you an idea of what's out there, the following was tweeted on December 6, 2019 by a yakker and self-described longtime Republican political strategist named Rick Wilson: "Anti-vaxxers are a scourge and a strong argument for re-education camps, the immediate seizure of their property, and putting their children into protective custody." Wilson is just a piece of flotsam, a Deep State barnacle. But there are more than a few judges, health officials, and goons with guns and badges who stand ready to carry out such actions, and *have* taken children from their parents often enough. Not all of them are consciously malevolent. Some are too stupid to be evil.

There's one more category into which those who are pro-vaccine fall. What separates them from the different kinds of people I've discussed above is that their conditioning is irrelevant. They are the genetically messed up. I turn once more to the excellent and suppressed book *The Eleventh Hour* by John Tyndall, from which I earlier extracted a quote about the media. In the chapter headed "The Cancer of Liberalism," Tyndall touches on an intriguing fact. The liberal, he asserted — meaning the truly committed liberal as distinct from the ordinary individual who has thoughtlessly adopted some liberal attitudes from watching too much television — is a biological type. "Assemble any group of liberals together on one side of any room," he wrote, "and then assemble on the other side a group of equal number and sexual composition but of outlook fundamentally opposed to liberalism. Most of us then entering the room would immediately be able to tell which group was which after the most cursory glance. Liberals, generally speaking, constitute the most unimpressive specimens of the white race...." Tyndall believed that an inborn weakness, as revealed in physique, posture and facial features, is inseparable from the liberal ideology.

This fascinated me. I reasoned that you can make the same distinction by glancing at any groups whose tastes, interests or beliefs are far apart. For example, take two groups of twenty men dressed in the same casual clothes. One group just went deer hunting, the other just went to the ballet. Could you differentiate them? Of course you could! The hunters will unmistakably appear stronger and tougher. Another example might be those who voted for Hillary Clinton as opposed to those who voted for Donald Trump, pretty much a straightforward contest between liberal and anti-liberal outlook, regardless of the fact that very few in either group knew about the squalid background of their preferred candidate. Just looking at the televised sea of faces, meaning white faces, at their campaign rallies, one could instantly distinguish them. This is very significant. It means that people have only limited control, if any, over how they feel and what they believe; in great part, it's determined in the womb. Of course, everyone is shaped to some degree by environment and fortuity in life, such as meeting someone who exerts a profound and lasting influence. But heredity, over which we have no control, is always the dominant factor. That is what I believe not only from reading but observation.

As a definitive example, take homosexuality. I'm one of those who believes that homosexuals are born, not made, even though the degeneracy which is all around us today might persuade some with weak instincts to go gay or bisexual, whereas fifty years ago that wouldn't have happened. It appears that homosexuals have existed in all ages and in every race, and by definition they are all genetically defective. Mother Nature can be a cruel mistress. To repeat our visual analysis, were you to compare, say, twenty homos of either sex to twenty heterosexuals, you'd instantly know which was which by their faces alone. Queer men tend to have weak or delicate features while most lesbians look sexless or downright mannish. All this is common knowledge and the butt of discreet jokes. Of course, there are plenty of exceptions in both men and women, but if the sample is large enough, there's no mistaking homo from hetero.

For the life of me, I can't imagine what it's like for a man to feel erotic pleasure while engaging in sodomy with another man — and most women must feel the same way about their own gender. And yet, homosexuals *do* feel this way because of their genetic make-up, of which their physical appearance is a manifestation. For them, their sexual behavior and the pleasure it gives them is "right" and the thought of having sex with a member of the opposite sex is "wrong" because it leaves them cold. No amount of logic, rational discourse or counseling will change them. They are what they are.

What I'm leading up to here is the fundamental genetic difference between those who are militantly pro-vaccine and those who are militantly anti-vaccine, or at least have expressed a deep concern about vaccine risks. It would be an interesting experiment to compare the faces of twenty or thirty proponents at each end of the spectrum. I have no doubt that, overall, the faces of the antis would clearly be more attractive and convey more kindness than the faces of the pros. In any event, there is something in the DNA of these people who have risen to the top levels of the public health bureaucracies and medical associations, about vaccine profiteers and pushers like Paul Offit and his admirers, about the columnists and commentators who incessantly denounce and ridicule anti-vaxxers, that makes them enemies of all good people. It's just who they are. Their attitudes are anchored in their protoplasm. They can't help themselves. They are destructive people who are driven to impose their destructiveness on others. They cannot and will not change. It's no more possible to convince them that their beliefs are wrong than it is to convince a homosexual that the kind of sex he or she indulges in is wrong. Naturally, the pro-vaxxers will say the same bad things about *our* DNA. So what we're dealing with here is two different life forms, two diametrically opposed natures for whom it is impossible to find any common ground or friendly divergence of opinion. Discussion or debate is a complete waste of time. So be it.

To clarify

IN THE introduction, I wrote that there is no proof whatsoever that anyone has ever benefited from a vaccination. That statement might have struck you as presumptuous, maybe even preposterous. I stand by it one hundred percent. There's no real science to refute my claim, only piles of crooked studies, the existence of which is one more line of evidence that the whole thing is a gigantic lie. But it must be admitted that, in the majority of cases, after a vaccine is injected, no one really knows what happens. And incidentally, I also must admit that in most cases, and especially when just a single dose is given, months or years apart from any other vaccination, it appears that no harm is done. However, I can't imagine that any youngster who undergoes the now standard regimen of fifty to sixty or more vaccine doses will go through life unscathed by this massive assault; my guess is that we're going to see shocking rates of cancer down the road — but again, I just don't know. Conversely, can there possibly be a benefit? Well, I suppose anything is possible. I consider it a benefit not if a vaccine prevents a disease that's nothing more than a temporary nuisance, but if it prevents a disease that would have caused permanent disability or death. And there's the catch — "would have." Preventative measures, by their nature, are impossible to calculate. How can you say that buying a new tire to replace a dangerously bald one prevented a blowout in heavy traffic that would have caused an accident resulting in serious injury or death? You can't.

Perhaps a better analogy can be made with cruciferous vegetables — broccoli, cauliflower, kale, Brussels sprouts, cabbage and others — all of which contain an organic compound called glucosinolate, said to have strong anti-carcinogenic properties. Is this true? I believe it is, and since I don't want to get cancer, I try to eat two or three servings of these vegetables a week, even though I don't care for them. But I really have no idea what benefit, if any, I get by eating them. Maybe what I've read about glucosinolate is a lot of baloney. And when you come right down to it, there's no way I can prove that I would've gotten cancer had I not eaten these vegetables on a regular basis for years. But there the similarity with vaccines ends. There's no multi-billion industry to buy influence with the media and politicians, no broccoli lobby that I've ever heard of. I don't know of anyone who has disputed the anti-cancer properties of glucosinolate. And even if sound science were to show that this substance has no benefits, I can't imagine anyone would ever claim that eating cruciferous vegetables can be hazardous to your health. To underscore, then, a belief that vaccines are effective is nothing more than that — a belief. People believe it simply because it's been repeated so often and so many other people believe it. There is simply no empirical evidence to support it.

❖

I reached the point years ago that I realized it's a waste of time to have any kind of dialogue or debate with pro-vaxxers. Long ago I stopped paying attention to the things they say. Nevertheless, I'd like to put paid to three assertions they often employ that, on the surface, may seem logical. The first is the concept of herd immunity, which holds that if a high percentage of the population is vaccinated against an infectious disease, the small unvaccinated percentage will be protected. This is pure deceit. Several infectious diseases in advanced countries, which still crop up in the Third World, dwindled to the vanishing point without any vaccination program. Throughout Europe, polio made a quiet exit in the mid-twentieth century even after health officials decided to forego the dangerous Salk vaccine. Besides, vaccines do not work, that is, the disease rates among the vaccinated and unvaccinated are about the same. Or are they? Middle-aged teachers and school nurses might know, even though the latter are likely to be strongly biased in favor of vaccines. They're the ones who see that kids are absent from school because they're sick at home. What can they tell us? Possibly that there's less "old style" measles or chickenpox around these days, but many more chronic and acute conditions that have brought about a higher absentee rate. Why not ask one?

Second is the line that the welfare of society takes precedence over individual freedom of choice — that no person should have the freedom of choice to remain unvaccinated when that choice threatens public health. This one goes back to the late nineteenth century; I've come across it many times. It answers itself on its face, because if vaccines worked, all people would need to do is get vaccinated to protect themselves from the unvaccinated. And what, after all, is "public health" but the health of all individuals that constitute a society? Well, it seems to me that when I was young, and there were far fewer vaccines in existence, there were also a hell of a lot fewer individuals with serious health issues than there are today.

Lastly, we hear that unvaccinated children endanger classmates who are "immuno-compromised." I never had an immuno-compromised classmate (though I did know two boys who died of leukemia), so if this is more common today, I would ask if vaccinations caused their condition in the first place. In fact, the Wikipedia page on this subject lists some strange diseases that are also spelled out as possible risks on vaccine product inserts, but no one who edits Wikipedia would dare to make that connection. Also, we know that during the Salk vaccine push of the 1950s, a number of parents came down with paralytic polio from changing the diapers of their recently vaccinated infants. Might not an immuno-compromised child, then, be infected with measles or mumps or chickenpox or whooping cough when a recently vaccinated classmate nearby coughs or sneezes? The vaccine zealots maintain that the live viruses in these vaccines can't shed and infect others, but I don't believe them. In addition, what these people are unwittingly admitting is that vaccines are too dangerous to give to those with an impaired immune system, who have no choice but to remain "unprotected" in their eyes. That being the case, it follows that they must also be risky to other children whose immune system is not that resilient, even if they are relatively healthy.

But all we do is go round and round with these arguments. The pro-vaxxer reading this mocks my logic and I mock his. Different brains are wired differently. Here we are, two thousand years after the fact, and there's still no broad agreement on whether Jesus was the incarnation of God. You say he was, I say he wasn't, I'm never going to convince you and you're never going to convince me, so what's the use of discussing it? I'll try to give an answer of sorts at the end of this book.

Vaccines cause autism. Repeat: Vaccines cause autism.

If they desire a thing, they declare it is true. If they desire it not, though that were death itself, they cry aloud, 'It has never been.' — Rudyard Kipling, on the intellectuals of his day (circa 1895)

IN MY JOB as a home heating oil truck driver on Long Island, I often make deliveries to houses where the oil tank is underground. The fill line of most of these tanks is at ground level, or just above it, and often sits in the front lawn. Most longtime oil drivers, myself included, have had a few accidental overfills where oil spills out onto the grass. When this happens, the grass takes on a waxy shine, and within a week or two turns an unsightly whitish yellow and becomes brittle. Needless to say, a patch of dead grass in the middle of a lush green lawn does not make for happy customers. It's also embarrassing to drivers who take pride in their work. If it's just a small area of grass that's been killed, the company usually sends an employee to dig up the contaminated soil, replace it, and plant new grass seed, or pays a landscaper to do it. Most people understand that accidents happen, that spilling a quart or two of oil on their lawn is not the end of the world, and if their company handles it properly, it's forgiven and forgotten.

Now can you imagine being a homeowner with a buried tank in the lawn, discovering that your oil driver had a little mishap, then calling up your company and being told, "Yes, our driver did overfill your tank by mistake but that has nothing to do with the fact that your grass is dead." What would you think? *Yet it is precisely this kind of denial, this brazen assault on the most elementary common sense, that has been employed a million times in this country when the behavior of a normally developing young child drastically changes within weeks, days, sometimes hours or even minutes of being vaccinated.*

Cause and effect is so obvious and plays out so often in daily life that we never give it a thought. Strong winds cause branches to fall from trees, ice on the road causes vehicles to skid, and so on *ad infinitum*. In matters of health, cause and effect also works in countless ways that are obvious, even when there's a long latency period. If you drink a pint of whiskey every day for a year and develop liver cirrhosis, the reason is immediately understood. Expectant mothers who took the drug thalidomide to ease their pregnancy in the 1950s gave birth to babies with profound deformities, as did Vietnamese women and wives of American soldiers who were exposed to the defoliant Agent Orange the following decade; the connection has never been questioned. Nor have I heard anyone deny that it was the toxic dust inhaled by those who worked at the site of the collapsed twin towers in New York after the September 11, 2001 terror attacks that provoked and continues to provoke thousands of chronic illnesses and fatal cancers. One can go on and on with examples of harmful substances wreaking havoc with the human body that everyone agrees upon. Yet the clearest illustration of cause and effect — the plain fact that vaccines cause autism — is met with the most emphatic denial by a vast multitude of health professionals and their friends in the print and electronic media.

❖

My daughter's autism was the main reason I wrote this book. My original intent was to make this discussion of the link between vaccination and autism a very long one. Then I realized that so many millions and billions of words have been written on both sides of this issue that all I could do to argue my point would be to recapitulate what others have already published. That's not what this book is about.

To me, this controversy of the role vaccines play in triggering autism is irrelevant. Rather the issue at hand is one of psychology, specifically the nature of the mind and the mechanics of denial. It's also an issue of mass insanity. How do you explain a group of highly educated people, probably numbering in the hundreds of thousands in this country alone, who actually believe that if you inject three or four or five doses of a mixture of chemicals and animal particles into the bloodstream of a small, healthy child, and his or her condition radically worsens in a short period of time, and there is no other plausible explanation for this change — how do you explain their insistence that the vaccine had nothing to do with it? I can think of three things: First, income. As Upton Sinclair once said, "It is difficult to get a man to understand something when his salary depends on his not understanding it." Since vaccinations are the bread and butter of pediatricians, this explains why they don't understand it. Second, mental stability. For most people whose minds have settled into a comfortable mold, it's too painful to change the way they think. Third, fear. I think a fair percentage of these people know very well that vaccines carry severe risks, including autism, and they harbor a secret fear that if enough people catch on to what's been happening all these years, they're going to be hunted down and killed on the spot.

There's a fourth factor as well, and it's a theme that surfaces again and again in this book. It's what I call the Jonestown syndrome, or what William Pierce called "the lemming factor" in human behavior — the compulsion to think and act like everyone else is thinking and acting, even when it makes absolutely no sense. And as always, we can thank our friends in the mass media for messing with our brains and leading us down the primrose path to destruction. Let me tell you about a personal experience that took place years ago. I decided it was time to draw up a will and establish a special needs trust for my children. My sister referred me to a friend of hers who is a lawyer in that field. I made an appointment with her over the phone and explained my situation, so she was familiar with my children's problems when we met at her office. As I expected of someone specializing in that modest area of law, she was courteous and intelligent, and also the mother of a teenage girl. Making small talk, and unable to resist getting in a jab at vaccines, I mentioned that the cause of my daughter's autism was her heavy brain bleeding at birth, but that in the overwhelming majority of cases, vaccines are to blame. "You really think so?" she said. "There's no doubt in my mind," I replied. I left it at that. From the inflection of her voice, I could tell she didn't believe me. I'm pretty sure she was thinking, "Well, all the experts agree that vaccines don't cause autism. Studies have proven it over and over. It's in the papers and on TV all the time. Are you one of those conspiracy theorists?" And if I didn't read her mind correctly, there are plenty of articulate people who have come right out and spoken almost these exact words.

I return to my statement in this work's opening pages, that I wish to accentuate what I have personally seen or experienced, and right behind that, what no one contests as being true. As I've already stated, the great majority of us who were born in the 1950s received no more than five or six vaccine doses growing up. I never heard the word *autism* until I was in my thirties. There has always been a striking correlation between vaccination rates and autism rates. The classic signs of autism — some combination of lack of speech, head banging, violent tantrums, withdrawal from parents and other people, self-stimulation, hypersensitivity to certain sounds, laughing or crying for no reason, and compulsive behavior patterns — were a phenomenon that first appeared in the early 1940s, with the rapid spread of the pertussis vaccine — for a long time considered the most dangerous vaccine by far — and was described in a famous paper by prominent child psychiatrist Leo Kanner, who popularized the word *autism*, as a "condition [that] differs markedly and uniquely from anything reported so far." For more than forty years, however, autism remained extremely rare. In the 1960s the rate was about 1 in 10,000. Today, depending on which source you believe, it's about 1 in 50.

No such estimate exists of parents of autistic children who believe that vaccination was the cause, but the number is surely great. In my reading of books, articles, websites and blogs I've come across several hundred, many providing their full names. But, it must be said, there are also parents who deny a connection, and others who are unsure. There is no other plausible explanation, however — just a bunch of silly ones — for the rate of autism that began to dramatically rise in the 1980s, then skyrocketed in the 1990s in tandem with a stack of new childhood vaccines, most or all of them containing the preservative thimerosal, which is about fifty percent mercury, one of the most toxic elements on the planet.

❖

Numerous studies have been conducted and have established beyond any doubt, we are told, that no causative link exists between vaccines and autism. Many of these studies focused on thimerosal, at one time an ingredient in most childhood vaccines, the use of which ignited a firestorm of controversy in the late 1990s, which spilled over into the new century. The reader will have surmised by now that I consider such studies not worth the paper they're written on. I have never examined any of them. Nor, for that matter, have I examined any studies that found that mercury in vaccines *has* played a major role in the autism epidemic. The most critical study implicating vaccines in autism — though it did not focus on mercury — was the one published in 1998 in *The Lancet* , the world's oldest and most respected mainstream medical journal. The study's leading author, British physician Andrew Wakefield, theorized that the measles virus in the measles-mumps-rubella (MMR) vaccine inflamed the small intestine, allowing harmful proteins to pass into the bloodstream and flow to the brain, causing autism. Wakefield's article, and the press conference that accompanied it, was a sensation and initiated a nationwide soul-searching on the issue of child vaccination in a country where, unlike this one, there is freedom of choice but, like everywhere else, the Dogma still holds sway. It should be added that Wakefield did not come to any hard and fast conclusions, nor was he even opposed to vaccination; he simply said that the MMR shot should be discontinued and the vaccine for each of the three diseases be given in separate doses. (Being 100% percent anti-vaccine, I totally disagree with that, though he may have hardened his position in recent years.)

Six years later, a hatchet job on Wakefield appeared in the London *Sunday Times*, written by an investigative reporter named Brian Deer. Numerous allegations of unethical conduct and borderline criminality were leveled at Wakefield, which touched off a cascade of events that included most of his co-authors and the editorial board of *The Lancet* repudiating him. His medical license was revoked as well. Overnight Wakefield went from international hero to international villain. It was just one more example in this world of lies we live in of the global media empire pre-empting the truth. Ever since then, the slander has followed him and all we hear about is the disgraced Wakefield whose study has been debunked, and this is cited as conclusive proof that vaccines don't cause autism. Like waving a magic wand. As if that's the final word. Wakefield ended up moving to Texas and starting a new life, educating the public with the help of a dynamic support team, through his excellent vaxxed.com web site.

Years ago I followed the Wakefield saga. I believe he is a man of integrity who was brought down by a pack of jackals who have never stopped yapping at his heels. I could be wrong. To read both sides of the story is to drown in a sea of conflicting information, which has been the case with everything about vaccines since day one. I have never read his article in *The Lancet*, nor do I intend to. It doesn't interest me. And it doesn't matter to me if Andrew Wakefield is a crafty and persuasive liar, if his study is a total fraud, and if he's an arsonist and bank robber to boot. Much more compelling to me than any study is his team's bus tours around the 48 lower states and the more than 7000 tragedies submitted by parents of vaccine-damaged children. Many were interviewed in their homes, the interviews posted on vaxxed.com. I've seen about twenty and most of these showed parents, often choking back tears, telling their stories of how their healthy children fell into the abyss of autism shortly after getting one or more vaccinations. You and I will never meet these people, but they're your next door neighbors, the nice family down the block with a quiet tragedy under their roof. To those who deny the connection, I ask: "Do you believe these people are paid actors? What the hell do you think is happening in the world around you?"

❖

Supposedly, thimerosal has now been removed from all routine childhood vaccines, except the influenza vaccine, which contains a trace amount. Trace amounts are still present in some of the vaccines given to children, though not routinely — that is, in vaccines that have been broken down into single or double doses, instead of "routine" three-in-one shots. No corresponding reduction in the autism rate has been seen. Therefore, we are told that the uproar over thimerosal, which dragged on for more than a decade, has proven to be a false lead. On this score I must say that after reading David Kirby's landmark book *Evidence of Harm*, I was thoroughly convinced of the corrosive effect of mercury on the brain, and that thimerosal *had* to be the culprit. What has transpired in recent years has made me take a step back and reconsider, but in light of the fact that so many vaccine pushers are liars I have by no means become unconvinced. Depending on what numbers you read, the autism rate has leveled off or is still increasing, but everyone seems to agree that it has not *decreased*, so it seems to me that one of three things must be true: 1) The pharmaceutical companies have lied about removing thimerosal from vaccines; 2) They did remove thimerosal, but stealthily and criminally continue to use a different compound that contains mercury; 3) The mercury is gone, but another chemical or biological agent added to many vaccines — possibly aluminum or formaldehyde — is still causing brain damage and autism.

I do think that in rare cases other causes can exist. My daughter was never vaccinated but her underdeveloped brain bled heavily at birth, and she also went into shock. Undoubtedly one or both events account for her autism. Because of the risk of premature birth associated with *in vitro* pregnancies — a burgeoning industry in recent decades, thanks to vaccine-induced fertility problems — autism that originates with pre-term brain damage might be more common than is generally acknowledged. Severe accidental injury to a full-term infant's brain is another obvious possibility. I have read that ultrasound imaging — a recent fad that so many couples indulge in to learn their baby's sex far in advance of birth, or to get a souvenir photograph of life in the womb — might cause a disturbance in the fetal brain leading to autism. The massive use of pesticides has been discussed as a possible factor. And in very rare instances, I suppose that an abnormally large dose of mercury absorbed by the body — say, from a dental filling or a coal firing plant — can initiate autism. But it seems to me that minor insults like these — if indeed they play any role at all in the autism explosion — pale in comparison to the effect of vaccines, much as drinking a small glass of beer each day might contribute minutely to liver disease in an alcoholic who downs a daily pint of whiskey. The fact remains that if autism ever existed prior to the mid-twentieth century — and I have come across no evidence that it did — it must have been regarded as a freak of nature, not worth contemplating as a condition that might reappear. It was only in 1943, a year when autism was still extremely rare, that eleven cases came to the attention of Leo Kanner, who felt it significant enough for the medical world to take note.

❖

Denial is a common phenomenon. To refuse to believe a painful but verifiable fact is a defensive mechanism to maintain psychic balance. It's part of the human condition — the excess evolutionary baggage that came with our big brains. It seldom hurts others. But denial carried to the extreme is one dictionary definition of insanity, that is, "extreme folly or unreasonableness." Those who are extremely unreasonable and wield power are a danger to others. Denial is part of the corporate mentality, and not only with vaccines, as we saw in a previous chapter. These people pretend not to see the elephant in the living room, but do see other things. They see autism as a genetic disorder. "Autism genes" has become the most popular subterfuge of the vaccine pushers. This reveals their mental bankruptcy, but as usual it's a matter of "believe what you want to believe."

Genetics comes into play when we talk about how mercury or any other toxic element is excreted from the body. Mercury elimination can be monitored by complex methods. Some children get rid of it easily, while others do not. Why is that? The answer is stamped in our genes — the mystery of the organism. The same amount of mercury that will kill one baby will make another blind, another deaf, another autistic, while having no immediate visible effect on the majority. Keep adding more mercury to that vaccine vial, however, and eventually you're going to kill every child.

An analogy can be made to altitude sickness. Everyone knows that the higher you ascend from the earth's surface the less oxygen there is in the atmosphere, but no one knows why some people will die at 15,000 feet while others don't feel a thing. I've had some personal experience with this. Once, while helping to push a vehicle stuck on a bad road in the Peruvian Andes at over 14,000 feet, I suddenly felt like I was hallucinating, and slowly sat down to regain my composure, which I did in about thirty seconds — a terrifying episode I'll never forget. At what in 1985 was the world's highest railway station, at 15,681 feet, in the town of Galera, also in Peru, I got off the train to take a quick photo and then my consciousness began to flicker from the mere exertion of climbing three steps back into the rail car. On board there was an attendant with an oxygen bag for those who needed it, as I did. Problems that high up — the highest I've ever been — are fairly common. Below 14,000 feet the rarefied air has never bothered me, though other travelers have told me of suffering headaches and dizziness at eleven or twelve thousand feet. Some say they can even feel the difference when climbing stairs in Denver, merely one mile high.

Gradual acclimatization — analogous to spreading out vaccinations over time — is a good idea at high altitudes and has enabled most, not all, adventurous hikers to make it to the top of Mt. Kilimanjaro, at 19,341 feet Africa's highest mountain. North of this mark a lot of people will die without the use of supplemental oxygen, an absolute necessity for those brave souls who take on the most challenging peaks in the Andes or Himalayas. The legendary exception is Reinhold Messner who twice made a successful ascent, without oxygen, of Mt. Everest, our planet's highest point at 29,029 feet. Only a handful of others have accomplished this superhuman feat. I suppose there are also a few people in this world who can break a thermometer, sprinkle the mercury on their breakfast cereal, and feel nothing at all.

And so autism marches on. How, when, and if this plague will end, no one knows. Every prospective parent should be alarmed. The worst prognosis I have read is that half of American children will be autistic by 2025. Frankly, I think this number is absurd. Long before it reaches that point, a social collapse must take place and we will revert to the law of the jungle. But even if it's 5% rather than 50%, how can civilization as we know it be sustained with so many helpless people to take care of? Things are bad enough now, with resources running dry in many states and bureaucratic indifference towards "special needs" children rampant — something I dealt with personally in my children's school system. Meanwhile, nothing at all is being done at the official level to address this scourge. To be sure, there must be hundreds of grass-roots organizations across the country pledged to get to the bottom of autism. Without a doubt, some of them are fighting the good fight, and fighting it hard, while many just go through the motions, like sponsoring walkathons and other useless fundraising events, sometimes with ulterior motives at work. Autism Speaks, perhaps the largest and most influential organization, is one of these sell-outs. It's a shame that so many parents of autistic children and other well-meaning people are led astray and throw their money down rat holes like this, but corruption and futile efforts have always hindered the pursuit of every worthy goal.

Elsewhere, in addition to destroying young minds, the public education system is training a new generation in the art of putting on a good show while evading the essential. With autism so prevalent now, nearly every seven-year-old has heard of it. Recently, at Seaford High School and MacArthur High School, near where I live, students placed blue plastic cups in a chain link fence adjoining school property, spelling "Autism Awareness" in words fifteen feet across. Thousands of motorists and pedestrians saw them. Two weeks later the cups were gone. What was accomplished?

❖

If those who deny the vaccine-autism connection were really interested in the truth, if they were really willing to back up what they claim to believe, they would welcome a study comparing autism rates among the vaccinated and unvaccinated. The fact that the medicrats have never shown any interest in such a study, have always balked and hedged at the suggestion of conducting one — the usual sanctimonious ploy of these bastards is that it would be "unethical" to withhold life-saving vaccines from children — is just one more piece of evidence indicting them for a crime that cries out for justice and action. The fact that a two-part newspaper article titled "The Amish Anomaly" that ran on April 18-19, 2005 in the *Washington Times* — which reported that not one case of autism existed in unvaccinated Amish communities near Lancaster, Pennsylvania — disappeared from that newspaper's website with the words "This story is no longer available on the site," tells us all we need to know. Repeat: *it tells us all we need to know.*

If you go on the Internet and type "Are there any autistic children who are unvaccinated?" you'll see a column of affirmative responses. Most of these responses have been deduced from studies — those eternal studies. There are vague discourses on the subject. I came across a few with names of parents and children — whether real or fictional, I don't know — but in not one of these stories that I've read do the details add up. There are plenty of shills and disinformation agents in cyberspace, and I suspect that's the case here.

As I said earlier, an honest study should be welcomed by every open-minded person. An honest study that seeks to establish the presence or absence of a connection between vaccination and autism should be performed by researchers on both sides of the issue in an atmosphere of mutual respect, with an agreement that all data be shared between themselves and with the public. This would keep it honest and neutralize any bias, letting the chips of good science fall where they may.

But this will not happen. Bogus studies are the cornerstone of the claims made by vaccine pushers. They've been lying and prevaricating since

1796. The thought of exposing truthful data to the light of day terrifies them. Deep down they know what snakes they are. Deep down, the thought of facing the wrath of an awakened and heavily armed people scares the hell out of them, as well it should.

Examining the other side

I AM committed to the truth, and I claim to know the truth about vaccines. But what would my belief be worth if I weren't willing to test it, if I weren't willing to risk being proved wrong by those who advocate for scores of childhood vaccinations? Why should I be afraid to hear out the other side, to give them a fair chance to prove me wrong and throw my mind into turmoil?

In a lifelong quest to learn the truth in many fields regarded as controversial, I have read thousands of books. Books that say something, that speak the truth, are easy to distinguish; they are clearly written and a pleasure to read. I have always enjoyed reading good books, even in the few cases where I was shell-shocked by the author's ideas, one book I remember well in this regard being *African Genesis* by Robert Ardrey. It becomes obvious that the author knows what he's talking about, even though no one is infallible and he might be wrong on some points. Occasionally, a valuable book might be tough to read due to a dense literary style — I think here of Charles Darwin's *On the Origin of Species* — but even in these cases the brilliance of the work shines through. On the other hand, as an inveterate reader, it doesn't take me long to determine that a book — or a website or a pamphlet or any other form of the printed word — is trash. And there's plenty of trash out there.

Back in the day when there was some sanity in our judicial system, and perhaps even now in some places, judges and juries, in trying to determine the truth, would take into account not only the testimony of those on the stand, but also their appearance and behavior. Is the person neatly dressed or does he look like a bum? Is he straightforward or evasive in answering crucial questions? Does he look uncomfortable, does he squirm or get angry, or does he stay calm? The *how* as much as the *what* is critical in deciding where the truth lies. Of course there are some persuasive liars in this world — just look at how our 45th president got elected — but they are far outnumbered by those whose mannerisms reveal the truth, or strongly hint at it. And in the end, the truth has an uncanny way of winning out.

Those who write the truth get to the point; what they write is satisfying to read and easy to understand because they have genuine insight. The fakers or enemies of truth have a way of filling pages with piles of words that add up to nothing; the method of their writing is as frustrating to deal with as the content, except when they write flaming nonsense, in which case it can be entertaining. As in the courtroom, appearance, that is, the *way* an author expresses himself, goes a long way in discerning truth or the lack of it.

I have read many books that come down hard on vaccines. I wanted to read at least one book on the other side, *just one*, that would challenge the certainty of my position on childhood vaccinations, that would make me take a step back and say, "Hmm, maybe I'm wrong about all this." I wanted to find a book whose author would write an introduction that would read something like this:

Vaccines have been the subject of intense debate for more than two hundred years. Understandably, they have been blamed for many tragedies involving small children. But I will present irrefutable evidence that this belief is totally wrong, and that vaccines are an absolutely essential measure to safeguard every child's health. The information presented in these pages should dispel any doubts parents may have as to the safety and efficacy of vaccines, and put an end to this controversy once and for all.

Not surprisingly, I never found such a book. Now I did say at the outset that my mind is made up and that I'm not interested in any further debate. Nevertheless, I gave the floor to the other side many times, reading twelve pro-vaccine books in all. For the remainder of this chapter I will review three of them, including the two worst ones. The next chapter is basically an extension of this one, in which I'll skim through five more.

❖

My first reading, *Magic Shots* by Allan Chase, is distinct for two reasons: I read it a long time ago, never having bothered to re-read it, and it's the only book I didn't read in its entirety. The title, I must say, whetted my curiosity. After all, magic is the opposite of science.

This is a strange book. It is exceptionally scatter-brained. What I remember is the many dull accounts of obscure scientists who lived and worked in centuries past, the long discussions about the properties of certain viruses, and the politically correct interpretations of social problems. Chase also has a "thing" about human excrement. I couldn't get a handle on this book. It had no center of gravity. There was little that I disagreed with, but only because I couldn't understand what the author was getting at.

Although I felt obligated to read this 508-page tome right through to the end, when I reached the halfway point I decided that I'd had enough. So I turned to the last sentence on the last page and this is what I read:

The continued failure to make intelligent and maximum use of available vaccines for people of all ages is therefore not only immoral — and for one third of the nation, it is actually genocidal — but it is also criminally wasteful of the tax monies our government collects from the hard-working people who pay most of those taxes.

So, it seems, one hundred million Americans are being killed off by a government that has neglected its duty by failing to vaccinate them. Of course!

Now I think it's bad form not to disprove, by rational argument, the viewpoints of those we disagree with, and to call them names instead. Nevertheless we must draw a line at times, and when someone claims, or implies, that one hundred million Americans are being killed off by not being vaccinated, it's time to walk out of the room. I'm not going to stand there and listen to a cuckoo bird.

❖

It's interesting that, even though more anti-vaccine than pro-vaccine books have been published, in keeping with the dictates of cultural Marxism, otherwise known as political correctness, there's a much wider choice of the latter at the two public libraries near my home, and I'm sure everywhere else. It's easy to pull these books off the shelf at random and thumb through them to get a gist of the content, which is how I picked out *Do Vaccines Cause That?!* by Martin Myers and Diego Pineda. Myers is a pediatrician, professor, and self-described internationally recognized vaccine expert with close ties to the federal health bureaucracy. Pineda is a medical writer. This book caters to those who read little beyond the comic strips and sports pages. Since these folks don't read books, I don't know why the authors went through the trouble of writing theirs. Then again, the foreword and back cover contain endorsements by doctors and pediatricians with glittering credentials, which is one more reason to keep your children away from these people.

Do Vaccines Cause That?! reminded me of Dr. Seuss's *The Cat in the Hat*, which I read when I was six. Consider the title — a giggly question on a subject of the gravest importance. Every chapter but the last one begins with the words "Do vaccines cause that?!" and ends with a cartoon. The front cover shows a silly signpost with signs reading "side effect," "disease" and "risk" pointing in different directions. This is a book for morons.

Most points the authors try to make are derived from studies. Studies prove this, studies disprove that. Here are some other snippets offered without comment: "Many media stories use faulty reports and parental concerns to depict a 'controversy' about vaccines, failing to mention that the scientific community does not feel that a controversy exists"; "Many misinformers charge federal employees and vaccine researchers with having conflicts of interest. The federal government has strict ethics rules that require divulging potential conflicts of interest for federal employees..."; "Often, it is easier to just start with reliable information than to try to sort out the misinformation from information. The first step is to start with an information source that you *know* is reliable [emphasis in original]. For many parents this is usually the pediatrician, family physician, or nurse practitioner..."; "Does the Web site feature anecdotes (stories about individuals) about purported serious adverse events instead of scientific evidence? If so it is likely not a credible source of information..."; "The vaccines routinely recommended for children are only a small portion of the immune system's 'memory.' Scientists estimate that based on the immune system's capacity to respond, a child could theoretically get 10,000 vaccines in one day and still not 'use up' his or her immune response or ability to respond..."

Here's one final quote, straight out of Lewis Carroll, as it appears under the chapter heading "Do Vaccines Cause Autism?!" on page 139:

Parents who cite the evidence of their own eyes that their children become autistic after receiving... vaccine are in the grip of one of the oldest of commonsensical fallacies — that association proves causation... It is the responsibility of scientists to tell them that intensive scientific research in a number of disciplines indicates that the evidence of their own eyes is not true. — Michael Fitzpatrick (physician and father of an autistic boy) Spiked Online November 11, 2005

I beg you: *Is any kind of evidence more undeniably true than the evidence we see with our own eyes?* For Mr. Fitzpatrick's sake, I do hope he employs the evidence of his own eyes the next time he crosses a busy street.

❖

I did read some pro-vaccine books that, unlike these two, showed some evidence of adult thinking, but for the most part still struck me as off the rails. I will not enumerate them here, just as I have not enumerated anywhere in these pages some of the anti-vaccine books I read. One book, however, must be singled out because of its influence over millions of parents, an influence that might be greater than that of all other pro-vaccine literature combined. That book is *The Vaccine Book: Making the Right Decision for Your Child* by Dr. Robert Sears. Sears's father, William Sears, is also a doctor and author, and his mother Martha a nurse; the Sears Parenting Library, with several titles, is a family business.

The Vaccine Book has become a bestselling "go-to" book for parents who, for the most part, want their children to get the shots but have heard bad things about vaccine reactions, are as concerned as they are confused, and desperately want advice from a doctor they can trust. Robert Sears, or "Doctor Bob" as he calls himself, fits the bill perfectly. Parents are very comfortable with his position, which can be distilled as promoting most of the vaccines, having reservations about a few, and delaying some of them. He is well-known for his alternative vaccination schedule, a marked deviation from the CDC recommendation. Also, Sears is for freedom of choice and does not think parents should be pressured or punished in any way if they don't want their children inoculated — and I give him a great deal of credit for that. All this has provoked the ire of many vaccination hardliners who have slapped him with the "anti-vaccine" label, which he most certainly is not, the very first words of the second edition of his book being "I am a pro-vaccine doctor." Some of his meaner opponents have called for the revocation of his medical license, and there was a flurry of related news articles in September 2016 in newspapers in California, where he practices, and in some of the big business magazines like *Forbes*. These creatures will settle for nothing less than the most draconian vaccine laws. They resent the fact that Sears has won over countless parents with his kindly face on his book's back cover, and his equally appealing personality which reveals itself in the way he writes. He's relaxed and informal. He wants to be your friend, and it's hard to resist. He ends his preface by inviting you to grab a cup of coffee and a snack, "sit back and put your feet up, and let's explore the vaccinations together." He comes across as the proverbial "nice guy" — a real life Dr. Kildare. And I'm sure he *is* a genuinely nice man. But he is also very ignorant and naïve, and this is what makes his book not only worthless but dangerous.

I should point out that the first edition of *The Vaccine Book* was published in 2007, and a second edition in 2011. I have read both editions, and while I didn't closely compare them, I didn't notice any major revisions. What I quote below is from the first book. Both contain a nearly identical and unprecedented statement on the title page; I'll quote here from the second edition: "This book was written in July 2007 and updated in 2011, and the information is correct as of these dates. As new information becomes available through research and experience, some of the data in this book may become invalid." In other words, "This information is correct today, but some of it might be incorrect tomorrow." How interesting.

The Vaccine Book contains much that is wrong and misleading. It's a terrible book, although, based on independent Internet reviews, ninety percent disagree with that assessment. Sears wrote it, he emphasizes in the preface, "because... *there is no other fully informed, unbiased book available*," adding further on, "I've done the research. I've spent the last thirteen years learning everything I could about vaccines and the diseases they are designed to prevent, and I've put it all together in one place so you can get all your questions answered." Yet earlier he announces that he's "not going to discuss every last conspiracy theory," meaning, I assume, that in his eyes only fools believe in the criminality of certain individuals in federal agencies and drug firms. He writes: "Doctors, myself included, learn a lot about diseases in medical school, but we learn very little about vaccines, other than the fact that the FDA and pharmaceutical companies do extensive research on vaccines to make sure they are safe and effective...We trust and take it for granted that the proper researchers are doing their jobs." *Oh really?* As to being unbiased, so many of his statements are subtly slanted in favor of vaccines and in violation of common sense that the opposite is true. This becomes evident in the first chapter, at which point I realized I would not be able to take this book seriously. Sears discusses serious side effects of the Haemophilus Influenza Type B (HIB) shot that came to light in vaccine trials. (Like nearly all who believe in vaccines, he prefers "trial" over "experiment.") This is what he writes:

In the vaccine trials, more babies who received the vaccine caught a serious HIB infection than those who didn't get the vaccine. However, these infants didn't get infected by the vaccine itself. Research shows that while the vaccine is beginning to work on the immune system, the immune system can't react to a natural HIB infection for a period of five days. An immunized infant is therefore susceptible to catching natural HIB for a few days while her immune system is "distracted" by the vaccine.

One does not need a degree in microbiology to detect a problem here. Either these babies *were* infected by the vaccine, which seems likely, or, as often happens, the vaccine suppressed the immune system to such a degree that Hib bacteria already existing in these babies' bodies, normally benign, became opportunistically virulent. (This can happen in other ways as well, such as getting meningitis after a polio shot.) This Houdini logic, so typical of vaccine advocates, is employed elsewhere in the book, as in his attempt to explain away seizures: "These can occur after any vaccine, usually from a high fever, not from the vaccine itself." Yet on another page, he lists fever as a common reaction to vaccines, which everyone acknowledges. Well, to my way of thinking, but not to his, if vaccines cause fevers, and these fevers cause seizures, this means that vaccines cause seizures.

Over and over he reminds us that vaccines do carry risks, but they're very rare; over and over he speaks of vaccine-preventable diseases, a misnomer because children often get diseases for which they've been vaccinated against; over and over he slips in sentences praising vaccines, while pretending to be objective.

I have a hard time believing some of his claims. For example, he considers rotavirus a common and serious disease in infants. I mentioned rotavirus in this book's opening pages. It's at the top of my list of diseases cooked up on Madison Avenue. I never heard of it until I was in my forties. What is it? Where did it come from? Is it a harmless virus turned invasive by one of the other new vaccines? Was it brought here by immigrants from Asia or Central America? Do you expect me to believe that, in America alone, rotavirus is responsible for half a million doctor visits each year? "During my training years," Sears writes, "I remember the hospital hallways would be overflowing with dehydrated babies and worried parents because of this bug." I'm sorry, but I live in a heavily populated area with several hospitals nearby, and I never heard of anything like this. According to the back cover of his book, Sears lives in Dana Point, California, which is about one hundred miles north of the Mexican border. Did he train in a hospital where illegal, malnourished invaders flocked for free medical care? Perhaps a hospital in Mexico itself? Or is he just exaggerating — or maybe even making the whole thing up?

Not once does Sears raise the question of vaccine failure. He just assumes they all work. To his credit, though, he does not sweep any of their potential dangers under the rug; he puts those dangers out there for all to see, though he downplays the risks. He *does* make an effort to give a balanced view. But then by manipulating words and twisting logic he usually comes up with something heavily tilted in favor of vaccines that contradicts what he just wrote.

Under the boldfaced question, "Is it your social responsibility to vaccinate your kids?" he writes:

This is one of the most controversial aspects of the vaccine debate. Obviously, the more kids who are vaccinated, the better our country is protected and the less likely it is that any child will die from a disease. Some parents, however, aren't willing to risk the very rare side effects of vaccines, so they choose to skip the shots. The children benefit from herd immunity (the protection of all the vaccinated kids around them) without risking the vaccines themselves...

Here he lures parents into a playpen of pretend words. Obviously, Dr. Sears? Obviously, my foot. This mentality — protecting the nation by ramping up vaccination rates — was unknown when I was young, when the proportion of kids with serious diseases — and the child mortality rate as well — was much lower than it is today . And the notion that unvaccinated kids are "freeloading" off their vaccinated peers is just one of numerous ploys invented by the opposition. I never saw it that way when I decided not to vaccinate my own children, and I've never heard of another anti-vaccine parent who has.

Sears says he wrote this book to take the confusion out of the childhood vaccine issue. Actually, he makes it more confusing than ever. Adding to the confusion are all the indecipherable statistics that he throws around, creating a confusion that he himself acknowledges. You can't get much more confusing than this: "Are vaccines safe? Yes. Do they have severe side effects? Yes. Are these severe side effects common? Not very. Is vaccinating to protect against all these diseases worth the risk of side effects? That's the million dollar question." How can you describe something as both being safe and having severe side effects, even if those side effects are not very common? And just what does "not very common" mean, especially after stating in the previous paragraph, "We don't know the exact number of serious vaccine reactions, as many are likely to go unreported or are not recognized. We also don't know how to factor in possible long-term hidden effects of vaccines."

❖

I'm sure Robert Sears, a father of three himself, has the best interests of children at heart. To repeat: his heart is in the right place, but his judgment most certainly is not. He can't see the forest for the trees. Had he come under a different influence while studying medicine, he might have ended up staunchly anti-vaccine. He can think outside the box, but just a little. Unfortunately, he's stuck in the tar pit of dogma and he can't get out. What seems incredible to me is that he apparently spent all those years of research reading nothing but medical studies, about sixty of which are listed as his resources — plus he said he read many more — yet not a single book of any kind about vaccines is mentioned (though he does obliquely refer to *A Shot in the Dark* as an important book that he did read, but has become outdated). All those endless studies, piled up like pick-up sticks, waiting for the wind to blow them down. No wonder he goes around and around and around and never gets anywhere! Incidentally, Sears has also authored *The Autism Book*, which I read — and learned virtually nothing of use. It's almost all filler. On only about twenty pages of this 350-page work does he bring up the crux of the matter, vaccines, and for the most part, peripherally. This book is about as stimulating as reading a dishwasher repair manual. Reading, however, is a subjective matter, and as with *The Vaccine Book*, a solid majority among the reading public gave it five stars, the highest rating. I'm with the six percent who gave it one star.

I was beginning to think that in trying to be fair, I was wasting my time reading all these books. Then I discovered Paul Offit.

The case of Paul Offit

I HAD COME across Paul Offit's name frequently in my study of the vaccine issue. Looking back, it's odd that I didn't read his works until I felt I was almost done examining the other side. About all I knew was that he was a zealous defender of childhood vaccination and, while he had his admirers, a lot of people hated him. In digging up more information, I discovered he is loaded with credentials, among them holding a top "immunization" post at the Children's Hospital of Philadelphia. Apparently he is quite renowned in that city, once even having thrown out the first pitch in a baseball game between the Philadelphia Phillies and the Arizona Diamondbacks. When I learned that he had written several books I began to look into them. I ended up reading five of the nine he has written to date. I first read *Deadly Choices: How the Anti-Vaccination Movement Threatens Us All*, then *Autism's False Prophets: Bad Science, Risky Medicine and the Search For A Cure*. I followed this up with *The Cutter Incident: How America's First Polio Vaccine Led to the Growing Vaccine Crisis*; *Vaccines and Your Child: Separating Fact From Fiction*; and *Vaccinated: One Man's Quest to Defeat the World's Deadliest Diseases*, on the life and career of Merck's foremost vaccine researcher, Maurice Hilleman.

Before I opened *Deadly Choices*, though, I asked myself if it was worth investing the time. I wanted to read a book written by someone with no vested interest — financial, emotional or otherwise — in vaccines. Offit, known to his detractors as "Offit for profit," had instantly disqualified himself. Anyone who could claim that it was "deadly" not to vaccinate, and that such a choice threatened me and everyone else, was impossible for me to take seriously. Just how am I, the people in my community, and the rest of the country "threatened" by unvaccinated children, and adults too for that matter? And to say that Offit has a vested interest in vaccines is a monumental understatement. He is the most prominent spokesman of those who passionately believe in the benefits of vaccines, the very face of compulsory childhood vaccination. It is an unattractive, slightly cruel face that made me recoil the first time I saw his photo. Offit has made a fortune patenting and pushing vaccines. Vaccines are his religion, his life blood. The dust jacket of *Deadly Choices* lists a dozen titles and awards, all connected with vaccines. So, my first thought was that it was a waste of time reading him, and in the end, my worst suspicions were confirmed. But what a reading experience in between!

Within an hour of opening *Deadly Choices*, I realized that this book was completely different from anything else I'd come across on the pro-vaccine side. Offit is an excellent writer. His ideas are well-organized, as are the chapters in this and his other works. The more I read, the more I marveled at his wide-ranging knowledge of every facet of vaccine history; he obviously had done an enormous amount of research. In fact, he filled in some gaps about which I was quite ignorant, particularly regarding the nineteenth century anti-vaccine movement in England, which was much more violent than I had realized. Furthermore, he was civil in his criticism of even his most vocal opponents; nowhere did he stoop to the name-calling and mud-slinging one often encounters in vaccine supporters, especially on the Internet. But most astounding of all, he neither concealed nor shied away from the strongest arguments we anti-vaxxers have put forth; on the contrary, he sought them out and addressed them head-on. This won my respect. Frankly, I was stunned by this approach, all the more so because, had I not already studied this subject exhaustively, some of his ideas would have made sense to me. In all my reading, I had never come across anyone so seemingly straightforward who stood for something so wrong.

And yet, at the same time, the more I read, the more irritated I became. Offit is just *too* authoritative, *too* sure of himself, *too* full of nervous energy and a weird sense of urgency in aiming his needle everywhere. He never pauses to express wonder or uncertainty; he is too quick to dismiss with a wave of his hand any inconvenient facts. This he does most often by citing studies. Studies of all kinds, specified and unspecified. Everything is studies. He makes this clear in the prologue of his autism book:

Since the late 1990s, many studies have shown that the rates of autism are the same in vaccinated and unvaccinated children. The CDC, the American Academy of Pediatrics, and the Institute of Medicine have all issued statements supporting these studies. So the notion that vaccines cause autism isn't a medical controversy.

So there you have it. Studies have shown that there's no connection between vaccines and autism. End of story. But not really. Some of the studies he cites are the very ones that former CDC researcher William Thompson confessed were faked. What many consider Offit's wildest assertion, however, is that a baby can endure 10,000 vaccinations which, as noted in the previous chapter, Myers and Pineda allude to. (Myers and Offit have been closely connected, as indeed many pro-vaccine authors and activists have been, in a well-heeled spread the praise club.) Many have lashed out at Offit for this alone, the 10,000 figure having circulated widely on the Internet among those who despise Offit, but it appears that a lot of people have missed a zero. This is what appears on page 42 of the paperback version, published in 2011, of *Vaccines and Your Child*, co-authored with a Charlotte Moser:

By calculating the number of [antibody-producing] *B cells in the bloodstream, the average number of epitopes contained in a vaccine, and the rapidity with which a critical quantity of antibodies could be made, we know that babies could theoretically respond to* a hundred thousand *vaccines at one time [emphasis added]. Of course, we're not saying that babies should get a hundred thousand vaccines at once. We're only saying that they could handle it.*

That, my friends, is the inexpressible madness of Paul Offit in a nutshell. Since *Vaccines* was the next to last Offit book I read, by the time I came across this line I was familiar with his unique methodology, which initially made him such a tough nut for me to crack. This is what Offit does: he combines accurate and fascinating historical narrative, forthright confrontation, honest and damning facts about vaccines and the people who made and tested them (though he doesn't find them damning), convoluted logic, dubious assertions, outright lies, and declarations — like the one just quoted — that could only originate in a mind so aberrant that it fits right in with the roughly three thousand Americans currently sitting on Death Row.

I could fill a hundred pages picking apart all the crazy things Offit writes, but for the purpose of this chapter just a few quotes will suffice. Here's one more:

In the 1960s, parents gave their children measles, mumps and rubella vaccines because they had witnessed firsthand the devastation wrought by those diseases: pneumonia and encephalitis from measles, deafness from mumps, and severe birth defects from rubella.

Actually, in the 1960s parents witnessed nothing of the kind. Offit and I are less than three years apart in age. He grew up in a suburb of Baltimore, I grew up 200 miles away in a suburb of New York. In other words, we grew up in the same society surrounded by the same everyday facts. How he came to perceive these facts the way he does is beyond my understanding. The best I can do is say that he is one of the filthiest liars who ever soiled this planet, or else he really believes what he writes, in which case I am at a loss for comment. I am inclined to think that, for the most part, he's a liar who knowingly imparts false information. (I am reminded here of an amusing quote I once read about the 1950s left-wing playwright, Lillian Hellman, "Every word she writes is a lie, including *and* and *the*," though in fairness to Offit he is at times astonishingly truthful.) This also applies to my final reading, *Vaccinated*, the strange title itself betraying something sinister about the man, implying something done by force, as in "there it is, done, you're branded." At least that's what I read into it. On the Big Lie index, this book, on the margins of which I penciled in countless exclamation points and question marks, is off the charts. It also contains eight photographs showing people young and old receiving injections of the wonderful vaccines developed by Maurice Hilleman, a criminal in his own right. Probably all of them were staged using empty needles.

❖

In the beginning, I mentioned that the meat of this work is based on evidence that is impossible to refute, mostly the evidence of what I have actually seen. Nevertheless, with so much juicy information flowing in all directions on the Internet (much of it admittedly suspect) I must relate an anecdote which I am convinced took place as told on ageofautism.com. Paul Offit has made a lot of enemies, and a few have made it their mission to expose him for what he is. One is a former college student named Jake Crosby whom Offit has called "my stalker." Crosby, it seems, has kept close tabs on Offit and follows him around. In May 2012, after Offit gave a lecture at the University of Pennsylvania, Crosby called him out during the question and answer period. The unfriendly conversation went back and forth for a few minutes, then Crosby really hit a raw nerve when he accused Offit of "making money off the backs of vaccine-injured children." Offit lost his temper, shouting, "No, that is bullshit! I don't do this for the money! Get out of here! Get the fuck out of here! Piece of shit!" The exchange was a long and engrossing one, causing commotion in the audience with others getting involved, and this continued even after both men left the building. Of course, I can't vouch for the accuracy of Crosby's account, but four years later a similar encounter was recorded on camera, proving that Offit does indeed have a recurring case of potty mouth when he's openly challenged. This brief clip can be seen on autisminvestigated.com. Since Offit is a darling of the controlled news media, neither outburst seems to have tarnished his public image.

So what are vaccines worth to Paul Offit? According to a link on the web page of Sharyl Attkinson, an investigative reporter formerly employed by CBS, between $29 and $46 million *alone* in royalty interests for the Rotateq vaccine he co-invented and which is marketed by the pharmaceutical giant Merck, with whom he has intimate business ties. Offit has several other vaccine irons in the financial fire, but has refused to talk about his income. He did, however, go after Attkinson in an August 4, 2008 hit piece that appeared in a California newspaper, the *Orange County Register*, which shortly thereafter published a clarification, stating, "It appears that a number of Dr. Offit's statements... were unsubstantiated and/or false."

Paul Offit has taken a lot of fire from people who don't like him. He's used to it. I think he even enjoys it. Parents, in particular, detest him. He has long been a lightning rod for their wrath. These are his opening lines in *Autism's False Prophets*:

I get a lot of hate mail. Every week people send letters and e-mails calling me "stupid," "callous," an "SOB," or "a prostitute." People ask, "How in the world can you put money before the health of someone's baby?" or "How can you sleep at night?" or "Why did you sell your soul to the devil?" They say I "don't have a conscience," am "directly responsible for the death and damage of hundreds of children," and have blood on [my] hands." They "pray that the love of Christ will one day flood [my] darkened heart." They warn that my "day of reckoning is coming."

Four pages later he mentions a specific death

threat, harsh words hurled at him as he walked through an anti-vaccine rally, other expressions of parental rage directed his way, and the precautions he had to take, including an armed guard posted for years at meetings he attended. This is good. I was happy to read these things. It encourages me to know that there are still Americans who are *alive*, whose instincts are still intact, telling them that this man is twisted and he's after their children. They might not be able to put their finger on it, but they sense, correctly, the evil lurking in his black heart.

Or am I wrong about all this? Well, it seems to me that if Offit is truly motivated by a desire to protect our children from terrible diseases, if he truly is a good, misunderstood man who knows a lot more than the rest of us, he would reach out to those who hate him the most — parents who just want to do what is best for their children — and try to set the record straight. But Offit has never done anything of the kind. Deep down inside, when he gets under the sheets at night, I can picture him thinking, "So all you scum out there hate me, do you? Well you just wait. I'll give you plenty more reasons to hate me — and screw your children and whatever happens to them." Of course, no one will ever know exactly what goes on in this man's brain, but implicit in much of what he's said and written is his wish to turn America into an outright vaccine police state. I can easily see him enforcing a nationwide vaccination decree issued by the CDC, on whose Advisory Committee for Immunization Practices he sits, and if you don't comply he'll send in federal marshals to kick in your door, tackle you, put a gun to your head, then strap your sobbing children to a board while a doctor forcibly inoculates them.

In addition to unflattering looks, Paul Offit has a genetic defect which he himself tells us about: he was born with two club feet. Such a deformity does not necessarily make a person evil, of course. But it could. Is Offit profoundly unhappy with who he is, and is his ambition to fully vaccinate every American child a way of avenging himself on the world? Does he have a secret animus towards healthy, attractive children? Again, I don't know, and only throw these out as possibilities. But how do you explain a man with an encyclopedic knowledge of vaccines, one of obviously superior intelligence, who has risen to the greatest heights of professional accomplishment and prestige, not to mention wealth — how do you explain why he writes the way he writes and does the things he does? Why is he so blind to all the evidence indicting vaccines, and so deaf to the lamentations of parents whose children were destroyed by them? How can he casually dismiss the innumerable accounts of severe reactions shortly after vaccination as just so many coincidences? What is wrong with him? What drives him?

To answer these questions, I can only speculate that Paul Offit is a mattoid. *Mattoid* is a rare word that you won't find in standard dictionaries, and the brief definition that appears on the Internet leaves a great deal unsaid. The late professor Revilo Oliver, writing in the 1960s, elaborates:

A mattoid is a person possessed of a mentality that is, in the strict sense of the word, unbalanced.... He exhibits an extremely high talent, often amounting to genius, in one kind of mental activity, such as poetry or mathematics, while the other parts of his mind are depressed to the level of imbecility or insanity. Nordau, who was an acutely observant physician, noted that such unbalanced beings are usually, if not invariably, "full of organic feelings of dislike" and tend to generalize their subjective state of resentment against the civilized world into some cleverly devised pseudo-philosophic or pseudo-aesthetic system that will erode the very foundations of civilized society.... It is clear that there is in the human species some biological strain of either atavism or degeneracy that manifests itself in a hatred of mankind and a lust for evil for its own sake.... It is probable that this appalling viciousness is transmitted by the organic mechanisms of heredity, and although no geneticist would now even speculate about what genes or lack of genes produce such biped terrors, I think it quite likely that the science of genetics, if study and research are permitted to continue, may identify the factors involved eventually.... It is quite likely that at the present rate, as eugenicists predict, civilization is going to collapse from sheer lack of brains to carry it on. But it is now collapsing faster and harder from a super-abundance of brains of the wrong kind. Granting that we can test intelligence, we must remember that at or near the top of the list, by any test that we can devise, will be a flock of diabolically ingenious degenerates.

That's one way to explain the case of Paul Offit. There is another, somewhat related explanation: Offit is a Jew of the worst kind, and if that statement repels you, you should think twice about reading the next chapter. But if you have an open mind, read on.

The Jewish Factor

The sun has never shone on a more bloodthirsty and vengeful people than they who imagine that they are God's people who have been commissioned and commanded to murder and slay the Gentiles. In fact, the most important thing that they expect of their Messiah is that he will murder and kill the entire world with their sword. They treated us Christians in this manner at the very beginning throughout all the world. They would still like to do this if they had the power, and often enough have made the attempt, for which they have got their snouts boxed lustily. — Martin Luther, 1543

We Jews, we, the destroyers, will remain destroyers forever. Nothing that you do will meet our needs and demands. We will forever destroy because we need a world of our own, a God-world which it is not in your nature to build. — Maurice Samuel, 1924

It is my firm belief that what I refer to as "The Problem of Israel" is one that threatens the very survival of life on Earth. — Michael Collins Piper, 2007

THROUGHOUT these pages I have made scattered and uncharitable references to various aspects and personalities of what has long been called "the Jewish question." The time has come to zero in on this subject because it clearly impinges on "the vaccine question" in this country. In writing and revising this book, no chapter troubled me more than this one — both because I know it will inevitably offend a lot of readers, and because it will be a waste of time, even for those who generally don't like Jews, to read it. Having adapted their attitudes to a lifetime of lies, most of them Jewish at their source, their minds will snap shut at every stage of this discussion which is surely the height of poor taste and, God forbid, anti-Semitism — whatever that is. This is emphatically true in a country that is not only the most Jewish in the world after Israel, but also has the world's highest percentage of Christian evangelicals or dispensationalists or fundamentalists, or whatever they prefer to call themselves — not only among the general population but at the highest levels of government. They are absolutely certain that Jews are God's chosen people, Israel can do no wrong, and that's the end of it. At this writing, this group includes Vice President Mike Pence, and former U.S. Army officer and CIA director and current Secretary of State Mike Pompeo, both of whom seem to be itching to start World War Three to fulfill biblical prophecy. Pompeo, a fervent hater of Iran and lover of Israel and Jesus, has said that he has a Holy Bible open on his office desk to remind him of the Word of God. He has hinted that the people of Iran should be starved into submission. On March 21, 2019, this psychotic Eagle Scout suggested that Trump might have been sent by God to save Israel from Iran, comparing him to Queen Esther in the Old Testament who saved the Jews from being killed by the Persian potentate Haman, after which the Jews slaughtered 75,000 Persians, Persia being the former name of modern Iran. Then there's John Bolton, former National Security Advisor and another rabid Zionist (fired by Trump, thank God), a study in pure evil who never met a war he didn't like, except the one in Vietnam he slunk away from when he was 21. Of course Donald Trump is also thoroughly Judaized in a secular sense — his ways shaped over decades of hobnobbing with Jewish real estate wheelers and dealers, and assorted Jewish perverts and shysters, not to mention his daughter's marriage to Jared Kushner. His sorry presidency caps a full century of formidable Jewish influence over America dating back to the disastrous era of Woodrow Wilson, another sad and tragic clown. Trump's slavish devotion to everything Jewish has not prevented his empty head from being knocked around like a hockey puck between bickering Jews whose opinions about him run the gamut, even as he does their dirty work — so far without starting another major war at their behest. As others have suggested, the pervasive media hostility towards Trump stems from his galvanizing of the white working class — the group that organized Jewry has always feared and hated — even though he has since utterly betrayed them. But he let the genie out of the bottle, especially by attacking some of the biggest fake news titans by name, an unpardonable sin. You will notice, however, that despite the constant media cannonades against Trump, they have never gone for his throat the way they did with Richard Nixon over his piddling cover-up of the amateur, unsuccessful Watergate burglary of which he knew nothing at the time — about as momentous on the scale of presidential crimes as jaywalking. The difference is that Nixon, as a young congressman, went after communists in government, a dastardly crime for which the media bosses never forgave him, and as revealed in tapes of private White House conversations with his aides — which he foolishly failed to destroy during all the Watergate nonsense — he was a man who really disliked and distrusted Jews, even though he was compelled by circumstances to appoint a few to important posts in his administration.

Suspecting that I've already lost a lot of readers, another problem confronted me. How do you trim forty years of study — poring over as many books and well over a thousand essays and articles culled from "hate literature" on a subject so vast and complex — to a chapter of reasonable length? The answer is, you can't. And yet to omit this subject would be to leave out a very important part of the vaccine story, and with it, the most pressing issue facing this country and the entire world. On the other hand, to incorporate it in this kind of book would require curtailing hundreds of facts that cry out for more detail. Well, that was the only course left to me. To those new to the Jewish question, reading this chapter will be like having bucket after bucket of ice water dumped on their heads. Several excellent books, nearly all of them effectively banned through organized Jewish pressure and Jewish control of the book marketing and distribution business — book burning without the flames — will be cited for the inquisitive few. But perhaps it would be best to soften the shock effect on those curious readers with fine sensibilities by beginning with a sketch of my introduction to the Jewish question many years ago.

❖

Although most people know that the New York metropolitan area has a huge Jewish population, there were very few Jews in my hometown, and I had almost no exposure to them growing up, other than a handful of Jewish teachers and students in elementary and junior high school. Like most people, the only distinction I made was that they practiced a different religion. My parents never talked about Jews, and until I was in my twenties I had no opinion about them either way other than an awareness that the orthodox ones stood apart in dress and religious customs, and generally speaking they were crafty in business and loved money. When I enrolled in college on Long Island at age 20 I became acquainted with quite a few Jewish students for the first time, but I liked or disliked them as individuals just as I would people of any other group. I knew nothing of their contentious place in history, and I was equally ignorant about everything regarding the modern state of Israel. In fact, one of my philosophy professors, a Gentile, a man with whom I had many stimulating conversations about world affairs over many a beer, sold me on the now stale Zionist myth: Israel, our one true friend in the Middle East, an island of freedom in a sea of Arab tyranny and terrorism, our gallant little ally risen from the ashes of near extermination at the hands of the Nazis, now defending herself against a new enemy malignantly jealous of her accomplishments and political freedom and bent on her destruction. Yes, I believed all this: for a few years I believed it passionately. It made me feel good. It put current events into focus and gave me a clear picture of what was happening in that troubled part of the world. So I can understand how appealing this myth is and why it still has a powerful hold on many Americans, especially those of my generation.

Three things happened in a short span of time in the late 1970s that, taken together, completely changed my outlook. I had a night job pumping gas and, being a red-blooded male, was thumbing through some skin magazines my boss kept hidden in a desk drawer. By chance, I came across an article by South Dakota senator James Abourezk — in *Oui* or *Penthouse*, I believe — about the Israeli attack on the *Liberty*, a U.S. Navy intelligence ship. This happened on June 8, 1967 during the Six-Day War. I had never heard of it, and to this day, I doubt one in a hundred Americans has. I was dismayed to read that the Israelis had shot up, napalmed, and torpedoed this American ship — which was monitoring, among other things, their land grabs in Syria — in a failed attempt to put it on the bottom of the Mediterranean Sea, an act which was sure to be blamed on the Egyptians. Furthermore, the attack was subsequently covered up at the highest levels in Washington D.C. and by the entire news media establishment; all accepted an official apology from Israel's leaders, and their explanation that it had been a case of mistaken identity, that they thought it was the Egyptian horse transport ship *El Quseir*, even though the

Liberty was much bigger, had a much different profile, and was flying a large American flag on a perfectly clear day. Much more evidence pointed to a planned, premeditated assault. Thirty-four men were killed, and 171 wounded during this massacre, which lasted two hours. Incredibly, although the ship was listing and had been riddled with more than 800 holes, it didn't sink, and was able to limp to port in the island nation of Malta. Every man on the *Liberty* knew the attack was deliberate, as did many military and government officials, but fear and intimidation prevented all but a few from speaking out. Books have been written by the survivors of this atrocity, one of whom, Phil Tourney, author of two works, I had the pleasure of meeting in 2017. At the time I read the Abourezk piece, I wanted to reject it, to write the man off as a typical liberal moron, but I couldn't: what he had written was calm, logical and thoroughly convincing. It confused and flustered me, because it was totally at odds with the image of Israel I had formed in my mind. It was like seeing a close friend sneak arsenic into your coffee, then when confronted, hearing him say he thought it was sugar. What kind of friend is that? (I have since learned that this "ally" of ours has carried out numerous acts of treachery against America.)

Shortly afterward, during a month long winter break at college, I decided to read Adolf Hitler's *Mein Kampf*. It was just a whim. Since I was about to take a course in modern European history, I felt it might be relevant, and I was simply curious about this work that everyone knows about, but almost no one ever reads, a book that I assumed would reveal the workings of a crazed mind belonging to the most evil man who ever lived. Like virtually everyone born and raised in America and throughout the Western world, that was how I looked upon Hitler. No bookstores or public libraries near where I lived carried *Mein Kampf*, but I found a used copy at the landmark Strand Bookstore in New York City. I hadn't read ten pages when I knew I wouldn't be able to put this book down. It was obvious that I was dealing not with a madman but a genius, and as I read further, a man with a sincere and burning ambition to lift up his people, to right all the barbaric wrongs that had been inflicted on Germany by the Versailles Treaty at the end of World War One, and to wipe out all traces of corruption and degeneracy. All these things he did, swiftly and ruthlessly, upon assuming political power eight years after *Mein Kampf* was completed, while he was still in prison. I was so taken by Hitler's clarity of vision that I began wondering if there was something wrong with me for finding myself in complete agreement with nearly everything he wrote! To this day I find it remarkable that in everyday life I've never met anyone who has read this book of almost 700 pages, in which Hitler openly and completely revealed himself, yet at the same time virtually everyone "knows" that he was the devil incarnate who was out to conquer the world. I find it equally remarkable that in all the thousands of damning references to Hitler I've encountered in my extensive reading, not one tried to refute a single sentence; in fact, I've come across only a handful who even bothered to quote what he had written. The mere thought of reading *Mein Kampf* and judging it on its merits or defects, like any other book, makes people tremble. Most people are incapable of reading serious literature anyway, and they can be excused, but it's about time that those who consider themselves pretty smart — while mouthing the usual platitudes about the satanic Hitler and his dreadful Nazis their whole lives — grow up, shut up, sit down and read the book.

I had only one reservation while reading *Mein Kampf* — the disturbing thought that, mixed in with Hitler's genius, his mind had run amok when he decided to biologically exterminate the Jews of Europe. This brings me to my third experience of more than forty years ago, which began with a day I remember well. I was standing on my front lawn when the postman came along, and in the day's mail was a flier from an outfit in California called the Noontide Press. The flier advertised two books, *The Myth of the Six Million*, anonymously written, and *The Hoax of the Twentieth Century*, by Arthur Butz, a professor of electrical engineering at Northwestern University. Not only did the advertisement deny that Hitler had tried to wipe out Europe's Jews, it openly ridiculed this tenet of established history. I stood there in disbelief; I'd never seen anything like this, and wondered how I got on their mailing list. "You gotta be kidding me," I said out loud. And yet, and yet... despite the tone, the content of the ad, like the Abourezk article about the *Liberty*, was totally rational. I knew I couldn't be honest with myself by laughing at this thing and ripping it up. I ordered both books.

In an earlier chapter I compared dogmas, among them vaccination and the Holocaust. *Myth* matched up exactly with the Neal Miller book on vaccination that instantly opened my eyes — both slender volumes of less than a hundred pages, easy to read, clear and concise. *Hoax* was much like Viera Scheibner's work *Vaccination* — more technical and tedious in places, but written in a tone that left no doubt in my mind that I was dealing with top-notch research. Butz's work cut through all the baloney and historical obfuscation like a laser beam. As I got deeper into it, I said to myself, "Nobody's gonna get anything past this guy." And since 1976, when the book was first published, no one has dared to even try to refute it; it's so much easier calling people names, or threatening them. What does that tell you? Within one week, the time I spent reading these two books, my belief in the so-called Holocaust was permanently demolished. In the years that followed I read more material that ripped the gassed six million Jews legend to shreds. But the icing on the cake was my visit to Auschwitz in 1991, which I wrote about earlier.

My introduction to Noontide Press opened the door to a whole new forbidden realm of knowledge that I had never known existed, and couldn't get enough of. In the pre-Internet age this meant subscribing to a few periodicals and reading many books which, while legally published thanks to our First Amendment, were censored and available only through the mail. A good deal of this "hate literature" dealt forthrightly with the Jewish question, and there were even a few renegade Jews who had written excellent books on this vital topic. Most memorable was *Antizion*, a compilation of documented quotes from more than 500 famous people, many of them names we're all familiar with, going way back to ancient Rome — philosophers, poets, composers, inventors, artists, statesmen, religious leaders and others — the best minds of Western civilization — who had spoken forcefully against Jewish character and influence, in some cases exceeding the most inflammatory passages in *Mein Kampf*. Why, I wondered, was this not common knowledge? Why, in a supposedly free country, are so few aware of books written by major figures like Martin Luther (*On the Jews and Their Lies*, 1543), Richard Wagner (*Judaism in Music*, 1850), and Henry Ford (*The International Jew — The World's Foremost Problem*, 1920)? And why are these books impossible to find in any bookstore or public library? Most notable was a speech given by Benjamin Franklin at the Constitutional convention of 1787 in Philadelphia (later claimed by Jewish groups to be a forgery) in which he referred to Jews as "Asiatics" and "vampires" among other things, and pleaded that they should be barred from America forever. Equally fascinating and suppressed were books written between the world wars by two European-born Jews who had emigrated to America, *You Gentiles* by Maurice Samuel and *Jews Must Live* by Samuel Roth. With blunt honesty, Samuel wrote that the fundamental differences in mentality between Jew and Gentile run so deep that no reconciliation would ever be possible. Roth, a publisher, apparently wrote to get even with Jewish competitors he believed had railroaded him into prison. He laid bare everything about his fellow Jews — their upbringing which instilled a contempt for physical labor, as well as a directive to cheat and distrust Gentiles, their lack of manners and crass behavior, their crowding and corrupting of the high-paying professions, and their overall parasitic existence. In recent times his words were echoed, in more fiery fashion, by Chicago native and world chess champion Bobby Fischer, who was half-Jewish himself. In a series of foreign radio interviews Fischer ripped into Jews every chance he got, calling them "a filthy, dirty, disgusting, vile, criminal people," "subhuman.... the scum of the Earth.... the bottom of the barrel of humanity." America, he said, was "a farce controlled by dirty, hook-nosed, circumcised Jew bastards." And he had much, much more to say on the subject. This was a man with an IQ of 187. To those who would see him as some kind of aberration, two other contemporary Americans with extraordinarily keen minds, William Pierce and Revilo Oliver, fully shared his sentiments about Jews, and as racial nationalists wrote a great deal about them — Oliver's *The Jewish Strategy* is a classic — the one difference being that their language was on a much higher intellectual plane.

Whatever one thinks of all this, the question again arises: Why are these facts unknown to the great majority of Americans, especially since this friction between Jew and Gentile goes back to the beginning of recorded history? And not only these facts, but many other incriminating ones, which resulted in their mass expulsion over and over from nearly every country or region of Europe? In simpler times, before radio and television, before the printing press was even invented, information and ideas were conveyed by word of mouth. In the Middle Ages, there was a much wider awareness of this seemingly eternal Jewish question. Men would gather around campfires, or flock to taverns and sit around rough wooden tables to discuss it over tankards of ale. Today, the reality outside of most people's immediate experience is filtered through the news and entertainment media — the most influential being the television screen. Reality is now falsified for ulterior purposes — a situation that was not nearly as potent two hundred years ago, and did not exist at all in medieval times — with the result that today, when the Jewish question is as critical as it has ever been, most Americans are oblivious to it, and would be aghast at the mention of such a thing. I have discussed the perfidy of the media enough in this book and wish to add nothing more here, other than to emphasize that Jews, through their remarkable talent for organization, exercise near total control over the fake news that blankets much of the world, and nowhere more than in this country. Only very rarely do awkward truths leak out. How they managed to achieve such international dominance over the past hundred years, despite their small numbers in every country except Israel (in the U.S. they comprise about three percent of the population), is astounding. And because Jews network so efficiently, the media work hand in glove with other censorship organizations like the powerful and sinister Anti-Defamation League (ADL) which, through its 29 regional U.S. offices, secretly monitors school curricula, library acquisition lists, and social activities of all kinds — not to mention their liaison with the FBI and numerous police departments — always on the lookout for any negative attitudes towards Jews and the state of Israel. These facts, I hope, provide an adequate answer to the question I posed at the beginning of this paragraph.

❖

By 1980, I was aware of much of the above, and realized that I'd been lied to my whole life. In October of that year, I arrived in Tel Aviv with my girlfriend on a cheap flight from Athens, Jew-wise, so to speak, and a totally different man from who I was five years earlier. We spent nine or ten days traveling around the country. Four impressions stay with me from my first trip to Israel: the astonishing sight of so many people, soldiers in uniform as well as plainclothes civilians, going about their business while toting Uzis or assault rifles; barbed wire everywhere; in Jerusalem the magical feeling of timeless, turbulent, pulsating history; and my last three days spent on a kibbutz called Gevulot, still in operation. This is worth going into.

In the mid-1970s, as part of my enchantment with everything about Israel, I was dazzled by the mystique of the kibbutz, a collective farm system which was supposedly "the only kind of socialism that works," proof that the Jews had come to a barren land and "made the desert bloom," as the myth went. What I saw in my three days there, and what I've believed ever since, is that this was a typical Jewish scam invented by a people who, in great part, are allergic to physical work, and the kibbutz myth was a way of attracting naïve *goyim* (plural of *goy*, a derogatory Jewish word for non-Jews) to perform the most menial tasks, while believing they were enjoying the great privilege of taking part in a progressive agricultural project. Life on the kibbutz for foreign volunteers amounted to eight hours of labor, spartan living quarters, three meals, and an income of about $80 a week. In reality, it's a voluntary form of an institution that exists throughout the world: prison. About fifteen Gentiles from various European countries, including a close-knit group of seven or eight Swedes in their late teens, were settled in as volunteers when we showed up. After a brief orientation, my girlfriend made a commitment, it being understood that volunteers could leave anytime they wanted. She ended up staying and working there for two months. In my brief stay I became acquainted with the other volunteers. One of them, a pretty Swedish girl, had a rash on her face from picking some kind of fruit from a tree with poisonous sap. The Jewish work assigner had failed to warn her about this. Did this faze her or her friends? Not at all. They were planning to stay indefinitely. I did, however, speak with a young English guy who told me he was fed up with varnishing chairs day after day, and intended to leave soon. For him the kibbutz mystique had worn off.

Carol, my impressionable girlfriend, was no doubt subjected to large doses of propaganda while she was there — just like our visiting congressmen, military personnel, and police chiefs trained by Israelis in "anti-terrorism" tactics (with some gas chamber yarns thrown in to stoke guilt feelings, as in "the world stood by and did nothing"). I remember picking her up at JFK Airport just before Christmas. She was gushing about what an "incredible experience" it had been working at Gevulot, though she never did explain what was so incredible about it. All of this, I believe, is a microcosm of the way Jews dupe and exploit dumb *goyim* while laughing behind their backs, life on the kibbutz being a comparatively mild example. It's a finely honed mind control technique that they employ most devastatingly through their control of the media.

Four years later I returned to Israel alone, having begun my trip in Turkey, then working my way down through Syria and Jordan before walking across the Allenby Bridge to the occupied West Bank. I had been apprehensive about visiting Syria, due to its scary image and the fact that there was no tourist information available, but I had applied for and received a visa with no problem from their embassy in Washington (since closed), and simply walked into the country from Nusaybin, Turkey at a hot, dusty, remote border post. The official on duty looked surprised when I handed him my U.S. passport, but stamped me in with no fuss and I sat down under a tree with a bunch of other people — men in traditional Arab dress, and exotic-looking women with gold filigree strung across their foreheads, a few with small children — waiting for some kind of transport, and purging my brain of the media poison that had instilled an irrational fear about coming here. Eventually a battered Buick Roadmaster station wagon pulled up, bags and suitcases were tied on the roof, the tailgate was opened to stuff in a pair of discouraged sheep, and off we went to Deir-ez-Zor, a small city recently wracked by war and suffering thanks to our Jewish neo-conservative Pentagon warlords. I found a shabby hotel, and spent the rest of the afternoon sipping tea in the lobby with the owner's two young sons, while learning a little Arabic and watching the Olympic games in Los Angeles on a fuzzy television screen. I had no problems at all traveling around the country on public transport even though I saw no other foreigners, except for a group of French tourists at the ruins of Palmyra.

My second trip to Israel lasted about ten days like the first, but this time I spent a few nights with a Palestinian family in the village of Sa'ir, near Hebron, and though it was uneventful, I got to sample some of the daily tension these people live with, which is surely much worse today. (Ten years later, on February 25, 1994 — which that year happened to be the day of Purim, a festival of hate and vengeance — one Baruch Goldstein, a settler from Brooklyn who was a medical doctor, walked into a mosque in Hebron and shot dead 29 worshippers, wounding 125, before being grabbed by survivors and beaten to death — one more reason to be wary of doctors, especially Jewish ones.) I also rented a car for a week and gave lifts to about a dozen hitchhiking and armed Israeli soldiers — they're everywhere on the road, or at least they were back then — just to chat with them and get inside their heads. Needless to say, I didn't reveal what made *me* tick, but I did have some interesting conversations, and even played devil's advocate on the subject of so-called Arab terrorism with the first guy I picked up. We talked a bit about Lebanon, which had suffered so much in that period from Israeli terror in the air and on the ground. I can still see him in the passenger seat, his rifle propped between his knees, grumbling about army service, and grimly telling me, "Nobody wants to go to Lebanon." It was interesting to hear that. Later I picked up another young conscript and his girlfriend, who had to report to a post near the Lebanese border. They sat in the back, from time to time speaking softly to each other in Hebrew, and I heard him take several deep breaths. He was obviously nervous about where he was going. They thanked me for the ride when I let them out. Just ordinary people, it seemed to me, who reminded me of the many ordinary Jews I've known on Long Island.

Here it might be a good idea to pause and clarify what I've written thus far for the sake of the perplexed reader, who may be acquainted only with Jews who seem like regular people, as many of them are, if a bit more neurotic on average than the rest of us. But there's a lot more to it than that. Jews are an ethnic group, or subrace if you will, whose origins are a matter of debate. Many are recognizable by that unappealing "Jewish look." Until the creation of Israel in recent times, they never had a true nation of their own. Their entire history is one of migrating in small numbers to host countries where the pickings are good, and rising to the heights of wealth and political power while supporting or engaging in all kinds of subversive activity, though this does not describe all Jews, of course. Alone among advanced peoples, in which there is a natural division of brains and brawn, Jews shun physical work. Granted that there are exceptions, this phenomenon is easily observed everywhere they live, including Israel, where most necessary labor is performed by 300,000 guest workers, mostly Asian. This is subsidized by American and German taxpayers, and infusions of cash from wealthy Jews living abroad, without which this nation would collapse in a week. Israel is the world's largest parasite colony. There's no country on earth like it.

There are good and bad in all peoples, as the saying goes. This includes Jews. The difference is that the percentage of bad Jews — those ranging from the generally unpleasant and unethical, to the predatory, to the truly monstrous, is astronomically higher when measured against any other people. By the same token there are Jews who live respectable lives and have honest jobs, and Gentiles who are born criminals or live parasitically, milking the System for all it's worth. This does not change the fact that, taken as a whole, Jews are biological parasites, as repugnant as that might sound. There's no other way to explain the fact that, collectively speaking, they have always, without exception, subsisted on the industry of others, while at the same time destroying them. Unlike every other white ethnic group, they are incapable of creating and sustaining a society of their own. I'll apologize for writing these words and destroy all copies of this book in my possession when I see evidence to the contrary.

If I were asked to offer one excerpt that best condenses my viewpoint, it would be the following from *The New Unhappy Lords: An Exposure of Power Politics* by British racial nationalist A.K. Chesterton, published in 1965:

World Jewry is the single most powerful force on earth and it follows that all the major policies which have been ruthlessly pursued through the last several decades must have had the stamp of Jewish approval. Indeed, common sense applied to such facts as have come to light must lead to the conclusion that the policies, directed against the most cherished Gentile values, were incubated by adroit Jewish brains and fulfilled, or carried to the verge of fulfillment, by the dynamism of the Jewish spirit. At the same time, so many Gentiles are associated with the conspiracy, both directly and through the formation of fronts, there are so many Gentile agents and agencies, and so many Gentile governments which have acquiesced in the conspiracy by falling into line with policies inimical to their own national interests, that it would be ludicrous to offload upon Jewish shoulders responsibility for the destruction, or near destruction, of Christendom and the Western world. Nevertheless it would be equally ludicrous to deny the Jewish part, especially where it is admitted.... Had we of the Gentile nations stood firm in defence of our own traditions and values, instead of cravenly capitulating, the Jews would have remained what they ought to be — a small sect living contentedly and at peace with their neighbours, exercising neither national nor international power and entertaining no inordinate ambitions. That, as I wrote at the outset, is how most of them actually do live. That a minority of them has been able to mount such a stupendous drive for world power is not their fault but ours alone, and it is we who must put things to rights — or perish.

As fair and balanced as this appraisal is, Chesterton, probably out of politeness, left a great deal unsaid, even though earlier in the same chapter he tackled the subject of Jewish subversion in some detail. What was implied in his choice of words like "destruction" and "perish" is a reality that, I'm afraid, most readers of these lines would prefer not to face. In the worst case scenario, if ultimate Jewish ambitions were to go unchecked, it would pave the way for the extinction of *Homo sapiens*, which Hitler foresaw — and he wasn't the first one. Short of extinction, it would mean the endless ruin of people in all forms and degrees, and frequently death on a cataclysmic scale, which is the ongoing world situation, the previous century being the most violent. For those interested in truth, I highly recommend any of the most up to date and incisive books on this topic, namely David Duke's *My Awakening*, *Jewish Supremacism* and *The Secret Behind Communism*, and Michael Collins Piper's *The Confessions of an Anti-Semite*, *The Golem*, *The Judas Goats*, *The New Jerusalem*, and *The High Priests of War*. Another excellent book I read in the midst of revising this chapter is *The Jewish Mafia: The Great International Predators* by French author and political activist Herve Ryssen, who has been jailed in France several times for his troubles. This work furnishes an overview of how legions of vile Jews, some recently departed, engage with impunity on a global scale in every imaginable sphere of criminal activity including, but not limited to, the heroin and ecstasy trade, prostitution and sex slavery, pornography, diamond and heavy weapons smuggling, trafficking in human body parts (often removed from murdered Palestinians), pirating of the aluminum and petroleum industries, money laundering, and banking and financial swindles of enormous proportions. One thinks here of the Rothschild dynasty and its affiliates which have financed and profited from so many wars; of Bernie Madoff, who stole $65 billion from investors, the biggest theft in the history of the world, and who probably went to prison only because some Jewish celebrities were among his victims; and of the Hungarian Jewish "philanthropist" George Soros, who likes to underwrite revolutions and who singlehandedly nearly destroyed the economies of four Asian countries in 1997, whose depredations are known throughout the world, and yet who somehow remains untouchable. If, however, you'd rather not read any of these books, and you're among the ninety percent of American Christians who not only has a Bible at home but might believe that some or all of it is true, check out a few of these chapters and verses from the Old Testament: Genesis 22:1-12, 45-47; Exodus 11:4-5; Leviticus 25:44-46; Numbers 21:2-3; Deuteronomy 2:33-35, 6:10-11, 7:1-8, 12:29-30, 20:16-17; Joshua 6:17-25, 8:24-28; Esther 9:5-6; Isaiah 13:15-16, 60:16, 61:5-6; Jeremiah 50:21. Here you'll get a bellyful of mass murder, cities totally destroyed, stealing land and dispossessing the inhabitants, enslaving people, living like kings without doing any work, maintaining racial purity — all of it directed by God and endorsed by Pence, Pompeo, and other Zionist lackeys scattered throughout the Deep State. I don't know of a more gory and psychotic work of literature. It doesn't matter much if most biblical stories are embellished or purely fictional; what matters is that in ancient times, as today, they define what it means to be Jewish. It says something that the world's number one runaway bestseller, which exposes how Jews think it's their divine right to do all these things, is rarely read by most Gentiles, who know nothing about it. And I've jotted down just a little bit; the Old Testament, which of course is the traditional Hebrew Bible, is chock full of these messages exposing how these people have been at war with the rest of the world for more than three thousand years.

Naturally, when such a people think they're morally superior to all other beings, you're bound to run into some real head cases, and we do indeed find that by their own admission, Jews have a much higher rate of clinical insanity than any other ethnic group — a lively topic among themselves on the Internet. Jews are a racial genotype, with individual variations, of course, as in all genotypes. Being Jewish has nothing to do with religion. The communist or Zionist Jew who never attends synagogue is the same fanatic as the most devout Hasidic Jew, or those of the slightly more mainstream Lubavitch-Chabad persuasion. The doctrinal core for most Jews, including those who aren't observant, is the Talmud, a voluminous literary work having many definitions that deliberately serve to confuse those who aren't supposed to read it. Basically, it's a series of rabbinical teachings first compiled around 200 A.D., and continually revised and expanded, explaining how the commandments of the Torah, the first five books of the Old Testament, are to be followed. The Talmud is taken very seriously by Jews with the strongest feeling of Jewish identity or religious conviction. It dwells on ways to cheat, rob and murder Gentiles; discusses sexual relations between adults and babies; calls Jesus Christ a sorcerer born to a whore during menstruation who is boiling for eternity in hot excrement (so much for our "Judeo-Christian heritage," just another hoax to be trotted out when it serves Jewish purposes). It lays down rules for disavowing solemn oaths before they're made, hitting one's parents after death, hair combing, pushing doors open, and other peculiar topics. The Talmud is so weird, so utterly demented, that it can actually be fun to read, and because of its ancient legacy, I suspect that some of it rubs off even on more sensible and moderate Jews. The mass expulsions of Jews in medieval times were often accompanied by public burnings of the Talmud by enraged Christians. Although most Christians today have never heard of the Talmud, it continues to wield plenty of influence over the minds of Jews. It is still studied and discussed in Jewish schools called yeshivas, which abound in the greater New York area, and its exhortations to murder along with its dripping hatred for all non-Jews resonate beyond the yeshiva circle. It has inspired boatloads of bloodcurdling quotes by prominent Jews, including quite a few from the lips of Israeli prime ministers. Modern editions of the Talmud are readily available, but shocking passages have been deleted, allowing Jews to deny that they ever existed — just another lie. Presumably Jews who study it closely use unexpurgated versions, difficult to obtain and very expensive.

I would be remiss here if I failed to mention some courageous Jews who have written books exposing the ugly truth about their own people, and the state of Israel in particular — foremost among them the late Alfred Lilienthal, author of *The Zionist Connection*, which was the first book I read, a long time ago, that reduced to rubble all the myths about this country and set the record straight for me. *By Way of Deception* by Victor Ostrovsky is one of the very few books mentioned in this chapter not to be buried by the far-reaching network of Jewish censorship. Not only was it printed, in 1990, by a major New York house, St.

Martin's Press, despite legal attempts by top Israeli officials to prevent publication, but amazingly it shot to the top of the *New York Times* bestseller list. Ostrovsky, born in Canada and raised in an ardent Zionist household, moved to Israel at a young age and was later recruited into the Mossad, the Israeli version of the CIA. He become a top level operative but, stung by conscience, reached a point where he could no longer endure the Mossad's murderous methods, and decided to reveal all their dirty secrets. Ostrovsky feared for his life and went into hiding even as his book sold like gangbusters. But far and away the most damning and sweeping work I've ever read, a book unprecedented in scope, is *Tell the Truth and Shame the Devil* by Gerard Menuhin, son of the late Yehudi Menuhin, the acclaimed conductor and world class violinist. This is the most no-holds-barred, information-packed, anti-Jewish book I have ever read — written by a Jew who totally renounced his Jewish identity. In the same vein, another admirable Jew, Oscar Levy, a renowned scholar and translator of German philosopher Friedrich Nietzsche's works, held nothing back in exposing organized Jewry, in the process disavowing his people forever. Asked by George Pitt-Rivers to review his manuscript *The World Significance of the Russian Revolution*, Levy wrote him a letter that the author made the preface of his short work, which exposed the real perpetrators of that apocalyptic event, and which I'll go into below. If anything, the following excerpt is more apropos today than when it appeared in print a century ago:

There is no race in the world more enigmatic, more fatal than the Jews. The question of the Jews and their influence on the world past and present, cuts to the root of all things, and should be discussed by every honest thinker....

You point out, with fine indignation, the great danger that springs from the prevalence of Jews in finance and industry, and from the preponderance of Jews in rebellion and revolution. You reveal the connection between the collectivism of the immensely rich International Finance — and the International Collectivism of Karl Marx and Trotsky [whose real name was Lev Bronstein] — the democracy of and by decoy cries. And all this evil and misery, the economic as well as the political, you trace back to one source — the Jews.

There is scarcely an event in modern Europe that cannot be traced back to the Jews. Take the Great War [World War One] that appears to have come to an end. Ask yourself what were its causes and reasons....

You have noted with alarm that the Jewish elements provide the driving forces for both Communism and Capitalism, for the material as well as the spiritual ruin of the world. From Moses to Marx, in practice and theory, in idealism and in materialism, in philosophy and in politics, they are today just what they have always been — passionately devoted to their aims and purposes....

If you are an Anti-Semite, I, the Semite am an Anti-Semite too, and a much more fervent one than ever you are. We have erred, my friend, we have grievously erred. We who have posed as saviors of the world; we who have even boasted of having given it "the Saviour" — we are today nothing else but the world's seducers, its destroyers, its incendiaries, its executioners. We, who have promised to lead you to a New Heaven, we have finally succeeded in leading you into a New Hell!

That was written in 1920. And what New Hell has Jewry led the world into today? Wars, upheaval and human suffering without end in the Middle East, millions dead, disfigured or displaced, tidal waves of refugees flooding into Europe to the infernal delight of Jews who are intent on conquering all the territory between the Nile and the Euphrates as ordained by God, and on ruling over a world in which all national boundaries and all human races, especially the white race, have been mongrelized, except, of course, theirs, the Jewish master race — a world in which everything that does not conform to Jewish law is torn down and trampled to dust.

It's depressing for me to sit here and write all this. As a live and let live kind of guy, nothing would make me happier than to see the leaders of Israel form a constructive, lasting relationship with the Palestinians and with neighboring Arab states — as indeed, most countries have with their immediate neighbors. Nothing would make me happier than to see the most aloof and insular type of Jew, like those in the various orthodox and Hasidic sects who constitute a large population in the New York City borough of Brooklyn, remain a tight-knit group and preserve their traditions — but also quit robbing the System blind by subsisting on undeserved welfare benefits or shady incomes — and become productive citizens while living on friendly terms with the outside world, like the Amish communities in Pennsylvania do. But I'm a realist. And being a realist means learning from history. In all history, there isn't a single instance of a Jewish population of any appreciable size living in harmony with other people. In every instance they have brought hatred and resentment down on their heads because of who they are and what they do.

❖

Despite the fact that Jews constitute only about three percent of the U.S. population, America is the most Jewish nation of all time after Israel, and in a sense it's even more Jewish. Numerically Israel ranks first, but America is the real citadel of Jewish wealth, power and influence. Many American Jews consider America the Promised Land, *their* country, because they have infused all our institutions with their character. The principles on which America was founded have nothing to do with Americanism today, which is represented by a flag that may as well have fifty Stars of David, so intimately tied to Israel and Jewish attitudes has this country become. Jews and Jewish organizations posturing as patriotic and champions of freedom deceive only the shallow, and this sadly includes many in law enforcement and the military. A good example is the FBI's groundless preoccupation with supposed Russian meddling in the 2016 presidential election, and the endless boring news coverage which followed it, while no one could possibly run for president in the first place without first proving his unconditional support for Israel, as Trump did and has never stopped doing. Apparently this isn't meddling. This whole Russia soap opera, followed by the Ukraine soap opera, appears to be as bogus as the pogroms of past ages and Soviet anti-Semitism in the 1970s.

The gigantic menorah erected near the White House every December with a ceremonial lighting, along with the conspicuous absence of Christmas decorations, which are banned, is a fitting symbol of our times. New York City is of paramount importance in this respect, and not only because so many Jews live in the area. Since well before the establishment of the state of Israel, New York has been the global hub of Jewish banking and media control. It was the staging ground of the 1917 overthrow of czarist Russia, a country violently hated by the Jewish financiers who bankrolled the revolution. Behind its pie-in-the-sky charter, the dysfunctional United Nations, now running out of money, was once the dreamed-of center of Jewish world government. These days, the governments of 27 states require public workers, as a condition of employment, to sign a pledge that they won't advocate any kind of boycott against Israel. Incredibly, or maybe not so incredibly, on March 23, 2017 Maryland senator Ben Cardin, who is Jewish, and Ohio senator Rob Portman, who isn't, introduced the Israel Anti-Boycott Act, which would have made verbal support of such a boycott an offense punishable by up to twenty years in prison and a one million dollar fine. Maybe they never heard of the First Amendment. Like quite a few Jews, particularly in the entertainment industry, Cardin's forebears anglicized the family name, Kardonsky, to conceal their Jewishness. Thankfully, the bill ran into some strong opposition, but at this writing its supporters are still trying to sneak in a watered-down version of it.

I have a two-volume set, *Encyclopedia of the Palestine Problem*. The author, a Palestinian refugee named Issa Nakhleh, compiled detailed reports of thousands of killings, bombings, beatings, groundless imprisonments, house demolitions, razing of villages, and other outrages committed by Jews against innocent civilians, mainly Palestinians, but also Arabs in Egypt, Jordan, Syria and Lebanon from 1939 to 1991, the year of publication. All of this continues, of course, as do worldwide assassinations of those considered enemies of Israel by roving Mossad hit squads. None of this makes the slightest impression on our politicians and Christian Zionist fanatics. Jews have such a stranglehold on the global media, and are so adept at turning reality inside out, that for a long time they convinced much of the Western world that they were the victims and their victims the terrorists. That perception has changed considerably in recent years, though not as much in the U.S. as in Europe, and not at all among a solid chunk of our populace and politicos, who interrupted a 39-minute speech given in the Senate chamber by Israeli Prime Minister Benjamin Netanyahu on March 3, 2015 with 26 standing ovations. Every Israeli leader has been a mass murderer, but Netanyahu has racked up one of the highest scores, and is so crooked his own citizens call him the "crime minister." Here he's given a free pass and worshiped in the halls of Congress.

Israel, and what amounts to its western province, the U.S., are the world's top two rogue nations. Jews, generally speaking, view the world in extremely subjective terms, and through their control of television have molded the minds of

Americans so thoroughly, seemingly to the point of genetic modification, that they too are unable to see reality beyond their own noses. Americans are the most lied-to and deluded of all people, the most diverse and thus the most divided, the most neurotically suspicious, as in the "If you see something say something" mantra and bag or metal detector checks that have become as familiar here as they are in Israel. Also like Israel, we bomb the hell out of everyone, then when a few fight back or avenge the loss of loved ones, we call them terrorists. Sadly, quite a few of our police, trained in Israel or by Israelis in this country, have become badge-wearing pit bulls, and Israeli influence on our military is even worse, as evidenced by the sickening atrocities carried out against helpless Iraqi soldiers, like the Highway of Death massacre bombing on February 26, 1991, a killing frenzy that would have done Genghis Khan and his Mongol hordes proud, the burying alive of surrendering troops in bulldozed trenches, and the tortures carried out for personal amusement at the Abu Ghraib prison, some sexual in nature, photographs of which can be seen on Wikipedia. Undoubtedly, Israeli advisors were on hand for all the fun. This carried over from the worse horrors inflicted on German civilians and soldiers before and after the end of the Second World War by American troops pumped full of propaganda spread by hate-crazed Jews. It's no coincidence that we Americans are at or near the top of the list in exporting terrorism, in the number of lawyers and stifling laws on the books, in our rate of incarceration in profit-run prisons, in our lack of medical freedom, in the number of over-vaccinated children, and, seen from abroad, in our barren culture, exemplified by the cinematic manure of Hollywood and the glitzy putrefaction of the Super Bowl halftime show.

There's a silver lining in all this. Thanks to the Internet, we're now aware that there's a growing number of Americans, and English speakers abroad, who know that Israel is a basket case and its American sponsor a Jew-run abomination. Cracks have formed in the global empire and widen all the time. On November 12, 2015, a Spanish judge, Jose de la Mata, signed a warrant for the arrest of Benjamin Netanyahu and six of his cabinet members, meaning that he would have been handcuffed and taken into custody had he set foot in Spain. This stemmed from the 2010 attack by Israeli naval forces on the *Mavi Marmara* in which some Spanish nationals were shot to death. This vessel, part of a six-ship flotilla, was attempting to break a blockade in order to deliver humanitarian aid to the Gaza strip. In 2001 a Belgian court issued a similar warrant against then Israeli head of state Ariel Sharon, another mad-dog killer, for his role in the September 1982 butchering of some 1500 Palestinians by Israel's Lebanese "Christian" allies at the Sabra and Shatila refugee camps in Beirut. Unfortunately, nothing came of these two injunctions, but they do point to an expanding awareness in the West of the nature of Israeli leadership. And while American politicians trip over themselves in condemning the movement to boycott the pariah state, the Irish parliament heads in the opposite direction, seeking to criminalize any transactions with businesses or individuals living on land seized by Israel after the 1967 Six-Day War. That legislation is pending at this writing. The 'Net also sizzles with scrutiny into Israeli involvement in the events of September 11, 2001. Having read and heard a great deal on what happened that day, and observing that there are Jewish fingerprints all over the crime scene, I'm convinced that this was a false flag operation conceived in Tel Aviv and carried out with the collaboration of some treasonous American military men. Israel was the only beneficiary of these terror attacks; everyone else lost.

It's most ironic that much of what I've discussed thus far is generally abhorrent to American Gentiles, who slavishly abide by Jewish-imposed taboos, while in Israel sensitive issues like Zionist control of "our" congress and even the plight of the Palestinians are more easily tossed around, though usually in the Hebrew language press. Very rarely — so rarely it goes unnoticed — the truth leaks out in the American mainstream media, as it did in the *Los Angeles Times* of December 19, 2008, in an opinion piece titled "How Jewish is Hollywood?" by a Joel Stein. Stein's candor was most refreshing. After expressing his amazement at how dumb the majority of Americans are for not knowing that Jews — in the words of a poll conducted by the Anti-Defamation League, America's *de facto* secret police — "pretty much run the movie and television industries," he reels off a list of eleven top executives in the business, all Jewish. Then, after brushing aside the advice of Abe Foxman, former ADL chief, to downplay the Jewishness of Hollywood, this arrogant buck wraps up by declaring: "I don't care if Americans think we're running the news media, Hollywood, Wall Street or the government. I just care that we get to keep running them."

❖

I introduced this chapter with some quotations from men who lived many years apart — the first of whom is quite famous — who intuited that a certain type of Jew, not at all rare, and conspiring with others of his type, is a menace to a large segment of humanity, if not all human life. This idea meshes with the title I chose for this book, and my purpose in writing this chapter will have misfired if by the end I have not proven to the reader that mass vaccination is part of the blind instinct in some Jews to destroy and murder, whether on a conscious or subconscious level. Let's descend a little further, one step at a time. Some readers who are still with me may think I'm overdoing it, but I would remind them that this is for all the marbles, as in "Will higher forms of man cease to exist," or even "Will some deranged Jew unleash Israel's rarely mentioned, burgeoning nuclear arsenal in an attempt to end all life on earth?" So bear with me as I drive my point home.

If there's any doubt as to how the top Israelis see the rest of the world, what follows might be illuminating. To my mind, Menachem Begin was the most savage of all that country's leaders, and that's saying something. That he received the Nobel Peace Prize in 1978 says much about how the world works. His term coincided with the Carter and Reagan years, but he had risen to fame much earlier as head of the underground Irgun, which waged a successful terror campaign to drive the British out of their Palestinian dependency and the local Arabs off their ancient homeland. I will not recount details of the rivers of blood this man let loose, other than to say he was the worst nightmare of the reincarnated Old Testament Jew. The following quote, attributed to him in a speech he made before the Knesset, the Israeli parliament, was published in the June 25, 1982 edition of *New Statesman*, a British political magazine:

Our race is the master race. We are divine gods on this planet. We are as different from the inferior races as they are from insects. In fact, compared to our race, other races are beasts and animals, cattle at best. Other races are considered as human excrement. Our destiny is to rule over the inferior races. Our earthly kingdom will be ruled by our leader with a rod of iron. The masses will lick our feet and serve us as our slaves.

Is a pattern beginning to emerge here? One among his few good friends, whom he undoubtedly regarded as a "useful idiot," as Lenin used to call influential Westerners who believed in communism, Begin counted the Reverend Jerry Falwell, head of the now defunct Moral Majority, a flag-waving, God-fearing outfit that had considerable clout in the seventies and eighties. A typical homegrown religious nut, Falwell preached about the end days when Jesus would float down to earth and zap the righteous into Heaven, while those who had turned their backs on Israel were in big trouble — or something along those lines. The late Falwell has his imitators, including Jerry Jr., who keep setting new "end of the world" dates that come and go — and when nothing happens they just move the date up a few years. And thirty or forty million Americans go right on believing these things.

Begin was an obvious mutant, even by Jewish standards, but you can find countless infernal quotes like this spoken or written by prominent Jews just in the books I've mentioned, and many more of recent vintage reported on truthful Web sites and other books and publications. Certain Jews have threatened to utterly destroy Iran, and even at this late date Germany, with nuclear weapons. Others in Israel have expressed a desire to blow up the entire planet if they feel their country is imperiled. Seymour Hersh, the veteran American investigative reporter, has written some excellent books on controversial topics. He delved into the danger of Israel's nuclear stockpile in *The Samson Option*, Samson being the biblical figure who, as related in the Book of Judges, pulled down the temple in a display of superhuman strength, killing all his Philistine enemies along with himself. While certainly a valuable book, I found it ambiguous in places, perhaps because Hersh is Jewish and retains some loyalty to his people. A much better book, in my opinion, is Piper's *The Golem*. Menuhin's *Tell the Truth* also touches upon aspects of this strange Jewish lust to exterminate all humanity if they themselves feel threatened with extermination. No other people entertain such thoughts. Between these three books, one will discover ample evidence of this, including deeply disturbing statements from leading lights in media, education and government, regarding Israel.

This troubling fixation on killing finds its expression in everything from shooting Palestinian children throwing stones to ending life on Earth.

Not long ago I read *The Making of the Atomic Bomb* by Richard Rhodes, and must confess that I was ignorant of the amazing preponderance of Jews engaged in the development of this hellish weapon, known as the Manhattan Project. The presumably Gentile Rhodes is one of those mawkish types who sees in Jews only an intelligent and industrious people with so much to offer the world, so unjustly persecuted through the ages, always wronged, never wrong. His stuffy book, a Pulitzer Prize winner, is shot through with myopic references to Hitler and his dreadful Nazis. Chapter 7, "Exodus," harps exclusively on the supposed maltreatment of Jews from the sixth century B.C. onward. The fact that so many Jews were involved in constructing a weapon of ultimate destruction, that he felt compelled to devote a full chapter to them, is lost on him.

None other than Albert Einstein was a central figure in this story. Of course everyone "knows" that Einstein was one of the greatest scientists of all time, just like everyone "knows" about Jonas Salk's life-saving polio vaccine. Indeed, thanks to the usual Jewish media brainwashing and habit of aggrandizing one of their own, the mention of his name invokes the word "genius," and the utterance of "genius" invokes Einstein. They're synonyms. Very few, however, including me, have any idea of what his theory of relativity is all about, and what he actually *did* in the realm of physics. There were other eminent twentieth century physicists, Max Planck, for example, who are never described as geniuses and are widely unknown. They weren't Jewish. Both Revilo Oliver and William Pierce (who was a physics professor at the University of Oregon) acknowledged that Einstein had made some noteworthy contributions but was vastly overrated. In recent years, however, there's been much buzz on the Internet claiming Einstein was a run-of-the-mill nobody who shoplifted the work of his peers and passed it off as his own, *a la* Salk. His contemporary, the inventor Nikola Tesla, called him "a long-haired crank."

Whatever Einstein did or did not accomplish, outside of his chosen field he had the mind of a third-grader. Professor Oliver considered him a borderline mattoid, and despite his mild praise, wrote that the world would be a better place had he never been born. Einstein appears to have been the first human being ever to conceive of nuclear weapons. In a letter of August 2, 1939 to President Franklin Roosevelt, while Europe was still at peace, co-signed by the Hungarian Jew Leo Szilard, he falsely insinuated that Germany, his native land which he had left shortly after Hitler assumed power, was already developing an atomic bomb, and persuaded FDR to begin funding similar research. Like the American Jew J. Robert Oppenheimer, who later voiced his regret that the bomb was not dropped on Germany, Einstein hated not only the National Socialist regime but the German people as a whole. He had, in fact, nursed an inexplicable contempt toward his country long before World War One and the subsequent rise of Hitler. In a letter written after the U.S. had entered the Second World War, he referred to their "wretched traditions," calling them "a badly messed-up people," adding, "I keep hoping that at the end of the war, with God's benevolent help, they will largely kill each other off." Behold the great genius, *Time* magazine's "Person of the Century."

Einstein got the ball rolling, but it was Oppenheimer who ran with it. The Manhattan Project's top scientist, rightly called "The Father of the Atomic Bomb," Oppenheimer, a tall, quiet chain smoker, given to bouts of depression and rage, was unstable to say the least. He was a pathological liar as well. He was also a strong communist sympathizer who surrounded himself with known communists during the wartime bomb research. Although it was never proven, many suspected him of being a Soviet agent: on his watch atomic secrets leaked out and made their way to the Soviet Union, which exploded its first atomic bomb on August 29, 1949. Eventually his security clearance was revoked by the U.S. Atomic Energy Commission. Upon witnessing the successful detonation of this horrific weapon at a test site in the New Mexico desert on July 16, 1945 (it would be dropped on Hiroshima three weeks later), Oppenheimer, as he later reminisced, thought of a scripture from the *Bhagavad Gita*: "Now I become Death, the destroyer of worlds." In a later interview, he said, "I hope they cannot see the limitless potential living inside of me to murder everything. I hope they cannot see I am the great destroyer."

While some Jews daydream about destroying the entire world, others would be content with blowing up just a continent or two. Introducing Pam Geller, a fellow Long Islander and a real conservative who, in past years, was a regular on Fox News and is still active on the Internet, doing her patriotic best to protect our American freedoms. Pam is a big Trump fan. She is also a deeply disturbed woman who has a very weak grip on reality and an unhealthy obsession with the religion of Islam. We are indebted to her for constantly warning us that Sharia law is coming soon to a neighborhood near you. Not many people take her seriously; even committed Zionists give her a wide berth. This is what she wrote on her bygone blog *Atlas Shrugged* on February 24, 2010: "I pray dearly that in the ungodly event that Tehran or its jihadi proxies target Israel with a nuke, that she retaliate with everything she has at Tehran, Mecca and Medina.... Not to mention Europe. They exterminated all their Jews, but that wasn't enough. Those monsters then went on to import the next generation of Jew killers."

❖

Zionism — presented to the world as a high-minded movement to establish a Jewish homeland in the Middle East — and now-obsolete communism — obsolete as an international conspiracy, but very much alive in the mentality of street scum like Antifa — are two sides of the same Jewish coin. The connection between them is historically fascinating but far too complicated to explore here. In a word, these international conspiracies are, or in the case of communism were, alternative paths to Jewish world domination, fluctuating between cooperation with and opposition to each other. Whether Israel became a formidable nuclear power primarily by way of the old Soviet Union or through its spies operating on American soil, by 1960 it was no longer a secret. David Ben-Gurion, the Mideast state's first prime minister, referred to his country's nuclear warheads as "holy." Ben-Gurion was a garden-variety messianic Jewish terrorist who once said, "I believe in our moral and intellectual superiority, in our capacity to serve as a model for the redemption of the human race." The January 16, 1962 issue of the popular American weekly *Look* magazine featured responses to a question put to a cross section of famous people, as to what the world would look like in 25 years. This, in part, was Ben-Gurion's vision of the world in 1987: "With the exception of the USSR as a federated Eurasian state, all other continents will become united in a world alliance, at whose disposal will be an international police force. All armies will be abolished, and there will be no more wars. In Jerusalem, the United Nations (a truly *United* Nations) will build a Shrine of the Prophets to serve the federated union of all continents; this will be the seat of the Supreme Court of Mankind, to settle all controversies among the federated continents, as prophesied by Isaiah."

President Kennedy was adamantly opposed to Israel acquiring nuclear weapons, and his acrimonious relationship with Ben-Gurion over this one issue is well documented. In his book *Final Judgment*, Michael Piper makes a strong case that this was one strand in the intricate conspiracy to assassinate the president, an element that no other historian has dared to touch. Also widely known was JFK's father's antipathy towards Jews and his admiration of Hitler's accomplishments in Germany. Others have pointed out that Kennedy was deemed a threat to the privately owned and very Jewish Federal Reserve Bank, which supplanted the nation's financial system in the fateful year of 1913, the same year the income tax was established and the sinister ADL was founded in reaction to the lynching of Leo Frank, an Atlanta Jew who had murdered a Christian girl named Mary Phagan. The president probably further incurred the wrath of Jewry by sacking the Jewish boss of the Joint Chiefs of Staff, Lyman Lemnitzer, after getting word of his false-flag plan Operation Northwoods, which called for the killing of American citizens and Cuban emigres in various ways, to be blamed on Fidel Castro. While a butcher in his own right, Castro had made even more enemies by becoming a "disobedient" communist and shutting down the gambling casinos in Havana, the playground of many Jewish gangsters, who handsomely profited from them. While no American hero, Kennedy did have a few good points.

I've written all this to impress upon you, my reader, the kind of mentality that I believe dwells, to a greater or lesser degree, in those Jews who push hard for vaccines. Now let's dig a little deeper into the subject of Jewish murder, taking quick note that, just as a remarkably high degree of Jews are found in the vaccine field, they also flock to another "medical" field, the murderous nature of which cannot be denied: about half the doctors who perform abortions in this country are Jewish.

❖

It was the misnamed Russian Revolution of 1917 and the forty years that followed that provides history's ghastliest example of Jewish tyranny. This event was soaked in vengeance due to many years of restrictive Russian laws that held Jews down, though daily life for them was not nearly as oppressive as what the Palestinians have long endured — laws that had been enacted due to Jewish exploitation and ruin of the Russian peasantry. There was nothing Russian about this revolution; it was the Jewish overthrow of czarist rule. The Jewish nature of this historical watershed — spattered by the shooting and bayonetting of Czar Nicholas, his wife and their five children on July 17, 1918 by Jewish Bolshevik animals, after they had spent months under house arrest — is openly admitted even in Jewish publications. That Lenin had only one Jewish grandparent and Stalin none, is as irrelevant as the fact that no Jew has ever been elected to the presidency here. It has always intrigued me that *Mein Kampf* is supposed to be proof of Hitler's diabolical madness, yet no one ever disputes what he wrote. Why is that? Because on every page he forcefully wrote the truth, as in his analysis of how Marxist Jews seize power, as they came very close to doing in Germany:

In the political field he refuses the state the means for its self-preservation, destroys the foundations of all national self-maintenance and defense, destroys faith in the leadership, scoffs at its history and past, and drags everything that is truly great into the gutter. Culturally he contaminates art, literature, the theater, makes a mockery of natural feeling, overthrows all concepts of beauty and sublimity, of the noble and the good, and instead drags men down into the sphere of his own base nature. Religion is ridiculed, ethics and morality represented as outmoded, until the last props of a nation in its struggle for existence in this world have fallen. Now begins the great last revolution. In gaining political power the Jew casts off the few cloaks that he still wears. The democratic people's Jew becomes the blood-Jew and tyrant over peoples. In a few years he tries to exterminate the national intelligentsia [in keeping with the Talmudic decree, "The best of the Goyim shall be killed"] and by robbing the peoples of their natural intellectual leadership makes them ripe for the slave's lot of permanent subjugation. The most frightful example of this kind is offered by Russia, where he killed or starved about thirty million people with positively fanatical savagery, in part amid inhuman tortures, in order to give a gang of Jewish journalists and stock exchange bandits domination over a great people.

It was Alexander Solzhenitsyn's opus *The Gulag Archipelago*, published in 1973, that brought to the world's attention the inferno of the Soviet prison system, which he knew from the inside. While not probing the Jewish nature of communism in *Gulag*, he did publish the names and photographs of the top six administrators of the labor camps where millions were worked to death. All six were Jews. He was expelled from the Soviet Union in 1974, and eventually settled in rural Vermont, where he lived for eighteen years. Although for a short time in the 1970s Solzhenitsyn was accorded hero worship, he never went along with the canard of "Soviet anti-Semitism" fashionable that entire decade, and was never really trusted by those in power. He alienated America's power brokers further by blasting the crass materialism and cultural degeneracy of his adopted country, reserving his harshest words for news media prostitutes, who promptly vanished him with the silent treatment, as they had done to Charles Lindbergh thirty years earlier.

The great writer returned to his beloved homeland in 1994. In his later years he researched and wrote what was to be his last book, *Two Hundred Years Together*, on the perpetual hostility between Jews and Russians. Published in 2002, it's the most detailed work on the Jewish role in the ascendancy of communism ever written. In it, Solzhenitsyn lays the death of 66 million Russians and kinfolk at the door of Jewry. How he arrived at this mind-boggling figure I have no idea, though I have no reason to believe a man of his character would lie about it. Because of the "fear of the Jews" syndrome, no publisher has ever translated it into English (it has been translated into French and German), though from what I understand, portions of it in English have recently been posted on the Internet. An uneven discussion of this book can be found on Wikipedia, which, as noted, consistently displays bias and distortion on controversial subjects, including vaccines. Solzhenitsyn succinctly stated his case elsewhere:

You must understand. The leading Bolsheviks who took over Russia were not Russians. They hated Russians. They hated Christians. Driven by ethnic hatred they tortured and slaughtered millions of Russians without a shred of remorse.... More of my countrymen suffered horrific crimes at their bloodstained hands than any people or nation ever suffered in the entirety of human history. It cannot be overstated. Bolshevism committed the greatest human slaughter of all time. The fact that most of the world is ignorant and uncaring about this enormous crime is proof that the global media are in the hands of the perpetrators.

Only a fool would deny the likelihood that this inner drive to attain political power and then to

You are in: World: Middle East

Front Page
World
Africa
Americas
Asia-Pacific
Europe
Middle East
South Asia
From Our Own Correspondent
Letter From America
UK
UK Politics
Business
Sci/Tech
Health
Education
Entertainment
Talking Point
In Depth
AudioVideo

Tuesday, 10 April, 2001, 16:01 GMT 17:01 UK

Rabbi calls for annihilation of Arabs

Rabbi Yosef is known for his outspoken comments

The spiritual leader of Israel's ultra-orthodox Shas party, Rabbi Ovadia Yosef, has provoked outrage with a sermon calling for the annihilation of Arabs.

"It is forbidden to be merciful to them. You must send missiles to them and annihilate them. They are evil and damnable," he was quoted as saying in a sermon delivered on Monday to mark the Jewish festival of Passover.

Rabbi Yosef is one of the most powerful religious figures in Israel, He is known for his outspoken comments and has in the past referred to the Arabs as "vipers".

> **The Lord shall return the Arabs' deeds on their own heads, waste their seed and exterminate them**
>
> **Rabbi Ovadia Yosef**

Through his influence over Shas, Israel's third largest political party, he is also a significant political figure.

As founder and spiritual leader of the political party Shas, Rabbi Yosef is held in almost saintly regard by hundreds of thousands of Jews of Middle Eastern and North African origin.

Search BBC News Online
GO
Advanced search options
Launch console for latest audio/video
BBC RADIO NEWS
BBC ONE TV NEWS
WORLD NEWS SUMMARY
BBC NEWS 24 BULLETIN
PROGRAMMES GUIDE

ISRAEL & the PALESTINIANS
FULL COVERAGE

Key stories
- Palestinian disarray
- Backing the bombers
- US demands
- Distant hope
- Arafat's successors?

Eyewitness
- Gazans kept apart
- Teenagers in uniform
- A family divided
- US Christians for Israel

Background
- Running Palestine
- Intifada Q&A
- History of bomb blasts
- Country profiles

FACTFILE
- Voices from the Conflict

TALKING POINT
- What is the alternative to Arafat?

AUDIO VIDEO
- TV and Radio reports

See also:
- 08 Aug 00 | Middle East
  Profile: Rabbi Ovadia Yosef

Part of a story from the BBC (still available in full on their Web site as of 2020)

commit mass murder — a biological impulse ineradicable in all too many Jews — has not spilled over into their drive to shoot American children full of vaccines.

❖

An affinity for mass murder has always resided in a striking percentage of Jews. It flourishes in modern times, as seen in the April 10, 2001 BBC headline featuring a photograph of the late rabbi, Ovadia Yosef, a major spiritual and political figure in Israel — the kind of headline rarely seen in Great Britain and never seen in the U.S. Although Mr. Yosef wanted all Palestinian Arabs dead, he was willing to spare other Gentiles, in another sermon referring to them as "donkeys" whose only purpose in life was to serve Jews. When the rabbi died on October 7, 2013 at age 93, 800,000 mourners — ten percent of the population — jammed the streets of Jerusalem for his funeral, the largest in Israel's history. The same impulse surfaced in the 1941 self-published book *Germany Must Perish!* by an American Jew named Theodore Kaufman. On the title page, Kaufman wrote, "This dynamic volume outlines a comprehensive plan for the

GERMANY MUST PERISH!

byword of science, as the best means of ridding the human race of its misfits: the degenerate, the insane, the hereditary criminal.

Sterilization is not to be confused with castration. It is a safe and simple operation, quite harmless and painless, neither mutilating nor unsexing the patient. Its effects are most often less distressing than vaccination and no more serious than a tooth extraction. Too, the operation is extremely rapid requiring no more than ten minutes to complete. The patient may resume his work immediately afterwards. Even in the case of the female the operation, though taking longer to perform, is as safe and simple. Performed thousands of times, no records indicate cases of complication or death. When one realizes that such health measures as vaccination and serum treatments are considered as direct benefits to the community, certainly sterilization of the German people cannot but be considered a great health measure promoted by humanity to immunize itself *forever* against the virus of Germanism.

The population of Germany, excluding conquered and annexed territories, is about 70,000,000, almost equally divided between male and female. To achieve the purpose of German extinction it would be necessary to only

• 87

A page from Kaufman's *Germany Must Perish!*

extinction of the German nation and the total eradication, from the earth, of all her people." On page 87, reproduced here, he recommends sterilization to accomplish this. The context is unclear as to whether he is mentioning vaccination as a less desirable sterilization procedure or just making a comparison. In either case, Jews who shout the loudest for wide vaccination coverage probably have a hidden agenda, and it should be noted that in the 1990s, laced vaccines *were* secretly used as an experiment to induce sterilization in tetanus vaccination campaigns sponsored by WHO in Kenya, Mexico, Nicaragua and the Philippines. A related item was reported in the London *Sunday Times* on November 15, 1998 under the heading "Israel planning 'ethnic' bomb as Saddam caves in." While I'm skeptical of some of the content, I would put absolutely nothing past these people. I'll reproduce a few lines here and invite the reader to examine the whole article and make up his own mind:

Israel is working on a biological weapon that would harm Arabs but not Jews, according to Israeli military and western intelligence sources. The weapon, targeting victims by ethnic origin, is seen as Israel's response to Iraq's threat of chemical and biological attacks.... In developing their "ethno-bomb," Israeli scientists are trying to exploit medical advances by identifying distinctive genes carried by some Arabs, then create a genetically modified bacterium or virus. The intention is to use the ability of viruses and certain bacteria to alter the DNA inside their host's living cells. The scientists are trying to engineer deadly micro-organisms that attack only those bearing the distinctive genes. The programme is based at the biological institute in Nes Tziyona, the main research facility for Israel's clandestine arsenal of chemical and biological weapons.... The disease could be spread by spraying the organisms into the air or putting them in water supplies.

❖

When I took the Trans-Siberian railway clear across Russia in 1980 — from the Pacific coast to the Polish border, with stopovers in five cities — the Red terror had long since abated, as had Jewish supremacism, though I'm sure Jews still had it better than most. It was a grim, shabby country, and a glimpse of everyday life confirmed that I would never want to live here. I'll never forget the awful food and the tired, musty smell of all the railroad stations, which a travel writer once described as a combination of body odor, rotten cabbage and cheap cigarette smoke. Russians have a reputation as a tough, humorless people, to which I can attest — understandable after all they've been through. In 1980 there was no fear in the air, though, and ordinary people who spoke some English were eager to strike up a conversation.

As the final decade of the twentieth century got underway, the house of communism, held together for so long by the crumbling plaster of big lies, came tumbling down in the Soviet Union and its east European satellites. In Russia, however, a new round of suffering descended on the common people who, for the most part under communism, were provided with the basic necessities of life if little else. With privatization of the economy, and a drunken court jester named Boris Yeltsin pretending to run the country, a handful of Jewish oligarchs went on a smash-and-grab spree in cahoots with their racial cousins abroad and domestic gangsters, plundering Russia's natural resources, taking over the media, and plunging most Russians to a level of poverty and despair they had never known. The elderly saw their pensions wiped out, fuel for heating in winter grew scarce, and there were extended power cuts. Violent crime ran amok, hunger and malnutrition stalked the land, and the only escape was alcohol or suicide. They were very bad years. In these desperate circumstances, thousands of young Russian and Ukrainian women, looking for jobs to support their families, were lured by newspaper advertisements promising *bona fide* employment in Israel. The ads were traps set by an international network of Jewish pimps. Once they arrived they were kidnapped, imprisoned in brothels or fortified apartments, and forced into prostitution under the most inhumane conditions. In Israel, the world's most ethnocentric state, one set of laws applies to Jews and another set to everyone else. Sexual slavery is legal in this country as long as the captive women are not Jewish, and the few cases that went through the courts in the 1990s were treated nonchalantly by Israeli judges. Amnesty International investigated this terrible situation, and it was even written up in an article headed "Contraband Women" in the very Jewish and pro-Israel *New York Times* of January 11, 1998. While the author used the code phrase "Russian crime gangs" and diluted the story with similar accounts in other countries, it reflected very poorly on Israel. But even though the *Times* is arguably the single most powerful media organ in the world, one article buried in one issue is not going to get noticed, let alone change anything. Have you heard of it? Ryssen's *The Jewish Mafia* devotes one long chapter to the subject in sickening detail, and reviews this distinctive worldwide Jewish tradition through the ages.

Unfortunately, these dastardly crimes committed against pretty, innocent Russian and Ukrainian women isn't the worst of it — far from it. I remembered reading in the underground press years back about an incident that sent shock waves through Italy. Unsuspecting television viewers switched on their sets to a regular news program only to see video footage of small children being sexually tortured to death. I recall that the TV station switchboard lit up with calls from people traumatized by what they had seen, many also enraged that it was shown without warning. But I don't recall reading about a Jewish angle. Just recently, however, I read a book titled *Blood Ritual* by Philip deVier which jogged my memory about this. The author tells us that he accessed a report from a now defunct web site, libertariansocialist.com, on December 9, 2000. Quoting the report from page 163 of *Blood Ritual*:

Rome, Italy — Italian and Russian police, working together, broke up a ring of Jewish gangsters who had been involved in the manufacture of child rape and snuff pornography. Three Russian Jews and eight Italian Jews were arrested after police discovered they had been kidnapping non-Jewish children between the ages of two and five years old from Russian orphanages, raping the children, and then murdering them on film. Mostly non-Jewish customers, including 1700 nationwide, 600 in Italy, and an unknown number in the United States, paid as much as $20,000 per film to watch little children being raped and murdered. Jewish officials in a major Italian news agency tried to cover the story up, but were circumvented by Italian news reporters, who broadcast scenes from the films live at prime time on Italian television to more than 11 million Italian viewers. Jewish officials then fired the executives responsible, claiming they were spreading "blood libel."

DeVier then observes: "Jews are a tiny minority in every country where they live, except in Israel. A minority of the customers might well be Jews, and at the same time the Jews could be vastly over-represented among them. Be that as it may, the *producers* of this hateful filth have been identified as Jews." While he is undoubtedly correct about this, one should not lose sight of the fact that Jews do not have a monopoly on extreme sexual deviancy. *The Franklin Cover-Up* by the late John DeCamp, a decorated Vietnam vet, lawyer, Nebraska state senator and father of four, is the most horrifying book I have ever read. Written at considerable risk to his own life, the author descends into the catacombs of the crimes mentioned in the previous paragraph, as they transpired, around the same time, in the corridors of power in Washington D.C. and the Omaha area, along with the incredibly corrupt court proceedings that accompanied them. If you want to know just how far gone America is, read this book — but I warn you, read it only if you can cope with depths of evil that don't seem possible. I might add that none of the three most notorious convicted sex monsters in this time frame — John Wayne Gacy, Ted Bundy, and Jeffrey Dahmer — were Jewish. Teenage boys were preferred by Gacy, while Dahmer liked them a little older; Bundy preyed on women. If you're curious you can read about them on Wikipedia or watch them interviewed on YouTube. It's particularly instructive to watch Bundy's last jailhouse interview, the day before he was electrocuted at Florida State Prison. Such a nice-looking, soft-spoken, all-American young man from a good Christian family. Golly, if he had been talking about something other than his addiction to pornography, and if you didn't know he had raped and murdered more than thirty young women — beheaded a dozen of them, and many times returned to their hidden, decomposing corpses to violate them again — you'd want your daughter to marry the guy. I say it's instructive because I believe that behind the caring façade of some of these vaccine pushers, who love to make pious noises about safeguarding the health of our children, there burns an unquenchable desire to kill them.

Blood Ritual is an admirably open-minded investigation into charges going back to biblical times that certain Jewish cults have engaged in the ritual sacrifice of Gentiles, mostly children, in ceremonies often tinged by atavistic sexual aberrations and a bitter hatred of Jesus Christ. The author traces the historical record right through to the twentieth century, uncovering some two hundred cases, and concludes that in many of these the evidence is weak, but in many others it's overwhelming, noting that vestiges of child sacrifice have existed in many cultures, and it's absurd to exempt Jews. Interestingly, he notes that there was an uptick in cases in the nineteenth century, as Europeans supposedly became more enlightened and their attitudes towards Jews became much more tolerant than what they were in the Middle Ages. The cover of deVier's lucid book, reproduced here, depicts the murder by three Jews of a young Polish woman named Agnes Hruza on March 29, 1899, for the purpose of collecting her blood. It was taken from a postcard widely circulated in eastern Europe at the time.

The most authoritative work on Jewish ritual sacrifice prior to *Blood Ritual* was Richard Burton's *The Jew, the Gypsy, and El Islam*, posthumously published in 1898. A larger-than-life figure — dauntless African explorer, disguised wanderer in Mideast bazaars, fluent in more than twenty languages, to name just a few marvels about him — Burton's name is recognized by practically no one today. Need I add that practically no one has heard of his book? Few men have been so finely attuned to the repulsive nature of the Talmudic Jew as he, and he analyzed this fact in all its nuances, caring nothing about etiquette. Burton lived and wrote in England's Victorian era. He called out the polite society of his day, forerunners of the modern "Let's all learn to live together" crowd, who were too fearful to reckon with the innate bloodthirstiness of a certain type of Jew.

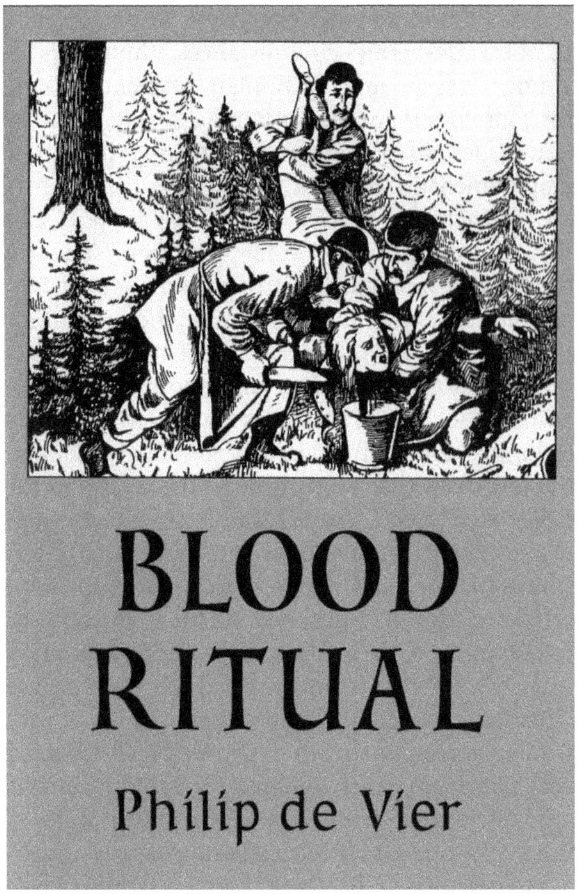

"Blood libel" is the cry certain Jews raise when they are accused of modern crimes, an allusion to what they want everyone to believe are laughably absurd medieval charges that they once used the blood of murdered Christian children in their secret ceremonies. Until recently, in fact, a statue depicting such a murder in 1286 and a mural illustrating a similar child sacrifice in 1475, could be seen by tourists at churches in Oberwesel, Germany and Trento, Italy, respectively. Undoubtedly these were the creations of the same kind of sick people among us today who are always ready to use the Jews as convenient scapegoats, because we all know that such things never happened, just like Benjamin Franklin never said what he said at the Constitutional Convention, just like *The Protocols of the Learned Elders of Zion* was a forgery written by some anti-Semitic screwball, just like Leo Frank was innocent, just like Israeli pilots mistook the *Liberty* for an Egyptian ship, just like Hitler gassed six million Jews, just like, just like.... Actually, this phenomenon is still with us. I call the reader's attention to, of all things, the Oprah Winfrey television show of May 1, 1989, which featured a panel discussion on child sacrifice. On this particular program, a homely Jewish woman, aged 29, discussed her upbringing in a family of ritual murderers, how this family custom dated back to the 1700s, and how there are many Jewish families practicing child sacrifice in America today. She also mentioned, without offering details, that she was forced by her parents to participate in it herself when she was young, and is now struggling to recover psychologically. The murmurs of a freaked out audience are barely audible. Her mother is an upstanding citizen living in the Chicago area today, she said. I never heard of a follow-up police investigation. How this program was allowed to air on nationwide TV — how it was taped in the first place — is beyond my comprehension. It can be seen on YouTube. A related item on YouTube can be found by typing "How to kill Goyim and influence people," an obvious lampoon of the title of the famous Dale Carnegie book. On August 18, 2010 a conference was held at the Ramada Hotel in Jerusalem where one old rabbi after another, speaking in Hebrew with English subtitles, takes the podium to defend a recent book published in Israel claiming that the Torah condones the killing of Gentile children. The video also shows a young Israeli poet in the audience, who had heard enough, shouting something, then being dragged out of the room by two burly men. Afterwards two American Jews are briefly interviewed in the lobby, the first condemning this interpretation of the wise rabbis — whether he was sincere is anyone's guess — and the second seemingly not bothered by it.

(February 2020 update: As seen on a stunning *60 Minutes* interview that aired on December 1, 2019, produced by Shachar Bar-On, who calls himself "a proud Jerusalemite," YouTube's Jewish CEO, Susan Wojcicki, told Lesley Stahl, also Jewish (as so many of that program's line-up have been over the years) that her extremely popular social media platform had taken down 70% of controversial content. This assignment was still being carried out by a detail of YouTube monitors, some of whom — get this — had "bought into" the content they were reviewing. Stahl is a typical bubblehead made for TV journalism, whose demeanor during this interview was such that she could've been holding a sign that read "It's about time we started this censorship." It's all about getting rid of misinformation, hate, and conspiracy theories, you see. Misinformation includes, in Stahl's own words, "medical quackery," and she even mentioned — can you imagine? — the assertion that vaccines cause autism. I did a quick check and found that the Oprah Winfrey program on child sacrifice has been removed, though there's still plenty of print discussion about it. At this writing, the YouTube censors haven't yet caught up with the Jerusalem conference.)

❖

Coming full circle, the new century began with the usually intoxicated Yeltsin stepping down and a former KGB man taking the reins. Within a few years, Vladimir Putin restored law and order and adequate living conditions, and jailed some of the top Jewish gangsters, while others fled the country, some to Israel, others to Great Britain and America. Putin, like Russia, has always been something of an enigma. On the whole I believe him to be a worthy person and a prudent statesman with a clear vision of the world scene — a rarity indeed among political leaders in our time. He is, in fact, the first leader since Czar Nicholas to personify the heart and soul of his people, building a new country on the ruins of the old. I'm aware that some American patriots, who point to his kowtowing to Israel and obsequious gestures towards Russian rabbis, see him as a weak-kneed figurehead, even a sell-out. That may be. There are certainly paradoxes about the man and he has deeply disappointed me at times, especially his refusal to budge one inch on even the most threadbare myths about the Second World War, and his signing into law a directive making Holocaust denial a criminal offense punishable by up to five years in jail. I would agree that in part he's under the thumb of organized Jewry, but also in part, and unlike any American president, he's his own man, trying to maintain a delicate balancing act on the world stage. To me he's the Russian version of Ron Paul. This is no place to analyze his many moves and the geopolitical complexities of the former Soviet Union. The fact that Putin has been continually disparaged here by bought lawmakers and the piranha press — in appraisals ranging from vague mistrust to foaming-at-the-mouth hatred — is evidence enough that there's some good in him. He has slammed the corporate profiteering mentality on several occasions, taking aim especially at mass vaccination and genetically modified food products, though trolls are always at work on the Internet and there are those who deny he has said, "We as a species have the choice to continue to develop our bodies and brains in a healthy upward trajectory, or we can follow the Western example of recent decades and intentionally poison our population with genetically altered food, pharmaceuticals, vaccinations, and fast food that should be classified as a dangerous, addictive drug." Statements like these, which dump sand into the gears of the New World Order, do not endear him to those who rule behind the scenes and their minions. Furthermore, he must know what the Jews did to his country in the twentieth century, even if at times he feels he must "play the game" to placate the Jewish global media nexus. A photograph of him shaking hands with a wheelchair-bound, beaming Alexander Solzhenitsyn, taken shortly before the latter's death in 2008, says all that needs to be said about Solzhenitsyn's trust in him. Did he deserve this trust? It goes without saying that Putin knows of *Two Hundred Years Together*, and one can only guess that he has read it. That he has also been photographed in the midst of a dozen rabbis adds to the mystery of the man and to the proof of global Jewish dominance.

❖

Hopefully a fair number of readers have found this discussion enlightening, and will want to read some of the suppressed books I've mentioned, because I haven't discussed one percent of the story. Now it's time to move on and apply these facts to the history of vaccination, and especially to the here and now. But before I introduce this Jewish rogues gallery, I'll pause and put things in their proper perspective, because this is not an open and shut case. To backtrack a bit, there are plenty of loathsome white Gentiles among us, from street level to Deep State level, and there are vaccine malefactors who are not Jewish. The most infamous, in my opinion, is Microsoft founder and multibillionaire Bill Gates, the wizard of worldwide population reduction under the guise of saving children through mass inoculation. Gates is truly twisted. He seems to think that being one of the world's richest men entitles him to play God, when he should be in the dock sweating for his life. (For the record, I too think the world would be a much better place with far fewer people, but would be very leery of any human being engineering such a project, least of all a mattoid like Gates.) Flimflam artists Edward Jenner and Louis Pasteur were not Jewish, nor are many mean and close-minded pediatricians, nurses and health officials scattered across fifty states. Furthermore, there are ordinary and harmless Jewish parents who have felt great pain and know firsthand the damage that vaccines do to children, and there are Jews, both doctors and laymen, who are fighting the good fight. The late Dr. Robert Mendelsohn is rightly regarded as a pioneer in the modern anti-vaccine movement. For years he wrote a syndicated column, *The People's Doctor*, featured in the *Chicago Daily News*, in which he frequently took aim at vaccines, expressing his skepticism about their efficacy and concern about their hidden risks. His articles were later compiled in a book. His insights and wry humor make reading them equally entertaining and informative. If he hadn't made a small joke somewhere about his Jewish background, I wouldn't have thought about it. Neil Miller's books have also been an invaluable contribution to the awareness of vaccine dangers. I don't know his ancestry, but judging from his name and the one photograph I've seen of him, my guess is that he too is Jewish. And there have been others.

So why am I singling out Jews as culprits in the vaccine racket? Because of the lopsided number of those involved in it. Because of their clannishness and talent for organization. Because behind the mask of lofty scientific pursuit what really drives most of them, I believe, is a genetic imperative to destroy other human beings while making big money, something they've been doing for a long time. Because it's no coincidence, as I've said, that children living in the most Jewish country in the world, after Israel, are also the most heavily vaccinated. Noteworthy also is the fact that in Israel, where needless to say Jewish officials establish public health policy, parents are free to decide whether or not their children get vaccinated. There are no Paul Offits running around in that country that I'm aware of. When people like Barbara Loe Fisher say that childhood vaccination is not a medical issue but a culture war, what they're really saying, whether they're aware of it or not, is that it's part of a broader Jewish war against their host society.

In the chapter on polio we saw some of the petty personal conflicts that characterized the search for a vaccine, the most celebrated being the long-standing rivalry between Jonas Salk and Albert Sabin. Both men were unconvicted felons, though Sabin appears to have been the more contemptible, at times even to his friends. A rank hypocrite who in 1951 had berated Hilary Koprowski — like himself, a Polish-born Jew — for testing a live-virus vaccine on institutionalized children in upstate New York, Sabin was elated to have his own dangerous vaccine tested in 1959 on 77 million young people living under communist tyranny in the former Soviet Union. What really happened as a result of that unprecedented experiment might never be known. This catty competition was more like a squabbling among Hasidic diamond merchants than anything related to real science.

Koprowski was up to his eyebrows in controversy. His disgraceful experiments on children in New York, Northern Ireland and the Congo were reviewed in the chapter on polio. Reading the many criticisms of him by others who worked on vaccines, Jew and Gentile alike, he comes across as a thief, a liar, a cutthroat — in short, a man without a conscience. Koprowski committed his third and most grave polio vaccination felony during his foray into the Congo, prior to that country's descent into anarchy upon independence in 1960. According to Paul Offit, he went there because "he was driven to prevent the horrors of polio." Oh, I'll bet he was, just like Offit says. Another Albert Schweitzer, without a doubt. Having traveled around rural Africa, and seen how these simple natives are so trusting of the white man, of his superior medicine and good intentions — a trust based on their past observations, because many sincere white people *have* tried to improve the lives of African natives — it enrages me to think of the Koprowskis of the world marching into African villages with their needles and flasks and causing babies to die like flies. This has happened many times and it still happens. It's how the world works, unfortunately. In any event, two investigative journalists, one British and one American, later wrote that Koprowski might have initiated the AIDS epidemic in Africa by administering doses of his polio vaccine that were contaminated with the virus that causes AIDS. Nutty idea? There have been so many allegations of experiments in population control through vaccination campaigns in impoverished countries, usually involving anti-fertility agents concealed in vaccines, like those I mentioned above, that only the woefully ignorant would laugh at such a possibility. And there have been as many accusations of sloppy procedures and gross indifference to human life in conducting these experiments. An American, Tom Curtis, interviewed Koprowski and several other aging veterans of the mid-twentieth-century polio vaccine campaign. He wrote what I thought was an excellent article that was published in the March 19, 1992 edition of the counterculture magazine *Rolling Stone*. Curtis did not draw any firm conclusions, nor did he assail anyone's character, but he did poke into a hornet's nest of troublesome issues and asked questions that the interviewees definitely did not like. Naïvely, Curtis held that these men had the best of intentions, and had, at worst, inadvertently contaminated the vaccine with AIDS in their haste to achieve a medical breakthrough. Koprowski issued a sharp reply and then did something that Jews are very good at doing: he sued Curtis and *Rolling Stone* — for defamation, no less. This sent waves of consternation through the scientific community, because Curtis had done nothing more than offer a hypothesis, and there truly was nothing defamatory about anyone in the article.

Rather than go into the back and forth verbal thrusts, I ask you to look at the big picture. A Polish Jew in his thirties emigrates to America, eventually sets up shop in Philadelphia (incidentally, Paul Offit's base and a city top-heavy in pharmaceuticals and suspect medicine), performs experiments on children, already mentioned, for which, in my opinion, he should've been strapped in the electric chair, then turns around and sues an American-born writer who did nothing more than state that Koprowski, and others, *might* have done something terribly wrong by accident. Koprowski won the lawsuit and was symbolically awarded $1 in damages, but he got what he probably wanted all along: a *mea culpa* published in *Rolling Stone*, telling readers what a great scientist he really is, and the satisfaction of dealing the magazine a heavy financial blow to the tune of half a million dollars in legal fees. Parenthetically, the British researcher Edward Hooper, who had spent years living and traveling around Africa, went on to write a very long and widely praised book on the subject titled *The River: A Journey Back to the Source of HIV and AIDS*. I haven't read it, but according to independent reviews, it implicates Koprowski and company in some very shady activities. A perusal of Hooper's website, aidsorigins.com, tells me that, unless I'm badly deceived, he is a very careful and meticulous researcher, and an honorable man.

Eclipsing even Koprowski's nefarious activities were the well-known experiments carried out by Saul Krugman at the Willowbrook State Hospital in Staten Island, New York, home to thousands of severely handicapped children, referred to in an earlier chapter. Older readers of these lines may recall the televised reports of Geraldo Rivera who, in 1972, long before sliding into his natural niche of trash TV, made his name by entering Willowbrook unannounced with a camera crew and filming the horrible conditions in that warehouse, which was eventually closed down. (Years earlier, Senator Robert Kennedy, who also made a surprise visit, called it a "snake pit.") In fact these videos, preserved on the Internet, were instrumental in the enactment of sweeping laws that forbade the use of children in medical experiments. But Rivera missed the bigger story, lost in the heart-rending images on the television screen — that the children of Willowbrook, most of whom were likely vaccine-injured to begin with, had been fed milkshakes that contained hepatitis B-infected human feces as part of Krugman's experiment. In a sane country, Krugman would've been thrown off a cliff. In this country, he won numerous awards and became president of the American Pediatric Society.

Maurice Hilleman, head of vaccine research at Merck Pharmaceuticals for many years, supposedly was of German stock and raised in an Episcopal household. Considering that he was born and raised in rural Montana, this may be true, but considering all the other lies in Offit's *Vaccinated* about him, he may have been hiding something. I suspect he was descended from *marranos*, Jews in medieval Spain and Portugal who converted to Christianity to avoid persecution or expulsion, while secretly adhering to Jewish beliefs. Whatever blood ran through his veins, with his abrasive personality and a mouth that constantly spat four-letter words, he fit right in with his many Jewish colleagues in what can best be called a mutual admiration/vilification society. In a colossal understatement he called Krugman's Willowbrook studies "the most unethical medical experiments performed on children in the United States." Yet in *Vaccinated* he appears in a photograph standing right next to Krugman in a group of five, winners of the prestigious Lasker Award, named after a Jewish advertising executive and his trendy philanthropist wife, an early advocate of abortion on demand. Actually, Hilleman himself was a proponent of testing vaccines on handicapped "throw away" children who were wards of the state, and he did so with his mumps vaccine. As a matter of fact, in all my reading on this subject, I have never come across a single person involved in the development of vaccines who raised any strong objections to this long-established practice — though a few felt uneasy about it — which is still carried out on helpless children in poor countries around the world.

The following statement attributed to Hilleman appears in *Vaccinated*: "Most children, mentally retarded or not, get the mumps. My vaccine gave all these children the chance to avoid the harm of that disease. Why should retarded children be denied that chance? Retarded children are perceived as more helpless; but the understanding of the shot they were given, and their willingness to participate in the trial, was no different than [that of] healthy babies and young children...." But I ask you: *the understanding of the shot they were given*_ and *the willingness to participate in the trial* on the part of retarded children and healthy babies! Did Hilleman really say that, or is this just one more Offit whopper?

For what was likely premeditated mass murder, we turn our attention to a different kind of victim, and yet another Jew from Poland who made his way to this country. The saga of Wolf Szmuness, born in 1919, from his native land to a Siberian prison during the war years, then to a Russian medical school, then back to Poland where he became a close friend of the future pope John Paul II, then to Italy, and finally to America, where in a few short years he became head of epidemiology at the New York City Blood Bank, is so bizarre that it calls to mind an issue that the great industrialist Henry Ford, in his book *The International Jew*, raised in the 1920s: "[The Jewish question] begins in very simple terms — How does the Jew so habitually and so resistlessly gravitate to the highest places? What puts him there? Why is he put there? What does he do there? What does the fact of his being there mean to the world?" Ford had in mind the shadowy Bernard Baruch, a man of no distinction other than his ill-earned fortune, who had wormed his way to the elbow of Woodrow Wilson, and remained an "adviser" without portfolio to the next four presidents. Another baleful, power-hungry Jew, the German-born Henry Kissinger, emulated Baruch in playing the globe like his personal chess board, plotting the downward destiny of nations, instigating war crime after war crime (see Christopher Hitchens's *The Trial of Henry Kissinger*), and telling half a dozen presidents what to do. Even at age 94 he sat down with Trump, who felt obligated to confer with him. Not to forget our first female secretary of state, the gargoylish Madeline Albright who, during a *60 Minutes* interview shown on May 12, 1996, once again with Lesley Stahl who this time was gently confrontational, said that the imposition of U.N. sanctions which had brought death to a half million Iraqi children, a war crime that no doubt delighted her, was "a very hard choice, but the price — we think the price is worth it." This was broadcast around the world to universal outrage, particularly in Arab and Islamic countries, but it went right over the heads of most Americans who saw the program. This woman actually is a Czech-born Jewess whose real name is Marie Korbelova. If the suffering in Iraq wasn't enough, she could barely contain her glee over the U.S.-led NATO attack on Yugoslavia in 1999, which included the Easter Sunday bombing of the capital city Belgrade. Nearly three months of air strikes killed between several hundred and several thousand civilians, and displaced hundreds of thousands, depending on whose statistics are most accurate. The creature will not be prosecuted for any of this. Instead, book signings — her latest is *Fascism: A Warning* — occupy her time these days.

Szmuness's important position allowed him to conceive and perform a large-scale experiment with a hepatitis B vaccine using homosexuals as subjects. However, he required that they be young, healthy and promiscuous, and that they agree to be vaccinated three times and donate blood ten times. Large ads were taken out in the queer newspaper *New York Native* seeking volunteers, and notice was circulated throughout Greenwich Village about the new untested vaccine. The response was favorable; 1083 homosexuals participated, though some believe the real number was much higher. The experiment was generously funded by various federal health agencies. It began in New York City in November 1978, ran for a year, then resumed in San Francisco, Los Angeles, Denver, Chicago and St. Louis, ending in 1981. The onset of a new and dreadful disease, AIDS, corresponds to this time period and these cities. Most people who have investigated AIDS beyond mainstream media bilge have read that it's caused by a manmade virus that was created by splicing the bovine leukemia virus and visna virus, found naturally in cows and sheep, respectively. They've read that it was created for possible use as a bioweapon that would work by destroying the human immune system. Not only do I find this wholly credible, I am also convinced that it was deliberately seeded in the hep B vaccine in Szmuness's trial to see how it would work. The fact that promiscuity was a requirement is compelling evidence that Szmuness wanted to kill a lot of people, or at the very least didn't care if a lot of people died.

Szmuness died of cancer in 1982, taking his secret to the grave. A doctor named Robert Strecker was the first to venture into the dark crypt of this satanic "science" and shed some light on it. You can watch *The Strecker Memorandum*, a 96-minute video, on the Internet, or read the details in *Emerging Viruses* by Leonard Horowitz or *AIDS and the Doctors of Death* by Alan Cantwell, both of which are summarized in this work's bibliography.

❖

Salk, Sabin, Koprowski, Krugman, Offit — these are the more famous Jews connected with vaccine production and experimentation. But there are many more players, some who developed vaccines and some who didn't, but whose work had a direct bearing on them. The head of the CDC, Rear Admiral Anne Schuchat, is Jewish, as is Walter Orenstein, the CDC's former longtime immunization program director. Leonard Hayflick, Stanley Plotkin, Samuel Katz, Max Tishler, Aaron Kellner, David Bodian, Baruch Blumberg — all these names appear in Offit's books and in other sources, and all are Jews. Nor is this list by any means complete. Many had a strong Jewish identity. Blumberg, who anointed himself inventor of the hepatits B vaccine, much to the chagrin of his competitors, is of special interest because he was a lifelong student of the Talmud, and according to his Wikipedia page, attended weekly discussions on it. No word can describe someone claiming to be a scientist and also a devoted student of the Talmud. At the same time, it underscores my suspicion that beneath the steady stream of pro-vaccine propaganda flows an undercurrent of Jewish ritual murder, especially targeting children. And if not a subliminal form of child sacrifice, some Jews might see in mass vaccination a dry run to exterminate, through sterilization or deliberate lethal infection, a group or an entire nation that they hate, such as the gang that tried to poison water supplies in Germany shortly after World War Two ended (see the Wikipedia page on "Nakam"). Or it might just satisfy a basic need to subjugate and destroy people, or perhaps offer an opportunity to become rich and famous and wallow in delusions of grandeur after doing fake and sloppy science, the consequences be damned. Whatever the case, the proportion of Jews among those who have taken up the banner for vaccines is truly remarkable.

Vaccinology is as Jewish as psychoanalysis, gun control, modern art, organized crime, the pornography industry, forced race mixing, and every other activity alien to the values and traditions of Americans of European descent. And in addition to those named above, there are many Jews on the sidelines, in the media and academia, cheering for vaccines and flinging mud at the opposition. One need only glance at the many Jewish names of those endorsing Paul Offit's views, like those on the back cover of his books. Other books, articles, web sites and blogs, with a decidedly nasty edge, reveal more of the same people, people like Alex Berezow, David Gorski, Seth Mnookin, Arthur Caplan and Dorit Rubenstein Reiss, some with big titles and connections who are all for cracking down on what irks them — like eliminating vaccine exemptions and revoking the medical licenses of anti-vaccine doctors. Berezow has suggested jail time for these doctors. Reiss has often teamed up with Offit, and like him, she is a repulsive creature that many people see through. Also like Offit, she has been smacked around pretty good on the Internet. She runs a site with the misleading name skepticalraptor.com. A Californian, she studied in Israel, her country of birth, and worked in that government's Ministry of Justice. From what I can tell, she never pushed hard for vaccination when she lived in Israel, if in fact she pushed for it at all. What made her come to this country with her loud vaccine bullhorn? In her mind, perhaps, vaccinations, like diversity and multiculturalism, should be forced on the children of *goyim*, but not on children of the Chosen People. Of late she has slapped the label of "terrorism" on anti-vax activists. The editorial in the May 8, 2017 issue of the *Boston Herald*, which advocated hanging people like me, was written by the unlovely Rachelle Cohen. Her face, and Reiss's too for that matter, betrays her tapeworm soul. Dishonorable mention also goes to another member of the tribe, Dr. Peter Hotez, president of the Sabin Vaccine Institute and honcho at the Baylor College of Medicine who, in an article that appeared in the March 3, 2017 *Scientific American*, wrote that the anti-vaccine movement in America ought to be "snuffed out" — an interesting choice of words. (For the uninformed, "snuff" is slang for murder.) Hotez received the distinguished achievement award from B'nai B'rith International that same month.

Considering all these facts, It would be interesting to find out how many Jews are involved in the most twisted medical projects in the maze of the Deep State, especially in Fort Detrick, Maryland. Undoubtedly, Gentiles are also to be found in these criminal enterprises. This invites a comparison with the inhuman medical experiments which allegedly took place in the wartime concentration camps connected with National Socialist Germany. Here, it's difficult to separate fact from fiction, for example with the gruesome experiments attributed to Dr. Josef Mengele, simply because so many Nazi atrocities have been exposed as Jewish fairy tales. However, we have it on the authority of Dr. Butz that some awful experiments did take place in the camps, presumably under the supervision of sadistic German doctors. I'm not aware of a definitive account of these activities, but I do know that Hitler was a humane man, and sought to act mercifully towards innocent people, even under emergency measures. Thus, I assume — and since I haven't investigated the subject I can only assume — that those in the camps selected for experiments, such as being immersed in ice water and put inside decompression chambers, to test the effect it would have on downed German fighter pilots — were hardened criminals, not children or inoffensive adults. In any event, as historian David Irving has pointed out, Hitler had far too much on his plate to concern himself with what was happening in the camps, which were not only detention centers but work stations vital to the German war effort. Certainly, Hitler never would allow helpless children, including Jewish children, to be used in cruel vaccine experiments, which in the twentieth century was common practice in America. One might object to this statement by pointing to the T4 program implemented in 1939, later abandoned, in which institutionalized Germans, including children, were euthanized. My reply would be that this was an impossible choice faced by Hitler: a choice between the lives of those with severe genetic defects — perhaps some even vaccine-damaged — abandoned by their families who were a burden to the state, and the lives of German soldiers seriously wounded in the early phase of the war, who were in desperate need of accommodation and care.

Finally, a word is in order about vaccination policy in Germany under Hitler — another subject about which there are information gaps. My understanding is that vaccinations, which in National Socialist Germany meant only smallpox and diphtheria, were required by law — nearly everyone got them. Like all social policies, this was overseen by the proper authorities; in this case, it was probably the personal decision of Dr. Leonard Conti, the Minister of Health. The top officials in Hitler's cabinet were very intelligent men, but apparently the vaccine dogma was so ingrained in their minds that childhood vaccinations were deemed beneficial without reservation. In *Table Talk* there is a passage in which Hitler expressed his awareness that some people did not believe in vaccines, but apparently, despite his extensive knowledge in many fields, including medicine, he had no opinion about them either way.

❖

As I update this in August 2019, I'm looking back at a tempest in a teapot earlier this year in Rockland County, just north of New York City. As a result of a lot of silly TV and newspaper coverage, New York became the fourth state to disallow religious exemptions, proving for the umpteenth time that what we have in this country is government by media. It seems an epidemic of measles broke out in March — 153 cases, we were told — with the parents of unvaccinated children in the orthodox Jewish community primarily to blame, though journalists knew to tread lightly here. Twice as many cases in a month were probably par for the course in the 1960s without anyone caring or calling it an epidemic. As always — going back in a straight line to Jenner's day — there was no mention of how many of the stricken had been vaccinated, and no interest in finding out. The county executive, Ed Day, started dancing as soon as the media started tuning up, declaring a state of emergency and banning all unvaccinated persons under the age of 18 from public places. I believe this was a first in American history. Not content with making a near total fool out of himself, in the next breath Day went all the way by saying there was no practical way to enforce his own ruling and he wouldn't even try! Since nothing in the newspapers around here can be believed, aside from the tide tables, and since this was merely the latest hyped story about measles, it doesn't interest me. I bring it up only because it raises the question as to the attitude of segregated Jews towards vaccinating their children. I looked into this a bit and also took a quick look at the issue in Israel. I'm guessing that the discussions posted on the Internet in English are sincere, and if so, there seems to be a division of opinion among these people, as everywhere else, over what's best for their children. There is full freedom of choice in Israel, children do not need any shots to attend school, yet the compliance rate is still 97% — or so the English-language press there tells us. That I don't believe.

Though very cunning, and generally intelligent in a pedantic sense, Jews are hardly a race of sages, and are vulnerable to medical dogmas like everyone else. Equally significant, as Hitler pointed out in *Mein Kampf*, while Jews become unified when they feel threatened by outer danger, left to themselves they have little sense of social cohesion. Under certain circumstances they have no compunction against harming and sometimes even killing each other. Numerous examples come to mind. A case in point is Israeli prime minister Yitzhak Rabin, assassinated on November 4, 1995 by a flaming radical student named Yigal Amir, who ludicrously perceived him as too moderate. When on July 22, 1946 the Irgun blew up the King David Hotel in Jerusalem, headquarters of the British Palestine administration, they surely knew that some ordinary Jews were employed there by the British, but they didn't care. In addition to 74 people killed in the blast, mostly British and Arab, 17 Jews died. And here's an interesting quote from Ariel Sharon, the Beirut butcher boy, who came right out with it in an interview with Israeli journalist Amos Oz originally published in the Israeli daily *Davar* on December 17, 1982: "Even today I am willing to volunteer to do the dirty work for Israel, to kill as many Arabs as necessary, to deport them, to expel and burn them, to have everyone hate us, to pull the rug from underneath the feet of the Diaspora Jews, so that they will be forced to run to us crying. Even if it means blowing up one or two synagogues here and there, I don't care...." The point I'm trying to make is that, broadly speaking, I doubt that Jewish vaccine pushers care much if Jewish children are victimized by vaccination, except for their own children and those of friends and relatives. Or else they're in denial about it, like most Gentile doctors and pediatricians. So even if the news reports were true, it was nothing more than a typical family spat, such as is prevalent among different classes of Jews in Israel. And my guess is that, now that everyone has forgotten about this "measles epidemic" of early 2019, it's business as usual, with the anti-vax Jews within the introverted orthodox community still refusing to vaccinate their kids and still sending them to school, in casual disregard of the new state law, with outsiders none the wiser.

❖

The title of this book asks if vaccines will be the end of us. A corollary question would be "Will Jews be the end of us?" As indicated by the quotations that introduced this chapter, it's something that thoughtful men through the ages have pondered. The late William Pierce, founder of the National Alliance, a racial nationalist organization still in operation despite several setbacks, pondered it long and hard. Although I don't agree with every idea Pierce espoused, I still regard him as the one of America's most farsighted thinkers. Pierce believed that the conflict between Jew and Aryan man was a battle of Nature beyond any moral considerations, such as takes place throughout the plant and animal kingdom, and ultimately one or the other will be totally killed off. Under the pen name Andrew MacDonald, he wrote and in 1978 published *The Turner Diaries*, a fantasy novel of nonstop national and global race war that featured, among much extreme violence, the nuking of New York City and Tel Aviv. The book is an underground classic that has sold close to a million copies. It's impossible to say how many readers have taken it to heart and shared it with others. I read it about ten years ago and can't say that it inspired me. I wish there was a better way. But that's not for me to decide. And let it be said that anyone with access to the Internet and a few hours to spare can find plenty of explicit death wishes, from Jews and others, directed at the entire white race. Here's a typical one from late Jewish professor Noel Ignatiev that appeared in the September/October 2002 issue of *Harvard* magazine: "The goal of abolishing the White race is, on its face, so desirable that some may find it hard to believe that it could incur any opposition other than from committed White supremacists.... Make no mistake about it: we intend to keep bashing the dead white males, and the live ones, and the females too, until the social construct known as 'the White race' is destroyed." I have no doubt that the great majority of Jews in positions of power worldwide think the same way, and their relentless push for childhood vaccination is a manifestation of their primal instinct to destroy, vaccines being one more weapon to weaken and enslave the hated white race, with the ultimate aim of destroying its will to survive and driving it to extinction. I can't prove that nor can anyone disprove it, but that's what I believe, and I rest my case.

Some may say I've gone way overboard in getting off on this tangent. In an important sense, though, it's not a tangent at all because I'm certain that the future of childhood vaccination is bound up with the future of Jewish domination in America, with all its inevitable tyranny and decay. And while I'm not a prophet and am not in the habit of making predictions, one thing is easy to predict: as long as the Jewish power structure remains intact — above all, as long as Jews maintain their death grip on the mass media — America will continue to go downhill, and there will be no let-up in the drive for child sacrifice through vaccination. I don't know how much resistance this drive will encounter. There's considerable resistance now, even if much of it is under the radar. Thus, despite published statistics, I don't think anyone has a clear idea of the percentage of those parents who blindly follow what the doctor says, those who have cut back on the number of shots, and those who have completely rejected them.

❖

2020 finds America and the world in a strange stasis, as the tectonic plates of reality and illusion, intelligence and stupidity, shift and grind against each other. Mainstream media lies have become so unrestrained that they're comical, while the Internet swells with an ever-growing community of truth seekers, at the same time the heavily Jewish social media and search engine companies go into attack mode to try to silence them. After two years of insufferable nonsense about Russia swinging the 2016 election, and not one word on how Trump's many Jewish connections made him a candidate in the first place, newshounds, straining at the ends of their leashes, are sniffing around for new red herrings. Meanwhile Trump continues to encourage the Mideast country he dearly loves to ratchet up their aggression and steal more land, as do the swamp creatures in his cabinet who surround him as part of the deal he made with the devil. But maybe tomorrow he'll spring something new on the world. Who knows with this guy?

The storm clouds of World War Three gather and disperse, gather and disperse, especially around Iran. More and more people everywhere have become aware of what the mostly Islamic inhabitants of the Middle East have known for a long time, that Israel is global cancer. But how many people now know this? Still far too few in America, but surely many more than fifty years ago. As more people recognize Zionism as the re-emergence of Bolshevism, as Israel becomes increasingly isolated, as reality closes in on its leaders and the unworkability of their delusions drives them closer to the edge, will they resort to the Samson Option? Will this swollen parasite try to destroy the entire planet?

Somewhere in his writings, Michael Piper called for a multinational army to invade Israel and oversee the dismantling of its stockpile of nuclear warheads. It's heartening to see that he's far from alone in recognizing the threat posed by Israel and its army of powerful American Jews and *shabbos-goy* disciples, particularly those believers in the apocalyptic death cult known as Christian Zionism. These insights, however, have been largely confined to the Internet, where they are usually conveyed and shared under the cloak of anonymity, and voiced by only a handful of foreign statesmen, like the leaders of Iran, Malaysia's Mahathir Mohamad, and the late Hugo Chavez of Venezuela. No Western leader as yet has had the guts to openly speak the truth. But some have lost their fear. A few good people, former members of the media, government and military establishments, have spoken out, and in turn have emboldened others. Courage can be contagious. I recently read *The Host and the Parasite: How Israel's Fifth Column Consumed America* by investigative journalist Greg Felton, a comprehensive study in treason and psychopathic personalities, mainly from the Clinton years to the early months of the Trump presidency. Although I have some criticisms of this book, among them his apparent unawareness that the Nuremberg war crimes trial was a Talmudic lynching bee dressed in judicial finery, and his mindless use of words like "fascism" and "democracy" which he never defines, these are minor distractions inside the framework of a most important work. Most striking, near the end, is his sketch of an ideal scenario to put things right — an overthrow of the regime in Washington D.C. by a man on horseback from within the U.S. military, followed by a series of forceful steps that he specifies. Could there possibly be one extraordinary man in our armed forces — a Caesar, Charlemagne, Napoleon or Hitler waiting in the wings — with the willingness and ability to take History into his hands and change its currently disastrous course? Anything is possible, but it would seem awfully difficult in our present circumstances. History is filled with surprises, though, and, having nearly reached the end of the road here on this unpleasant subject, in the last chapter I'll reflect a little more on what may lie ahead.

❖

I am not worthy to speak aloud of Adolf Hitler, and his life and work are not suited to any sentimental talk. He was a champion of mankind, and a herald of the message of justice for all nations. He was a reforming figure of the highest rank, and his

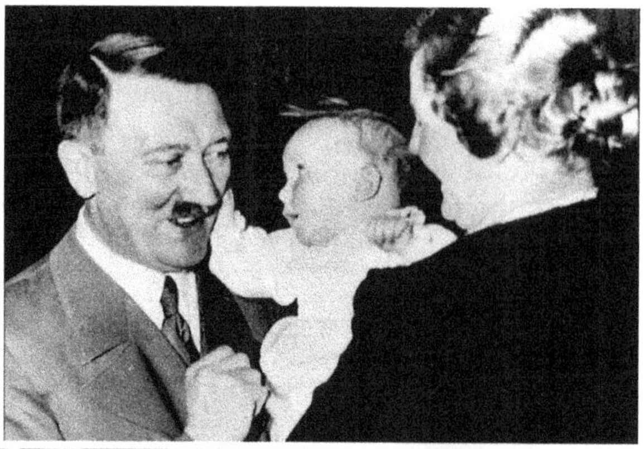

historic fate was that he had to work at a time of unparalleled perfidy that ultimately beat him down. — Knut Hamsun, Norwegian writer and Nobel Laureate

Adolf Hitler is the most maligned and misunderstood historical figure of all time, the object of so many outrageous lies that no one should waste his time refuting any of them. A cultured, emotionally rich man of galactic-level intelligence, he singlehandedly brought happiness and prosperity to a great nation that had been plunged into the depths of poverty and chaos at the end of World War One. An ordinary soldier in the German army during that conflict, Hitler was wounded by shellfire, and after recovering and returning to the front lines, he was temporarily blinded in a British gas attack. To his rougher comrades there was an air of mystery about this brooding loner, this sensitive genius who kept to himself, his faithful fox terrier always at his side. Yet they became very fond of him as a man of exceptional bravery and generosity, who always volunteered for the most dangerous missions and never hesitated to share his food while going hungry myself.

Knowing the horrors of war firsthand, carrying the pain of the world in his heart, Hitler strove for peace in Europe after becoming chancellor of Germany in 1933. He was the first man ever to not only intuit the parasitic, destructive nature of the Jew, but to do something about it. Nevertheless, despite his stridently anti-Jewish policies, he never ordered a single innocent Jew put to death, and in fact thousands of ordinary Jews continued to live quietly in Germany, though as second-class citizens, right through the war years. Nor did Hitler forget the kindness of Dr. Eduard Bloch, the Jewish physician who had treated his cancer-stricken mother when he was a boy, personally intervening to make sure he received special protection. The "Final Solution" that he approved was a plan — put on hold during the war — to deport most of Europe's Jews to east Russia or Madagascar.

Hitler was a true hero who embodied every virtue that good people hold dear, a beloved leader worshiped in his country and greatly admired throughout the continent of Europe. His pledge to stamp out forever the menace of Bolshevism, which had begun to seep westward from Russia, inspired the formation of the largest volunteer army in world history, the Waffen SS, comprised of a half million men from every nation in Europe, who fought alongside German units. This suppressed fact alone blows the myth of "The Good War" — a phrase coined by Jewish author Studs Terkel — out of the water. Although Hitler believed in the superiority of Aryan man, he was a friend to all decent people of all races. Naturally he felt closest to his own countrymen, and he especially loved the children of Germany. Had there been hundreds of highly placed vaccine profiteers getting rich on the dead and broken bodies of children, which is routine business practice in America, and had they been exposed, he undoubtedly would have had every one of them shot. The image of Adolf Hitler as a mass-murdering monster has nothing to do with reality and everything to do with Jewish control of the education system and the mass media everywhere in the Western world, and with it the incorrigible Jewish habit of pounding the wildest lies, mostly through the hypnotic effect of television, into the minds of the masses.

❖

Most American veterans of World War Two went to their graves never doubting the righteousness of the cause they fought for, never questioning the Big Lie that Hitler was out to exterminate the Jews and conquer the world. Most of them were irreversibly brainwashed. But there were exceptions. The most outstanding was the charismatic George Lincoln Rockwell, a man of multiple talents — like Hitler, a gifted artist as well as a spellbinding orator — who dropped out of Brown University to enlist in the Navy shortly before Pearl Harbor, becoming a fighter pilot in World War Two and eventually rising to the rank of commander. Throughout his military career, which included service in the Korean War, he despised Franklin Roosevelt and became alarmed by the communist infiltration of the government in Washington, D.C. But it was his reading of *Mein Kampf* in 1951 that changed his life, and led him, like some other American and British airmen, to feel the deepest remorse for bombing Germany during the Second World War. A genius in his own right, Rockwell instantly grasped Hitler's insight that the ultimate aim of organized Jewry, whether under the banner of Communism or Zionism, is to destroy the white race by any means, including fratricidal wars and intermarriage with those of other races. We see this process

at work today in all the wars and misery across the Islamic world, fomented behind the scenes by Jewish brains and money, with endless hordes of uprooted refugees, mixed in with miscellaneous human refuse, swarming into an enfeebled Europe. In America we have the same deliberately unchecked inundation of non-white immigrants from around the world, especially from south of the border, which has been going on for decades and has permanently disfigured this country, all of it supported by influential Jews and Jewish organizations.

Rockwell went on to found the American Nazi Party, declaring his goal of winning the U.S. presidency in 1972. Heartbroken by his beloved wife and their four children leaving him, financially depleted by the loss of his Navy pension, Rockwell took to the streets to get his message across, fearlessly hurling himself at the enemy at every opportunity, undaunted by never-ending hecklers, assaults, death threats, unjust arrests and lock-ups, often spicing his activism with pranks that would make the gods laugh. A tall, handsome man, he spoke at numerous colleges in the 1960s, quickly winning over hostile students with the soundness of his ideas, his priceless sense of humor, and the typhoon force of his personality. Tragically, his wild and woolly tactics attracted some unbalanced followers, and on August 25, 1967 one of them gunned him down in cold blood. In the photo below, Rockwell protests a local ordinance against "hate groups" by picketing outside the Dallas office of the Anti-Defamation League. On the previous page is the front cover of the autobiography of an extraordinary twentieth-century American whose character will inspire all noble souls who wish to save Western civilization and the race that created it.

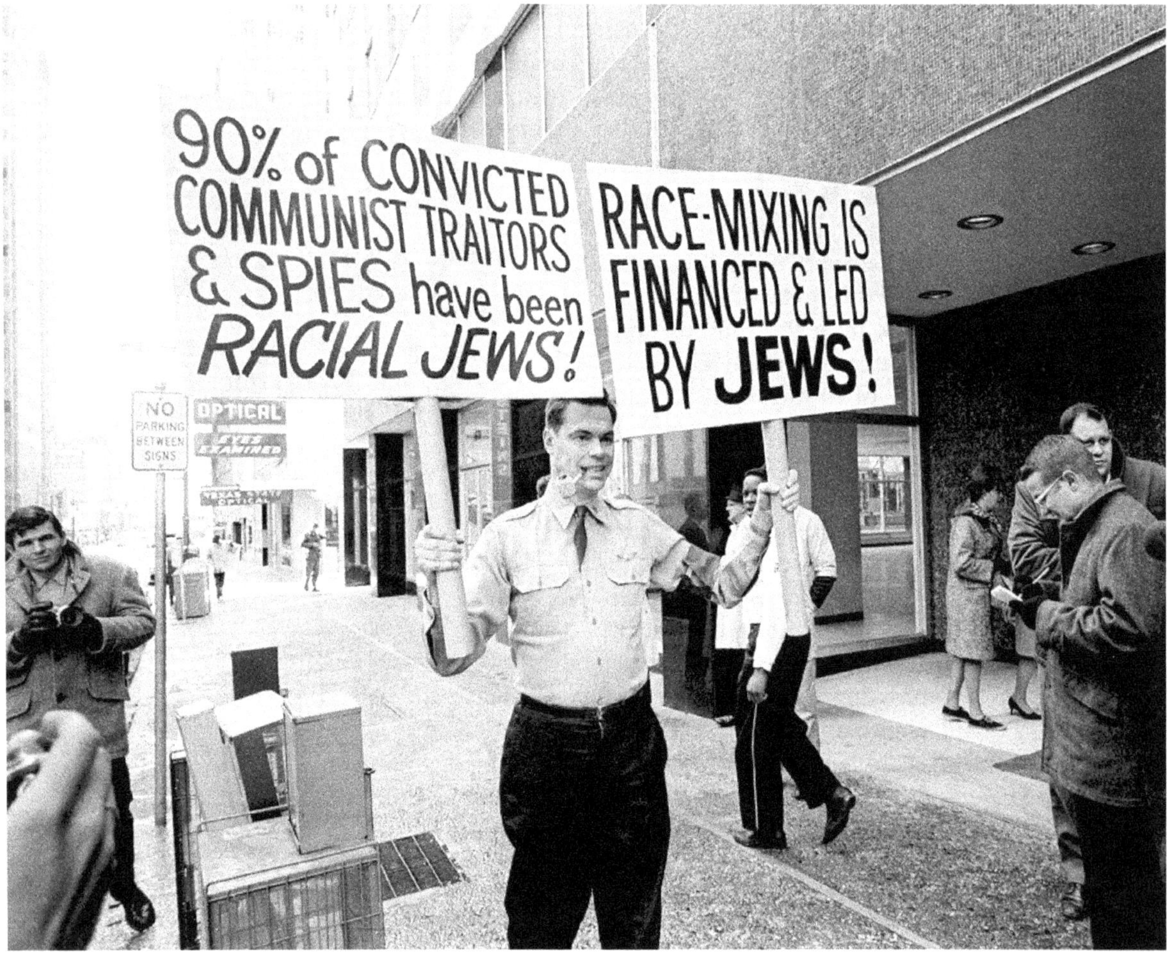

An extremely important book

IN AN opening page I explained that, due to the ocean of information I had to navigate, it's unlikely that this work is free of error, as desirable as that would be. On some minor points I'd just have to wing it, relying on memory or on something I'd read long ago. Re-reading everything I'd written, there was only one really sore spot, a discrepancy that puzzled me and I felt had to be put right. I had made conflicting statements as to how the worldwide AIDS epidemic began. There was the Strecker memorandum, which pointed to the diabolical invention of a new virus by splicing cow and sheep viruses in the U.S. Army laboratory at Fort Detrick, Maryland in the 1970s; the idea that AIDS was somehow connected to the oral polio vaccine experiments in Africa around 1960; the allegation made by Dr. William Douglass, and duplicated in less seditious terms on the front page of the May 11, 1987 London *Sunday Times*, that the smallpox vaccine campaign in Africa in the 1970s was the starting point; that AIDS first came on the scene in the early 1980s in the wake of Wolf Szmuness's hepatitis B vaccine trials with homosexuals. Clearly, one or more of these suppositions had to be wrong. I wanted answers. Whom better to turn to, I thought, than Edward Hooper, who had not only written an acclaimed book on the subject, but also ran the web site aidsorigins.com. There was a contact tab. On April 26, 2018 I fired off an email telling him a little about me and my book, and asking him if he could straighten all this out for me. I had no idea if I'd get a response.

To my surprise, I received a cordial reply from Ed that very day, in which he untangled the problem to some degree, adding that a full explanation could be found in Chapter 10 of his book. Although I'm a lifelong bookworm, it's not often that I read a book 500 pages, let alone 1073, in length. Nevertheless I decided to have a look at *The River: A Journey to the Source of HIV and AIDS*, and since the tenth chapter began on page 151, not too far in, why not start from the beginning?

I quickly realized that this was a very special book that I'd have to read through to the end. It was the combination of a spirit of open-minded scientific investigation, even though the author wasn't a scientist; an uncanny ability to synthesize boatloads of disorderly data; a staggering amount of research, including more than 600 interviews conducted across Europe, Africa and North America; an air of transparent intellectual honesty; and a lyrical style that made reading a pleasure, in places offsetting some dense scientific terminology. Here was a man who embodied the very best of the British, whose intellect and contribution to the treasury of knowledge ranked right up there with men like Edward Gibbon, Charles Darwin, Sir Arthur Keith and David Irving. On top of all that, Hooper had spent several years in Africa, an intrepid traveler whose adventures on that continent far outdistanced mine, as incredible as they were. As expected, *The River* brought to light a few factual errors on my part, but I considered them so minor that they're not worth mentioning. I didn't bother correcting them either, nor did I revise the comments that accompany my list of sources, because I like how they show the continually evolving nature of this book. I didn't correct my contradictory accounts of the origin of AIDS either, which was my sole reason for writing to Ed, but these I *will* discuss.

Chapter 10 solved the mystery. Hooper tells us that in 1992, not a year after the collapse of the Soviet empire, a former Soviet counterintelligence official named Yevgeni Primakov publicly admitted that "the KGB planted stories in the late 1980s which alleged that the HIV virus was the result of a Pentagon experiment." But more convincingly, in my eyes, was the fact that scattered reports of symptoms in African natives that would later be attributed to human immunodeficiency virus, or HIV, the precursor of AIDS, had entered the medical literature in the 1960s, and a blood sample from as early as 1959 had tested positive for HIV. This blew Strecker, Douglass, and the front page of the *Sunday Times* out of the water. I went back to that video of the Strecker memorandum that I'd seen on the Internet — Robert Strecker, serious, intelligent, handsome, athletic build, all-American scientist. He had it all figured out, I thought, but he was wrong — though sincere, I'm sure — and I was wrong to believe him. (Strangely, in everything that followed, the Szmuness hep B vaccine matter was neither addressed nor resolved; the name appears only once in *The River*, in a footnote. This vaccine was likely grown on the same substrate as the Koprowski vaccine, which I will come to.)

On a deeper level, *The River* touches on the mysteries of the Life Force — mysteries that are as intriguing as they are humbling insofar as they reduce man to his animal essence and expose his hubris — the attitude of some that they can manipulate Nature without ripping the lid off Pandora's Box. I wondered, above all, about the relationship between viruses, the tiniest living organisms, and higher animals, especially primates. In this regard, I remembered something I'd learned long ago that fascinated me — I presume it's true — about how the aboriginal Indians in the Americas, who had never been exposed to the smallpox virus, were decimated by this disease after European explorers and settlers arrived. Hooper's theory — that an oral polio vaccine (OPV) campaign in central Africa in the late 1950s spawned the worldwide AIDS epidemic — incorporates these profound issues, and attracted the attention of Professor William Hamilton of Oxford University, a highly renowned evolutionary biologist — considered Darwin's heir by some. Hamilton broke ranks with his chickenhearted colleagues and became Hooper's friend and mentor. He wrote the foreword to *The River*. So consumed was he by the possibility that HIV had jumped from chimpanzee to man by way of mass vaccination and initiated the catastrophe of AIDS, and so concerned by the grave implications, that he accompanied Hooper on a fact-finding journey to the regressive and perpetually unstable Congo in 1999. He returned six months later with two companions, for the lowly task of gathering chimp feces in the wild for laboratory analysis, which he felt was the key to it all. He had confided to Hooper that he would risk anything, even his life, to learn the truth. Sadly, his words were prophetic. On this, his second trip to the Congo, Hamilton contracted malaria and died on March 7, 2000, shortly after returning to England.

In *The River*, Ed Hooper presents a nearly airtight case pinpointing central Africa as the source of this dreadful disease — a disease that has thus far killed something like fifty million people globally, with an equal number walking around infected today. AIDS is in retreat in some parts of the world, but continues to spread in parts of others. No one has suffered more than the people of Africa. It's a very involved story, as one would expect of a book this long, but I'll distill it as best I can: Chimpanzee kidneys, which naturally harbor HIV, harmless to chimps but deadly to man, were used by Hilary Koprowski to grow his experimental oral polio vaccine in the 1950s in what was then the Belgian Congo (not to be confused with the much smaller former French Congo to the west, both countries now independent and both informally called "the Congo"). HIV (technically HIV-1) infected this vaccine, which was administered to about one million people in the Congo, as well as in neighboring Ruanda-Urundi (also a territory of Belgium at the time, today the independent countries of Rwanda and Burundi). This vaccination campaign fell under the auspices of the Belgian government, and was the brainchild of Koprowski who supervised it with his partner Ghislain Courtois. HIV is *not* a man-made virus, nor was it deliberately unleashed on hapless African natives, but rather it was the unforeseen outcome of devil-may-care risk taking, Koprowski's trademark. Through promiscuity and prostitution, mainly in urban areas, but also through normal heterosexual activity and human contact, HIV spread throughout Africa and beyond. In Africa, the latency period prior to full-blown AIDS or AIDS-related terminal illness can be several years but small children and babies, who can be infected through the placenta or from breast milk, die faster because of their immature immune systems. Itinerant homosexuals and bisexuals were primary carriers and Haiti was likely a major transmission route. (Many Haitians had gone to work in the Congo when the white population fled after independence in 1960, but they later returned home. Haiti was a popular playground for American homosexuals in the 1970s.) The OPV/AIDS hypothesis is obviously earth-shaking in its repercussions, and must be kept under wraps or "discredited" by the pro-vaccine crowd and every institution connected with it.

The publication of *The River*, I later learned, was a milestone event among AIDS and vaccine researchers around the world. You'll find discussions about it up and down the Internet. The book's biggest detractors, not surprisingly, were and are the usual authorities in organized medicine, pharmaceuticals, and the ivory towers of academia. Despite their clever arguments and attempts to punch holes in Hooper's character, not one has accepted his challenge to submit solid material or documentary evidence demonstrating that his theory is wrong, which evidence, he asserts, he would gladly acknowledge and admit he is wrong. Flipping through books I'd already read, I noticed that *The River* had been mentioned, but the recognition was so brief that it hadn't stuck. In two paragraphs in *Vaccinated*, Paul Offit dismisses the book on laughably shallow grounds — no surprise there — as does David Oshinsky in *Polio*. Neither claimed to have read it.

Long before I finished reading *The River*, I knew I'd be contacting Ed to express my unstinting praise for his accomplishment. I also wanted to address what I felt was a huge flaw in his work even though it came down to one page. But there was more to it. Looking back, it was both impractical and imposing on my part, but I wanted to enlist his support in inaugurating a study of vaccinated versus unvaccinated children, and in addition get his opinion on my own book by sending him some chapters. This touched off a flurry of emails across the Atlantic, which lasted all of eleven days. Since I was made to understand that these were private comments, I promised Ed that I would neither quote from them nor refer to their contents. All I will say is that there were some irreconcilable differences between us, in both personality and viewpoints, that put a quick end to our dialogue. Despite this, it was very kind of him to take the time to compose thoughtful and lengthy replies to each email I sent. How many people are like that? This is a man who had connected and often locked horns with some of the world's top specialists in the virus and vaccine fields, most famously at a widely publicized conference held on September 11-12, 2000 at the Royal Society in London, where his theory was unceremoniously rejected. He probably knows more about the biochemistry of AIDS than anyone alive, and has conducted what might be the most exhaustive scientific investigation ever undertaken. He has written a book for the ages on it, a book I put right up there with Charles Darwin's *On the Origin of Species* and Robert Ardrey's *African Genesis*. It was a privilege to exchange ideas with Edward Hooper, if only through the medium of the Internet, and even though it didn't go very far.

❖

Validating my beliefs about vaccines was the last thing on my mind when I began reading *The River*, but by the time I finished reading it every one of them had been validated. One that really stood out was my conviction that the essence of vaccination is child and animal sacrifice — George Bernard Shaw had called it "a peculiarly filthy piece of witchcraft" — this in connection with the gruesome experiments secretly carried out on chimpanzees at a secluded jungle facility, and the removal of their internal organs while still alive, with all the physical and emotional distress that entails. Something else was also confirmed: the spooky feeling I'd had about Philadelphia when I was there for one day. I'd been to Philly only once in my life, on an eighth grade field trip, but seeing the iconic Liberty Bell had been on my son's wish list for a long time, so we took the bus from New York City and stayed one night. Aside from the famous historical sights, Philadelphia is home to the Wistar Institute, which crops up nearly a hundred times in *The River*. The Wistar web site bills itself as "a global leader in biomedical research with special expertise in cancer, immunology, infectious diseases and vaccine development." Actually, it has hosted some strange and clandestine activities, and in the late 1950s several shipments of chimpanzee kidneys were flown there from the Congolese city of Stanleyville (now called Kisangani) via Brussels, the Belgian capital, to make polio vaccines, then sent back to the Congo. Furthermore, even back then, Wistar had close ties to the Children's Hospital of Philadelphia (CHOP), longtime headquarters of Paul Offit, who succeeded Hilary Koprowski as vaccination's Prince of Darkness. Koprowski served as director of Wistar from 1957 to 1991, when he was ousted for reasons that are not difficult to imagine, given his inbred criminality.

My son and I visited Philadelphia on July 1, 2015. The city was gearing up for the fiftieth anniversary, on the Fourth of July, of what the local homo guild was boasting as the first coming out of the closet in America. Every lamppost in downtown Philly was decked out in the rainbow colors of sissyhood. We waited on line with other tourists to see the Liberty Bell. Just steps away from it was a photograph of smiling communist Nelson Mandela, and another, taken in 1965, of picketing sodomites. They actually looked normal back then, even corny. One man, in jacket and tie, held a sign that read "Homosexuals are Americans too." I repeat: these large photographs were hanging on the wall just a few feet from the Liberty Bell! Outside the building there was a display of panels telling the story of the American Revolution. It was openly disrespectful of the Founding Fathers, in particular their supposed hypocrisy in striving towards political freedom while condoning slavery. The depredations of various African tribes as a result of the slave trade took up a lot of space. I was shocked by all this. Just who is running this city, I wondered, who is responsible for all this poison which had no purpose other than to agitate blacks and glorify homosexuality? A hint was provided right across the street by the presence of the National Museum of American Jewish History. In front of it was a billboard advertising some current exhibits: Albert Einstein's pipe, Jonas Salk's polio vaccine flask, Irving Berlin's piano. I hadn't heard of this museum, but I did know that some bad race stuff had gone down in Philly, that it was a simmering hotbed of black racial hatred. I had no intention of venturing beyond the tourist area, especially with my son. I could write much more about my impressions of the city of brotherly love, but I'll sum up by saying that my sixth sense was aroused by all these elements of subversion. I could almost smell the evil hanging in the air. This was Bolshevism poised to strike. And three years later, reading about the sinister activities that had taken place at the Wistar Institute — although Hooper doesn't quite describe them that way — added a creepy dimension to my memory of this city. Philadelphia: the cradle of America and the deathbed of America. Enough said.

❖

My conversation with Ed Hooper was quite illuminating, and I regret that I can't reproduce any of it, but for my purpose I think I can do just as well examining what he states in his book and on his web site. (Incidentally, for those who have no inclination to read *The River*, I highly recommend watching the 90-minute documentary film "The Origin of Aids," a synopsis of the book, by clicking the link on the home page of aidsorigins.com.) I've already given a capsule review of this work, and cannot emphasize too strongly the importance of its implications as they bear on the future of man. Nevertheless, I wish to point out some major flaws in Hooper's analysis, even though they constitute just a few fragments of what he has published.

The most amazing thing I learned about Ed is that he's what I would call a flaming liberal. I really couldn't grasp this at first. Here was a man whose thoughts flowed with impeccable logic, who expressed himself so clearly and beautifully, who was so admirably open-minded, testing his theory against every possible scientific hitch, and often traveling great distances to investigate these hitches. How then, could such a perceptive man make such a sappy comment about "white supremacist intransigence" in Birmingham, Alabama (where he had gone to interview a prominent AIDS researcher) in the 1960s (p. 647), or write a phrase like "rancid examples of racism and sexism," such as appears in an essay on his site? Actually, a brief background of the author on the back cover of *The River* gives advance notice of this, noting that he had served as a United Nations official and a BBC correspondent in Africa. Just above that was a glowing excerpt from the reviewer of an openly homosexual book club. All the liberal credentials were there. This flabbergasted me because, aside from just two or three signs of the mental disorder that is liberalism, I felt, while reading the book, that I was perfectly attuned to Ed's way of thinking; I felt that I was dealing with a mind of the highest development. I've often observed how people can be so right about some things and so wrong about others, that their brains can be divided into compartments that work independently and very differently to each other, but this was the most extreme case I'd come across.

I bring this up because it helped me to understand that, in some matters, vaccines among them, Ed Hooper doesn't get it and he'll probably never get it — and I don't mean that as an insult, just an observation. Although, as I've said, his knowledge of HIV and the polio vaccine that caused it is unmatched, he seems to know very little about the history and properties of vaccines, nor have I seen any evidence that he wants to learn. This includes his avoidance of the issue on his occasionally updated web site, except for one flat article titled "As Far as is Known, Modern Polio Vaccines are Safe." In the 181 pages of fine print notes with designated sources that fill out *The River*, which reveal an astounding amount of research, not a single book that points to the failures and dangers of vaccines is cited. In fairness, this isn't what his book is about, but since it's a closely related topic, I think that anyone writing about any aspect of vaccines ought to educate himself on it. In reading *The River*, one would get the impression that the author is totally unaware of any controversy surrounding vaccines, or if he is, he considers the position of anti-vaxxers like me so weak that it's not worth mentioning.

There were only a few passages in *The River* that raised my hackles, that seemed totally out of alignment with Ed's brilliance, and since this was such an enjoyable read, I tried to forget about them. Then, on page 812, came the bombshell:

Vaccines are one of the greatest triumphs of medical science — some would say the greatest. Every year they save the lives of tens of millions of humans and hundreds of millions of animals. The vast majority of vaccines are safe, and have been proven so over the course of many years. And perhaps the brightest jewel in the crown is the polio vaccine.

The race to conquer poliomyelitis and develop a vaccine was a wonderful, imaginative, and heroic scientific enterprise. Those fumbling beginnings back in the dawn of virology led directly to the situation we have today, where the breaking of the chain of wild poliovirus transmission is a realistic goal for the near future. Already the virus has been eradicated from the Americas, and only one case was reported from Europe in 1997. Ironically, because the recent political upheavals have interrupted polio vaccination initiatives, the Congo is now one of the strongholds of the virus. However, it is now almost certain that once "National Immunization Days" are staged there and in other vulnerable countries, poliomyelitis will become the second viral disease — after smallpox — to be eradicated from the human race through medical intervention.

This will be a magnificent achievement, for which the great pioneers of tissue culture development and polio research, such as John Enders, Renato Dulbecco, Joseph Melnick, David Bodian, Jonas Salk, Sven Gard, Albert Sabin, Herald Cox, and Hilary Koprowski deserve enormous credit.

This statement comes out of nowhere. It has no relevance whatsoever to the book's theme and to what directly preceded it. Every assertion in it is false. How does one explain it, then, other than to presume that the author is shackled by one of the major dogmas of our time without knowing it, without realizing that he has just forgotten the discerning words he had written much earlier, at the beginning of Chapter 2: "Sometimes a statement or a version of events gets repeated so often that it becomes accepted by a large part of humanity as fact, even when the supporting evidence has been disproved long before. This is one of the ways in which myth can take over from reality." Concerning Koprowski, just four pages earlier, in discussing the aftermath of the lawsuit filed by him against Tom Curtis and *Rolling Stone*, which sent a chill through the halls of science, and which I have already mentioned, Hooper had written, "I have already been threatened with litigation by a lawyer representing Koprowski and [his associate Stanley] Plotkin, even before this book is published. I say shame on them, for trying to conduct their science through the courts." And now he's heaping "enormous credit" on the man? This is what's called doublethink.

One thing that makes *The River* such a great read is that we hear, as if we're in the same room, from so many people who were involved, wittingly or unwittingly, in one of the greatest scandals and ensuing cover-ups of all time, including several Europeans who had lived in the Congo and worked at Camp Lindi, the sequestered laboratory where hundreds of female and young chimpanzees — captured by hunters in the rain forest in raids that often involved shooting male adult chimps defending them — were housed and underwent vivisection, their carcasses later buried or given to locals to eat. *The River* was researched and written in the 1990s, when most of these folks were quietly living out their sunset years in retirement, the majority in France and Belgium. There's a wide range of personalities in every field, and in response to Hooper's probing questions, some became reflective, others uneasy, still others agitated. These conversations were taped and the quotes in the book are verbatim. The author also describes their hand gestures, facial expressions, long silences or angry outbursts as the case may be. So much more truth, I believe, is extracted this way than in poring over studies or statistics, even those that are reliable. We learn a great deal about Koprowski from these reminiscences. Many who knew him well spoke of him in terms of utter revulsion, though he did have some friends and supporters. What we see is a man of wit, intelligence and charm, a cultured man who collected art and played classical piano, but under the skin a dishonest and ambitious cutthroat who could turn vicious in a heartbeat, a chameleon always ready to change his stance when it was expedient to do so, and one who was considered unethical even by his unethical colleagues. In his interviews with Hooper, recounted in two riveting chapters, Koprowski becomes a babbler of lies, contradictions, evasions and convenient memory lapses, whose mood changes according to the nature of the question. He reveals a lot about himself in these interviews, which he was hesitant to grant, but probably figured he'd better grant in a clumsy attempt to clear his name.

Hilary Koprowski died in 2013 at the age of 96. Earlier I wrote that he deserved to end his life in the electric chair. After finishing *The River*, I can add that I would've been happy to throw the switch myself. Koprowski was a virus as deadly as any he ever saw under an electron microscope. He was driven by some demonic urge that kept his black heart beating far too long, an urge that could only be satisfied by destroying other human beings. My conjecture is that his OPV campaign in Africa was a dry run for a possible future campaign of sterilizing or maybe even killing large numbers of, in his view, undesirable people by loading vaccines with poisonous ingredients. Unable to go that far, he still had a lot of fun sitting on the throne of fake science and gambling with millions of lives, confident that the global Big Lie machine would certify him as a great man, perhaps just a notch below the immortal Salk. That posterity might curse him as an ogre a million times worse than Typhoid Mary was not something he had bargained for. What I've just written is my own take on the man, but his track record is undeniable. Yet Ed Hooper, despite exposing fresh slime dripping from this serpent, dresses him up in pretty words like "wonderful," "imaginative" and "heroic." Whatever his reasons were for writing what appears on page 812 — perhaps he was just trying to give him every possible benefit of the doubt — we clearly had some temperamental differences. (A few years after his book was published, Hooper updated aidsorigins.com with new and damning material on Koprowski that had come to light. Also, he relates that Stanley Plotkin, one of the interviewees in *The River*, mailed him a postcard with the heading "Punishment of the Apple Stealers," an illustration of one figure clubbing another over the head, and the cryptic message, "Dear Ed, Thinking of You." Whatever Plotkin had in mind only he knows, but that's pretty bizarre behavior coming from a former prominent vaccine developer and director of Sanofi Pasteur, one of the world's largest pharmaceutical companies.)

Remarkably, right after the excerpt I quoted, and still on page 812, Hooper does a U-turn and in the fourteen remaining pages of this chapter presents a sobering account of disasters wrought by human tampering with Nature's laws, vaccines included, plus an insane new proposal to test a live-virus AIDS vaccine and a few dubious assertions from the usual "experts" shuffled in. I learned a lot in these pages. Although I knew about vaccine damage to our dogs and cats, I was unaware of the extent of it, and inquiring further I learned that there have been many vaccine-associated problems with other domesticated animals, notably horses and cattle. I had read plenty about Gulf War syndrome, and the nefarious experiments staged by the U.S. military, but I was completely ignorant about the myxoma virus disaster in Australia, in which the introduction of foreign rabbits as an experiment to control soil erosion nearly wiped out the local rabbit species and wrecked the ecosystem. Nor had I ever given a thought to the grave risks involved in trying to save the lives of a few human beings by transplanting organs from other primates and from pigs. These are profound issues that made me pause and reflect on the deepest mysteries of Life. On page 824, Hooper asks, "Do we, as a species, have sufficient safeguards in place, enough *independent* checks and balances, to ensure that when the men in white coats, or those in gray suits, get it wrong — as they surely sometimes do — that their mistakes will not be the end of us?" That question struck me as so crucial, as *the* ultimate question, that I decided to make it the title of my own book.

It's a pity that so few people outside of a couple of specialized fields have read *The River*, or even know about it. Hooper tells an engaging story about how, concerned that his book was dying on the vine shortly after publication, he met in the *New York Times* cafeteria with science writer Larry Altman and persuaded him to read it. The global influence and political power of this newspaper is awesome — and almost always in a destructive sense. In any event, Altman was impressed by the book and wrote what Hooper says was a "very positive" review of it. I read this review of November 30, 1999 on the Internet after reading the book and found it positive, though I wouldn't say very. The saying "damn with faint praise" came to mind. Had I read it without knowing about the book, I would've yawned and turned the page. But reading is subjective, and others read it differently. Since the *Times* is the Orwellian Ministry of Truth to so many journalists and other literati in the U.S., Hooper was soon contacted by quite a few of them. I noted the excerpts of sixteen other enthusiastic reviews on the first two pages of the Back Bay edition, but wonder how many, in this country at least, had played "follow the leader" after the almighty *Times* had given *The River* the thumbs up. Unfortunately, though, what little momentum these reviews created quickly faded, and the book has had very little impact compared to, say, Rachel Carson's 1962 classic work *Silent Spring*, which documented how the indiscriminate use of pesticides was destroying the natural environment. If not for my research, I would know nothing of it. *The River* was never displayed, and I believe never stocked, at any of the local chain bookstores. Of the 107 public libraries on Long Island, a computer search shows that exactly two have it on the shelf. I have never seen it mentioned on the five or six anti-vaccine websites I occasionally browse. I would think the anti-vax movement would make it their bible despite Hooper's misplaced faith in vaccines. In an email discussing an unrelated vaccine matter sent last year to a woman in Andrew Wakefield's group in Texas, I incidentally mentioned *The River* and strongly recommended she read it if she hadn't already. Her only response on this point was that she had heard of it.

Even before I plowed through the 1073 pages of *The River*, I'd reached a point, having studied vaccines off and on for nearly half my life, where I just didn't feel like reading any more. But of all things, a new surge of curiosity came over me after I finished it. I wanted to dig deeper into polio, even though I'd read much more about this disease and its associated vaccines than any other, with smallpox a close second. So I obtained a copy of *A History of Poliomyelitis* by John Paul (1971), a title I'd occasionally come across, and the only book of its kind. Paul, a medical doctor who had done some field work in Connecticut with polio in the 1930s, was apparently a bland background figure during the years when the race was on for a vaccine, a sort of moderator who got along well with all the famous researchers. Naturally he supported their efforts. I found his book rather dull, and skimmed much of it. But what got my attention is that only the first 27 of its 469 pages covers the period from ancient times to the mid-nineteenth century, when the history of polio began in earnest, albeit slowly. We learn that a British physician named Michael Underwood was the very first to write of cases involving weakness or paralysis of the legs, which *might* have been polio, in his *Disorders of Children*, published in 1784. This was a widely read textbook of the period that went through several editions and was translated into German. Yet this extremely rare condition aroused no interest among doctors, and Underwood admitted that he himself had never seen a single case. Paul writes:

For the twenty-five years following Underwood's account, with the exception of a description by the Italian surgeon Monteggia, we hear very little about either debility of the lower extremities or paralysis in infants following fever. It was just as if the disease did not exist or had never been described. None of thirteen classifications of illnesses published between 1793 and 1818, which listed all the diseases known to the medical world, included a reference to it.

And twelve pages further: "It was not until the mid-1830s that physicians began to report small groups of poliomyelitis cases which had occurred almost simultaneously. Probably the earliest of these on record is the well-known outbreak recorded by Baham in England in 1835."

England again — the birthplace of vaccination — and like Down Syndrome, a mere two generations or so after Edward Jenner started the whole business. And right around the time this practice was slowly catching on throughout England. How interesting. In the decades that followed, clusters of paralysis also began appearing in America and a few European countries. Questions come to mind that Paul didn't ask. Did this come about as a result of smallpox vaccination? Did Jenner's various concoctions using cow and horse lymph introduce something new into the human organism? Was polio a benign virus that had adapted to our systems millennia ago, until the smallpox vaccine somehow "awakened" it (perhaps in the same way that rotavirus, if it really exists, has been provoked by all the new childhood vaccines)? Is this just one more instance of medicine based on greed and ignorance piling one disaster on top of another — first the smallpox vaccine, then polio, then the polio vaccine, then strange new cancers and AIDS, then — an AIDS vaccine?

I then wondered why it had never occurred to me to ask if vaccination played a role in the 1918 outbreak of the Spanish Flu which, along with AIDS and the Black Death of the fourteenth century, ranks in the top three deadliest epidemics of all time. I looked into this, and I won't dwell on it because the data is sketchy. To this day, mainstream medicine considers this disease a mystery. It abounds in debatable facts and unfounded theories, so that each account seems to be guesswork based on individual prejudices, as do the number of dead, which ranges between 50 and 100 million worldwide. Only a few mavericks, defying the great taboo, have suggested a vaccine connection. Nevertheless, several clues point in this direction, beginning with the name; although the disease ravaged Spain, as it did so many other countries, it did not originate there, so that the term "Spanish Flu" seems to have been a smokescreen, typical of the name games vaccine proponents have always played. Healthy young men were the most susceptible, while the young and old were largely spared, which is highly unusual for respiratory diseases. Most significantly, the pandemic began during a war in which unprecedented numbers of soldiers, not only from Europe but from Asian and African colonies of France and Great Britain, and of course the U.S. and British realms, took part. Then, as now, soldiers were treated like pin cushions and many likely received a stiff cocktail of all the vaccines that were being dabbled in at the time but were unknown to the general public, like diphtheria, tetanus and typhoid fever, in addition to smallpox boosters. Since much is conjecture, I won't go further other than to say that it seems likely that, just like AIDS, the Spanish Flu was caused by mass vaccination.

Then there's foot and mouth disease, a consequence of the cruel procedure whereby smallpox vaccine was manufactured in times past. Calves were strapped down, their bellies shaved and slashed multiple times, and viral matter smeared into the wounds to grow more virus, which would be scraped off and ground into powder to make vaccine. Released back into the herd, this infection would spread to other animals. In addition to inflaming internal organs, vesicles swelled and ruptured in the hooves and mouths of these animals, causing intense pain. Epidemics of this disease occurred in the decades overlapping the nineteenth and twentieth centuries, here and in Europe; those of 1902, 1908 and 1914 in the U.S. were particularly devastating. All told, millions of cows, infected and uninfected, were slaughtered to prevent the disease from spreading. Here, I think, the correlation between vaccine virus contamination and mass infection was crystal clear. From a biological standpoint, it's identical to AIDS sweeping through whole African families and villages merely from proximity.

I strongly suspect that infectious polio, Down Syndrome, foot and mouth disease, Spanish Flu and AIDS all originated with vaccines. The evidence is very compelling. What's missing is a smoking gun. But suppose vaccines were the source of just Spanish Flu and AIDS. This means that in the twentieth century the number of people killed through vaccination approaches or exceeds the number of people killed in all wars.

One final thought. When I was a schoolboy, I learned that when Christopher Columbus made landfall on his epic voyages, he was certain he had reached Asia and discovered some of its offshore islands. I later learned in my adult reading that this is indeed true, that he had made some faulty geographical assumptions and clung to some mistaken beliefs in the face of opposing evidence. But despite some errors in judgment, he is rightly regarded as one of the greatest mariners of all time, whose discoveries had a profound influence on the course of history. *The River* is to modern scientific investigation what Columbus's voyages were to the age of exploration. Edward Hooper's extraordinary effort in trying to solve a great medical mystery — even though he made some incorrect inferences — have shown that vaccination — which usually involves inserting the organic matter of different species into the human bloodstream — carries risks of the greatest magnitude, risks that just might be "the end of us." The significance of his findings cannot be overstated, and I can only hope that somewhere in the world there exists a true leader and man of destiny who will read this great but deplorably neglected book and act on it.

Ramblings, reflections, and conclusions

EVERY author who writes a book with a vital message, real or imagined, hopes to set the world on fire, so to speak, or at least to win the hearts and minds of millions of readers. I'm no exception. Yet, as I get near the end, I really wonder if this book, on its own, will put the slightest dent in the vaccination madness. And, as I debated with myself all along, I wonder if I did the right thing by often digressing from the main subject, and by taking on too many sacred cows. But either way, the written word alone has never moved mountains, or even hills. Every intelligent human being should be aware of the life-or-death conclusion reached by Edward Hooper in *The River*, but look at the obscurity into which this extremely important book has sunk. I see some of the most pivotal and inspirational works on my bookshelf, like Revilo Oliver's *America's Decline: The Education of a Conservative*, and William Gayley Simpson's *Which Way Western Man?* — the latter the greatest book I've ever read — and sadly ponder that almost no one I know has ever heard of them, much less read them. Indeed, I myself had never heard of *Nutrition and Physical Degeneration* by Weston A. Price — a monumental work first published in 1939, which is not the least bit political — until I was about fifty, well after I began investigating vaccines. Over ten years, Dr. Price, a dentist and dental researcher, traveled to the ends of the earth on his own dime to study the differences in skeletal and dental structure between mostly primitive people who ate food in its natural state on the one hand, and on the other hand those in more advanced nations and primitives under the domination of Europeans who ate refined foods loaded with sugar and starch. The book contains hundreds of photographs of people from both groups showing their teeth, and what these photos reveal is so significant and revolutionary that this book has the power to put both the dental and processed food industries out of business. (I might add, incidentally, that I saw plenty of corroborating evidence of Price's findings in rural Africa where so many natives have beautiful, straight, pearly white teeth, even though their diet is limited to bush meat and a few edible fruits and vegetables. In fact, having no utensils, many times I saw them use their strong teeth to uncap a bottle of beer — something I certainly wouldn't try!) Yet what influence has this book had? I doubt one in a hundred would even recognize the title. What has everyone heard of, what is far and away the bestseller of all time with five billion copies in print, what will you find in 90% of American households even though nearly all of it is deathly boring, absurd or downright psychotic and very few have read more than ten pages? The Holy Babble.

Even though baboons can't read and write, can't drive cars, can't build roads, houses and cities, they have a highly developed social structure, and are the most fascinating of primates. They are keenly aware of the dangers that surround them and they know how to meet those dangers. They have been known to kill leopards, their mortal enemy, in defense of the troop even if one or two of them get disemboweled in the process. They've been around for eight million years, a lot longer than any human race, and are clearly very adept at surviving. Dare we say that we're more intelligent? May we not be just another failed primate, like the gorilla, who seems to be well on the way to extinction, not only because of human encroachment, but because he is a confused creature whose day-to-day behavior is detrimental to his survival?

In my Africa traveling days, I recall a story about a tourist, camera in hand, who approached a baby elephant that had strayed from its family and was then promptly charged, stomped and tusked to death by the herd matriarch. Closer to home, most of us have read, at one time or another, about a hiker who suddenly came upon a grizzly sow and her cubs, and was badly mauled. Those who live to tell about these encounters are the lucky ones. Stories like these evoke our sympathy, especially when a man or woman who meant no harm is killed, but no sensible person is revolted by animals attacking in these circumstances. It's plainly understood, no explanation required, that such aggression is obedience to the most basic law of life, protecting one's young. Such aggression, even when it results in a tragic death, commands quiet respect. How perverse would it be for a grizzly bear to just sit on her haunches, unconcerned, and watch a human being walk right up to her cubs? About as perverse, in my opinion, as human beings who passively watch their children and grandchildren getting vaccinated over and over and over again. It would seem that, despite our unique capabilities, humans, far from being the most intelligent species in the animal kingdom, are, in the way that counts most, one of the dumbest.

❖

I can't imagine a country such as this one, with a semblance of peace and stability, going downhill faster than from what I've seen in my lifetime. The 1960s were my formative years. I was six at the beginning at that decade and sixteen at the end of it. It was in 1958 that Ezra Pound declared that "all America is an insane asylum." God, what would he say about this country today! I don't remember much from 1958, but the sixties were unforgettable, truly a fascinating time to grow up. Compared to the situation today, those times, which seem so pure and innocent looking back — though ominous things were happening below the surface — were a highwater mark in American history. Had I been taken in 1965 to a remote tropical island and lived there fifty years, then transported back here, I'd swear I was on some godforsaken planet. I know now that the 1960s was a transitional decade. From 1967 I sensed that the social fabric was starting to unravel. Hippies, drugs, violent protests against the Vietnam War, and leftwing radicalism on college campuses had all become familiar amid the first glimmerings of what was called the sexual revolution. But to me the real event of 1967 was the incredibly destructive race riots that exploded in so many cities. The late sixties really began the downward trajectory to where we are today, but even then, only two or three vaccine doses had been added to what we got in the 1950s, and America was still very much a white country — or better said, a white and black country with the racial lines clearly drawn and easy to see. The invasion from south of the border and from all points east was still years away, as were hypnotic words like "diversity" and "multiculturalism." Another bright spot of that era, an era that seems so normal now, was the popular music that flourished — so many catchy tunes and pleasant melodies and songs that touched deep emotions. I spend more time than I should in front of the computer screen with headphones on, trying to recapture my youth. Just yesterday I listened to Judy Collins's *Both Sides Now* over and over. What a lovely song and what a lovely lady, even if, like most musicians back then, she was not what I would call a role model. From reading the nostalgic comments of others around my age, who compare that music with today's spaced-out crud, I know I have plenty of company.

Superficially, everyday life in America is as normal as when I was young. Things are orderly, no one is starving, and there's little serious crime in white neighborhoods. When on the road making oil deliveries, I often drive past schools, and in the spring and fall I see children playing outside during recess, and high school sports teams practicing. I see groups of students getting out of high school around three o'clock, and they always look healthy and happy, horsing around like my friends and I used to do. And sometimes I've asked myself, "Could I be way off the mark to think that the lives of so many youngsters have been shattered by vaccines?" But I know that appearances can be deceiving. In addition to what I've read, there are many little things that tell a different story. One thing is all those signs on residential streets that I photographed, which represent just a fraction of maimed children. I remember my father, when his hearing started to fail him in old age, coming home from a visit to the ear, nose and throat doctor and saying that he was astounded at how many small children were in the waiting room with their parents. I've spoken to three teachers and one school principal around my age who totally agree that young people today have so many more mental problems than when we were young. One is my brother, a retired high school math teacher and athletic coach, who certainly doesn't think like me. He once told me that he had never seen so many whiny, sickly, maladjusted kids as what he dealt with every day. In September 2019 I made a trip to Wyoming — a fine state which comes in last place with a population of less than 600,000 but rates near the top in human genetic quality, a state blessed with little diversity, a state where one still finds the best old American values — and stopped to have a look at the Cheyenne Depot Museum, inside the now inactive railroad station. I took a self-guided tour behind a small group of first or second graders who were there with four teachers and aides on a field trip. Three wore glasses — none of my classmates did when I was their age — and two were in specially equipped wheelchairs. The girl, detached from her surroundings, stared silently at the ceiling, her mouth wide open. The boy screamed from time to time until one of the aides took him from his wheelchair and held him, which calmed him down. It was a sad sight, and one that was unimaginable two generations ago.

It's hard to say how much of the emotional problems so prevalent in children today stem from being over-vaccinated and how much from the social environment, which has changed so radically in the last fifty years. Children growing up in a world of security cameras, metal detectors, MTV, rampant divorce, prescription drug use, of electronic gadgets which reduce natural social interaction and allow instant access to hardcore pornography and all kinds of perversions, children conditioned to believe that "diversity" and "multiculturalism," euphemisms for national suicide which they see in the alien faces all around them, are a positive social force, that transgenderism and homosexuality are normal, as evidenced by the many "gay-straight clubs" in public schools today — children exposed to all these corrosive stimuli, which we baby boomers never knew, are bound to be much more messed up than we were, totally screwed up. Combine this with all the chemicals and biological junk that has been injected into their bodies, and I suspect that over the course of two generations, the System that rules America will prove more deadly than the Khmer Rouge's reign of terror in Cambodia.

I'll say again that I generally don't trust statistics — whether of war dead, the AIDS infected, unemployed adults or vaccine casualties — which are too often juggled by people trying to get their point across and are subject to the wildest swings. To quote Mark Twain: "There are three kinds of lies: lies, damned lies, and statistics." But for the sake of a rough guide, let me quote some numbers that were posted on the valuable ageofautism.com

Web site on September 16, 2016. In American children, 1 in 9 have asthma; 1 in 3 allergies, 1 in 10 ADHD; 1 in 6 learning disorders; 1 in 40 autism; 1 in 10 bowel disease; 1 in 20 seizure disorders; 1 in 50 a peanut allergy; 1 in 200 diabetes. Hearing and visual handicaps were not listed. Some of these numbers seem inflated to me, and I have no idea how the correspondent crunched them to conclude that every other American child suffers from one or more of the above. Then again, this might be in the ballpark. Allergies, incidentally, can be much more debilitating than people imagine. A wide variety of foods, airborne particles and even aromas, for instance, can cause swelling or hives, and a bout of itching as severe as contact with poison ivy. Peanut allergies, from what I've read, can be life-threatening; this has come on the scene only recently and the incidence is increasing. The writer is correct, of course, in pointing out, as many others have, that these problems were nearly nonexistent not long ago.

I like to base my beliefs on my own experience and observation, but I have no contact with groups of children or adolescents. But just this once, let me use those 17 or 18 children in Cheyenne as a sample group. When one considers the likelihood that some of them had conditions that weren't apparent, like those listed above, and that most if not all of them were caused by vaccines, then a fifty percent casualty rate is not at all far-fetched. It's all very hypothetical, of course. But suppose it's one in three, or even one in four. Why in God's name are we doing this to our children?

❖

I return to the question posed by Harold Buttram which I printed at the beginning of this book, and raised also by Edward Hooper in the preceding chapter, which I'll embellish here. Is man — or rather the wrong kind of man, who so often dominates his fellow man — so greedy, rotten, uncaring or ignorant that he will put an end to our species, or at best reduce us to a scattering of feeble primates floating aimlessly in the current of evolution? I have to say yes. Just look at how we've wrecked the natural environment in much of the world, and, if you've been following recent events, how birds and insects are dying or disappearing in large areas of America, very likely due to the atmospheric effects of the insatiable drive for more and more pointless technology. My gut feeling is that any young person who received even half the vaccines on the current CDC schedule is going to pay a dear price, if not right away, then later in life. Looking at the big picture, it comes down to a quote by the immortal British historian Edward Gibbon: "History is indeed little more than the register of the crimes, follies and misfortunes of mankind."

The question is, will vaccination be the final entry on the register? I again express my regret with my inconsistent method of updating this book over the course of ten years. During the last stage, I read a bunch of articles on the Internet that bring Harold Buttram's main concern, the meltdown of human reproduction, into sharp focus. I haven't incorporated them in this book. Some of it is shocking stuff that reads like science fiction. You can find the latest developments by using search terms like "epigenetics" and "DNA sequencing." I'll only touch on a bit of it here, while admitting that I know very little about microbiology beyond what I learned in high school biology class. But I think that's sufficient for a basic understanding of the grave problems we're up against. The underlying issue is the widespread injection, through vaccination, of traces of foreign animal matter into the human bloodstream, and the consequential alteration of the recipient's DNA — with that possibility obviously increasing with each additional vaccine dose. If this results in a change in the number of chromosomes, or some other chromosomal mismatch, no pregnancy will occur. We see something like this in nature, or rather man's tampering with nature, when a horse mates with a donkey, or a lion with a tiger. The offspring is sterile — ligers usually, mules always. It seems to me, that with the ever increasing rates of infertility, a parallel process may be at work in heavily vaccinated human beings. I don't know. Do you? Do you think the mattoids at Merck and the CDC know?

Recently there's been increasing speculation that vaccines are scrambling the sexual identity codes imprinted on genes, which has brought us new multitudes of not only homosexuals but pathetic freaks who can't figure out if they're boys or girls. This transgender rights movement came out of nowhere only about twenty years ago. Today, words like transgender and other linguistic abominations are in common use, with Facebook recognizing 58 designations of male, female and whatnot as user identities. What a world we live in. The vaccine explanation is somewhat esoteric, but who knows that it isn't the cause? Modern medicine has more cutting-edge horrors in its bag of needles, already in the works or pending experimental results with laboratory animals, including monkeys, but I won't go into it here. As things already stand, tens of millions of over-vaccinated young Americans may be the human equivalent of genetically modified corn, and at this point, no one really knows what this bodes for the future. But I'm sure it won't be good.

❖

Anthropologists tell us that eons ago, when our ancient forebears in Europe migrated north across that continent, those who lacked the foresight to prepare for the harsh winter ahead, who failed to build adequate shelters, gather firewood, and lay in a plentiful supply of dried fish, meat, honey and nuts, died of starvation and exposure, leaving no ancestors behind. Only the more intelligent survived, and they reproduced, and over generations their ancestors continued to be tested by Nature with the less mentally alert and physically fit continuing to die out. I sometimes wonder if this vaccination insanity amounts to the same evolutionary culling in a much shorter time span, if it will be left to the unvaccinated descendants of the smarter ones to carry on the race, while the children of the obtuse majority — those who blindly trust their doctors, television, and the pronouncements of government agencies — will grow up to be such wrecks physically and emotionally, that their lines are headed for extinction. Is all this fanciful nonsense on my part? Time will tell. In any case, it's impossible to deny the dysgenic trend of our times.

The survival and progress of European man and Western civilization is my foremost concern. The reader, I'm sure, has detected my racial prejudices at various points in this book. The white man, here and in his other homelands, has more on his plate than just the danger posed by vaccines. Let me make one thing perfectly clear if I haven't already: there is a race war being waged, both underground and in the open, the driving force of which is Jewish, with the ultimate aim of biological white extinction. I believe this is linked, however loosely, and whether consciously or subconsciously, with the drive to vaccinate — which doesn't mean that other races haven't also paid a heavy price as a consequence of mass vaccination. But we, and only we, have been conditioned to abhor the very thought of racial pride, identity and survival, even as the cold, hard facts of racial differences scream in our faces. Untold millions of central Americans and other nonwhites in the western hemisphere, among them a large criminal underclass, pour into the country that whites created to escape the terrible living conditions of their native lands. They want what the white man created and what they are unable to create. The flow is always in one direction. Asians, particularly Chinese and Indians, have swarmed into America in vast numbers to make a better life, and have been quietly networking and crowding out whites here in the universities, in certain businesses, and in upper middle class jobs, especially in medicine and technology. And again, it's a battle between weak and strong that takes place everywhere in the natural world. It appears that only the white race, for centuries the master of the planet, is the only race susceptible to manipulation of feelings of guilt, charity and fairness at the expense of his survival, the only race confused and weakened by meaningless words, allowing other races to assert themselves, sometimes violently, in the white man's own countries, while the media forever lecture us on the danger and evil of what they call white supremacy, which in reality is an instinct still alive in some whites for their survival as a people. I'm all for diversity and multiculturalism among the countries of the world, with their separate identities — not within one country because then it becomes a wobbly, unnatural patchwork of disunity. Just look at what a chaotic dump America has become, especially in large sections of all our major metropolitan areas. I admire other cultures — my walls and bookshelves are decorated with folk art from around the world that I've picked up in my travels — and I'm even fond of certain other races. I'm appalled at how criminal vaccine manufacturers use Third World children to test their newest products. But that doesn't mean that they and their families should be permitted to gate crash white countries, nor that they shouldn't be deported back where they came from. But it's going to take a major unforeseeable event to make that happen.

❖

As a politically incorrect malcontent who has subscribed to various hate sheets for nearly forty years, I've gotten on all kinds of unusual mailing lists, and frequently get unsolicited literature in the mail from people looking for new subscribers. Most of it is related to the faltering economy. I usually take a quick look at it before throwing it out. I've been reading these gloom and doom predictions of economic meltdown for so long now that I just can't be bothered with them anymore, which is not to say that the economy can't crash tomorrow. Furthermore, I think it's wise to be prepared for a nationwide bank failure and ensuing social chaos. It has happened in other countries. But I just can't dwell on it, mainly because I've never been able to understand the dynamics of inflation, national debt, balancing the budget, deficit spending and all the rest. It's over my head. What I do understand is that the wirepullers of international finance, who control nearly all the world's banking systems, are the most powerful and rapacious parasites on earth. And I know their tentacles are intertwined with those of the world's wealthiest corporations, including the pharmaceutical firms that make vaccines.

But there's a force in the world infinitely more powerful than them, and that force is what we call Nature. Just how subservient we are to Nature's ways, I saw for the first time in my life in the aftermath of Hurricane Sandy, which struck the northeast on October 29-30, 2012, devastating the New Jersey coast, the waterfront areas of New York City, and the south shore of Long Island. I live on the south shore, less than a mile from the bay where, when the area was developed, thousands of canals were cut and houses built. In my lifetime we've had some hurricanes that caused substantial erosion and property damage, but aside from the 100-mile-per-hour winds, what made Sandy unique was the rising ocean and bay tides that poured inland, flooding tens of thousands of homes and disabling electrical systems everywhere. Sandy is the fourth costliest hurricane in U.S. history. I know there have been worse natural disasters in America, but this was by far the worst I've ever witnessed. There were 285 fatalities, most from drowning. I saw and heard of things that I never could have imagined. Not far from my home, which escaped damage, three houses next to each other burned down because there was six feet of standing salt water in the streets and fire engines couldn't get through. Yachts were torn loose from their moorings, washed out to sea and vanished. The large sewage treatment plant in East Rockaway broke down, and shockingly, human waste began backing up through the underground plumbing and in several homes came out of sink and bathtub faucets. You couldn't buy gasoline at most gas stations because there was no electrical power to the pumps, and when it was restored, lines of eighty or ninety cars were a common sight. I delivered oil in some areas near the water a few weeks later and saw boats strewn about everywhere on dry land. There was some looting in the more affluent areas. For a long time I saw convoy after convoy of out-of-state utility trucks from as far west as Wisconsin that had come to repair the electricity grids. Lindenhurst, two towns to my east, got pounded especially hard because so many developments are right on the water. For months afterward I saw piles of debris — furniture, sheetrock, wooden fixtures ruined by all the water — brought out to the curb to be picked up by the overworked sanitation crews. Some houses were knocked off their foundations; it was much worse to the west, near Kennedy Airport. God knows how many vehicles were destroyed by high water and falling trees. Needless to say, it was financially devastating for those who, in the best of times, live paycheck to paycheck. One thing I remember well was the sad faces of people walking around in the hardest hit neighborhoods, expressions that seemed to be asking, "How could this have happened here?"

It all got me to thinking: What if the Atlantic Ocean and the bays fed by it had risen just a few feet higher? What if Merrick Road, the southernmost east-west business route that runs along the western half of Long Island, dotted with supermarkets its entire length, became impassable, and delivery trucks couldn't get through? And food suddenly became scarce? And there was no electricity and running water? What if, what if? It would be a matter of just two or three days before the tougher boys and men started smashing in supermarket windows and clearing food and drink from the shelves for themselves and their families, fending off competitors the way brown bears do in Alaskan rivers during the salmon run. The more aggressive blacks, of which there are plenty in several Long Island towns, would quickly revert to cannibalism. Think I'm exaggerating? Well, it happened all over the Congo in the early 1960s, right after Belgium granted that colony its independence and the new country went haywire. Civilization is much more fragile and precious than you might think. If and when law and order break down around here, due to factors beyond human control, we'll probably see something along the lines of what happened in prehistoric times — the survivors will be those with the foresight to have adequately armed themselves and stockpiled plenty of ammunition along with food and water, and perhaps bought a place to escape to upstate. The fashionable and socially respectable, trapped in their mental playpens of liberalism and conservatism, won't last long. They'll learn the hard way that this planet is not a playground. They and their faulty genes will disappear from the earth. The stronger and more intelligent will survive and propagate.

Am I being too melodramatic? The people of Banda Aceh might not think so. Banda Aceh is a city on the northern tip of Sumatra, the largest island in the archipelago nation of Indonesia. On December 26, 2004, an earthquake erupted on the floor of the Indian Ocean, creating a tsunami that devastated coastal regions of south Asia. A wall of water ninety feet high slammed into Banda Aceh and rushed inland more than two miles. When the water receded, 167,000 decomposing corpses awaited burial or cremation. It's possible that the more intuitive fled from the coast while there was still time, but I really don't know. Can something like that happen here? The "experts" tell us that it almost certainly can't, that this part of the world is much less prone to severe geological disturbances. But no one had ever spoken about something like Sandy happening here. And if Sandy could happen, why couldn't something worse happen? What's to prevent a future storm, packing a little more wallop, from totally inundating the island of Manhattan, which is only about two miles wide? As it was, the subway lines and most road tunnels into Manhattan were flooded during Sandy, and several square miles in the vicinity of the shoreline were under three to fifteen feet of water. What if another catastrophic storm, with just half the force of the 2004 Indian Ocean tsunami, hits the New York area? Can anyone picture the chaos and savagery that would explode when "the greatest city in the world" goes down? And not 167,000 but one or two million rotting corpses in all five boroughs litter the streets? Can you imagine how quickly the world would change with the sudden destruction of New York City?

Now, let me be realistic. What I just wrote *could* happen, but it's unlikely. But I'll tell you what *is* happening. Diseases like tuberculosis and hepatitis A, brought here by Third World invaders, legal and illegal, are becoming fairly common in urban areas and in the southern states, though outbreaks have affected only a small number of native-born Americans. So far. Piles of feces, along with discarded hypodermic needles, left behind by homeless human dregs, are now a common sight on the streets and sidewalks of downtown areas in several western cities, and in California, especially San Francisco, the situation is reputed to be worse than anything I've seen in any Third World city. Malaria, prevalent in colonial times, and totally eradicated only about seventy years ago, has also reappeared. I'm surprised it hasn't become pandemic, since it's transmitted so easily. It might yet. A host of other imported diseases have arrived here with the deluge of dark skins, but I doubt they can spawn anything like the Black Death, which may have wiped out half the population of Europe in the fourteenth century. If such a plague should take hold, however, puny man will have no more control over it than he has over the recurring fury of wind and waves.

Speaking of uncontrollable forces of Nature, this is a final update on this final chapter in the middle of March 2020, well over a month into incessant news coverage of the spreading corona virus. My God, if claims about vaccines are so conflicting and contradictory, what is one to make of the bedlam of opinions and interpretations surrounding this latest international health issue, even among sincere truth seekers? I don't know what it's all about, and even if I did, this isn't the place to dive into it. One thing I know with certainty: there is no more glaring proof that the Jewish global media machine can lead the governments of the world around by the nose by deciding what is news, slanting it accordingly, and drenching everyone with it. Whether there's some ulterior motive at work here, such as a conspiracy to impose medical martial law on the U.S. and other nations, perhaps involving a trial balloon campaign of forced vaccination, or whether the media masters are really frightened about a worldwide pandemic that might sweep them away along with a sizable chunk of the world population, or whether the truth lies somewhere else, I cannot say. What I will say is that it would be nice if deaths by vaccination around the world, which probably far exceed deaths by corona virus in the last six weeks, got one millionth of the air time and column space. I can only judge the present by the past and make a good guess about the current situation. I think the whole thing is much ado about very little, very few will be affected, and by the end of the year this big scare will die down and fade out.

❖

Kill or Let Your Children Be Killed is the title I originally chose for this book. I decided to change it, one reason being that I didn't want to give anyone the impression that the book's message was an imminent call to violence, which is a crime, the distinction between legally protected speech and criminal speech having been codified in the 1969 Supreme Court ruling on *Brandenburg vs. Ohio*. So, I should make it clear that *my top priority is simply to change the law without resorting to violence to allow unconditional freedom of choice for parents to vaccinate or not vaccinate their children.* If this doesn't work, then somewhere down the road it just might take bullets and bombs to settle the issue. But for now, no one should be afraid to exercise the little freedom we still have in defense of our children, and push that freedom to its legal limit. Can anyone claim that we don't owe them that?

The first step towards accomplishing anything is to face the fact that, as I've repeatedly stated, we live under a System that's rotten to the core and cannot last. I hope it has sunk in by now: *America is death.* Death and suffering, inflicted on foreign populations as well as our own, make up a substantial portion of America's gross national product. It emanates from the top tiers of media, finance, government, pharmaceuticals, medicine, law enforcement and the military, and trickles down to perhaps twenty or thirty million in the public and private sectors who cling to the status quo — people whose salaries are hooked into human misery, however remote, like the factory worker who makes syringes and vials, or the bartender at a tavern frequented by Army personnel. Most of these people are not evil, of course, but it goes without saying that the top one percenters are. Among them must be several thousand, at least, whose work involves childhood vaccines and who are well aware of their dangers, but pretend otherwise. Every last one of them should be sitting in the dock and facing the death penalty. Instead they live well, and in many cases have grown fat off the System.

A few far-sighted men have written extremely violent novels, depicting all-out war waged by homegrown rebels against America's rulers and their flunkies. I've already mentioned *The Iron Heel* and *The Turner Diaries*; Frank Weltner's *My Dead America* also comes to mind. It's difficult to estimate how many Americans have been inspired by these books, or who otherwise have become totally radicalized in their thinking — and how many are not far behind them. Not long ago, on a Web site geared towards disgruntled military veterans, there was an open call for an armed million-man march through Washington D.C. followed by the storming of federal buildings, dragging bureaucrats out into the streets, and hanging them. Does this, and some scenes in the aforementioned books, sound naughty? Just what do you think America has been doing to the rest of the world for the past 75 years?

Fiction aside, there's an historical precedent to the guerilla warfare depicted in the novels named above. That precedent is the Ku Klux Klan, which was founded in Tennessee in 1865 amidst the intolerable Federal tyranny known as the Reconstruction era which crushed the South following its defeat in the Civil War. I am neither praising nor denigrating the violence of the Klan, which in any case is an historically obsolete organization; I'm simply relating the plain fact that the original Klan was an anti-terrorist guerilla force fighting against a horrible regime, and contrary to the Hollywood image, the KKK was not motivated by racial hatred or an itch to lynch innocent blacks. Much of their wrath, in fact, was directed at the white carpetbaggers and scalawags who had empowered the worst black elements. With few exceptions, beatings and killings were meted out to those who deserved it. This had major repercussions within the federal government, and after twelve years of insanity, the Klan restored civilization to the southern states. I'm simplifying here, but that's the gist of what happened, which you won't read on Wikipedia or in any other politically correct source. Most Americans, who have been robbed of their own history, know nothing of this inspirational story. They have never heard of Thomas Dixon, who grew up in North Carolina during the darkest post-Civil War years, nor his novels *The Leopard's Spots*, *The Clansman*, and *The Traitor*, pinnacles of American literature. They have no idea that in 1915, D.W. Griffith's *The Birth of a Nation*, a movie considered a technological landmark, glorified the Klan's triumph of good over evil, to the dismay of northern audiences. Even in the South today, perhaps, the word "Reconstruction" is likely to draw a blank among the younger generation, though books by local authors about those shameful years can be found at Civil War battlefield gift shops. Dixon's works, however, have vanished down the Orwellian memory hole, even though they sold millions of copies in the early twentieth century.

Let's talk some more about violence, since we have visited so much of it on the rest of the world, and since it's an everyday feature of the American landscape, especially in the big cities. What do you think Tom Jefferson had in mind when he wrote that "the tree of liberty must be refreshed from time to time with the blood of patriots and tyrants"? The immortal German composer Richard Wagner constructed bombs used in a political uprising in Dresden in 1849 — an uprising which for him had cultural overtones, and after which he fled his native land and lived in exile for several years — proving that genius does not necessarily exclude revolutionary violence. That there are times when killing becomes imperative — and in one instance when deadly force could have forestalled history's greatest genocide — was something Alexander Solzhenitsyn thought about as he ground out his eight-year prison sentence:

And how we burned in the camps later thinking: What would things have been like if every Security operative, when he went out at night to make an arrest, had been uncertain whether he would return alive and had to say goodbye to his family? Or if, during periods of mass arrests, as for example in Leningrad, when they arrested a quarter of the entire city, people had not simply sat there in their lairs paling with terror at every bang of the downstairs door and at every step on the staircase, but had understood they had nothing left to lose and had boldly set up in the downstairs hall an ambush of half a dozen people with axes, hammers, pokers, or whatever else was at hand? After all, you knew ahead of time that those bluecaps were out at night for no good purpose. And you could be sure ahead of time that you'd be cracking the skull of a cutthroat. Or what about the Black Maria sitting out there on the street with one lonely chauffeur — what if it had been driven off or its tires spiked? The Organs would very quickly have suffered a shortage of officers and transport and, notwithstanding all of Stalin's thirst, the cursed machine would have ground to a halt!

If... if.... We didn't love freedom enough. And even more — we had no awareness of the real situation. We spent ourselves in one unrestrained outburst in 1917, and then we hurried *to submit. We submitted with* pleasure! *We purely and simply* deserved *everything that happened afterward.*

I used to wonder how the people of Russia — unrefined by European standards, but still a civilized country under the czars — allowed themselves to be led deeper and deeper into a situation which, for sheer terror, has few parallels. Well, I see the same process at work here. Every day, well over a million Americans walk through a security gauntlet at the nation's airports, submitting to the commands of TSA dullards, without the slightest sense of violation. Every day, more than 10,000 newborn babies are needlessly injected with hepatitis B vaccine without so much as a whimper from their parents. It's clear to me that most Americans — especially those who live in the cities and suburbs and wouldn't think of owning a gun — embody the same passive stupidity of all those hapless Russians whom Solzhenitsyn wrote about. And while it's great that there are so many privately owned firearms in America, without sufficient organization and numbers, an isolated shooting of a state or federal vaccine enforcement goon who knocks on your door is unlikely to benefit others.

America, it bears repeating, is a medical dictatorship, probably as oppressive in regard to childhood vaccination as the worst countries out there. You can forget about your constitutional rights because this is not a nation of laws, but a nation of bureaucratic vagaries. As I've noted, if you don't want your child vaccinated as a prerequisite for attending school, in all but four states you have the right to file a philosophical or religious exemption explaining your position. The superintendent or school nurse reviewing it also has the right to deny it, and to take the matter further by reporting you to the police. And the police have the right to go further still so that you end up in court. And the judge has the right to take your child away from you. So any talk about "rights" is just hot air. So far as I know, no state mandates a hepatitis B shot before a newborn baby leaves the hospital, but it's routinely given just the same. Yet even requesting that the shot not be given can arouse suspicion and lead to undesirable consequences. In my research I randomly came across about fifteen recent cases in which parents had broken no law, yet were fed through the judicial wringer. Some were locked up for a few days for disobeying a court order to have their children vaccinated. Much worse are incidents, by no means rare, where parents have been charged, under the guise of "shaken baby syndrome," with grave felonies to cover a medicrat or doctor after an infant suffers seizures or dies after a vaccination. You can find plenty of detailed cases on the Internet. What a truly horrible country we live in when authorities can stoop to such crimes — against innocent, grieving parents no less — and get away with it.

Needless to say, Congress will never put an end to this. Ninety-five percent of our elected representatives are scum to begin with or they wouldn't be where they are. It's an axiom that to be elected to public office at any level of government you have to sell your soul to the highest bidder, and the pharmaceutical lobby has more money to buy political whores than it takes to buy all the gold in Fort Knox. Literally. Of our 535 senators and congressmen, can anyone name five with any honor and integrity, five with the backbone to swim against the political tides on a contentious issue and do the right thing? And what do congressional investigations ever accomplish? In 1999, Dan Burton, a Republican congressman from Indiana, who chaired the Committee on Government Reform, held the first of nearly twenty hearings over the years on Capitol Hill, on the issue of mercury in vaccines and autism, at which heartbroken parents from around the country testified. Burton put a few top CDC and FDA officials in the hot seat and grilled them mercilessly. He had his own ax to grind. He had seen his grandson sink into the abyss of autism almost immediately after being injected with nine vaccine doses at one routine doctor visit. When he was done, Helen Chenoweth, a feisty congresswoman who also had an autistic grandchild, reamed them out some more. And what good came out of all this? None, absolutely none.

In human events, when two irreconcilable ideas clash, superior force, or the threat of it, usually wins. In America, the anti-vaxxers are mindful of the gun and handcuffs the policeman carries, because experience has taught them to be. The pro-vaxxers never worry about these things because the State, which stands behind the vaccine dogma, is on their side. But the balance of firepower in a country like America, where so many citizens distrust or flat out hate the federal government, is very tricky. Clearly, the police and the military are aligned with the State. The System takes good care of them, and anything that threatens it is likely to be perceived as a personal threat. Upton Sinclair's quotation is worth repeating: "It is difficult to get a man to understand something when his salary depends upon him not understanding it." For most of those who make a career of the police or military at all departments and levels, it's a nice salary, plus a generous benefits package and pension, that molds their thinking. They owe it all to the System. And furthermore, there's plenty of indoctrination from Jewish organizations, especially the Anti-Defamation League, which influences a large segment of the military and law enforcement community, aided and abetted by Israeli trainers who instill brutal methods of cracking down on dissidents.

There are ever-widening cracks in the edifice of the System, however. Discord within the military, especially between the top and bottom castes, which has always prevailed, is probably at its most critical today. How can it be otherwise when, as we have seen, Air Force pilots are compelled to get a vaccination they're all but told to their faces may cause grave harm? After every foreign war American combat veterans have been treated like disposable livestock, but in 2009 Janet Napolitano, the bull-dyke Secretary of Homeland Security under Barack Obama, showed the elitists' true colors when she alluded to soldiers returning from Iraq and Afghanistan as potential terrorists. In 1994 a questionnaire was distributed to Marines at Twenty-Nine Palms combat training base in southern California, which asked if they would shoot American civilians who refused to surrender guns deemed illegal by the government. And there have been several polls since then to get our troops' opinions on this matter, undertaken by those in the higher echelons of the armed forces, where most military gangsters and traitors are found. The general consensus, from what I have gathered, is that most would *not* shoot at their countrymen, and find the very idea abhorrent. But with so much "diversity" in the armed forces these days, I'm not too sure about that anymore. Turmoil in the military is on full display on the gritty Web site militarycorruption.com, where servicemen who have honor expose rotten career men in straightforward and often rough language.

The situation, I'm sure, is much the same with the police. As with the military, America's police forces have their share of bullies and thugs, and also as with the military, one is likely to find the most corrupt and pliable at the higher levels. I also believe, along with many other dissidents, that with the drift towards more security and less freedom, the police have become more militarized and in many cases more vicious. Nevertheless, having been acquainted with about twenty cops over the years, I think a solid majority of them are basically good guys who try to do their jobs the right way. I also think that this percentage shrinks the higher you go in government, and at the federal level there's a sizable criminal element, especially in the FBI, an agency with a checkered history which, more than ever, resembles the feared and hated secret police outfits in the formerly captive nations of the Soviet bloc. In our increasingly upside down country, and in the increasingly twisted worldview of high-level FBI officials, any nonconformist, anyone who holds politically incorrect opinions, is a possible terrorist. How much of this mentality trickles down to the rank and file is impossible for me to say. I'm sure there are good people in the FBI. I'm equally certain that you can find the same kind of thugs once found in the Stasi and the KGB. This has always been the case, but it has gotten considerably worse in recent years, with more *agents provocateurs*, and more vermin skilled in cover-ups and frame-ups — not to mention a homosexual clique that shields pedophiles, as was made all too clear in John DeCamp's *The Franklin Cover-Up*. I suspect, and I can only hope, that there is plenty of dissension in the ranks.

Returning to police officers at the local level, with whom I've had a fair amount of contact, like nearly all men of force they are not the type to find fault with the System, nor sympathize with those of any political ideology who do. Few are deep thinkers or book readers. They're flag wavers and TV watchers, who for decades have absorbed all they've seen on the evening news and other programs. But unless they work in a white enclave, they know that the System, which treats them well financially, can instantly turn on them. They know, for example, that if they make a split-second decision to shoot some black animal brandishing a knife or a gun, they might very well be sacrificed on the altar of political correctness; their life can suddenly become a judicial nightmare, with the looming possibility of a long prison term. They are much less likely to know, however, that Jewish entertainment executives are the ones who produce and profit from the recordings of black rap musicians whose lyrics egg on their fellow blacks to commit violent crimes, including explicit messages to murder cops. In lapping up all the bilge about ISIS and al-Qaeda and Islamic terrorism, spouted by all the talking heads on the television screen, they are totally unaware of the bulk of evidence indicating that the Israelis and their minions in the U.S. played a key role in the September 11 attacks.

Most important, as individuals who are rarely intellectual, few police, and troops for that matter, have studied the vaccine issue, and overwhelmingly they are the type who leave all that to the "experts". But cops are fathers too, who want to do what's best for their children, and there's no doubt that their children have been victimized by vaccines in the same proportion as the children of those in other professions. What is all the financial security the System provides worth compared to a lifelong affliction of his child, or grandchild, brought on by a tainted vaccine? Despite their general satisfaction with the status quo, most rank and file members of the police and military know in their hearts that things are not right in America, even if they can't articulate it. If the few issues I've just raised ever come into the open, there could well be an abrupt change in the direction weapons of all kinds are pointed. No wonder Jews are so determined to keep an iron grip on the news and entertainment media.

That the police and military can never be trusted by the American ruling class, which includes top military officials and all media executives, was made abundantly clear in the failed Hungarian revolt of 1956. It wasn't until the standalone historian David Irving published *Uprising!* in 1981 that readers outside of Hungary learned of the anti-Jewish nature of the bloody insurrection in a nation that had seethed in the yoke of total communism where, as in all east European nations in the years after World War Two, Jews had coasted to power on the wave of Soviet victory. But what was remarkable about this revolt, and shocking to the overlords in Budapest and Moscow, was that many ordinary police and soldiers, in the pay of the System, hated the regime so bitterly themselves that they raided their armories and distributed guns to the workers and peasants who, of course, were strictly forbidden to own them. The question is, can a full-scale uprising like this happen here? In comparing Hungary in 1956 to America today there are similarities as well as differences. Hungary is a small country, the size of Indiana, with a homogeneous population then and now, both features of which offer a great advantage. America'a advantage would be that there is no foreign enemy force to worry about, like the Soviet tanks that rolled in and ultimately crushed the rebellion. On the other hand, Americans as a whole have never known anything like the repression that was clamped down on Hungary, where most men had been thrown into prison, and often beaten or tortured, on absurd pretexts. Then again, what could more justifiably incite a people to rise in holy rage than what has been done, and continues to be done, to American children?

We live in the here and now, and despite unprecedented levels of discontent, all signs at present point to a continuation of the slow process of social disintegration that will not reach critical mass in the near future. We are not near the point where Hungary was in 1956 and Romania was in 1989, so for now it behooves knowing, caring parents to do things quietly, calmly and legally. Let me tell you what I would do if I were a parent of a young child, or had a baby on the way. First, if possible, give birth at home with the aid of a midwife, or, if you're fortunate to have one, a sympathetic obstetrician. This way you won't have to worry about the obscene and disgusting ritual of your baby being welcomed to the world with a toxic hepatitis B jab, as well as the recently ordained and equally cruel vitamin K injection. Of course, as everywhere else, you'll find different personalities working in neonatal wards, but why even deal with the anxiety of raising a red flag with some bitch nurse or bastard doctor by saying no to the needle? Second, if you live in a state where schools do not allow religious or philosophical exemptions to vaccination, keep your child out of school and opt for home schooling. Home-schooled children are invariably healthier and brighter, although this may in large part be attributed to genetic factors, namely having parents who reflect the dictum of George Bernard Shaw: "Intelligent people don't vaccinate their children." The question of depriving children of social interaction often comes up, and it's a fair question. I look back at my own elementary school education, from 1958 to 1965, as a happy and innocent time among my fifty or so classmates, divided into two classrooms. Even though there were as many petty animosities as there were close friendships, I cherish that experience. Most important, with two exceptions, we were all of European ancestry and whatever one's ethnic background, and whether one was Catholic or Protestant, counted for almost nothing. We were like a big family without even thinking about it. But in much of America today, this homogeneity, this natural feeling of kinship and belonging, no longer exists in the classroom. In my view, this inner isolation that white children must feel when going through primary school in the company of mestizos or blacks or Chinese or whatever is worse than staying out of school altogether, especially if they have ample opportunity to take part in social activities with their own kind. As for education, the public schools are an unmitigated disaster, the initial phase in creating a population of zombies. I have already commented on this and need not repeat myself here. In my day it was not as bad, and nothing sexually weird was pushed on us, though the process of being brainwashed by teachers who themselves had been brainwashed on certain issues, notably World War Two, was in full swing. Beyond the three Rs, taught in the first grade, I learned little of lasting value all the way through college. To an inquisitive boy like me, school was an unending exercise in frustration, with just a few bright spots. If your child has a natural thirst for knowledge, encourage him to read good books. If he's not the studious type, encourage him to pursue things that interest him. Either way, he'll learn more and undergo less brain damage than if he attends school. Every intelligent American's motto should be: "No shots, no school? Okay then. The hell with school!"

I should point out that, at present, so far as I know, home birth and home schooling are legal in all fifty states, though you may have to meet certain conditions under the squinting eyes of local bureaucrats. Also, if you choose to keep your baby out of the hospital or your child out of school, he or she is not required by law to get a single vaccination in any state — although, as I've emphasized, this is a precarious freedom that exists only on paper, in some states more than others, and it's most wise to sidestep the System quietly. Should the CDC usurp state vaccine laws, or some states on their own declare a medical emergency and try to enforce compulsory vaccination by going door to door, then the time will have come to uncase the guns, join with like-minded citizens and informed local police, and declare war on Vaccination Nation Jonestown.

❖

During the latter stages of writing and revising this book, I've often glanced at a small photograph of my daughter that I keep on my desk. It reminds me of the reason I've spent the last ten years, off and on, discouraged then determined, drained then inspired, at work on a project that, were it a lighthearted subject, would've taken me only a year to complete. Her photo tells me to stay in my seat and get this done, instead of relaxing with a book or putting on some music and pouring myself a glass of wine — though I've often given in to that temptation. But most of all it speaks for all the young lives wrecked by vaccines. My daughter is fidgety when you try to make her sit still for a picture, and this one, taken by a professional photographer at her activity center, is not my favorite. Yet her facial expression in this photo is unlike any other, and when I look at it long enough my eyes well up because it's an expression that seems to say, "Why did this happen to me?" And every time that happens, it reinforces my sense of purpose, which is to spare as many parents as I can the heartache of asking, "Why did this happen to my child?"

I gave some serious thought to writing a full account of my daughter's condition, from the early days when she was not progressing and my wife and I feared the worst, to the present when, soon to turn 26, she plods on day to day, cared for and drugged against my wishes in a home with two older men, less disabled but like her lacking the faculty of speech. I could write a very poignant account of the day the bottom fell out of my life, the day, at my wife's insistence — I didn't want to face the dreaded diagnosis — I took my daughter to be evaluated by a specialist. Near the end of our appointment, after close observation, I was told, in words that are probably often used to soften the blow, that Mary was "on the autistic spectrum." This is nicer than hearing "your daughter is autistic," but of course it means the same thing. I could make you cry by telling you what it was like driving home, a numb shell of a man with nothing to live for, then walking through the door, holding my daughter's hand, my wife's eyes silently searching mine for a few moments, then breaking the news to her. I decided against it because it's a long story that merits a book in itself, and it's so closely connected with another long story — the wrong fork I took in life by marrying the wrong woman, and all the marital strife and harrowing divorce litigation that followed. So, the only comment I wish to make here is that, in the society that eventually emerges from the mess that is now America, the judicial system will be as radically transformed as the medical system.

Another reason for not writing in detail about my daughter is that I don't want to convey the idea, as promoted by the media spotlight, that autism is the only tragedy for which people have claimed a connection with vaccines. Autism is only the tip of the iceberg. Vaccines have triggered disorders of all kinds affecting the brain, skin, blood, intestinal tract, joints and nervous system; they're the primary cause of Sudden Infant Death Syndrome (SIDS), and also have induced asthma, seizures, learning disabilities, visual and auditory problems, diabetes, meningitis, various types of cancer and more. These children, who far outnumber the autistic population, have been robbed of the well-being that should have been their birthright. Routine vaccinations have cursed their lives too, as well as the lives of their parents and grandparents.

I've often reflected on how my life today — and what I hope to make of it by writing this book and raising the sword against vaccines — is the sum total of chance events, wrong decisions and bad luck. I have no complaints, as I wrote at the beginning, and I explained why. But here I'll add another dimension to show just how fluky life can be, and how a providential stroke of luck is behind the creation of this book.

In 1988, three years before we got married, my wife and I traveled for two months around five countries in southern Africa. Leaving South Africa, our starting point, we hitchhiked and cadged lifts north through sparsely populated Botswana, pulling up in the small, dusty town of Maun, gateway to the far north, famous for its wildlife and a vast waterway of channels and tributaries known as the Okavango Delta, through which a river of the same name flows. Aside from game drives in the Chobe and Moremi reserves, the thing to do here if you're looking for adventure is to take an excursion in a dugout canoe, known locally as a *mokoro*, poled along by a native, in the hope of seeing some of Africa's wild animals which are prolific here. Africa lurks with the unexpected and unpleasant, so a *mokoro* ride carries with it a whiff of danger. There were a few white-run, fly-by-night travel storefronts in Maun; I walked into one and worked out a two-day itinerary in which we'd get a ride deep into Moremi, where we'd camp for the night, then in the morning our *mokoro* guide would meet us and take us out for the day on the Okavango. When I think of how some people are a bundle of contradictions, I always think of my ex-wife, who was and still is the classic nervous wreck. Not long after this trip she was prescribed Xanax for her anxiety, and later she became too stressed to drive her car on the interstate, sticking only to local roads. Yet I doubt one woman in a hundred has the spunk to pitch a tent in a place like the Moremi, which swarms with big game of all kinds. We passed a fitful night in our sleeping bags, punctuated by those unforgettable African night sounds, and at eight in the morning, as planned, Samuel, our *mokoro* man, showed up. We got into a beat-up Land Rover with him, and after a short ride, followed him down to a lagoon where he kept his *mokoro*.

A *mokoro* is long, slender, and very low to the water. I settled into the front, my wife in the middle, and Samuel in the rear with his long pole, guiding us through a maze of inlets by pushing against the bottom. We kept our eyes open for wildlife coming to drink, but saw just a few antelope. Still, it was a beautiful day, this was relaxing and unique, and I was glad to be doing it. We had been gliding along for twenty minutes or so when I felt a slight wobble and then saw the front of the *mokoro* dip below the surface. Water rushed in — so silently, so peacefully — and within seconds the *mokoro* sank from under us and we were standing in chest-deep water. The few things we had taken with us, including my camera, still in its nylon carrying bag (I would later discover it was ruined) slowly drifted away on the gentle current. My wife, nearly hysterical, threw her arms around my neck and gasped, "John, are there crocodiles here?" I knew there were — plenty of them — but I told her to stay still and keep quiet so as not to attract any crocs that might be close by. We had not seen any, but that didn't mean we hadn't passed any swimming along the bottom or concealed in the reeds at the water's edge. This animal is the deadliest predator on earth, and we were now helpless, dangling bait. As one African big game hunter put it, "*Crocodylus niliticus* is the one man-killer who, if he's big enough and you're available enough, will eat you every time he gets a chance." Neither of us had ever been so terrified — not only of dying, but of dying in a way that was too horrible to contemplate. But there was nothing we could do but wait as we watched Samuel wade through the water to retrieve our belongings that had floated downstream, then return and upright our *mokoro*. It seemed like an eternity, and I don't even remember if we continued our trip or called it a day and went back. Samuel told us that his pole had gotten caught in some vegetation, and his yanking it free caused the *mokoro* to capsize. We were both just greatly relieved to get back on dry land and live to tell about it.

The following week we checked into a cheap hotel in Bulawayo, Zimbabwe. Pinned to the bulletin board in the lobby was a flier with a photograph and a story about a nice English family of six that had come to Zimbabwe with some church group. On the far right was a handsome boy about twelve years old. They had gone rafting on the Zambezi River, the boy had fallen overboard, and he was taken by a crocodile. The flier implored the reader to pray for his family. Pray for what? How do you continue believing in a personal God after seeing something like that? How do you even go on living?

Years later, when the Internet and YouTube came along, I typed in "Okavango mokoro crocodile" to revisit our misadventure. I watched four video clips of close encounters with enormous crocodiles, one of which ended as the croc appeared to attack the *mokoro*. All I could do was look upward, close my eyes, and exhale deeply. But for the throw of the dice, I, or my wife, or both of us, could've ended up like that poor boy who fell into the Zambezi, could've been grabbed and eaten by one of these engrossingly horrid reptiles, living relics of the age of dinosaurs. But for the throw of the dice, my children would never have been born, this book would never have been written, and I would never have gone to war against childhood vaccination — and maybe, just maybe, the world would have been poorer for it. So who am I to complain about what I've gone through, and what I might yet go through, since that day in July 1988 in the Okavango Delta of Botswana, when what easily could've happened didn't happen?

❖

I'll reiterate here that not only is a country as good as its people, but a country *is* its people. Certain areas within the recognized boundaries of the United States are so dissimilar in human composition that we're not really talking about the same country. The leaders of the Second American Revolution will have to take this into account. Large areas of every major American city are now Third World slums where English is a second language. As many areas are black jungles where few outsiders dare to venture. Much of southern California and the southwest are more Mexican than European. How is it possible to clear up this demographic disaster, unprecedented in history, without provoking anarchy and bloodshed on a huge scale? How do you even begin to reclaim human cesspools like New York City, San Francisco, and every other megalopolis for that matter? My answer is you can't, so just cut them loose. The only solution I see, the only glimmer of hope for the future, is for those areas of the country least blessed by all this diversity, areas where the good old values and traditions still endure — I have in mind the three northern New England states and most of the northwest quadrant — to secede. This has happened repeatedly in nations fractured by human differentiation, and nowhere do these fault lines run so deep as in America today. Although the noxious fumes of television and the public education system permeate American households everywhere, there are still regions far removed — geographically, culturally, racially, spiritually — from New York, Washington and Hollywood. Another aspect of these areas is the largely unspoiled natural environments, which goes hand in hand with small populations, among which there is, I believe, a higher quotient of people with ingenuity and creativity, compared to the white population at large. It follows that the vaccination laws in these states are more lax than the national average, and from what I've read, North Dakota is the most lax of them all, granting exemptions for schoolchildren without the slightest fuss. But there are still rotten people out there — I've read a few disturbing anecdotes out of New Hampshire and Idaho — and no one should let their guard down until these stupid laws are swept clean from the books.

It's amazing how a change of scenery can lift me out of the funk of living thirty miles from the world's most loathsome city. In September 2017 I packed up my car and drove with my son to Maine for a week of camping and exploring. I hadn't been to Maine in nearly thirty years, and had forgotten how everything about this state is invigorating — not just the beauty of the pristine seashore and forests but the cleanliness everywhere, the pleasant little towns and shops, and of course the people, courteous, civilized, and on the whole attractive and healthy looking, as one would expect where opportunities for rugged outdoor activities abound — *and white*. Very little diversity in Maine, and none at all that I saw, except for seven or eight blacks in Portland, the state's largest city with a population of 67,000. This was my first time in Portland, and it struck me as everything an American city should be. Perhaps I'm romanticizing here; I may have missed some depressed towns and villages, like several I've seen in rural upstate New York, but Maine gave me hope for America, and I've felt that same little surge in my more frequent visits to Vermont and New Hampshire. Montpelier, Vermont — leafy little capital of 7,535; imagine that! (Since writing these lines, hundreds of gate crashers from Angola and the Congo, and thousands of Somalis were dumped in Portland, straining that city's resources. I also learned that, years earlier, the same faceless, nation-wrecking termites quietly transplanted about 5,000 Somalis into Lewiston, Maine's second-largest city, bringing with them the usual violence and social tension.)

This takes me back to the second-smallest state capital, one of only four bypassed by the Interstate highway system. Pierre, South Dakota is 32 miles from I-90, there's not much to see there, hardly anyone makes it a destination, and these were all excellent reasons to have a look at it during a meandering trip by car with my son in 2012. We've taken several such trips, and one of our whimsies is visiting state capitols, or at least driving right by them. After settling in in Pierre, we rode a few blocks to the sedate structure, parked, and walked up the steps. A sign indicated that the building would be closing at 7 PM. It was 6:15, a Friday. We went through the main entrance. There were no cameras, no metal detectors, no security guards. The place was deserted and we had unrestricted freedom to wander. I couldn't believe it. We walked around, peered into the legislative chamber, poked into a hearing room, stepped into the offices of state officials. Every door, except the one to the governor's office, was unlocked. I thought, "Can you picture something like this in Albany, Atlanta, Austin or any other capital city drowning in diversity?" The twenty minutes we spent in that empty building left a lasting impression on me — an impression of trust, decorum and social harmony. It exemplified the kind of America our children and grandchildren should inherit. This is what America should be, and still is, in some states.

A carping critic may point out that South Dakota and the other states comprising the northwest parcel of the lower 48 are hardly examples of racial homogeneity, given the 20 or so Indian reservations spread across them. True enough, but pestholes like Portland (Oregon) and Seattle, crawling with liberals and multiculturalism, seem much more problematic to me — hopeless, in fact. I'm no authority on Indian affairs, and I'm aware of mutual feelings of resentment between Indians and whites that have flickered for a long time, but through the years the relationship has been tranquil for the most part, and shows the wisdom of keeping different races physically separated. I've driven through four reservations without the slightest fear, and in fact on our visit to South Dakota we attended the annual Oglala tribe Powwow on the Pine Ridge reservation, spending the night there in our tent amid hundreds of Indian tents and teepees. Most Indians we dealt with were friendly in a reserved kind of way, and there's every reason to believe this peaceful relationship can last indefinitely.

To offer a final illustration of what a reborn America could be and should be, I go back to my 1986 visit to Iceland, where 99% of the 330,000 inhabitants are white, the overwhelming majority of Scandinavian stock. Iceland is as progressive a nation as you'll find on this planet. I happened to arrive by air in Reykjavik on June 16th which was, unknown to me, the day before Icelandic National Day, the equivalent of our Fourth of July. As I strolled the quaint, compact capital I noticed a crowd gathering outside the Parliament building. Curious, I joined them. Shortly afterward, the president of Iceland, a woman named Vigdis Finnbogadottir, came out with a small entourage and walked right past me, not thirty feet away. There were no bodyguards, no police, no uniforms. In 2008 and successive years the government there set an example for the world by arresting and prosecuting corrupt bank officials whose globally connected swindles caused a financial meltdown in the country, and severe economic hardship for a substantial percentage of the population. Twenty-nine Icelandic bankers went to jail, and the country's massive debt was written off with the slash of a pen; the vultures of international finance were simply told to go to hell. More recently, the country underwent an emotional crisis of sorts when a violent criminal was shot and killed by police — unbelievably, the first time this had ever happened in Icelandic history, which predates the invention of guns. The literacy rate in this country is 100%, and it seems unnecessary to add that there are no compulsory vaccination laws.

❖

I have exhausted every argument, trod every avenue, veered down many a side street in my endeavor to expose vaccination for the monumental and deadly fraud it is. This book has now reached its last page, and it's up to the reader to decide if I've succeeded. From the beginning the only reward I've sought is the satisfaction of applying my power of perception to the real world so that children do not needlessly suffer and die, of providing sharper talons and a greater wing span to a vital movement that other good people have started and built up. And if I my efforts fail, I still have a pipedream that, 150 years from now, people will read about me in the same vein that people today read about, say, Ignaz Semmelweis. For those who would accuse me of vanity for saying that, all I can do is smile and plead no contest.

I'm 66 as I write these words. Although I've enjoyed good health in body and soul, with a few bumps, pretty much my whole life, psychologically a change has come over me. At this age, you're well past the halfway point in the journey and you realize you're not going to live forever. I suppose most people who had a happy childhood become nostalgic in later life, looking back at their youth when the idea of growing old was inconceivable. That's certainly true in my case. Where we're born, the era we're born into, and the conditions in which we're raised are beyond our control, totally accidental. Looking back, it was really something to grow up in the 1960s. Ezra Pound called America an insane asylum in 1958, and he was right, but compared to today the sixties were so much more sane. Despite all the assassinations, the destructive race riots, the Vietnam War, despite the social decline that picked up steam towards the end of that decade and has accelerated ever since, it was a good time to be young. In so many ways, America seemed like a better country back then. One way is that we only got a few shots when we were kids. No one ever gave a thought to doubling that number, let alone multiplying it tenfold, and no one could have foreseen that such a disaster would take place in our lifetimes. Children never needed all those horrible vaccinations when I was young. Children don't need any of them today.

Bibliography

BELOW IS a partial list of books I have read over the years which, more than any other source of information, educated me about vaccines. They are listed in chronological order of their first printing. All are well worth reading, though with a few of them, as noted, I sharply disagree in places. What's nice these days is that, with the Internet, one can bypass the arbiters of culture in the media and instead read many independent reviews written by sensible people, and some people not so sensible, to gauge the value of any book. What struck me when I checked the publication years, since I hadn't thought about it, was that, aside from the Cantwell book which doesn't deal with childhood shots, in six of seven published prior to 1994 autism is never mentioned (*A Shot in the Dark*, the one exception, devotes only two pages to it), whereas every book from 1994 onward mentions this condition, and autism is the central subject in two of these. There were 22 vaccine doses on the CDC schedule from birth to four years in 1990, which gradually increased to 43 doses in 2001. How much more correlation does one need to show?

To avoid confusing the reader, I'll repeat that I have not changed my notes on the Cantwell and Horowitz books from the way I wrote them before reading Edward Hooper's *The River*, which invalidates a fair amount of their content. If one sets aside the major errors regarding Robert Strecker's theory of the source of AIDS, however, they're still valuable books.

Many other excellent books on the perils of vaccination have been published in recent years. No part-time researcher can possibly read all of them, but in my opinion that's not necessary because they keep reinforcing the same facts that have been known for twenty, fifty, a hundred years. Because those who wield power and influence in the U.S. are so hostile to the anti-vax movement, most books critical of vaccines will not be found in bookstores and public libraries. A few will. All of them, however, can be purchased through the Internet.

Speaking of the Internet, I have mixed feelings about buying books this way. It's great that we all have the opportunity to circumvent the censors, but the Internet has destroyed many legitimate businesses, book retailers included. Amazon immediately comes to mind. Amazon is a monopoly which operates like a monopoly. I say this from firsthand knowledge, having written and self-published a humorous book about my experiences as a baseball umpire in 1999, and having dealt with this leviathan personally. I did get exposure and sold hundreds of books through Amazon, but I was paid a pittance, and their employees, who toil miserably under the lash of Jeff Bezos, the world's richest man, are impossible to communicate with on a human level. I know that it's easy, efficient and economical to purchase books from Amazon, and I confess I've done so myself many times, but it's much more fair to the self-published author, and also to the publisher if one is involved, to buy books from them directly, or to buy from brick and mortar stores, both independents and chains. Unless the author is famous, in which case big sales are virtually guaranteed, writing and selling books is an arduous task, and often a financial loss. The old-fashioned transactions I've mentioned assure an equitable share of the profits. However, the remaining chains, like Barnes & Noble — and again, I speak from experience — frown upon self-published authors because of their stupid corporate rules, and anything that smacks of genuine controversy scares them to death. I have little sympathy for them.

A distinction should be made between faceless and often shady Internet booksellers that hawk anything between two covers, and Web sites dedicated to specific issues or subjects that offer books for sale. For example, Neil Miller's books are published by New Atlantean Press (affiliated with the site thinktwice.com), a small company in Santa Fe, New Mexico which, despite its obscurity, is not exactly "underground" as they work with mainstream book distributors. New Atlantean specializes in the vaccine issue, natural health and alternative medicine, selling their own books and those of other publishers. Some of the titles below, other than Miller's, might be available from them. This is the kind of company people should patronize.

There's another reason to buy controversial books from an upfront source whenever possible. 2017 witnessed a corporate crackdown on books that disseminate politically incorrect ideas. This crackdown has not affected anti-vaccine books — not yet at least — but it *has* affected other areas of knowledge, in particular the forbidden facts surrounding the Holocaust racket. A notable development was the revocation of credit card purchases, in which Wells Fargo Bank and PayPal, likely pressured by the Anti-Defamation League, played a prominent role. This battle continues. Among the targets were the Institute for Historical Review (ihr.org) and the Barnes Review (barnesreview.org). These firms sell suppressed books that should be read by every thinking man and woman. (Parenthetically, quite a few spineless directors of public libraries, including some near where I live, have installed censorship filters on their computers, so that when one tries to log on to these and other dissident Web sites, a page appears telling the user that library policy blocks access to sites promoting "intolerance and hate," words that always conveniently go undefined.). Also in 2017, for the first time, Amazon, which since its formation had at least held to an exemplary policy of selling controversial books of all kinds, caved in to Jewish pressure and quietly removed numerous titles from their inventory — that is, they banned them. On this list is what I consider the best book about World War Two ever written — a fast-paced chronicle of mini-chapters packed with photographs that I had never seen, and historical facts I had not been aware of, even though I'd already read a great deal of censored literature on that conflict, upon which the entire edifice of lies and false values of our age is constructed. That book is *The Bad War* by M.S. King, whose Web site is tomatobubble.com. I have never met nor spoken with Mr. King, but I'm most happy to give his book a plug. With these final remarks, I've deviated from the subject of this book one last time. For the last time, I will remind the reader that *it all ties together*. A dumbed-down nation of mental slaves who think we "won" the Second World War, who know and care nothing about history, and the lessons it teaches, is the same sorry bunch that allows their children to be injected with scores of poisonous vaccines. The first step out of this rut is the acquisition of knowledge which the following books provide.

• *Horrors of Vaccination Exposed and Illustrated* by Charles Higgins (1920). Reading this impassioned book, one comes to realize that all the angles of vaccination today — fraud, corruption, dogma, police state intimidation, experimentation on soldiers, serious injuries, death, denials, alibis — were commonplace a century ago, and people are none the wiser today. This book centers on the smallpox vaccine, the only one in wide use in the early 1900s, but it covers every conceivable aspect, including case histories of children who died after vaccination, photographs of horrific skin eruptions in victims young and old, and notes on con artist Edward Jenner sourced from 19th century British literature. It also includes descriptions and photographs of cattle suffering from an epidemic of foot-and-mouth disease resulting from the cruel practice of planting virus in the bellies of calves to make the vaccine.

• *The Poisoned Needle* by Eleanor McBean (1957). Long out of print, though reissued in 2005 — unfortunately in very small type — this hard-hitting book was a voice in the wilderness warning about the danger of the Salk polio vaccine amidst the roaring news media festivities. It tells "the other side of the story" that so few Americans have heard. The first part of the book covers the smallpox vaccine, and includes many case histories and additional damning facts about Jenner that would be difficult to find elsewhere. A most valuable book and historical document.

• *Every Second Child* by Archie Kalokerinos (1974). The story of a doctor who treated Aborigine babies in the remote Australian outback. Living in filth and subsisting on a terrible diet, many of these babies would die soon after public "health" teams descended on them and lined them up for vaccinations, paying no heed to their individual medical histories and living conditions. Dr. Kalokerinos figured out that these shots rapidly depleted their tiny bodies of Vitamin C reserves, and saved many on the brink of death with injections of this essential vitamin. The book also shows that Australian medicrats are every bit as ignorant and close-minded as their American counterparts.

• *A Shot in the Dark* by Harris Coulter and Barbara Loe Fisher (1985). A well-researched and sobering exposition of the whole-cell pertussis vaccine — the "P" in the DPT (or DTP) shot — long considered by far the most dangerous of all the childhood vaccinations. Although this vaccine was completely phased out in 1997 and replaced by the less risky but still unsafe acellular version, it shouldn't be forgotten that for more than 50 years its alarmingly high toxicity was common knowledge, yet nobody did anything about it. That attitude has never changed, nor has the unbelievably reckless and callous behavior of those who profit from vaccines, much of which this work documents.

• *Immunizations: The Terrible Risks Your Children Face That Your Doctor Won't Reveal* by Robert Mendelsohn (1988). The late Dr. Mendelsohn was an avowed enemy of the medical establishment, but as author of *Confessions of a Medical Heretic* and *How to Raise a Healthy Child... In Spite of your Doctor* (two books I highly recommend) and *The People's Doctor*, a newsletter that doubled as a syndicated column in the *Chicago Daily News* — probably the only column of its kind in America — he was a folk hero to informed parents. *Immunizations* is a selection of his articles from 1976 to 1988, all infused with his signature wit and wisdom.

• *AIDS and the Doctors of Death* by Alan Cantwell (1988). Subtitled "An Inquiry into the Origin of the AIDS Epidemic," this book is more or less a condensed narrative of Leonard Horowitz's *Emerging Viruses* (see below), but at 230 pages it's less than half the length, focuses only on AIDS, and is an easier read. It presents hard evidence that AIDS is a man-made disease and was deliberately inflicted on homosexuals by way of Hepatitis B vaccination campaigns in selected American cities. The author is a homosexual, though he doesn't advertise the fact, and the book is shot through with sexual relativism and gooey liberalism. Nevertheless, Cantwell's intentions are pure, his research is sound, and he avoids the kind of wayward speculations that occasionally mar *Emerging Viruses*. This book could not have been written without the earlier detective work of Dr. Robert Strecker, a very conscientious researcher, not given to wild allegations, who first came up with the hypothesis of an AIDS-Hepatitis B vaccine connection. (A bombshell 96-minute video, *The Strecker Memorandum*, can be watched on the Internet.) Cantwell consulted with Strecker and properly credits him.

• *Vaccines: Are They Really Safe and Effective?* by Neil Miller (1992). This is the first book I ever read about vaccines that began my long education. It's a basic, easy-to-read work, only 70 pages long, that I still consider the best introduction to the subject. In retrospect, my only criticism is that it relies too heavily on graphs, statistics and studies. Nevertheless, I would still recommend it to a general reading audience, and especially to those who don't usually read books, but might be persuaded to read just one. It just might get them to start thinking.

• *Vaccination: 100 Years of Orthodox Research Shows That Vaccines Represent a Medical Assault on the Immune System* by Viera Scheibner (1993). This work is unique. Having plowed through some 30,000 pages in major medical journals going back nearly a century, the author discovered that mainstream researchers have consistently reported on the failure of vaccines to prevent diseases and on their harmful side effects — often alongside papers claiming vaccine victory over the same diseases! Thus, the pro-vaxxers stand refuted by their own colleagues. About 500 studies are referenced. Although this book contains the most medical terminology of any in this bibliography, it's not that complicated, and Dr. Scheibner's no-nonsense approach and uncanny ability to pick apart and swiftly demolish specious arguments make it a straightforward read.

• *The Medical Mafia* by Ghislaine Lanctot (1994). Dr. Lanctot is a French-Canadian wild woman, a firebreathing rebel who attacked the Quebec medical establishment with a pen she wielded like a sledgehammer. Vaccines are just one of the many targets she smashed. This book has an unusual format and many illustrations. It also has a New Age veneer and is downright silly in places. Nevertheless, I include it here because Lanctot is such a colorful character and we need not agree with everything a person believes to acknowledge that he or she can be dead right on certain issues. *The Medical Mafia* actually attracted quite a bit of media attention in Quebec and created such a furor that Lanctot went on trial before the Quebec College of Physicians. After three years of deliberations, she was barred from practicing medicine for life — a verdict that she celebrated! All this is recounted in a follow-up book, *The Trial of the Medical Mafia* by Joachim Schafer, also an enjoyable work and perhaps more pertinent in that a greater part of it deals with the subject of vaccines, as batted around in courtroom testimony.

• *Immunization: Theory vs. Reality* by Neil Miller (1995). This is Miller at his best, a wide-ranging survey on all aspects of the vaccination fraud, mostly focused on childhood shots, but including a chapter on the outrageous experiments carried out on our troops during the Gulf War. I've read two other books by the author, *Immunization: The People Speak!* and *Vaccines, Autism and Childhood Disorders*, and they're all brilliant, but I consider this his most important.

• *Emerging Viruses* by Leonard Horowitz (1996). Explores the origins of the AIDS and Ebola viruses, and the various theories as to whether they occur in nature (unlikely), escaped accidentally from experimental laboratories (more likely), or were deliberately seeded in human populations through vaccination campaigns (most likely). There is solid documentation throughout the work connecting the dots between the CIA, military, scientists, medicrats, and even presidents. Unfortunately, the author, who is Jewish, is obsessed with Nazis and their nonexistent extermination program and is very confused politically, sometimes weaving far-fetched ideas into his account. But these are minor defects that pale in comparison to the book's overall importance. One chapter is a fascinating interview with the aforementioned Robert Strecker, who raises a specter of unspeakable evil at high levels of government that few people are wiling to contemplate.

• *The Virus and the Vaccine* by Debbi Bookchin and Jim Schumacher (2004). Moderate in tone but shocking in message, this book zones in on the frenzied "polio vaccine decade" of 1954 to 1963, especially the unsettling discovery of vaccine contamination with the cancer-causing monkey virus called SV-40, and how the many people involved — including those who determined national vaccination policy — reacted to it. It's an untold, very human story of bureaucratic infighting that pitted a few good people against a lot of mediocre or irresponsible people who went into hot denial and looked the other way rather than face cold, hard facts and see their world crumble.

• *A Stolen Life* by Marge Grant (2005). No one knows how many parents there are in this country whose lives have become a daily trial caring for a vaccine-damaged child, but the number is surely great. This is the story of one woman, an ordinary housewife from Wisconsin, whose son Scott was destroyed mentally and physically in 1961 by a then supposedly new and improved four-in-one (DPT plus polio) vaccine called Quadrigen. She then went on to educate herself and became an activist. With her husband Jim, who died in 2001, she sacrificed everything to take care of her son. Every parent in her situation will find solace in her story, and every human being with heart and spirit will be inspired by this fine lady.

• *Evidence of Harm* by David Kirby (2005). A work of immense importance on the connection between mercury in vaccines and autism, and the only one mentioned in this work to appear on the parody known as the *New York Times* bestseller list. What makes this book special is that the author writes as a moderator throughout, letting the controversy speak for itself through the words of parents of autistic children, their scientific allies, and a few politicians on one side (notably Dan Burton, whose grandson became autistic after receiving a barrage of shots, touched on earlier), and public health officials, scientists and politicians on the other side. Remaining neutral from beginning to end, Kirby does not draw any firm conclusions, but it's difficult for me to imagine that this book would not have a profound impact on any intelligent reader.

• *Polio: An American Story* by David Oshinsky (2005). An indispensable book from which I borrowed generously in writing the chapter on the polio vaccine. In conducting his research, the author spent years examining the personal papers of the major protagonists, and interviewing some of their associates and family members. The portrait he paints is one of a bunch of rats scurrying for cheap fame over the prevention of a disease that endangered comparatively few, although he apparently didn't see it that way. On the back cover, an advance reviewer wrote that *Polio* "vividly retells one of the greatest of all American success stories," which makes me wonder if this gentleman also considers the maiden voyage of the *Titanic* one of the greatest of maritime success stories. I was dumbfounded when I saw Oshinsky's name on the back of Paul Offit's *The Cutter Incident*, attached to laudatory remarks about the book and a solidly pro-vaccine stance. I will never understand how some people think.

• *Dissolving Illusions: Disease, Vaccines, and the Forgotten History* by Suzanne Humphries and Roman Bystrianyk (2013). This is probably the most comprehensive review of the entire history of vaccination ever written. It is extremely enlightening. It references more than 800 books, medical journals, newspaper articles and other literature going back to the early 1800s, which have been swept under the rug by the apostles of modern medicine and their accomplices in the media. In a later book, *Rising from the Dead*, Dr. Humphries, a nephrologist who served as a hospital intern, relates her personal interactions with patients who suffered kidney damage from routine shots, and with colleagues who were arrogant defenders of the vaccine dogma and others who agreed with her but were too scared to speak out.

• *The Vaccine Court* by Wayne Rohde (2014). Mr. Rohde is the father of a boy who regressed into autism after his first childhood vaccinations. His book is based on interviews and case histories of parents like him whose children were gravely injured by vaccines and struggle, often unsuccessfully, to win compensation from the federal government to cover some of the lifelong expenses incurred by their children's disabilities. It exposes the 1986 National Vaccine Injury Compensation Program as a bureaucratic train wreck, and makes one wonder if there is a government on planet Earth that treats its citizens worse than the one in Washington D.C. does.

• *Vaccine Whistleblower* by Kevin Barry (2015). Hearing a story from the inside is always edifying. This is the story of a conscience-stricken CDC employee, William Thompson, who could no longer take the rampant fraud and corruption that pervades that agency. "Senior people just do completely unethical, vile things and no one holds them accountable," he states. The meat of this book is a transcript of four legally recorded telephone calls (Thompson wasn't aware they were being recorded) between Thompson and Dr. Brian Hooker, an autism researcher. Thompson's revelations are absolutely shocking. If anyone has any doubts that the top health officials in this country are not the worst of criminals, and are responsible for untold human suffering, let him read this book.

Just say no to censorship

THE BATTLE for free speech in America today has much in common with the aspirations of east Europeans in the years leading up to the collapse of communism. With the invention of the Internet, thinking Americans can instantly access facts and ideas suppressed by our controlled media. At this writing, tech giants like Facebook, Twitter and YouTube are hard at work scrubbing politically incorrect content from the World Wide Web. It's difficult to predict how this will play out — if, for example, government or corporate censors will go further and make a move to pull down Web sites that speak truth to power, or block them with cyber filters, as is done in China and even in quite a few American public libraries. So far, they haven't put much of a crimp in the ability of intelligent people to express themselves and communicate with each other.

The first section below lists Web sites on health matters that I have found useful. Most deal solely with vaccines, while others deal with a variety of health issues that overlap vaccines. In the second section I've listed sites that cover politics, history and current news, both domestic and international, some from a racial perspective. Two are foreign (Iranian and Russian) English-language sites. Some occasionally discuss vaccines, and some are connected with printed publications. All are concerned with the future of America and Western civilization. There's a wide range of viewpoints spread across them, not all of which I agree with, and a few can sometimes be annoyingly sensationalist, but everyone involved has striven to see things clearly and retain the capacity to *think*.

Shut off the TV. Throw the newspaper out. Stop reading *New York Times* bestsellers. Open your mind.

- ageofautism.com
- greatmothersquestioningvaccines.com
- nvic.org
- ritalindeath.com
- thinktwice.com
- vaccineliberationarmy.com
- vaccineresistancemovement.org
- vaccineriskawareness.com
- vaccines.news
- vaclib.org
- vactruth.com
- vaxxed.com
- whale.to

- americanfreedomunion.com
- americanfreepress.net
- barnesreview.com
- codoh.com
- davidduke.com
- globalresearch.ca
- ihr.org
- kevinalfredstrom.com
- natall.com
- national-justice.com
- nationalvanguard.org
- newobserveronline.com
- presstv.ir
- renegadetribune.com
- rense.com
- rt.com
- stormfront.org
- tomatobubble.com
- unz.com
- whatreallyhappened.com

2020-2021: Journal of a world gone mad

HAD SOMEONE with extraordinary prophetic powers predicted, at the end of 2019, what would happen in 2020, and what the world situation would be like today, I would've written him off as a colorful lunatic or an aspiring horror story novelist. Even I, fully aware of the evil inherent in most of these vaccine pushers, never imagined that they would implement a program so nakedly aggressive and murderous as what has recently taken place and continues to take place far and wide — and as clearly as anywhere, right here in America. It never occurred to me that it would come to this. But really, why should I be shocked? This is the logical culmination of the vaccine dogma, the end of the road after 225 years of vaccine lies. This is it. It's all aboard for Jonestown, last stop the Kool-Aid dispenser, the point of a needle filled with a concoction its leading pushers know will sicken, sterilize or kill millions or billions of people. It has already wreaked an incredible amount of havoc on humanity — in a mere seven months, as I write, possibly more than any other vaccine ever developed, though technically it's not a vaccine at all but a totally reckless gene-manipulating experiment.

Before getting into the global contractions that began in early 2020, I'd like to look back at a few things I wrote. There was one factual error in the chapter on autism. I had written that the exceptionally dangerous whole-cell pertussis vaccine came into widespread use in the early 1940s, and that the child psychologist Leo Kanner had first observed and written about the typical traits of autistic behavior around this time. That is incorrect. While Kanner did indeed write the first paper on autism in 1943, in which he coined that word, the pertussis vaccine came on the scene in 1914 — though the original vaccine seems to have been used hardly at all, and it underwent further development, with the first mass guinea pig experiments (called "clinical trials" by the vaccine cultists) beginning in 1934 — and it was in 1938 that Kanner first noticed young children behaving abnormally. As far as cause and effect goes, however, this still makes perfect sense. Other than that, I'm not aware of any mistaken assertions, and there's certainly nothing substantial I would change, though I've been enlightened about a few things. I always knew that the Democrats were the party of the Left, but until 2020, I didn't realize how far gone so many governors and big city mayors are. On average, I've learned, they're much worse than Republican politicians, the great majority of whom are despicable in their own right. To me these Democrats are communists, by which I mean they're indistinguishable from the squalid office holders who ran things in the former communist dictatorships in eastern Europe. I also was unaware of their partners in crime, the many radical prosecutors and attorneys general, mostly black, who came out of the woodwork following the nationwide race riots instigated by the news media, and whose appointments were said to be funded by George Soros, which I have no reason to doubt. I was too generous in my appraisal of Vladimir Putin and a few others. Donald Trump is gone, having betrayed his deluded supporters, whose numbers have happily dwindled. I do believe the election was stolen from him, but why should any patriot care? Does it really matter that Biden is headed for a cliff driving at 100 mph, whereas Trump was only doing 70 mph? Trump's one enduring legacy will be the absolute disaster of Operation Warp Speed, which unsurprisingly earned him the idiotic praise of conservatives near and far, though by now some are having second thoughts.

Since March 2020, so much has happened and continues to happen on a daily basis, and I have read so much and watched so many videos on the Internet, that it's impossible to present more than the tiniest fraction of it while keeping this afterword to a reasonable length. Some aspects of the Covid event can only be surmised, but most of it is clear. It's not that complicated. The lessons of history are the primary guide, to which I've applied my intuition and everything I've learned in life. The priceless Internet continues to be practically the only source of truth, but as always, one must aim to distinguish the certain or nearly certain, the probable, the plausible, the unlikely, and the ludicrous. One thing is clear to me: we are in the midst of a global Bolshevik revolution.

The one foundational truth we should never lose sight of is this: From the beginning, everything about Covid-19 has been a fake news event. There was absolutely no evidence that SARS-CoV-2 was anything more than a boogeyman virus like SARS-CoV-1 (2003), H1N1 (2009), West

Nile (2012), and Zika (2016).

In other words, had the news media never reported a word about it, no one would've noticed anything out of the ordinary. no one would've noticed anything unusual in the incidence of respiratory disease compared to 2019 or any previous year.

Let that sink in. But even in February, before the first reported U.S. Covid death on the last day of that month, I detected a building fear campaign on WCBS, the news radio program I listen to for ten minutes each morning — my only exposure to the mainstream fake news, other than a quick glance at the headlines on one or two websites run by the major television networks. After March 13, the day Trump declared a national emergency, WCBS went into overdrive. On some days I turned the radio on again, later in the day and at night, and listened for a full hour. Even in Russia under Stalin during the Second World War, I can't imagine such wild, nonstop lies calculated to generate such intense fear and panic. In addition to skyrocketing numbers of deaths and infections, obviously pulled out of a hat, it was as if the announcers had been instructed on how to pause and inflect to achieve the maximum emotional impact on listeners. This frenzied, round-the-clock "news" reporting was all a sane man could take. And the flip side of all the "fear porn" to which the media's favorite experts contributed was the usual Soviet censorship of thousands of brave, dissenting doctors and other independent researchers worldwide, who from the beginning saw this scam for what it was. Here is the stifled voice of just one, which I came across in November 2020, a Dr. Roger Hodkinson of the Royal College of Physicians and Surgeons in Canada:

There is utterly unfounded public hysteria driven by the media and politicians. This is the biggest hoax ever perpetrated on an unsuspecting public. There is absolutely nothing that can be done to contain this virus. This is nothing more than a bad flu season. It's politics playing medicine, and that's a very dangerous game.... There is no action needed. Masks are utterly useless. There is no evidence whatsoever that they are even effective. It is utterly ridiculous seeing these unfortunate, uneducated people walking around like lemmings obeying without any evidence. Social distancing is also useless.... Positive testing results do NOT indicate clinical infection. It is simply driving public hysteria and ALL testing should STOP immediately.... Using the [Alberta] province's own statistics the risk of death under 65 is 1 in 300,000. The scale of the response is utterly ridiculous.... all kinds of business closures, suicides.... you're being led down the garden path.

To expand on this, let's review some salient facts. One is that a fake pandemic, or plandemic, as others have aptly called it, has been in the works for a long time. The most exhaustive timeline on this, going back to the late twentieth century, is found on the invaluable website covid19propaganda.com. I'll mention four recent milestones. On January 10, 2017, ten days before Trump took office, Anthony Fauci — no introduction needed — gave a lecture on infectious diseases and pandemic preparedness at Georgetown University, in which he said, "There will be a surprise outbreak" confronting the incoming administration. That gives the game away right there. On April 26, 2018, twenty months before anyone heard of Covid, the European Commission, an executive body, first proposed vaccine passports for crossing borders in a document entitled "Strengthened Cooperation Against Vaccine Preventable Diseases." This occasion was the first time the phrase "vaccine hesitancy" was heard. On September 12, 2019, a "Global Vaccination Summit" was held in Brussels, Belgium, again by the European Commission, in conjunction with WHO, attended by big shots from many countries. Three round table discussions were held, entitled "In Vaccines we Trust," "The Magic of Science," and "Vaccines Protecting Everyone, Everywhere." On October 18, 2019, "Event 201" took place at the Pierre Hotel in New York City. This was a "tabletop exercise," a simulation of a global pandemic and the conspirators' response to it, on which the Covid pandemic, which kicked into high gear five months later, was closely modeled.

The first reports of a novel corona virus, SARS-CoV-2, came out of Wuhan, China in late December 2019. I don't know what happened, or didn't happen, in Wuhan, though it appears certain that the slimy Fauci, who has flip-flopped at least fifteen times in his recommendations, and whom even Trump had the brains to call "an idiot and a disaster," was involved in research at the virology lab there going back to 2014. I suspect that secretive biowarfare research took place at this lab, as has long taken place in our own facilities, especially at Fort Detrick, Maryland. I also believe there may be a connection between possible radiation sickness in that city and cities in other countries where 5G wireless communications had recently been rolled out. Nevertheless, this obsession with China, this business of solely blaming China for the origin of Covid, is a huge and mendacious distraction from the brutal lockdowns in 2020, and the vaccination juggernaut in 2021. If SARS-CoV-2 really exists, and whether or not it's natural or manmade, it's quite harmless to nearly everyone except the aged and those with co-morbidities, who are more vulnerable to any kind of viral infection.

Jon Rappoport, who moderates the hard-hitting blog nomorefakenews.com, maintains that the Covid-19 virus does not exist, that is, no scientist in the world has ever isolated and identified it as a distinct virus in the corona family. Inquiries on this subject sent to the top health departments in several Western countries were ignored or answered evasively. Rappoport has also said that the PCR test, used to detect infection, is utterly useless and rigged to spit out false positives like a fire hose. This man is an independent investigator who has been delving into government-medical-pharmaceutical chicanery and analyzing so-called pandemics for more than thirty years. He is unsurpassed at stripping away dense scientific jargon, reducing it to simple terms, and exposing "experts" like Fauci as bullshit artists. The late Kary Mullis, the inventor of the PCR test, was another man who despised Fauci as a know-nothing career bureaucrat. Mullis never intended the PCR test to be a diagnostic tool, which is exactly what it became.

So, through 2020, tens of millions of asymptomatic Americans drove to designated Covid testing sites or stood in lines outside walk-in clinics, waiting for a bogus and painful nasal swab test for a likely nonexistent virus to be administered by strangers in bizarre protective attire, which added to the drama. Nothing like this has ever happened before. It was given impetus by an apocalyptic warning, issued by Neil Ferguson, a technocrat based at Imperial College in London (which is not a traditional college but rather a hotbed of social activism, connected with the UN, WHO, and global policy makers), that the world was on the verge of the worst epidemic since the 1918 Spanish Flu. Ferguson's track record was an absolute joke. Using computer models, which is all this geek knows how to do, he had also made scary predictions about swine flu in 2009, dengue fever in 2015, and Zika virus in 2016 that never materialized. That he was taken seriously in 2020 by so many heads of state and other movers and shakers does not reflect well on their IQs. It was Ferguson who, on March 16th, released a doomsday report predicting massive Covid death tolls in the U.K. and the U.S., and after seeing what the Chinese and then the Italian governments were capable of pulling off, called for draconian lockdown measures. He's the one responsible for all the school and business closures, social distancing, face masks, isolation, quarantining and contact tracing crusades. In May he was caught having an extramarital affair, setting a pattern of stinking hypocrisy for all the petty tyrant politicians around America who were seen countless times with family and friends at social gatherings, ignoring their own harsh edicts.

Right through 2020 I didn't know anyone who claimed to be sick with Covid, and most people I spoke to didn't know anyone who knew anyone who supposedly had it. I don't even remember hearing anyone cough in 2020! Oh, there were a few who told me that they knew people who died of Covid, or knew doctors or nurses who worked in hospitals that were overwhelmed by Covid patients. I don't doubt that they knew people who died, and it's tactless to press them for details when they tell you that. But fraud, a specialty at the CDC, was in the air, reflected in the huge drop in mortality statistics for flu and pneumonia. Deaths from many different causes, respiratory and otherwise, were being recorded as Covid deaths. Hospitals had financial incentives to do this, and it was a great way to inflate numbers and ratchet up public fear. Some brave doctors, like Scott Jensen in Minnesota, and Annie Bukacek in Montana, who often fill out death certificates, went public with these shenanigans. In recorded phone conversations made by James O'Keefe, who runs projectveritas.com, four funeral directors in the New York metro area, who routinely view these documents, said they were seeing Covid listed as the cause of death all the time, even though they and the families they were serving knew damn well that the deceased had not died of Covid. As for hospitals inundated with Covid patients, I'm sure they were swamped in some cases, but like everyone else, doctors and nurses fall for the power of suggestion and the pressures of conformity, and with the media blaring its fear porn 24 hours a day, I'm sure they were convinced that anyone with respiratory symptoms was infected with Covid.

This was in April 2020, the peak of hysteria, when the global epicenter of Covid had moved from China to Italy and Spain to New York City, with the Queens enclave of Elmhurst "the epicenter of the epicenter." I lived 32 miles from Elmhurst. A cousin who lived on eastern Long Island, 50 miles out, called me to see if I was okay. I wasn't sure if he was serious. He was. A traveling companion from way back when emailed me from Australia with the same concern. Since I like to confirm things with my own eyes, on April 11th I took a ride to Good Samaritan Hospital in West Islip, seven miles east. The visitor's parking lot was only twenty percent full, there was no activity at the emergency room entrance, and there wasn't a soul in the spacious waiting room. Five days later I drove in the other direction to the Nassau County Medical Center in East Meadow, about 20 miles from Elmhurst. Here too, nothing was happening at the ER entrance and the parking lot was mostly empty, but as I pulled into the driveway, I saw a security guard and a large, ominous sign ahead that read "All Visitation Suspended." I got the impression that I was most unwelcome here, so I made a U-turn and left. Unlike Good Sam, the NCMC is a public hospital, which means if you're poor you can come here and get treated for free.

My take on what happened at Elmhurst Hospital, and at other public hospitals in New York and in big cities around the country, is as follows. With the constant media fear mongering, I think a great number of black and Hispanic deadbeats, unhealthy to begin with and not the brightest people, who experienced any respiratory symptoms at all, even a simple cough or cold, panicked and went to the hospital. Whether these patients were lured into hospitals by design, or an opportunity arose after they were admitted, they were deliberately killed off on orders from above — by over-sedation, improper use of ventilators, or gross neglect. This was done because these people are a drain on Medicaid, which is close to going broke. That's my theory. Two out-of-state nurses, Nicole Sirotek and Erin Olszewski, who went to New York hospitals separately on a mission of mercy, were aghast at what they saw. They made videos, emotional in places, still up on the net, and both point-blank called it *murder*. All these victims conveniently became Covid death statistics.

Medicare funds, which provide for old folks who have lived productive lives, also are drying up, so they have to be bumped off too. I refer you to M.S. King's *The Morphine Genocide: How the Fed-Med Mafia Kills Our Elderly with Palliative Care*, published in 2017. Like the rest of Mike King's mostly excellent books, it was axed by the censors at Amazon. When it was still sold there, three or four reviewers attested that their own mother, father or other relative was murdered in the hospital. This policy of empowering medical "death panels" snuck in under Obama's Affordable Care Act. King cites some articles in the American mainstream press that hint at the crimes, but the best one is from England, where the National Health Service is up against the same financial crunch — one of those rare nuggets that appear in the controlled press, in this case the *London Daily Mail*. The article, headed "Top doctor's chilling claim: The NHS kills off 130,000 elderly patients every year" appeared in the June 19, 2012 edition, and can be read online.

This takes us to the tragic nursing home deaths that occurred in several states, some 12,000 in New York alone, under Governor Andrew Cuomo. Cuomo is such an odious creature and is so widely detested here, it's amazing he's still in office. He's one of many Democratic politicians who, with his brutal lockdown decrees, took pleasure in inflicting psychological pain on ordinary people, just to flex his muscles. For the past three or four months, the media masters have been playing a cat and mouse game with him for reasons only they know, turning up the heat, then turning it down, then up and down again. Will they force him to resign or will they let his crimes slide? What so many forget is that back in April 2020, when his executive orders were being carried out and he was preening his chest hairs at his daily press conferences, the media were puffing him up as a superstar, as a take-charge leader making the tough decisions needed to hold the deadly Corona virus at bay. Maybe he should even think about running for president, they hinted. And the boobs lapped it up. After all, didn't TV and the papers claim that his approval rating had soared to 85%? If 85% agree on something, it has to be right, right? Well, there were more than 200,000 New Yorkers who had parents in nursing homes, and they weren't too impressed with their governor, whose policies were as cruel as they were illogical. There's no need to go into the details here, which are widely known and undisputed. The only remaining question is whether, acting out of malice or from hidden pressure exerted from above, he deliberately killed all those old people. I think the evidence points in that direction. In a sane society, he, along with his state health commissioner Howard Zucker, would be tried in a court of law for premeditated mass murder, and if found guilty, executed.

There's no doubt in my mind that all this business is being orchestrated by sinister, unseen personalities, and that maggoty politicians like Cuomo, along with heads of state around the world, are doing what's expected of them. Those who don't toe the line are taking big risks. Tanzania's president John Magufuli, suspicious of PCR test kits sent to his country by WHO — as well he might have been, considering the past exploitation and killing of African children in vaccine experiments — tested a variety of inanimate and living subjects. In April 2020 he announced his findings to the world: a goat, a papaya, and motor oil tested positive for Covid, prompting laughter from many quarters. Magufuli told WHO to pack up and get out of Tanzania. In January 2021 he warned his countrymen to stay away from the Covid vaccine. Two months later he was dead, supposedly of a heart attack. Pierre Nkurunziza, the president of neighboring Burundi also opposed lockdowns, was a vaccine skeptic, and expelled WHO officials from his country. He succumbed to a heart attack as well. He was 55, Magufuli 61. Were these men murdered? I'd bet on it.

On an impulse, I recently skimmed through A.K. Chesterton's *The New Unhappy Lords*, which I read many years ago, and which I quoted from earlier. I've read several fine books on the conspiracy, or better said conspiracies, among the rich and powerful, especially since World War Two, to rule the planet — to impose globalism, international communism, one world tyranny, a new world order, whatever you want to call it. To my mind, Chesterton's book is the best of them, because he is eminently logical and doesn't duck the taboo issues of Jewish subversion and racial mongrelization, especially in his native England, where even in the early 1960s, when he was writing, nonwhite immigration was well underway, as was the humiliation and destruction of the white man in Africa, as some British colonies buckled under to lunatic black rulers, while others dug in against the betrayal of worthless politicians in London — a precursor of our current Black Lives Matter movement. The role of international finance, the behind-the-scenes pressure applied to spineless world leaders, the perfidy of the United Nations, the hidden hand directing global affairs — nothing has changed. In Chapter 24, he lists the mostly obscure names and positions of nearly a hundred men from North America and Europe established in government, media, academia, banking, foundations, big business and the military who had met in secret more than once, in lavish surroundings, with strict security precautions, to discuss — what? As Chesterton makes clear, only a fool would deny that there's a conspiracy at work here. The faces have changed in the last sixty years, of course, but they are the same unfeeling, mechanical creatures conspiring to rule over a global plantation, to control the lives of a raceless, spiritless, degraded humanity. They have morphed into what today appears to be the foremost organization of power addicts, an elitist club known as the World Economic Forum (WEF). The only differences today are the ascendance of technocracy, the push to vaccinate, and the fact that the conspiracy is out in the open, right in your face if you care to open your eyes.

In late 2020 I read two books co-authored by Klaus Schwab, a half-Jewish German national and the founder and executive chairman of the WEF, which was formed in 1971. These books are *Shaping the Future of the Fourth Industrial Revolution*, published in 2018, and *Covid-19: The Great Reset* which came out in July 2020. These books have to be read to be believed. I had occasionally come across the ideas he expounds, but they seemed so far out that I didn't take them seriously and never looked into them. That was a mistake. He is for real with his "trans-human" agenda of merging man with machine, literally creating a new being who will be "a fusion of our physical, digital and biological identity." The technology of injecting sensors into the body, which monitor vital signs and convey them to cell towers, is already here and marketed on profusa.com. Recently I came across an interview Schwab did on Swiss television on January 10, 2016 in which he discussed planting microchips in our brains. Naturally, he's big on vaccines, and has said, "People who refuse to be vaccinated are a health threat to everyone else." Like Bill Gates, a close colleague, he comes across as a reincarnation of Pol Pot. He envisions a world of total surveillance where everything is electronically connected with everything else, thoughts can be read by police, and get this, human organs can be printed by 3D technology (*Shaping the Future*, page 145). His two greatest mentors are the death merchants Henry Kissinger (who in addition to his war crimes in Southeast Asia, pressured the government of Rhodesia to take the primrose path to black Marxist rule, with the inevitable dispossession and killings of white farmers after the country became Zimbabwe), and Nelson Mandela, about whom I've written enough. It would seem that, as part of his dystopian plan to phase in a new version of man, the white race needs to be phased out. In fact, Schwab claims he was deeply affected by the death of George Floyd, and the WEF website, weforum.org, has much boilerplate discussion about ending racism — only white racism, that is.

The media-instigated anarchy and race riots, which tore through nearly all major cities and many large towns in America, was, of course, the other major event of 2020. Whether George Floyd died of a fentanyl-induced heart attack or of suffocation from the knee of a rogue white cop, is irrelevant. Floyd was just another black bum in a country whose cities are overflowing with black bums, a drug addict and a career criminal, who once did five years in a Texas prison for sticking a gun in a woman's stomach during a home invasion, which in itself justified an appointment with Old Sparky. What is relevant is that the Jewish cabal that decides what is "news" used this incident to ignite the American racial powder keg, and touch off more racial explosions around the world. Once again, the awesome, unrestrained power of the media was on full display. This, of course, was just one in a long series of media attacks against white police officers working in the hopelessly crime-ridden big cities who have to deal with black animals all the time. A racial explosion has been brewing for decades, and we've seen it happen before, but as someone who lived through the summer of 1967, I can attest that, even though the scale of mayhem and destruction in 2020 was about the same, back in '67 there was no toppling and defacing of statues, no invasions of white suburbs by menacing black mobs (some of which took place very close to my home), no "cancel culture," and no shaming or persecution of any whites who deviated one inch from the party line of "systemic racism" and "white supremacy," which are as fictitious as the Covid pandemic. And this time the raw hatred and contempt for the white race was thrown like acid in our faces, including implicit and even explicit calls for murder in both the news media and social media. On October 3rd, 400 members of the No Fucking Around Coalition, a black paramilitary group, marched heavily-armed through Lafayette, Louisiana, a city of 120,000 with a white majority. A similar march took place in Tulsa, Oklahoma, on the hundredth anniversary, on June 1, 2021, attended by Biden, of what the media ludicrously described as a senseless white-on-black racial massacre, which ended with the tribal chief whipping up the crowd with the usual megaphone rhetoric and exhorting them to "kill everything white in sight." I thought of the Hitler quote I reproduced on page 232, in which he describes how the Jew gradually erodes all the foundations of a civilized society until he feels confident enough to drop the pretense of social justice and begin the butchery. Crushing lockdowns that temporarily or permanently ruined so many lives, combined with a determined attack on orderly white society, employing Antifa and Black Lives Matter anthropoids as shock troops — what we had here was a classic Bolshevik revolution, calling to mind George Lincoln Rockwell's novel definition of communism as "the organized mutiny, led by the Jews, of the biologically inferior people of the world against those who have created civilization." Any doubt about the Jewish role should have been dispelled by a brazen full-page ad that appeared in the *New York Times* on August 28th, 2020, sponsored by more than 600 Jewish organizations, from the mighty Anti-Defamation League down to the local activist synagogue. The eye was drawn to "We speak with one voice when we say, unequivocally" — then below that, in gigantic block letters — BLACK LIVES MATTER. And it soon became clear that this latest planned racial uprising was tied in with the Covid hoax, as in New York, where black throngs were exempted from the face mask and social distancing requirements that were being zealously enforced by that city's politicized cops. Around the country, thousands of violent black felons were released from prison, ostensibly to prevent Covid from spreading in those facilities. And the connection was also evident in the corporate world, especially professional sports teams, many Jewish-owned, so proud to proclaim "Black Lives Matter" on billboards and stadium signs, while requiring all non-playing personnel to participate in the senseless obedience ritual of wearing face masks. For not wearing masks on the sidelines, several NFL coaches were fined $100,000, their teams $250,000.

As one would expect, the thrust behind the increasingly fierce injection campaign in the U.S. is heavily though not entirely Jewish. If it weren't for Bill Gates, a gentile, we would not be in the extreme situation we're in today. Over the past twenty years, no man or woman comes close to his leadership in worldwide vaccination efforts, always with a philanthropic spin. Gates is like an alien from another planet, possessed of a mentality that I cannot fathom. He apparently sees himself as ringmaster of the solar system. He is crazy and dangerous. He has other projects too, like having most of us in advanced nations eating synthetic beef by 2030, and sending a fleet of specially equipped aircraft into outer space to spray particles that would dim the sun's rays in order to cool the stratosphere, his remedy for global warming. That one is on hold for now as he sorts out life after divorce, while still buying up and owning more American farmland than anyone else. Behind his facade of caring for Third World children, he sees vaccines as one way to depopulate the world. (Other rich and powerful men, Warren Buffet, Henry Kissinger, and the late David Rockefeller among them, have also favored depopulation, Kissinger having hinted at enforced starvation.) These were Gates's exact words during a 2010 "TED talk" to a private audience, widely circulated on the Internet: "The world today has 6.8 billion people. That's headed up to about nine billion. Now if we do a really great job on new vaccines, health care, reproductive health services, we could lower that by perhaps 10 or 15 percent." More than once he has publicly stated his ambition to deliver Covid vaccines to everyone on the planet. He has been implicated in many criminal vaccine schemes, especially in African countries and in India where he has been accused of maiming 496,000 children and adults in a polio eradication campaign from 2000 to 2017. I've come across no facts that refute this charge. People around the world can see into his dark soul. I've read at least a hundred comments on the Internet calling for his death. I've never heard him express any remorse for the great harm he has caused. In his mind, perhaps, he has never harmed anyone. It happens that Nelson Mandela was one of his heroes, and he even had the honor of meeting the world's most beloved terrorist.

Aside from being a shrewd and unprincipled businessman, Bill Gates is a nobody — an unappealing nerd of no creative talent, no moral sense, no character. Had he not been born into wealth and privilege, which gave him a huge advantage in life, he might have ended up with some dull technocrat job making $25 an hour. Several people have tried to determine what makes him tick; a thorough analysis, including a two-hour documentary titled "Who is Bill Gates?," worth watching, can be found on corbettreport.com. Psychopaths have always been part of the human condition, so Gates is far from unique in that category. What makes him unique is his wealth. I, for one, don't know what it means to have $130 billion in the bank, or in the Bill and Melinda Gates Foundation, but with that kind of money you can obviously be a big spender and walk through almost any door. You can purchase a pet weasel named Anthony Fauci and have him follow you around and eat out of your hand. You can meet with a bonehead president in the White House, convince him that vaccines are great, and get him all excited about Operation Warp Speed. You can throw hundreds of millions here, hundreds of millions there at academicians to win their support. You can hire the very best public relations firm to make you look like something you're not. You can pal around with pedophile and fellow billionaire Jeffrey Epstein and fly on his private jet. You can fund Imperial College to keep it afloat, and get another computer worm, Neil Ferguson, to ruin people's lives. You can become chummy with airhead celebrities and appear on TV news and entertainment programs. You can buy favors with the media masters, who are naturally drawn to your genocidal plans. You can pump endless streams of cash and wrap your tentacles around all kinds of global agencies, organizations and corporations that push for total population control through vaccination and digital tracking devices. You can bribe government and health officials everywhere to carry out vaccination campaigns guaranteed to kill and maim. And you can plan all these things while swanning around in a $2 million a week rented yacht or hanging out in one of your five mansions. The one thing you can't do is buy off the Truth, which is that you're a worthless little turd whose name will become an ugly, permanent stain when the history of this age is written.

As I mentioned, the wheels of a global vaccination conspiracy had been turning for years before the jabs actually began — on December 14, 2020, here in the U.S. I must admit that, despite my extensive research, I was unaware of this. Conspiracies by their nature are hatched in secrecy. It was in April 2020 that Trump announced Operation Warp Speed with his usual oily pitch of great things to come. The shots were eight months away, but the wheeling and dealing behind closed doors, with billions of dollars up for grabs, which always accompanies major vaccination campaigns, had begun, with Pfizer and Moderna winning the big prize, and their lawyers hammering out a clause removing any liability for the vaccine deaths and injuries they knew were inevitable. As 2020 wore on, I had an uneasy feeling about what was coming. There were few tangible hints, but what there was, was ominous. In May, the utterly repulsive Alan Dershowitz said "If you refuse to be vaccinated, the state has the power to literally take you to a doctor's office and plunge a needle into your arm." I came across a 2017 clip of Tal Zaks, the chief medical officer of Moderna and an Israeli, discussing exciting new mRNA technologies and how the next class of vaccines would "hack the software of life." Those words sent chills down my spine. He had a little diagram of how it would work. Playing God, taking Nature into his own hands. Absolute madness. But naturally Moderna was playing it safe, insulated against any lawsuits, and as I later learned, after winning the government contract, Zaks dumped his company stock in structured sell-offs. Technically it was not illegal, but it was the epitome of sleaze, and made him very wealthy. Word is that he plans to leave Moderna in September 2021 for greener pastures. On October 23, 2020, seven weeks before the injection campaign hit stride in England, the British government posted this notice on ted.europa.eu: "The MHRA [Medicines and Health Care Products Regulatory Agency] urgently seeks an Artificial Intelligence (AI) software tool to process the expected high volume of Covid-19 vaccine Adverse Drug Reactions (ADRs) and ensure that no details from the ADR's reaction text are missed." Here was proof that the top politicians in London knew in advance how dangerous this vaccine was, but approved its use in a mass vaccination offensive anyway.

I wrote earlier about the over-representation of Jews in the vaccine field, both at the pharmaceutical companies and public health agencies, which in practice are one and the same. I could have mentioned more Jews in the top slots at the CDC. The new director under Biden is Rochelle Walensky, member in good standing of Temple Emanuel in Newton, Massachusetts, conveniently just a few miles from Moderna headquarters in Cambridge, who on April 8, 2021 declared racism "a serious public health threat." This is the mentality of the people who rule over us. Perhaps the CDC will soon be pushing an anti-racism vaccine. But when I looked into the advisory committee dredged up from the CDC and FDA, which met to discuss emergency authorization of the Covid-19 vaccines in late 2020 — which needless to say was already a done deal — I saw so many Jewish names and faces that I thought I might've stumbled on the Knesset website. Further investigation revealed the kind of conflicts of interest you always find with these people. But that's just half the story. The CEO of Pfizer is a Greek Jew, Albert Bourla, and the chief scientific officer is Mikael Dolsten, a Jew born and raised in Sweden. The billionaire CEO of Moderna is Stephane Bancel, a French Jew. Wikipedia states, "Bancel has been described as having a secretive approach to Moderna, and as being a tough operator." That's an understatement. Moderna is an extremely shady start-up company and Bancel a horrible person whom many found impossible to work for. Like Zaks, he unloaded a ton of Moderna shares through 2020 at great personal profit. The chairman and CEO of Johnson & Johnson is Alex Gorsky, who also is Jewish. Pfizer and J&J have rap sheets a foot long. It's always the same with these pharmaceutical companies. They commit crime after crime, destroy millions of lives, get hit with big fines and pay out a billion here, three hundred million there — just the cost of doing business. Then they go on their merry way to commit more crimes. No one ever goes to prison, much less to the gallows. And these are the companies that have injected 180 million American sheep, caused tens of thousands of tragedies already, yet cannot be sued. And their tribal cousins in the media just keep covering up for them. It's time to realize that these are criminal syndicates every bit as brutal as the most infamous Mexican drug cartels. In fact, they're more to be feared. No one is forced to use cocaine or heroin, but the vaccine manufacturers, with their huge stable of bought politicians who pass laws mandating their products, are a different story. And only a fool would deny that they employ assassins to get rid of people they regard as a threat. Their latest victim was Brandy Vaughan, a Merck sales executive turned anti-vax activist, who ran the website learntherisk.org. She knew she was being targeted, told her friends about it, and was found dead by her nine-year-old son in her Santa Barbara, California home on December 7, 2020. Dead from natural causes, the coroner dutifully reported.

Predictably, some vicious and deranged Jews were unable to control themselves. In a *Wash-*

ington Post piece of April 22, 2021 headed "We Should Soon Stop Catering to the Vaccine Holdouts," Jennifer Rubin advocated making life impossible for those who cannot present secure proof of vaccination. Rubin has written so many knucklehead columns over the years that even her colleagues have long wondered out loud why the *Post* doesn't fire her. Seventeen days later, another unhinged Jewess, the washed-up entertainer Bette Midler, threatened unvaccinated children with this tweeted gem: "If my kid can't bring peanut butter to school then yours can't bring the deathly plague. Vaccinate or I'm bringing the Jiffy." How charming. She was referring to peanut allergies, a by-product of the flood of new childhood vaccines in the 1990s, which are potentially fatal, so this was a thinly disguised death wish, not to mention just another mindless rant: if vaccines protect against disease, then vaccinated children need not worry about any "deathly plague" brought by unvaccinated children. Lastly, there's our old friend Peter Hotez, who went bonkers in an April 27, 2021 article, "Covid Vaccines: Time to Confront Anti-Vax Aggression," published in *Nature*, a print journal and online magazine that calls itself scientific. Hotez wants to see anti-vaccine groups dismantled, and wrote darkly of terrorism, nuclear armament, far-right extremist groups and Russian-sponsored misinformation, while plugging the London-based Center for Countering Digital Hate, as well as a Virginia-based rumor-tracking analytics company. He wants to empower WHO, the UN, cyber security, law enforcement, public education, and strengthen international relations. He is one very sick puppy who recently got himself into a heap of trouble for making outrageous allegations against Sharyl Attkisson, the former CBS correspondent who often writes articles mildly critical of the vaccine establishment on her website. At this writing, Hotez has a defamation lawsuit hanging over his head.

I don't want to oversimplify. The Jewish role in this whole Covid-19 saga, as in most of their destructive activities, is not black and white. There's some gray. There are plenty of gentile Cromwell types involved in this. Another gray area is Israel itself, where the government imposed lockdown rules that were harsher than in most Western nations, then followed up with an aggressive vaccination campaign. Some have suggested that the Israeli population was injected with saline solution, and the Jew duped the dumb *goyim* yet again. I don't believe that. I've seen enough evidence on Israeli websites to convince me that Israelis have had their share of casualties, in addition to the fact that Jews, like whites, are divided among themselves in getting or not getting the shot. Furthermore Netanyahu (now out of office), may have received a backhander from Pfizer and, gangster that he is, may not care if his own people are harmed. Alternatively, maybe he wanted to make everybody forget about all his crimes and make himself look like a strong leader in a futile attempt to get re-elected. I don't know. Maybe also, he, or whoever makes the top public health decisions there, genuinely believes that mass vaccination is a good thing. I believe this to be true of other world leaders, like those of Russia and Iran, who have foolishly bought into the vaccine dogma and the whole Covid scamdemic and, while wisely shunning the poisons used here and in Europe, are using vaccines like the Sputnik V which, from what I've read, employs the same dangerous biotechnology.

But let's not avoid the million dollar question. We have this nexus of Jews who sit at the top of the CDC, the FDA, the vaccine manufacturing firms, the television networks and social media, not to mention those swarming around the vegetable Biden. In fact, the real acting president is Jewish chief of staff Ron Klain, who daily tells his mumbling boss what to say and do. Only a blind man could fail to see that there's tribal solidarity at work here, even if there are some personality conflicts. Every one of them is on board with the plandemic and the vaccination campaign. It goes without saying that they all have criminal minds. All of them are engaged in a cover-up of an immense tragedy which has been going on for seven months now. Why don't any of them raise their voices about it? Why isn't Biden's Coronavirus Response Coordinator, the Jew Jeffrey Zients, addressing the situation? What are their motives for not stopping it? The polio and swine flu vaccination campaigns of 1955 and 1976 were halted in much less time with only a minute fraction of the casualties, when it became clear that the vaccines were backfiring. Why in 2021 is it full speed ahead, fueled now by a fresh round of scaremongering nonsense about the Delta variant? I answered these questions the best I could in the long chapter titled "The Jewish Factor," so I'll try to answer them in one sentence here: A sizable number of Jews are genetically driven to destroy civilization and murder people on a huge scale, even though it may ultimately lead to human extinction, including their own.

❖

As I've said, I always like an insider's view of things. Thankfully, there are a few honorable people who work, or have worked, in the top public health agencies and in the big pharmaceutical firms. Michael Yeadon, former chief of allergy and respiratory research at Pfizer, is the most outspoken. He has been sounding the alarm for several months now, as in this statement:

It's become absolutely clear to me, even when I talk to intelligent people, friends, acquaintances.... and they can tell I'm telling them something important, but they get to the point where I say, "Your government is lying to you in a way that could lead to your death and that of your children," and they can't begin to engage with it. And I think maybe ten percent of them understood what I said, and ninety percent of those blank their understanding of it because it's too difficult. And my concern is, we're gonna lose this because people will not deal with the possibility that anyone is so evil.... I've talked to lots of people, and some of them have said, "I don't want to believe that you're right, so I'm just gonna put it away because if it's true I can't handle it."

Yeadon strongly suspects that this global vaccination push is a conspiracy to depopulate the planet. So do I — among some of the conspirators anyway. But they could not do their infernal work without the support of a vast army of pimps and toadies in government, journalism, NGOs, scientific research institutions, large corporations and academia.

We are in the midst of an unprecedented event. As with the lockdown measures that became increasingly harsh and ludicrous as 2020 wore on, once the vaccination steamroller started up and went forward, it never stopped. The goalposts kept moving, again and again, until the steamroller knocked them down and rolled right over them. It's still rolling. Last month I read that they're now after children as young as two, and they're even talking about booster shots. Judging by the VAERS index of adverse event reports, which probably represents at most ten percent of such events, this is the most dangerous substance ever put into a syringe, aside from lethal injections given to condemned prisoners.

From January 1st to June 26th, VAERS compiled 376,300 reports. A quick scrolling on vaers.hhs.gov shows that nearly all of them are from the Pfizer or Moderna Covid shot. According to healthimpactnews.com (a superb site if you can overlook the quotes from scripture), during this time frame, plus the latter half of December, 6113 deaths, 5172 permanent disabilities, 6435 life threatening events, and 51,588 emergency room visits were reported to VAERS. There have been more reports filed with VAERS in the first six months of 2021 than in all the years from 2012 to 2020 combined. I might add that I've mentioned VAERS to at least twenty people over the years, and no one had heard of it. Since you can examine the data for yourself, I won't bother reproducing any of it, as I did for previous years earlier in this book. Were I to reproduce the first half of 2021, it would add 80,064 pages to the book you're holding. And we continue to be told this is a safe shot. In fact, according to the CDC's own website, cdc.gov, as posted on May 27th, VAERS (which the CDC and FDA jointly supervise) had received 4863 reports of Covid-19 vaccine deaths up to May 24th. Yet at the top of this same page, under "What You Need to Know," the first bullet point statement was "Covid-19 vaccines are safe and effective," with the words "safe and effective" in boldface! The madness of this whole thing is beyond description.

The situation in Europe is no better. There's a system there, similar to VAERS, called EudroVigilance, which keeps track of Covid-19 vaccine deaths and injuries, and breaks them down into 27 categories (cardiac, gastrointestinal, respiratory, etc.). Tabulations are regularly updated on healthimpactnews.com. The casualties reported in Europe exceed ours, and there are many more besides, because several European countries do not participate in the system.

It's impossible for me to go into detail on all this harm, and I don't want to get bogged down in biological processes and terminology that I don't understand, but it's well worth passing on some information that I've come across. From what I gather, these experimental injections permanently change the DNA, which means that it can't be undone in future offspring — if it doesn't destroy the fertility of recipients to begin with. It seems that these people are now the equivalent of genetically modified corn. I've read that 82% of women who were injected during the first or second trimester of pregnancy have had miscarriages. In an interview conducted with Jeff Rense, who runs the excellent aggregate website rense.com, the aforementioned Dr. Hodkinson issued a grave warning about the unknown future effects of these shots. He was particularly concerned with myocarditis, inflammation of the heart, which unlike the liver and kidneys cannot repair itself. What this means in plain English is that millions of young Americans who got this injection may die of heart attacks in their thirties or forties, though no one knows yet what the future holds.

Other dissident doctors have come right out and said there's going to be massive death and suffering within two or three years as a result of "cytokine storms," meaning, as I understand it, the immune system going haywire in fighting common transient illnesses, or in the continuing replication of spike proteins, that have a notably harmful effect on the circulatory system. I've read numerous accounts of shedding, which we know happened with the Salk polio vaccine, and likely with others as well. Vaccinated people can infect others through physical contact or exhalation. Vaccinated women have reported serious disruptions of their menstrual cycles, or unusually heavy blood flow, or passing blood clots. I've even read of post-menopausal women having periods again. And from simply being around vaccinated women, unvaccinated women have experienced the same symptoms. There are also many cases of metal objects sticking to the skin of vaccine recipients, usually at the injection site, which means that an electromagnetic or magnetic agent was stealthily put in at least some of these vials. I've seen about ten videos of this, along with one of a man in India with an LED bulb which lit up as he held it near his shoulder, then went off when he moved it away. Is all of this an elaborate collaborative hoax? I doubt it. There's so much more information out there, but at this point I'll refer the reader to the web sites I've mentioned.

The majority of Americans, including a high percentage of the 180 million who were foolish enough to be injected, are oblivious to everything I've written in this afterword, and in the whole book for that matter. My best guess, and I'm being conservative here, is that in the seven months since its inception, about 50,000 Americans have been killed by this shot, and about 4 million have had serious reactions, many of a permanent, life-altering nature. But that's not even two percent of the total population — not yet, anyway. So those who don't know anyone in this small, hypothetical minority will go on living in their alternate reality. Although many have become increasingly distrustful of the mainstream media, they still cling to it.

In my social circle, about seventy percent got injected, including some who are quite intelligent. That really amazed me. It just amazes me that intelligent, responsible people would not suspect that all along there's been something fishy about this whole Covid business, would get this shot without any hesitation and without investigating it first. Early this year I emailed my brother and sister long lists of testimonies, taken from the Internet, of ordinary people detailing the worrisome symptoms they've been experiencing, some beginning within minutes of inoculation. They went out and got their shots anyway. So many people really are sheep, lemmings, zombies, who mindlessly go with the flow, who automatically bow to the stronger opinion, who actually believe that the CDC and the media would inform us if these shots carry serious risks that are far from being "extremely rare," that pet phrase so beloved of the vaccine pushers. After all, haven't they been conditioned for years, like good little Pavlov dogs, to yap every time they hear the word "anti-vaxxer," and wag their tails when they're told how wonderful vaccines are? I've noticed also how happy some people are to let others know they've gotten their shots, to show the world they're with it, what these days is called "virtue signaling." One would think that the flaming idiocy of the relentless campaign to vaccinate America — as seen, for example, by enticing people with free beer, hamburgers, donuts, marijuana, lottery tickets, Uber rides and the like — was reason enough to opt out. But no. And there are still many, no doubt vaccinated, who buy into the Delta variant scaremongering. Just yesterday, seven weeks after the "Mask Required for Entry" signs came down all across New York, I observed that almost half the shoppers in my local supermarket were still wearing them.

Still, it's encouraging that 150 million Americans have not been injected, that the adults among them are instinctively wary, if not downright hostile, to any message force-fed them by the media and government. And cracks in the façade are spreading among the vaccinated as well, even as the social media giants are hard at work scrubbing messages that are at odds with the proclamations of WHO and the CDC. Just recently I discovered a new website, c19vaxreactions, the brainchild of Ken Ruettgers, former NFL veteran, an offensive lineman for the Green Bay Packers. His wife has been suffering every day for five months with an array of symptoms, beginning the day after she got her first Moderna shot. The many heartbreaking personal stories on this site are powerful stuff that will get around, and I'm sure there are many other grassroots efforts like this. The home page reads: "We are a large and ever-growing group of Americans who were previously healthy and have been seriously injured by the Covid vaccines (Pfizer, Moderna, J & J, as well as Astra-Zeneca in the clinical stage in the United States.)" But here's the next sentence: "We are pro-vaccine, pro-science, and were excited for the opportunity to be vaccinated and to do our part in helping to end the pandemic." And while Wisconsin Republican senator Ron Johnson deserves credit for appearing on a public platform with Ruettgers, he starts off by going into contortions over how everyone assembled is pro-vaccine, he's gotten every flu shot since Swine Flu (sic!), he's a huge supporter of Operation Warp Speed, he's happy to report that 300 million doses have been given, the vaccine has saved countless lives, and it has contributed to the end of the pandemic. It's hard for me to picture a more disgusting performance.

Can this battle be won when well-intentioned public figures, out of fear or ignorance, distort the truth like this? I suppose we should be grateful that the truth is getting out in drips and drabs. I've made my aversion to Fox News known in this book, and I'm no fan of Tucker Carlson, who manages to slip in a pro-vaccine lie or two each time he expounds on the subject, but sometimes he makes a critical point. On his May 5th program he announced the VAERS Covid-19 injection death count, which on that date was 3362. To my knowledge, no one else on television has reported news like this.

Unless I missed something, so far, in seven months, a grand total of two deaths has been reported on WCBS news radio. If there are other conservatives enlightening the public, well, that's good. But the only hard-hitting, uncompromising source of truth remains the Internet. I don't know a single red state governor who has spoken disapprovingly about the injection effort. This includes Florida's Ron DeSantis and South Dakota's Kristi Noem, both of whom stood their ground against the lockdown militants and banned vaccine passports, but think the Covid-19 vaccines themselves are just fine, and claim to have gotten theirs. It's as if the label "anti-vaxxer" is a scarlet letter. Can you imagine a security guard whose job it is to apprehend shoplifters insist that he's not "anti-klepto"? I've read hundreds of posts of family members of injection victims, or the victims themselves, who feel obligated to write, "I'm not anti-vax, but...." I was dumbstruck when I heard even Dr. Hodkinson deny the charge. To the contrary, he thinks vaccines in general have done much good; it's just that this one is a killer. What is it with him, and other fine people like him? I tried to answer that in the chapter on dogma, but I still find it perplexing that they have to bow low like this. One thing that all these otherwise sensible people have in common is that they never, ever single out any vaccine and explain in detail why they think it's so good. My reply to the accusation of anti-vaxxer is this: "You're damn right I'm an anti-vaxxer! Who in his right mind wouldn't be?"

❖

Like many other Americans, and Europeans too I'm sure, I've fantasized about cleansing this planet of all the human bacteria that have proliferated far too long. While writing this book, I often daydreamed about writing another, my own version of *The Turner Diaries*, that legendary underground novel of total race war, which saw Tel Aviv and New York City disappear under a mushroom cloud. I'd take a more humane approach. I wouldn't nuke Israel, I'd simply cut off all aid and watch it sink like a stone. I'd have my protagonist, an Air Force fighter pilot of genius and audacity like George Lincoln Rockwell, stage a mutiny, having won over a dozen fellow pilots with his charismatic leadership. I'd have him serve notice on all news media bases in New York, and the dozen or so vaccine production plants in the country, that they were in imminent danger of attack, just as the German government placed newspaper ads in 1915 warning prospective passengers not to sail on the *Lusitania*, which was secretly carrying munitions and thus a fair target for German U-boat torpedoes, which did in fact sink it on May 7th of that year with great loss of life. This will allow any innocents to quit their jobs and stay away from these buildings; they've been warned and now it's their decision. New York City is the global news media nerve center. It's from here, primarily, that disinformation and deviant ideas, like this whole Covid-19 clown show, flow into homes across America, and stream out to the world's most distant outposts. And, as I believe the evidence shows, at its core this is an attempted worldwide

Bolshevik revolution, the latest eruption of the Jewish genocidal mentality openly proclaimed in the Old Testament.

In the book I've daydreamed about, in advance of twelve Apache helicopters, armed to capacity, taking off from McGuire AFB near Trenton, 300 commandos will storm the *New York Times* building, and the television studios of ABC, CBS, CNN and NBC. All executives, news producers, columnists and script writers will be arrested and marched down to the Javits convention center on 11th Avenue. Within days they will be joined by thousands more journalists, politicians, pharmaceutical firm and health department officials at all levels, rounded up by local militia units from around the country. Now the choppers come swooping in over Manhattan. Within minutes, a salvo of air-to-surface Hellfire missiles reduces the five above-mentioned media septic tanks to piles of smoking rubble. The aircraft split up and head out to vaccine factories in Baltimore and Rockville in Maryland, Andover and Norwood in Massachusetts, and Marietta, West Point and Swiftwater in Pennsylvania, leveling these structures with their remaining payload, before returning to base to refuel, rearm and head off to a few midwestern states to finish the job of demolishing all remaining vaccine manufacturing facilities, after again refueling at Wright AFB in Dayton. Soon, with the media government gone, the one in Washington implodes, and puppet governments in Europe topple one by one. Within days, "reality" begins to change for the world's normies. The convention center will become a makeshift courthouse and all the prisoners will be given a fair trial. Those found to have misled the public about any aspect of Covid-19, or of vaccines in general, which I would guess to be eighty percent of them, will, in line with the Alexis Carrel proposal quoted on page 11, be immediately executed by firing squad on the banks of the Hudson River. Of course a great deal more would need to be done, but this is a crude, preliminary sketch of the book I'd love to write but probably won't.

Crazy, you say? Well, we all have our fantasies. And logistically, it may well be impossible to pull off. I don't know; I'm not a military man. But really, are not the events of the last sixteen months the true measure of insanity? Isn't what I've just described a better alternative than watching the vaccine killing fields fill up while doing nothing? As the vaccination steamroller keeps forging ahead, I'm almost beginning to believe that some people involved in this scheme will have big smiles on their faces as they watch all humanity go extinct. So if this condensed plot line plants a seed somewhere, great. Now back to the real world.

I cannot overemphasize the avalanche of events taking place without letup around the world, as people rebel against what has become medical martial law. London, Athens, Paris and smaller European cities have seen huge, raging protest marches in recent weeks. At the moment, life in France seems to be on the verge of becoming unbearable, as the most repressive Covid laws ever go into effect. Canada and Great Britain are also sliding backwards, as is Australia, where the police have often descended to sickening levels of brutality, and protests banned outright in some cities. Four British Airways and five Air India pilots died in May and June, some of them undoubtedly from the injection. And now, a whistleblower from Jet Blue has just revealed that five pilots who flew for that airline — which like most, pressures or requires all their employees to submit to the shot -— died in recent months. I have seen the names and ages of all but one of these men, and photographs of most of them. When in the history of aviation has anything like this ever happened? Across the world, the focus has shifted from inhumane mask and social distancing laws to the specter of mandatory vaccination, even as sad reports of death and serious injuries, often accompanied by graphic photos and videos, pile up on social media faster than the censors at Facebook can delete them. I also cannot overemphasize the indispensability of alternative news websites in these times, nor the scale of mainstream news censorship, which is nearly absolute. Let no one criticize the government of North Korea for depriving its citizens of information.

Thousands of lawyers and doctors in America and Europe — there are that many good ones, I was happy to learn — have spoken out in the strongest language against the fake pandemic and the global Covid-19 vaccination campaign — the media pretending they don't exist. One of them, a German lawyer named Reiner Fuellmich, who has years of experience suing large corporations, has initiated legal proceedings against WHO, the CDC, and what he calls the Davos group (meaning the World Economic Forum, which meets annually at a resort in Davos, Switzerland) for crimes against humanity. I wish him success, but I don't see how he can prevail without military support. This is war, and in war history shows that superior force, not superior morality, has the last say.

With the loosening of restrictions in late May and a seeming return to normalcy here in New York state, I hoped that things were finally winding down, that the System would be content with jabbing about 60% of the population after interest quickly dropped off. I was wrong. In mid-June, Andrew Cuomo announced plans to go door-to-door in towns where zip codes showed low injection rates, to coax residents into getting a shot. So far I haven't seen it. Now, in July, Biden's controllers are promoting the same activity. In the two videos I saw, residents had sharp words for the door knockers, who walked away. How far they will go with these intrusions remains to be seen. Will it fizzle out or will it intensify? I wouldn't be surprised if these canvassers were instructed to report the addresses of belligerent citizens for inclusion in an "extremist" or "anti-vaxxer" database, to be singled out for a special visit by FEMA goons for a forced injection, a trip to a detention camp, or worse. After seeing how events have unfolded since March 2020, I now realize that anything, absolutely anything, here and abroad, is possible if the ruling scum think they can get away with it. I've written more than enough in this book about the historical lessons of government terror, here and in Europe. In America, at least, goons have to reckon with the possibility of community defense groups springing up and spraying them with gunfire.

I have a sobering question: What is going to happen to the 180 million people in this country alone who received this injection, and particularly to those of child-bearing age or younger? It goes without saying that there will be much weeping and gnashing of teeth — there has been already. But really, how bad will it be in the years ahead? Will it be the worst event in the history of our species? Or not nearly as bad as some are predicting? My gut feeling is that it's going to be very bad, much worse than any previous manmade disaster. I do believe that hundreds of millions, if not billions, will suffer and die prematurely, or will not be able to have children, because of what a handful of twisted creatures are doing to us. But I just don't know. And I don't know how long it will take to assess the damage.

A few times I've run into a depressing quote from the pen of Edward Gibbon, the eighteenth-century English historian whose six-volume masterpiece, *The Decline and Fall of the Roman Empire*, is widely regarded as the greatest historical chronicle ever written. Casting his eye over 1500 years of society and civilization, Gibbon wrote: "History is indeed little more than the register of the crimes, follies and misfortunes of mankind." I believe in evolution, and that using our brains constructively for two million years is what allowed us to advance from primates on the African savannah to where we are today, meaning above all the many splendid achievements of the European peoples. So why have we done so many stupid, barbaric things over the centuries, like burning people at the stake for heresy, and slaughtering millions of people in distant lands who were bothering no one? The events of the last eighteen months have shown, in so many ways, just how fundamentally primitive and irrational most human beings are, and how fragile civilization is. Why don't we ever learn? Why, now, are we in the midst of what may well prove to be the greatest crime, folly and misfortune in the history of our species? Is there any hope for us?

I have to believe there is, despite the terrible trial we're now going through on the never-ending upward path.

Purchasing information

By special arrangement, this book is offered for sale by some organizations that the author has supported over the years. If you ordered this copy from one of them, you are encouraged to do so again. Or you may order directly from the author, by cash, check or money order. COST: $20 each for 1 or 2 copies, $15 each for 3 copies or more. These prices include postage and handling, and any applicable sales tax. Send order and payment to:

John Massaro
P.O. Box 45
Jeffersonville, NY 12748

www.ingramcontent.com/pod-product-compliance
Lightning Source LLC
Chambersburg PA
CBHW081239220326
41597CB00023BA/4160
9781737966005